T0201629

Electromigration in Metals

Learn to assess electromigration reliability and design more resilient chips in this comprehensive and practical resource. Beginning with fundamental physics and building to advanced methodologies, this book enables the reader to develop highly reliable on-chip wiring stacks and power grids. Through a detailed review on the role of microstructure, interfaces, and processing in electromigration reliability, as well as characterization, testing, and analysis, the book follows the development of on-chip interconnects from microscale to nanoscale. Practical modeling methodologies for statistical analysis, from simple 1D approximation to complex 3D description, can be used for step-by-step development of reliable on-chip wiring stacks and industrial-grade power/ground grids. This is an ideal resource for materials scientists and reliability and chip design engineers.

Paul S. Ho is Professor Emeritus in the Department of Mechanical Engineering and the Texas Materials Institute at the University of Texas at Austin. He has received research awards from the Electrochemical Society, IEEE, IITC and Semiconductor Industry Association, among others.

Chao-Kun Hu has recently retired as a research staff member in the reliability department at the Thomas J. Watson Research Center of IBM. He has received IBM Corporate awards, the IEEE Cledo Brunetti Award, the EDS Recognition and IITC Best Paper awards, and Invention of the Year from the New York Intellectual Property Law Association.

Martin Gall is the director of the US Operations Reliability Engineering Department at GlobalFoundries Inc. He is an IEEE *Transactions on Device and Materials Reliability* (TDMR) editor and the recipient of IEEE IITC Best Paper and SRC Mentor of the Year awards.

Valeriy Sukharev is principal engineer for Calibre Design Solutions, Siemens EDA, Siemens Digital Industries Software. He is a recipient of the 2014, 2018, and 2021 Mahboob Khan Outstanding Industry Liaison Award (SRC) and the Best Paper awards from ICCAD in 2016, 2019, and 2020.

Electromigration in Metals

Fundamentals to Nano-Interconnects

PAUL S. HO
University of Texas at Austin

CHAO-KUN HU
IBM Thomas J. Watson Research Center

MARTIN GALL
GlobalFoundries Inc.

VALERIY SUKHAREV
Siemens EDA, Siemens Digital Industries Software

CAMBRIDGE
UNIVERSITY PRESS

University Printing House, Cambridge CB2 8BS, United Kingdom

One Liberty Plaza, 20th Floor, New York, NY 10006, USA

477 Williamstown Road, Port Melbourne, VIC 3207, Australia

314–321, 3rd Floor, Plot 3, Splendor Forum, Jasola District Centre, New Delhi – 110025, India

103 Penang Road, #05–06/07, Visioncrest Commercial, Singapore 238467

Cambridge University Press is part of the University of Cambridge.

It furthers the University's mission by disseminating knowledge in the pursuit of education, learning, and research at the highest international levels of excellence.

www.cambridge.org
Information on this title: www.cambridge.org/9781107032385
DOI: 10.1017/9781139505819

First published 2022

Printed in the United Kingdom by TJ Books Limited, Padstow Cornwall

A catalogue record for this publication is available from the British Library.

Library of Congress Cataloging-in-Publication Data
Names: Ho, P. S., author. | Hu, Chao-Kun, 1946– author. | Gall, Martin, author. | Sukharev, Valeriy, 1952– author.
Title: Electromigration in metals : fundamentals to nano-interconnects / Paul Ho, Chao-Kun Hu, Martin Gall, Valeriy Sukharev.
Description: Cambridge ; New York, NY : Cambridge University Pres, 2022. | Includes bibliographical references and index.
Identifiers: LCCN 2021050167 (print) | LCCN 2021050168 (ebook) | ISBN 9781107032385 (hardback) | ISBN 9781139505819 (epub)
Subjects: LCSH: Interconnects (Integrated circuit technology)–Materials. | Metals–Electric properties. | Electrodiffusion.
Classification: LCC TK7874.53 .H62 2022 (print) | LCC TK7874.53 (ebook) | DDC 621.3815–dc23/eng/20220110
LC record available at https://lccn.loc.gov/2021050167
LC ebook record available at https://lccn.loc.gov/2021050168

ISBN 978-1-107-03238-5 Hardback

To Hillard B. Huntington
for his pioneering studies of electromigration

Contents

Colour plates section to be found between pp. 210 and 211

Preface

The study of electromigration (EM) can be traced to the mid-1900s with experiments to observe the effect of electrical current on mass transport in solid and liquid metals. This was followed by theoretical studies to analyze the "electron wind" driving force and its effect on diffusion due to the scattering of the electrons or, more generally, the charge carriers with the mobile ions. The subject took a drastic turn in the 1960s when EM was found to induce crack formation, leading to failure of aluminum conductor lines in silicon chips. This was a serious reliability problem for the microelectronics industry, coming at a time when the industry was at the beginning to develop very-large-scale integration (VLSI) of integrated circuits. This has generated great interests to study EM in thin films and metal lines, which has distinct characteristics. For metal lines on a chip, the current density can be much higher (100× or more) than in bulk metals due to substrate cooling, but diffusion occurs at a lower temperature, about 0.5 of the absolute melting point T_m, where it is dominated by structural defects, such as surfaces, interfaces, and grain boundaries. In addition, the confinement of the inter-level dielectrics and the silicon substrate can sustain mechanical stresses to counteract EM to reduce mass transport and damage formation.

As semiconductor technology continued to develop, Cu damascene structures were introduced to replace Al interconnects in 1997, and subsequently low permittivity (or "low k") materials were implemented as interlevel dielectrics. This changed the basic material, structure, and process of on-chip interconnects, with EM lifetimes being predicted to degrade by half for every technology node, even with a constant current density. With device scaling, the dimensions of the Cu lines continue to reduce to the nanoscale, significantly affecting the Cu microstructure and adding complexity and the demand for EM reliability. Responding to the technological needs, EM studies of Cu interconnects were greatly extended to provide better understanding and to develop effective approaches such as interfacial engineering and alternate metallization for improving the EM reliability at the nanoscale.

The study of EM includes a wide range of topics, from materials science and semiconductor device physics to statistics and probability, making it a truly multi-disciplinary subject. The physics of failure and the stochastic process are topics of active research by experts in different areas of science and engineering. The aim of this book is to provide a comprehensive resource for those interested in the fundamentals of EM as a physical phenomenon and in its role as a major problem in chip reliability. We hope that this book will be useful for materials scientists, reliability engineers, and chip physical design professionals, as well as for the increasing number of graduate students pursuing interdisciplinary research of materials science and electrical engineering.

This book is organized into nine chapters extending from the basic studies in bulk metals to on-chip interconnects from the microscale to the nanoscale. The group of the first four chapters reviews the basic studies, starting from an introduction to EM in Chapter 1, followed by a review of the theory and the studies of bulk metals in Chapter 2, including a kinetic analysis of the solute effect. In Chapter 3, the discussion is focused on the analysis and X-ray diffraction measurements of thermal stresses induced by the dielectric and substrate confinement in Al and Cu lines, and on showing how substantial hydrostatic stresses can be induced, leading to void formation. This is followed by Chapter 4, analyzing the kinetics of mass transport under EM and thermal stresses, leading to Korhonen's equation and the Blech "short length" effect. Here, analytical solutions and simulations are presented in one dimension (1D) to analyze the microstructure and stress effects on damage formation and early failures under EM.

The second part of this book contains five chapters focusing on advanced topics relating to the scaling effects on EM at the nanoscale, including microstructure evolution, massive-scale statistical tests, and power grid applications. Starting from Chapter 5, the EM characteristics and reliability studies of Cu conductor lines are reviewed and assessed as scaling continues to the nanoscale. Several innovative approaches have been developed to improve EM reliability for Cu interconnects, including the use of cap layers and alloying effects. In Chapter 6, we investigate the scaling effect on microstructure evolution using a high-resolution electron diffraction technique to examine Cu interconnects to 22 nm linewidth and Co interconnects to 26 nm linewidth. Based on the observed microstructures, a Monte Carlo simulation was developed based on total energy minimization to project the scaling effect on grain growth and the implication on EM reliability for future technology nodes. In Chapter 7, the analysis of EM and stress evolution is generalized from 1D to 3D damascene interconnects, where a physics-based simulation is set up to analyze the microstructure effects on stress evolution and void formation. Results from the simulation are combined with scanning and transmission electron microscopy (SEM/TEM) experiments to study void nucleation, migration, growth, and shape evolution leading to interconnect degradation.

With continued scaling to the nanoscale, the statistical nature of damage formation becomes important to assess EM reliability, which is dominated by the early failures of the weakest links. This topic is discussed in Chapter 8, presenting a novel approach based on the Wheatstone Bridge technique to perform statistical EM tests on a massive scale to detect early failures. In Chapter 9, we conclude with a discussion of a novel approach to assess EM reliability for power grid systems based on a mesh model to account for system redundancy and to track EM degradation in multibranch interconnect trees across the die. Such problems are important for future development of reliable Cu interconnects as scaling continues to expand the wiring structure to the material limit at the nanoscale.

Over the years, all of us have been fortunate to work and interact with many colleagues who helped to broaden and improve our understanding of the fundamentals of EM and its role in interconnect degradation, notably Robert Rosenberg, Matt A. Korhonen, Tony Oates, King-Ning Tu, and William D. Nix. We acknowledge the fruitful collaborations with many brilliant scientists and engineers; in particular, Hisao Kawasaki, Ehrenfried Zschech, Ennis Ogawa, Steve Anderson, Carl V. Thompson, Farid N. Najm, Armen Kteyan, and Junjun Liu have provided discussions and ideas to generate many interesting and important experimental and theoretical results described in this book. We thank the members of the University of Texas at Austin group, particularly Steve S. T. Hu, D. W. Gan, S. H. Rhee, L. J. Cao, M. Hauschildt, J. Kasthurirangan, and I. S. Yeoh, who have contributed to the research results cited in this book. The research support from the National Science Foundation, the Semiconductor Research Corporation, SEMATECH, IBM, Motorola, Intel, Siemens EDA (formerly the Mentor Graphics Corporation), and the University of Texas at Austin is gratefully acknowledged. Part of the work for Chapter 5 was performed by Research Alliance Teams at various IBM Research and Development Facilities, which is gratefully acknowledged. We thank especially Sarah Strange and Julia Ford of the Cambridge University Press for their continuous help with the book preparation. Finally, and most importantly, we thank our wives for their patience and understanding while we worked to complete this book.

1   Introduction to Electromigration

Electromigration, or electrotransport, describes the phenomenon of atomic diffusion in metal driven by an electric current. Its study can be traced back more than a century, starting with the first observation reported by Geradin [1] in molten alloys of lead-tin and mercury-sodium. Systematic studies did not start until the early 1950s with the work of Seith and Wever at the University of Muenster, Germany [2]. By measuring the mass transport across the phase diagram of some Hume–Rothery alloys, they observed that the direction of mass transport can be reversed and is correlated with the type of the majority charge carriers, i.e., electrons or holes, in the specific alloy phase. This observation provided the first evidence concerning the nature of the driving force for electromigration. It showed that the atomic motion is not determined solely by the electrostatic force imposed by the applied electric field; instead, it depends on the direction of motion of the charge carriers. This prompted Seith to introduce the idea originated by Skaupy [3] of "electron wind" to account for the induced mass transport, an idea that laid the foundation for the basic understanding of electromigration. Seith and Wever [4] also introduced the method of using the displacement of an indentation on a metal wire to measure the induced mass transport. This "marker motion" technique, now called the vacancy flux method, has become one of the standard measurements of electromigration in bulk metals.

The concept of the "electron wind" driving force was first formulated by Fiks [5] and Huntington and Grone [6]. Working independently, these authors employed a semiclassical "ballistic" approach to treat the collision of the moving atom by the charge carriers. In a more rigorous treatment, Huntington and Grone considered the initial and final quantum states of the charge carriers involved in the collision process, where the "wind" force experienced by the moving atom is derived by integrating the spatial variation of the interaction with the charge carriers. These ideas yielded a driving force depending on the type of defect, the interaction with the charge carriers, and the atomic configuration of the jumping path.

The formulation of the driving force was a major contribution to the study of electromigration, as it demonstrated the possibility of using electromigration to probe the interaction of mobile defects with charge carriers in metals. This has stimulated considerable interest starting in the 1960s in theoretical and experimental studies on electromigration in metals. In theoretical studies, Bosvieux and Friedel [7] developed the first quantum mechanical formulation of the driving force in terms of electron–ion interactions for a simple case of impurity atoms in a "jellium." With some simple model assumptions, these authors have formulated an approach that can be extended to metals with more complex electronic structures. This has stimulated a series of subsequent theoretical studies with focus on the physical basis of the driving force, particularly regarding the dynamic screening of the moving ion and the effect of the band structure on the electron–ion scattering. (See reviews by Verbruggen [8] and Sorbello [9].) While the theory continued to develop, experimental data have accumulated to show systematic trends to collaborate the theoretical results. In the experimental studies, the interest was initially concentrated on self-electromigration in pure metals using primarily the vacancy flux method and with a few using radioactive isotopes. These studies were first focused on metals with electron charge carriers and then extended to metals with hole carriers (see the reviews by Huntington [10]). The experimental study was later extended to alloys and liquid metals by Rigney [11].

Until that time, the application of electromigration (EM) was rather limited, with efforts largely directed toward the purification of interstitial, particularly gaseous, impurities in refractory metals [12]. The interest and the direction of electromigration study took a drastic turn in 1966 when Blech and Sello [13] reported that the crack formation in aluminum conductor lines on silicon chips was induced by electromigration. This was a serious reliability problem for the microelectronics industry, coming at a time when the microelectronics industry was starting the development of very-large-scale integration (VLSI) of integrated circuits [14]. This has generated significant interest to study electromigration, not only for its basic physics, but as an important reliability problem for the integrated circuits. Most of the studies were aimed at electromigration in thin film conductors used for multilevel on-chip interconnects with linewidths of submillimeters and operated at moderate temperatures of about half the absolute melting point. Under these conditions, the current density driving the mass transport increases by more than two orders of magnitude, while electromigration occurs primarily along structural defects, such as surfaces, interfaces, and grain boundaries. Particularly relevant to damage formation is the flux divergence site where damages in the form of voids or hillocks can form because of the local imbalance of atomic flux induced by material and microstructure inhomogeneity. Such characteristics of EM damage have shifted the focus of the investigations from bulk materials to thin films and conductor lines [15–17].

Until the early 1990s, interest in EM was primarily focused on aluminum lines, the dominant material used for on-chip interconnects at that time. As the technology evolved, the Al on-chip interconnects became more complex with multiple metal levels integrated with SiO_2 interlayered dielectrics. Following Moore's law [18], device scaling continues to reduce the size of the conductor lines by a factor of ~0.7 every technology generation, with corresponding increase in its number to sustain the improvement of device density and chip performance. This cumulated to the development of Al/SiO_2 interconnects with 5–6 metal levels and a minimum linewidth of ~0.4 μm for high-performance microprocessors of the 0.25 μm technology node. At this time, the study of electromigration has advanced well into the microscale domain with materials and microstructure characteristics distinctly different from bulk metals or thin films.

One distinct aspect of EM in conductor lines arises from the thermal stresses generated in the line due to the confinement by the Si substrate and the interlevel dielectrics and barriers to contain the metal line. The thermal stress provides an additional driving force to counter the electrical current for mass transport and can substantially affect the kinetics and mechanism for damage formation. The stress effect on EM was first observed by Blech in 1976 in a study of EM in Al conductor lines [19] and established the concept of a critical current density-length (jLc) product with a "critical length" Lc where the stress gradient can completely counteract the current j driving force to stop EM [20, 21]. The stress effect took a drastic turn in the 1980s to become an important reliability problem with the finding of stress-induced void formation in passivated Al lines that can cause line failure without EM [22, 23]. This phenomenon has generated great interest at that time to investigate the thermal stress characteristics and the stress voiding phenomenon. It soon became clear that stress voiding is synergetic to electromigration, and together they form a framework for analyzing the kinetics of damage formation under EM. This has led to the formulation of the Korhonen equation [24, 25] and subsequently more general 3D analyses.

This can be considered the first major period of EM studies on conductor lines driven by technological needs to develop Al-based on-chip interconnects for large-scale integration. Significant interest has been generated from the industry and university laboratories with research topics focusing on effects of materials, microstructure, and line dimensions on mass transport and the development of reliable EM structures. These included the solute effect [15], the intermetallic layer effect [26], and the development of statistical studies [27], among others.

Since then, the study of electromigration has been sustained by unprecedented advances of the IC miniaturization from the micro- to the nanoscale, with continuing improvement of the device density, chip performance, and manufacturing cost as predicted by Moore's law. This has prompted the development of Cu interconnects to replace Al interconnects in 1997 [28]. and subsequently the implementation of low k dielectrics as interlevel dielectrics [29]. The development of Cu damascene structures brought new focus into the study of EM and thermal stresses in Cu interconnects. This began the second major period of electromigration studies where not only the basic material set has changed but also the interconnect structure with the development of the dual damascene process for Cu interconnects [30]. While Cu has better intrinsic reliability than Al, electromigration continued to be a major reliability concern with distinct characteristics and damage mechanisms. The degradation of EM reliability for Cu damascene lines depends on materials and processing, particularly relating to defects induced by chemical-mechanical polishing (CMP) at the interface and the plasma processing of the weak low k dielectrics. Several important studies have been reported, including the analysis by Lane et al. showing that the activation energy for void growth under EM can be directly correlated to the bond energy of the Cu interface [31]. Also important was the prediction by Hu et al. that the EM lifetime would degrade by half for every technology node due to dimensional scaling even with a constant current density [32]. Results of these studies clearly showed that interfacial engineering is an effective approach for improving the EM reliability for Cu damascene interconnects. This has led to several approaches to improve EM reliability for Cu metallization, including metal cap layers [33] and solute alloying [34], among others.

With continued scaling, the dimensions of the Cu lines reach to the nanoscale, which significantly increases the number and complexity of the interconnects together with changes in the Cu microstructure and increases in the line resistivity. Responding to the technological needs, the EM studies of Cu damascene structures were extended to the nanoscale and the development of alternate metallization. These included the formulation of general 3D kinetic analyses [35, 36], the development of statistical tests at a massive scale [37, 38], the scaling effect on microstructure [39, 40], and electrical resistivity [41, 42].

This book aims to provide a comprehensive review of the current understanding of electromigration and interconnect reliability from fundamentals to nanointerconnects. It is organized into nine chapters and largely follows the chronological order of electromigration studies, starting from the basic studies in metals and extending to interconnects from the microscale to the nanoscale. First, the fundamentals of electromigration and the studies of bulk metals are reviewed in Chapter 2, including a formulation of electromigration based on irreversible thermodynamics, a kinetic analysis of the solute effect and a review of the theory and bulk materials studies. In Chapter 3, the discussion on EM in metal lines starts with thermal stresses induced by dielectric and substrate confinement with distinctive characteristics to interact with EM to control the mass transport and void formation. Here we describe the X-ray diffraction technique used to determine the triaxial stress state in Al and Cu lines and show how the structural confinement can give rise to substantial hydrostatic stresses to induce void formation leading to line failure. This is followed by Chapter 4 analyzing the kinetics of mass transport under EM and thermal stresses leading to the Korhonen equation and the Blech "short-length" effect. Here analytical solutions and simulations are presented in 1D to analyze the microstructure and stress effects on damage formation and early failures under EM.

Starting from Chapter 5, the discussion is focused on EM in Cu interconnects and topics related to the scaling effects at nanoscale, including microstructure evolution, electrical resistivity, massive-scale statistical tests, and power grid applications. The discussion proceeds first in Chapter 5 on EM characteristics and reliability studies of Cu conductor lines as scaling continues to the nanoscale for ultra large-scale integration (ULSI). The Cu interconnects fabricated by the dual damascene process have structural features leading to mass transport and failure mechanisms distinctly different from the Al interconnects. Several innovative approaches have been developed to improve EM reliability for Cu interconnects, including the use of cap layers and alloying effects, which will be discussed. In Chapter 6, we investigate the scaling effect on microstructure evolution and the implication on EM reliability of Cu and Co interconnects. The scaling effect on microstructure was measured using a high-resolution transmission electron microscopy (TEM)-based precession electron diffraction (PED) technique. With this technique, microstructure evolution was measured to

22 nm linewidth for Cu interconnects and to 26 nm linewidth for Co interconnects. Based on the PED results, Monte Carlo simulation was developed based on total energy minimization to project the scaling effect on grain growth and EM reliability for future technology nodes.

In Chapter 7, the analysis of EM and stress evolution is generalized from 1D to 3D damascene interconnects. This chapter starts with a general 3D analysis of stress evolution by treating the vacancies and displaced atoms generated under EM as local volumetric strains in the grain interior and at grain boundaries (GBs). Then a physics-based simulation is set up to analyze the microstructure effects on stress evolution and void formation, considering the variations of the GB diffusivities and elastic properties of individual grains in the interconnect. Results from the simulation are combined with scanning and transmission electron microscopies to study void nucleation, migration, growth, and shape evolution, leading to interconnect degradation.

With continued scaling to the nanoscale, the statistical nature of damage formation becomes important to assess EM reliability, which is dominated by the early failures of the weakest links. This topic is discussed in Chapter 8, presenting a novel approach based on the Wheatstone Bridge technique to perform statistical EM tests at a massive scale to detect early failures in very large systems. In contrast, we show in Chapter 9 that recent measurements showed that the weak link approach cannot accurately predict the lifetime for power grid–like structures. A novel approach is presented for power grid EM assessment using a physics-based approach combined with a mesh model to account for redundancy while still fast enough to be practically useful. Existing physical models for EM in metal branches are extended to track EM degradation in multibranch interconnect trees across the die. These problems are important for future development of reliable Cu interconnects as scaling continues to expand the wiring structure to the material limit at the nanoscale.

References

1. M. Gerardin, *Compt. rend.* **53** (1861), 727.
2. W. Seith and H. Wever, *Z. Elecktrochem.* **59** (1953), 891–900.
3. F. Skaupy, *Verhandl. Deut. Physik. Ges.* **16** (1914), 156.
4. W. Seith and H. Wever, *Z. Elecktrochem.* **59** (1955), 942.
5. V. B. Fiks, Forces produced by conduction electrons in metals located in external fields, *Soviet Physics–Solid State* **1** (1959), 14.
6. H. B. Huntington and A. R. Grone, Current-induced marker motion in gold wires, *Journal of Physics and Chemistry of Solids* **20** (1961), 76–81.
7. C. Bosvieux and J. Friedel, Sur l'electrolyse des alliages metalliques, *Journal of Physics and Chemistry of Solids* **23** (1962), 123–136.
8. A. H. Verbruggen, Unpublished PhD thesis, Free University of Amsterdam (1985).
9. R. S. Sorbello, Theory of the direct force in electromigration, *Physical Review B* **31** (1985), 798.
10. H. B. Huntington, Electromigration in metals. *Diffusion in Solids: Recent Developments,* ed. A. S. Nowick and J. J. Burton (New York: AIME, 1974), 303–352.

11. D. A. Rigney, Electromigration in alloys and liquid metals. *Diffusion in Solids: Recent Developments,* ed. A. S. Nowick and J. J. Burton (New York: AIME, 1974), 140–159.

12. D. T. Peterson, Electromigration of hydrogen, deuterium in vanadium, niobium and tantalum. *Electro- and Thermo-Transport in Metals,* ed. R. E. Hummel and H. B. Huntington (New York: AIME, 1977), 54–67.

13. I. A. Blech and H. Sello, The failure of thin aluminum current carrying strips on oxidized Silicon, *Fifth Annual Symposium on the Physics of Failure in Electronics*, (1966), 496–505.

14. P. S. Ho, *Proceedings of the 20th IEEE Symposium on Reliability in Physics* (New York: IEEE, 1982), 284.

15. F. M. d'Heurle and R. Rosenberg, *Physics of Thin Films 7* (New York: Academic Press, 1972), 257.

16. F. M. d'Heurle and P. S. Ho, Electromigration in metals. *Thin Films: Interdifusion and Reactions,* ed. J. M. Poate et al. (New York: Wiley-Interscience, 1978), 183.

17. P. S. Ho, F. M. d'Heurle, and A. Gangulee, Implications of electromigration on device reliability. *Electro- and Thermo-Transport in Metals and Alloys,* ed. R. E. Hummel and H. B. Huntington (New York: AIME, 1977), 108–139.

18. G. E. Moore. Cramming more components onto integrated circuits, *Electronics* **38** (8) (1965), 114–118.

19. I. A. Blech and E. Kinsbron, Electromigration in thin gold films on molybdenum surfaces, *Thin Solid Films* **25** (1975), 327–334.

20. I. A. Blech, Electromigration in thin aluminum films on titanium nitride, *Journal of Applied Physics* **47** (1976) 1203–1208.

21. I. A. Blech and K. L. Tai, Measurement of stress gradients generated by electromigration, *Applied Physics Letters* **30** (1977), 387.

22. J. Klema, R. Pyle, and E. Domangue, Reliability implications of nitrogen contamination during deposition of sputtered aluminum/silicon metal films, *IEEE Proceedings International. Reliability Physics Symposium* **22** (1984), 1–5.

23. J. Curry, G. Fitzgibbon, Y. Guan, R. Muollo, G. Nelson, and A. Thomas, New failure mechanisms in sputtered aluminum-silicon films, *IEEE Proceedings of the International Reliability Physics Symposium* **22** (1984), 6–10.

24. M. A. Korhonen, P. Borgesen, and Che-Yu Li, Mechanisms of stress-induced and electromigration-induced damage in passivated narrow metallization on rigid substrates, *Materials Research Society Bulletin* **17** (1992), 61–69.

25. M. A. Korhonen, R. D. Black, and C-Y. Li, Stress relaxation of passivated aluminum line metallizations on silicon substrates, *Journal of Applied. Physics* **69** (1991), 1748.

26. J. K. Howard, J. F. White, and P. S. Ho, Intermetallic compounds of Al and transitions metals: effect of electromigration in 1–2-μm-wide lines, *Journal of Applied Physics* **49** (1978), 4083.

27. M. Gall, M. Gall, C. Capasso, et al., Statistical analysis of early failures in electromigration, *Journal of Applied Physics* **90** (2001), 732–740, doi:10.1063/1.1377304.

28. D. Edelstein, .J Heidenreich, R. Goldblatt et al. Full copper wiring in a sub-0.25 micron CMOS ULSI technology, *Tech. Dig. IEEE International Electron Devices Conference* (1997), 773–776.

29. W. Lee, J. Leu, and P. S. Ho, Low dielectric constant materials for ULSI interlayer dielectric application, *Materials Research Bulletin* **22** (1997), 19–27.

30. R. Rosenberg, D. C. Edelstein, C.-K. Hu, and K. P. Rodbell, Copper metallization for high performance silicon technology, *Annual Review of. Materials Science* **30** (2000), 229–262.

31. M. W. Lane, E. G. Liniger, and J. R. Lloyd, Relationship between interfacial adhesion and electromigration in Cu metallization, *Journal of Applied Physics* **93** (2003) 1417.
32. C. K. Hu, R. Rosenberg, and K. Y. Lee, Electromigration path in Cu thin film lines, *Applied Physics Letters* **74** (1999), 2945.
33. C. K. Hu, L. Gignac, R. Rosenberg, et al., Reduced electromigration of Cu wires by surface coating, *Applied Physics Letters* **81** (2002), 1782.
34. C. K. Hu, J. Ohm, L. M. Gignac et al., Electromigration in Cu(Al) and Cu(Mn) damascene lines, *Journal of Applied Physics* **111** (2012), 093722.
35. V. Sukharev, E. Zschech, and W. D. Nix, A model for electromigration- induced degradation mechanisms in dual-inlaid copper interconnects: Effect of microstructure, *Journal of Applied Physics* **102** (2007), 053505.
36. R. J. Gleixner and W. D. Nix, A physically based model for electromigration and stress-induced void formation in microelectronic interconnects, *Journal of Applied Physics* **86** (1999), 1932–1944.
37. A. Volinsky, M. Hauschildt, J. B. Vella, et al., Residual stress and microstructure of electroplated Cu films on different barrier layers, *Materials Research Society Symposium Proceedings* **695** (2001): https://doi.org/10.1557/PROC-695-L1.11.1.
38. A. S. Oates, Strategies to ensure electromigration reliability of Cu/Low-*k* interconnects at 10 nm, *ECS Journal of Solid State Science and Technology* **4** (2015), N3168–N3176.
39. K. J. Ganesh, A. D. Darbal, S. Rajasekhara, et al., Effect of downscaling nano-copper interconnects on the microstructure revealed by high resolution TEM-orientation-mapping, *Nanotechnology* **23** (2012), 135702.
40. L. Cao, L. Zhang, P. S. Ho, P. Justison, and M. Hauschildt, Scaling effects on microstructure and electromigration reliability for Cu and Cu(Mn) interconnects, 2014 IEEE International Reliability Physics Symposium, Waikoloa, HI. (2014), 5A.5.1–5.5.
41. K. Croes, C. Adelmann, C. J. Wilson, et al., Interconnect metals beyond copper: reliability challenges and opportunities, *2018 IEEE International Electron Devices Meeting (IEDM), San Francisco, CA* (2018), 5.3.1–5.3.4, doi: 10.1109/IEDM.2018.8614695.
42. D. Gall, The search for the most conductive metal for narrow interconnect lines, *Journal of Applied Physics* **127** (2020), 050901.

2 Fundamentals of Electromigration

2.1 Introduction

In this chapter, a formal description of electromigration is presented by treating electromigration as a phenomenon of mass transport under a driving force in metals within the framework of irreversible thermodynamics. Although the formulation as presented is formal, it is useful in clarifying the thermodynamic nature of electromigration as arising from the interaction of the moving atoms and the charge carriers. Following this formulation, an example is given to describe electromigration in dilute alloys as a correlated atomic diffusion phenomenon, which is used to analyze the solute effect on electromigration in Section 2.3. This example shows that by considering the details of the correlation in the atomic jumping process, the thermodynamic parameters can be expressed in terms of the atomic jumping frequencies and the correlation between the solute and the solvent atoms. In this way, the solute effect on electromigration can be formulated and analyzed with results to show how the solute diffusion can affect the solvent diffusion in electromigration. In Section 2.4, the theory on the driving force is summarized with a discussion on the controversy of the screening effect on the direct force. Then we present the results from experiments designed to measure electromigration of hydrogen impurities to resolve this question. This is followed by reviewing in Section 2.5 the EM studies carried out in two systems of bulk metals: substitutional and interstitial. The chapter then concludes with a summary.

2.2 Thermodynamic Description of Electromigration

Electromigration can be treated as a general phenomenon of forced atomic motion in a solid with several interacting components within the framework of irreversible thermodynamics (see, for example Adda and Philibert [1]). Here we follow the approach of Huntington [2] to study the solute effect on electromigration in dilute alloys.

In a multicomponent system, the atomic motion of the ith component depends on its interaction with all the other elements in the system. This can be expressed in general by writing the atomic fluxes J_i as

$$J_i = \sum_j L_{ij} X_j \qquad (i = 1, 2, \ldots, n), \qquad (2.1)$$

where L_{ij} is the phenomenological coefficient correlating the flux J_i of the ith component to the driving force X_j of the jth component. In general, the driving force X_j is derived from the gradient of the chemical potential μ_i and other types of potential such as the electrical potential and thermal and stress gradients. In electromigration, the driving forces from the thermal and stress gradients can also be important as discussed in later chapters, depending on the system and the boundary conditions. Here we express the flux equations separately for the motion of the atoms and the charge carriers and assume uniform temperature and stress distributions as follows:

$$J_i = T^{-1} \sum_j L_{ij} \nabla \left(\mu_j + q_j \phi \right) - T^{-1} L_{ie} q_e \nabla \phi, \qquad (2.2a)$$

$$J_e = T^{-1} \sum_j L_{ej} \nabla \left(\mu_j + q_j \phi \right) - T^{-1} L_{ee} q_e \nabla \phi. \qquad (2.2b)$$

Here the diagonal elements L_{ij} are related to the diffusion of the ith element, and L_{ee} is related to the flow of the charge carriers. There are two types of off-diagonal elements: L_{ij}, relating to the effect on the atomic jumps of the ith element due to the correlated jumps of the jth element, and L_{ie}, relating to the interaction between the atom and the electrons. The first type of these elements, L_{ij}, applies to both the chemical potential μ_j and the electrical potential $q_j \phi$, where the correlation in atomic jumps affects not only the diffusion process relating to μ_j but also the electromigration process relating to $q_j \phi$. The second type of the off-diagonal elements, L_{ie}, constitutes the "electron-wind" force due to the interaction of the charge carriers with the moving ion. This term is usually combined with the direct electrostatic force, i.e., the $q_j \nabla \phi$ term, to yield the electromigration driving force in terms of an effective charge $Z^* e$ as

$$\begin{aligned} F_{eff} &= |e| Z^* E \\ &= |e| (Z^w + Z^e) E, \end{aligned} \qquad (2.3)$$

where $Z^e e E$ is the electrostatic force and $Z^w e E$ the "electron-wind" force and $|e|$ the absolute value of the electronic charge and E, equal to $-\nabla \phi$, is the electric field. In this way, the formulation can analyze the solute effects on electromigration in dilute alloys.

At this point, not all of the nondiagonal coefficients are independent so that the Onsager relation cannot be applied to reduce the number of independent coefficients [2]. Before this simplification is carried out, the mass conservation condition $\sum J_i = 0$ and the mechanical equilibrium condition $\sum \nabla (\mu_j + q_i \phi) = 0$ can be imposed to ensure an equilibrium state of the system. These conditions reduce the number of the independent coefficients L_{ij}. For a system where mass transport involves diffusion via vacancies, it is convenient to introduce the vacancy as an additional component to balance the mass transport of the moving atoms as

$$J_v = -\sum_i J_i. \tag{2.4}$$

Using this condition, one can transfer (2.2a) and (2.2b) from the laboratory reference frame to a lattice reference frame where the atomic motion can be observed as relative movement of the lattice planes. This approach of tracking the mass transport in electromigration is one of the commonly used methods in measurement and is called the *vacancy flux method* [2], which will be discussed in Section 2.4.

After transposing the flux equation (2.2a) and (2.2b) to the lattice frame, the system is under mechanical equilibrium where the flux and the force are conjugate variables and the Onsager relation can be applied to yield $L_{ij} = L_{ji}$. After this simplification, a basic formalism is established to treat electromigration in multicomponent systems. Several applications of common interest can be found in [2, 3].

2.3   Kinetic Analysis of Solute Effect on EM in Binary Alloys

In this section, the thermodynamic formulation is used to analyze the solute effect on EM in binary alloys [4, 5]. This problem was of considerable interest in the early EM studies since the addition of copper was found to be effective to improve EM lifetime of the aluminum lines [6]. In a dilute alloy, the total atomic flux comes mainly from the diffusion of the solvent atoms, and so to understand the solute effect, one must analyze how the diffusion of the solvent atoms can be affected by the solute atoms and change the overall atomic transport. This problem was first analyzed by Doan [7], who investigated the atomic process in impurity diffusion and derived the various ratios of the jumping frequencies. The solute effect presented here is based on irreversible thermodynamics and analyzing the kinetics to examine how the interactions of the electrons with the solute atoms can affect the transport of the solvent atoms under EM [5]. Once the problem is formulated, the analysis is applied to several alloy systems to evaluate the solute effect and to search for solute additions that can reduce the overall electromigration.

Consider a homogeneous binary alloy of atomic species a and b with atomic concentrations of c_a and c_b, respectively. Following thermodynamics, the atomic fluxes under an electric field E at temperature T can be related to the driving forces through the phenomenological coefficients L_{ij} as

$$
\begin{aligned}
J_a &= -(L_{aa}\nabla\mu_a + L_{ab}\nabla\mu_b)/T + (L_{aa}q_a + L_{ae}q_e + L_{ab}q_b)E/T \\
J_b &= -(L_{bb}\nabla\mu_b + L_{ba}\nabla\mu_a)/T + (L_{ba}q_a + L_{be}q_e + L_{bb}q_b)E/T,
\end{aligned} \tag{2.5}
$$

and the vacancy flux is

$$J_v = -(J_a + J_b). \tag{2.6}$$

In (2.5), the flux consists of two parts: the first two terms on the right are the diffusion contributions due to the concentration gradients, and the othes are the EM induced by the field E. Here μ_a and μ_b are the chemical potentials associated with the solute and the solvent atoms, respectively, and $Z_a^* e$ and $Z_b^* e$ are the respective effective charges as defined in (2.3). It is clear that the electromigration of both solvent and solute is affected by their mutual interaction as indicated by the terms containing L_{ab} and L_{ba}. The phenomenological approach can be correlated to the atomistic details of the diffusion process by expressing the coefficients L_{ab} and L_{ba} in terms of the frequencies of the jumping processes involved in the atomic transport. Such an analysis provides a link to examine the kinetics in electromigration by considering the interactions between the moving species on the overall atomic transport. The analysis was first formulated by Manning [8] and is used here to study the solute effect on electromigration.

Following Manning's approach to formulate the solute effect, (2.5) is first simplified by applying the Gibbs–Duhem relations for a homogeneous binary dilute alloy under equilibrium and by defining

$$L_{ae}q_e = L_{aa}|e|Z_a^w + L_{ab}|e|Z_b^w$$
$$L_{be}q_e = L_{ab}|e|Z_a^w + L_{bb}|e|Z_b^w. \tag{2.7}$$

Then (2.5) can be simplified as

$$J_a = -k\left(\frac{L_{aa}}{c_a} - \frac{L_{bb}}{c_b}\right)\nabla c_a + L_{aa}Z_a^* eE + L_{ab}Z_b^* eE$$
$$J_b = -k\left(\frac{L_{bb}}{c_b}\right)\nabla c_b + L_{ab}Z_a^* eE + L_{bb}Z_b^* eE, \tag{2.8}$$

where c_a and c_b are the solvent and solute concentrations and k the Boltzmann constant. For dilute alloys, Manning [8] shows that

$$L_{aa} \simeq \left(\frac{L_{bb}}{c_b}\right)D_a \quad \text{and} \quad L_{bb} \simeq \left(\frac{Nc_b}{kT}\right)D_b f_b, \tag{2.9}$$

where D_a and D_b are the uncorrelated diffusion coefficients of the solvent and solute atoms, respectively; N is the atomic density; and f_b is the solute correlation factor. (The correlation factor is related to the return probability of the diffusing atom after exchange with a vacancy [9].) Using these expressions, the electromigration components of the atomic flux can be written as

$$J_a^E = \left(\frac{Nc_a D_a}{kT}\right)Z_a^{**} eE \quad \text{and} \quad J_b^E = \left(\frac{Nc_b D_b f_b}{kT}\right)Z_b^{**} eE. \tag{2.10}$$

Here Z_a^{**} and Z_b^{**} are the apparent effective charges and related to the actual effective charges by the following relationships:

$$Z_a^{**} = Z_a^*\left(1 + \frac{L_{ab}Z_b^*}{L_{aa}Z_a^*}\right) \quad \text{and} \quad Z_b^{**} = Z_b^*\left(1 + \frac{L_{ba}Z_a^*}{L_{bb}Z_b^*}\right). \tag{2.11}$$

In (2.8), the solute effect can be traced to the additional contribution to the flux of the solvent a in the alloy due to the mass transport of the solute b through the off-diagonal coefficient L_{ab} from both diffusion and electromigration. In a dilute alloy, there is extra mass transport of the solvent atoms induced by vacancies pushed to them in an opposite direction due to the flow of the solute atoms. Manning [8] called such cross-term effect on EM the "vacancy-flow" effect.

To evaluate the solute effect, it is convenient to express the diffusivity in the alloy as

$$D_a = D_a(0)(1 + b'c_b),\qquad(2.12)$$

where $D_a(0)$ is the uncorrelated diffusivity of the pure solvent and b' is called the solute enhancement factor on solvent diffusion. Combining (2.8)–(2.12), one obtains

$$J_a^E = \frac{Nc_a D_a(0) Z_a^* eE}{kT}\left[1 + \left(b' + \frac{D_b^*}{D_a(0)}\frac{Z_b^* L_{ab}}{Z_a^* L_{bb}}\right)c_b\right]$$

$$J_b^E = \frac{Nc_b D_b^* Z_b^* eE}{kT}\left[1 + \frac{L_{ab} Z_a^*}{L_{bb} Z_b^*}\right].$$

(2.13)

These expressions indicate that solute addition affects both solvent and solute EM, which can be traced to the off-diagonal term L_{ab} due to the interaction of the diffusing atoms with the charge carriers. For the solute atom, the driving force is modified by a constant factor, while for the solvent atoms, the effect is linearly proportional to the solute concentration. For most substitutional alloy, the sign of Z_b^* is expected to be the same as Z_a^*, so the solute effect depends primarily on the sign of L_{ab}/L_{bb}. This ratio can be positive or negative, depending on the interaction between the solute and the solvent atoms in the jumping process.

To evaluate the solute effect, one requires data for Z_a^*, Z_b^*, D_b^*, $D_a(0)$ together with b' and the ratio of L_{ab}/L_{bb}. While the values of Z_a^*, Z_b^*, D_b^*, $D_a(0)$ can be measured directly, the values of b' and L_{ab}/L_{bb} have to be deduced depending on the kinetic model used to represent solute diffusion. One such model was set up by Howard and Manning [10] for metals with face-center cubic (fcc) crystalline structures by considering only the nearest-neighbor atomic jumps as illustrated in Figure 2.1.

In this model, only the nearest-neighbor jumps for a vacancy associated or dissociated from a solute atom are considered. Based on this model, Manning deduced the following expression:

$$\frac{L_{ab}}{L_{bb}} = \frac{-2 + \phi\omega_3/\omega_1}{1 + \tfrac{7}{2}F\omega_3/\omega_1},\qquad(2.14)$$

where ϕ and F are both function of ω_4/ω_0. Doan [11] showed that the enhancement factor b' can be expressed in terms of the atomic jumps as follows:

$$b' = -18 + 4\frac{\omega_4}{\omega_0}\left(\frac{\omega_1}{\omega_3} + \frac{7}{2}\right).\qquad(2.15)$$

The formulation was applied to analyze the solute effect for EM in Al, Cu, Ag, and Au where detailed analyses have been carried out based on existing diffusion data to extract the ratios of atomic jumps required for evaluating the solute effect [5]. Results

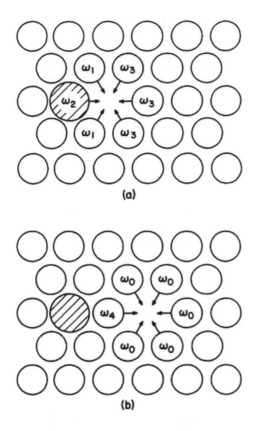

Figure 2.1 Vacancy-jumping frequencies for jumps in the (111) plane of an fcc lattice: (a) vacancy next to a solute atom; (b) vacancy dissociated from a solute atom. Reprinted from [5], with the permission of APS Physics.

of the analysis are summarized in Table 2.1 in terms of three α parameters, which are defined as follows:

$$\frac{J_a^e}{J_a^e(0)} = 1 + \alpha_a c_b; \quad \alpha_a = b' + \frac{D_b^*}{D_a(0)} \frac{L_{ab} Z_b^*}{L_{bb} Z_a^*}$$

$$\frac{J_b^e}{J_b^e(0)} = \alpha_b = \frac{D_b^*}{D_a(0)} \left(\frac{Z_b^*}{Z_a^*} + \frac{L_{ab}}{L_{bb}} \right) \tag{2.16}$$

$$\frac{J_v^e}{J_v^e(0)} = 1 + \alpha_v c_b; \quad \alpha_v = \alpha_a + \alpha_b - 1.$$

In this way, the solute effect is expressed in terms of α_a for solvent EM as a function of the solute concentration, α_b for the effect on solute EM, and α_v for the effect on the overall mass transport.

Results as summarized in Table 2.1 show two general features about solute effects on EM. First and rather surprising, most of the solutes are expected to enhance the EM of the

Table 2.1. Results from kinetic analysis of solute effects on electromigration.

Solvent	Solute	Z_b^*	Z_b^*/Z_a^*	L_{ab}/L_{bb}	α_b	b'	α_a	α_v
	Cu	−6.8	0.4	−0.4	~0	~100	~100	~100
Al	Zn	−33.9	2.0	−0.79	2.4	19.5	16.3	17.7
	Mg	−52.4	3.1	−0.79	8.3	23.0	14.2	21.5
	Zn	−23.3	2.9	−0.29	7.3	8.4	6.1	12.4
Cu	Fe	−61.3	7.7	−1.2	5.4	−1.2	−8.9	−4.5
	Cu	−16.1	1.3	−0.65	0.56	9.8	9.1	8.7
	Zn	−27.6	2.2	−0.53	4.9	13.5	10.1	14.0
Ag	Cd	−29.1	2.3	−0.73	4.5	11.4	6.5	10.0
	In	−48.9	3.9	−0.71	13.2	17.5	6.0	18.2
	Sb	−115.0	9.2	−1.2	47.5	73.1	7.5	54.0
	Fe	−58.0	4.6	−0.44	2.2	−12.6	−13.7	−12.5
Au	In	−22.2	2.4	−1.3	7.7	84.4	62.5	69.2
	Sn	−32.6	3.5	−1.4	25.9	162.3	101.8	126.7

Source: Reprinted from [5], with the permission of APS Physics.

solvent, i.e., $\alpha_a > 0$, except for Fe in Cu and Ag. It is convenient to separate the contributions to α_a due to enhanced diffusivity (b') and the vacancy flow factor (the second term in α_a). Judging from the values in Table 2.1, the positive enhancement in diffusivity generally overwhelms the negative vacancy flow correction, resulting in the solvent EM enhanced by the solute. This is because the large positive b' term can overcome a retarding vacancy flow effect with a universal negative sign of L_{ab}/L_{bb}. Second, α_b is universally positive and larger than 1, despite the negative vacancy flow effect. This indicates that the solute migration is generally faster than the solvent atoms because it is enhanced by the migration of the solvent atoms (Z_b^*/Z_a^* is larger than L_{ab}/L_{bb}). However, this effect can vanish since the solute atoms migrate faster than the solvent atoms and can be depleted, so a minimum solute concentration is required to maintain this effect.

It is interesting to trace these effects by examining the relative rates of atom jumps shown in Figure 2.1. Here the negative vacancy flow effect can be traced to the fact that $\omega_3/\omega_1 < 1$ for all the solutes analyzed. As the ω_3 jump dissociates the vacancy from the solute and the ω_1 jump maintains the vacancy–solute pair with the vacancy jumping around the solute, $\omega_3/\omega_1 < 1$ indicates that the solute tends to trap the vacancy around itself instead of releasing it to move along with the current flow. The exceptions in the solute effect of Fe in Cu and Ag can be traced to a rather unique combination of $\omega_4/\omega_0 < 1$ and $\omega_3/\omega_1 < 1$, indicating an interesting jumping process for the vacancy around the Fe atom. The value of $\omega_4/\omega_0 < 1$ implies that the vacancy does not prefer the associative jump to near the Fe atom, but the fact $\omega_3/\omega_1 < 1$ indicates that once the vacancy–solute pair is formed, the vacancy tends to stay bound. Energetically, this suggests the existence of a higher potential barrier for the dissociative jump ω_3 than for ω_1, and for the associative jump ω_4 than for ω_0. The origin of such characteristics is not clear but may be traced to the nature of the interaction between Fe and Cu or Ag atoms.

One significant effect derived from this analysis is that the addition of most solutes will not reduce the overall rate of electromigration. This raises a basic question regarding the mechanism of using alloy addition to improve EM lifetime of conductor lines, as observed for the case of Cu addition in Al [12] and later the use of Mn addition in Cu [13]. The difficulty suggests that since EM in conductor lines is dominated by grain boundaries and interfaces, the kinetics of solute addition on EM must be different from the bulk lattice. The solute effect in Al and Cu lines has been extensively studied by experiments and modeling and will be discussed later in Chapters 5 and 7.

2.4   Theory of Electromigration

The development of electromigration theory has evolved around the two components of the driving force: electrostatic and electron-wind. The early development in theory includes the semiclassical ballistic model by Fiks [14] and by Huntington and cow-orkers [2, 15], and the quantum-mechanical formulation by Bosvieux and Friedel [16]. Although those early models are relatively simple formalism, most of the key concepts of the driving force have been incorporated. The theoretical development has continued for more than two decades, with new formulations and refinements. The course of the theoretical developments has been reviewed by Sorbello [17, 18] and Verbruggen [19].

In the ballistic model, the electron-wind force is formulated from the momentum transfer to the jumping atom due to its collision with the charge carriers, and the electrostatic force is taken to be the force exerted by the field on the bare ion. The original formulation by Fiks [14] is simple and intuitive, where the wind force is calculated from the total momentum transfer per unit time due to collision with the electrons and expressed in terms of an effective charge Z^w defined as

$$Z^w = -n_e \lambda_e \sigma_e, \qquad (2.17a)$$

where n_e is the electron density, λ_e the mean free path between collisions, and σ_e the cross section of the electron–ion collision. This expression was subsequently extended to include the contribution from the positive holes [20] as

$$Z^w = -n_e \lambda_e \sigma_e + n_h \lambda_h \sigma_h. \qquad (2.17b)$$

Huntington and Grone [15] treated the electron–ion collision by considering the change of the electronic states caused by the collision. In this semiquantum mechanical treatment, a probability matrix was set up to account for the transition from an initial to a final state of the electron. The state of the electron and its momentum were initially taken to be those of the free electron. Subsequently, this was modified to include the scattering of the holes by considering the momentum transfer coming from the pseudomomentum of the electronic states, i.e., the momentum of the electron plus the lattice [21]. It was also pointed out that the force is not constant during jumping and should be deduced by integrating the transition matrix over the jumping path of the moving atom. Such an approach is important for treating different types of jumping processes. For example, while jumping of the interstitial atom is not expected to affect its charge state due to interacting with electrons throughout the jumping path, the jumping of the substitutional atom is quite different since the electron scattering at the saddle point can be considerably higher than at the lattice sites of the starting and final positions due to the different electron screening. This effect is incorporated by means of a parameter of specific defect resistivity. Accordingly, the effective charge Z^w can be expressed as

$$Z^w = -Z\left(\frac{\rho_d}{N_d}\right)\left(\frac{N}{\rho}\right)\frac{|m^*|}{m^*}$$

$$= {}^K\!/_\rho(T), \tag{2.18}$$

where the ratio of the specific resistivities between ρ_d/N_d of the moving defect and ρ/N of the lattice atom provides a measure of the electron–ion scattering throughout the jumping process, and the $|m^*|/m^*$ term is related to the sign of the charge carrier with an effective mass m^*. The value of ρ_d/N_d varies according to the jumping mechanism. For the case of substitutional self-diffusion, where the scattering power varies approximately in a sinusoidal manner from the equilibrium lattice position to the saddle point, it can be taken to be about half of that at the saddle point; for impurity diffusion, its value can be appreciably higher even for substitutional jumping.

Bosvieux and Friedel [16] developed the first quantum-mechanical formulation of the driving force. They deduced the wind force by calculating the electrostatic force exerted by the electron charge density on the bare ion in the presence of the electric field. Sorbello [18] showed that this approach can be justified by Oppenheimer's adiabatic approximation and is equivalent to the use of the Feymann–Hellman theory to treat a closed system. In this formalism, the wind force is derived by applying the operator equation of motion for the ion momentum as

$$F_i^w = \int n(r)\left(-\frac{\partial V_b}{\partial R_i}\right)d^3r, \tag{2.19}$$

where $n(r)$ is the electron density in the presence of the impurity and electric field, and V_b is the bare impurity potential located at position R_i. The electric field and current have two effects on the electrons: first, the wave function is modified because of charge polarization; and second, the Fermi surface is displaced due to the current flow. In this way, the direct force comes from the sum of the polarization effect on the charge plus the electrostatic force on a bare ion, where the wind force is derived from the second effect due to the interaction with the current flow. These two effects are assumed to be independent and were evaluated by a self-consistent, time-independent perturbation method based on free electron wavefunctions. Interestingly, one can show that by carrying out the calculation to the first order, as in the original work, the wind force can be reduced to a form identical to that of the ballistic model.

One controversy arose from the result derived for the direct force by Bosvieux and Friedel [16]. The calculation yielded a result that an interstitial impurity would be completely screened, and thus encounters no direct force, while a substitutional impurity would experience a force identical to the lattice atom. The complete screening of the interstitial impurity is surprising and contradicts the result of the ballistic model, which has a direct force corresponding to a bare ion. Even though the magnitude of the direct force is small in comparison with the wind force for most metals, the controversy regarding the extent of screening, particularly for the interstitials, has lasted for a long time. This problem has since been resolved as discussed later in Section 2.5 for EM of interstitial atoms.

The pioneering work of Bosvieux and Friedel has led to considerable interest in subsequent development of theories for electromigration. In the development of formal theories, Das and Peierls [22] formulated the problem using a semiclassical approach based on the Boltzmann transport equation. This formulation included for the first time the many-body nature of the scattering problem using an electron distribution function to calculate the scattering probability. Following the self-consistent approach of Bosvieux and Friedel [16], these authors calculated the scattering using a stationary transport equation and a perturbation method to treat the change of the distribution function and the impurity potential due to ion scattering. The final expression for the wind force in the lowest-order approximation is similar to that derived by Bosvieux and Friedel except for a parameter that depends on the electron mean free path. As a result, the wind force becomes dependent on the mean free path of the electrons. However, since the final state is not restricted to an upper bound of $2k_F$ at the Fermi surface in this semiclassical treatment, the integration to yield the wind force diverges. This difficulty was resolved by Sorbello and Dasgupta [23] and Schaich [24] using a quantum-mechanical generalization of the formulation based on the density matrix theory. Using a Liouville equation and a density matrix operator (the quantum-mechanical equivalents of the Boltzmann transport equation and electron distribution function, respectively), this approach yielded a similar expression for the wind force except that the divergence is removed because of the cutoff of the momentum vector at k_F. Other than that, the expression is the same as that derived by Das and Peierls [22] for a weak scatterer.

Subsequently, a linear response theory was developed for calculating the electromigration force on impurity based on a quantum field theory by Kumar and Sorbello [25]. This was first applied to the case of a dilute distribution of weak scatterers interacting with nearly free electrons by Sham [26] and Schaich [27], which was subsequently extended to treat strong scatterers by Rimbey and Sorbello [28]. Although the formulation is complex, this approach has two advantages: first, it starts from a general expression of the force in the presence of current and field; second, the many-body, quantum-mechanical nature of the scattering process is treated in a self-consistent manner. In the weak scattering limit, this approach gives a wind force for an impurity atom the same as obtained by Bosvieux and Friedel in the lowest-order approximation. However, the direct force was found not corresponding to the complete screening as deduced by Bosvieux and Friedel. The nature of the difference was clarified by Sorbello [18] as due to an additional contribution arising from the dynamic nature of the scattering process due to the interaction with the current flow. Sorbello showed that the magnitude of this contribution depends on the extent of the interaction between the charge carrier and the impurity. For weak scatterers, the direct force corresponds to the unscreened case of the bare ionic charge, thus confirming the result of the classical ballistic model. With this result, the controversy regarding the extent of screening on the direct force appears to have been resolved for the case of weak scatterers. For strong scatterers, the ionic charge is in general not completely screened. Thus, the magnitude of the effective charge depends on the extent of the dynamic scattering. Under certain circumstances, the dynamic effect can give rise to an effective charge of the direct force exceeding the ionic charge. The magnitude of this correction term is about 10–30% as estimated for a square-well model of a strong scatterer [18].

Two additional contributions to the driving force have been formulated by Landauer and co-workers. The first is called the residual resistivity dipole and is related to the wind force [29]. This term arises from the voltage increase required to maintain a constant current because of the increase in the resistivity due to the impurity. It is essentially a second-order correction to the wind force, although its magnitude may be comparable to that obtained by Bosvieux and Friedel for a weak scatterer. The other is called the carrier-density modulation effect and is related to the direct force [30]. This term originates from the decrease in the local electric field due to the increase of the electron density near the impurity in the presence of current flow. It is related to the dynamic origin of the direct force in electromigration, similar in nature to that described in the framework of the linear response theory. The effect, as formulated in a semiclassical manner, generally reduces the magnitude of the direct force.

Following the theoretical development, experiments have been carried out to directly measure the direct force and the wind force of interstitials by Verbruggen et al. [31]. This finally resolves the direct force controversy, and the results will be discussed in Section 2.5.

Since the Bosvieux–Friedel approach incorporates the impurity potential in a self-consistent manner, the formalism is useful as it can be adopted to calculate the wind force for real metals. For this purpose, the calculation becomes a matter of determining the total charge density $n(r)$ arising from the interaction between the charge carriers with the impurity and its surrounding atoms in the presence of electric field and current. For defects with complicated jumping configurations, there can be a substantial contribution to the wind force from scattering with the surrounding atoms. It has been suggested that such a contribution may be important for understanding electromigration for fast diffusers [32] as well as for electromigration in grain boundaries [33].

The calculation of the effective charge for real metals began initially for simple metals with a nearly full band of conduction electrons. Sorbello [34] adapted a pseudopotential approach to calculate the driving force based on a self-consistent approach. Using a local pseudopotential, the driving force on different types of defects can be calculated by taking the appropriate structure form factors. The resultant force was found from two contributions: first, a one-body interaction coming from the scattering of the electrons by isolated defects, and the other, two-body interactions arising from interference effects in scattering. Numerical results show an agreement within 50% of the measured effective charges for simple monovalent (e.g., Na and K) and divalent metals (e.g., Be and Mg) as well as their alloys. For metals with more complex electron structures, such as the noble metals, the agreement is less satisfactory since those metals are not as well described by pseudopotentials. Another calculation treated the anisotropy in the driving force for divalent metals of Zn and Cd with hexagonal crystalline structures. This work was first developed using Bloch wave functions and a perturbation method to treat specifically the effect on scattering due to the anisotropy of the Fermi surface at the Brillouin zone boundary by Feit and Huntington [35] and Huntington et al. [36]. Later this was generalized by using a dielectric matrix to treat the effect on electron screening due to the band structure [37]. In this calculation, a semiempirical pseudopotential form factor was used, and the results were able to account for the observed anisotropy in the driving force for Zn.

Later works on numerical calculations include one for liquid binary alloys by Jones and Dunleavy [38] and another for transition metals by Gupta et al. [39]. The work on liquid binary alloys [38] was based on the linear response theory. Using simple model potentials, the calculation appears to yield a reasonable semiquantitative description for alkali metal alloys but not for heavy metal alloys, such as Pb-Sn. The study on transition metals [39] uses a formal scattering theory to treat electron–ion interactions based on a self-consistent potential of the moving atom. The calculated effective charge for several transition impurities in Nb was found to be relatively small and did not seem to agree well with experiments.

2.5 Electromigration in Bulk Materials

In this section, the results of experimental studies of electromigration on bulk materials are reviewed. Most of these results were obtained at an early stage of electromigration studies up to about 1985 before the studies on thin films. Here the discussion starts with a summary on the measurement techniques. It is followed by a review on the results from two key areas of investigation on bulk materials, namely substitutional and interstitial electromigration.

2.5.1 Techniques of Measurement

The techniques of measurement for electromigration have been reviewed previously for bulk materials by Huntington [2] and for thin films by d'Heurle and Ho [40]. Essentially, there are two basic types of measurements: one determines the compositional change induced by the current flow, and the other measures the mass transport

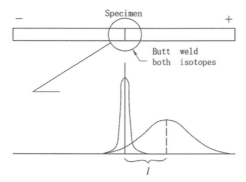

Figure 2.2 The isothermal isotope method: (a) inert marker concentrations; (b) concentration of matrix isotope, where l is the displacement and t the duration of the experiment. Temperature is determined from the diffusional spreading profile (b).

relative to a fixed reference frame. The principles of these two types of measurements apply to bulk materials as well as thin films.

The first type of measurement is similar as the usual diffusion measurement except that the sample must be designed to allow current flow with a high current density to yield a measurable amount of electromigration. The principle of this type of measurement is shown in Figure 2.2.

For bulk studies, the sample is usually made into a cylindrical form with liquid cooling of both ends. The element to be measured can be incorporated uniformly into the sample or inserted near the center of the sample. Upon passage of the current, the composition distribution will change due to electromigration. For the case with a uniform initial distribution, a steady-state distribution can be established after sufficient time, which can be derived from (2.8) by setting $J_b = 0$ as

$$\nabla \ln c_b = Z_b^{**} eE/f_b kT$$
$$= \frac{Z_b eE}{f_b kT} \left(1 + \frac{L_{ba} Z_a^*}{L_{bb} Z_b^*} \right). \tag{2.20}$$

Thus, a measurement of the profile of c_b can yield directly Z_b^{**}/f_b without knowing the diffusivity of the moving species. In the case of self-electromigration, the vacancy flow factor inside the bracket simplifies to unity, then the measurement gives Z_b directly. Since the diffusivity is usually difficult to measure precisely, this technique provides a simple means to measure the effective charge.

For the case with a sharply defined source, its effective charge can be measured from the displacement of the center of the source [2] according to the following expression:

$$Z_b^{**} = -\frac{l}{t} \left(\frac{f_b kT}{DeE} \right), \tag{2.21}$$

where l is the displacement of the center and t the electromigration time. In this case, the determination of Z_b^{**} requires the measurement of D. This can be done by

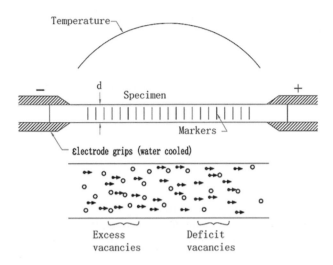

Figure 2.3 Schematic electromigration experiment for vacancy flux method: • electron, ▸ vacancy.

measuring the thermal spreading of the profile c_b without the passage of current. Since D is difficult to measure, this method is more tedious and less accurate than the steady-state method. Again, for self-electromigration, one can measure Z_b directly.

Several methods of chemical analysis can be used to measure the composition profile, with the most common being the isotope-sectioning technique and the ion or electron microprobe technique. Other techniques based on measurements of properties depending on the composition, such as electrical resistance and mechanical hardness, have also been used.

The second type of measurement is called the "vacancy flux" method, which measures the change in the sample dimension due to the creation and annihilation of vacancies along the sample during electromigration. The principle of this measurement is illustrated in Figure 2.3.

The early experiments simply measured the displacement of surface markers (scratches or indentations) separated by a uniform spacing along the length of a wire sample [15]. Later, in a study of Al [41], it became apparent that the change in the dimension is not confined to being along the length of the sample only; there are also transverse dimensional changes. The latter was found to depend on the aspect ratio of the sample and should be included to account for the total vacancy flux. The simplicity of the measurement has made this method one of the principal techniques in electromigration studies, and it has been applied to measure bulk electromigration for several pure metals. Results of these measurements are in general agreement, although not with the same accuracy, as those obtained from chemical analysis.

The general principles of these two types of measurements have been extended to thin film studies, although the difference in the characteristics of electromigration necessitates changes in the experimental condition as well as in the analysis of the composition profile [42]. While the details of the thin film studies can be found in a review article [40], two salient features are worth noting. First, those studies are usually carried out on films with a supporting substrate, which significantly increases the dissipation of the Joule heating generated by the electrical current. As a result, the current density applied can be increased by more than an order of magnitude, i.e., from 10^3–10^4 A cm^{-2} to 10^4–10^5 A cm^{-2}. This gives rise to a corresponding increase in the driving force, thus improving the accuracy of the measurement. Second, in thin films the temperature range where electromigration of interest is lower than in bulk materials is at about half of the absolute melting point, where the mass transport is dominated by grain boundaries. Thus, the effect of microstructure becomes important in controlling the mass transport and has to be considered, which requires a different analysis for electromigration at grain boundaries [43–45]. In spite of these differences, the two basic types of measurement for bulk materials have been applied for studying thin film electromigration. For measurement of the composition distribution, chemical analyses using electron microprobe [43] and radioactive isotopes [44] have been employed.

Other measurement techniques have been developed to measure the mass flux, specifically the mass accumulation and depletion along thin film stripes during the course of electromigration [46]. Among this type of measurement, of particular interest is the technique of drift velocity measurement [46, 47]. In this technique, one can measure the mass transport as well as the stress generated in the film stripe due to electromigration. Such measurements on mass flux are of basic interest to EM studies of conductor lines since they provide a direct measure of the rate of damage formation due to mass depletion under electromigration.

2.5.2   Electromigration in Interstitial Systems

Studies of electromigration of interstitials can be traced to the pioneer work in the early 1930s on H in Pd [48] and C in Fe [49]. However, the interest in this area of study was not generated until the late 1960s, when the research group in the Ames Laboratory found electromigration to be effective in purifying interstitial impurities such as H, C, O, and N in refractory and rare-earth metals [50–52]. This was later stimulated by the basic interest in studying the quantum effect on the isotope dependence of Z^* [53, 54]. Since then, interstitial electromigration has been investigated in a number of metals, particularly for elements in columns I11 to VI11 in the periodic table. Results of these studies have been reviewed by Huntington [2] and Wever [55]. Later in the 1980s, there was renewed interest in investigating interstitial electromigration due to the interest in measuring the direct driving force to resolve the controversy regarding the extent of ionic screening [56, 57]. Studies have also been carried out to probe the behavior of the fast diffusing impurities in Pb alloys [58].

Conceptually, interstitial electromigration is appealing because of its simplicity. Unlike the vacancy, an interstitial can diffuse without the complication of the saddle-point configuration associated with the atomic exchange with the vacancy. Therefore, the driving force is nearly constant throughout the entire jump, simplifying the theoretical treatment of the scattering process. In addition, the fast transport of interstitials reduces the measurement difficulty since the steady-state approach can be used to yield a reliable measurement of Z^*. However, these advantages are offset by the fact that few of the impurities can be classified as interstitials and only in metals with relatively complex electronic structures, such as the refractory metals. The latter factor complicates the interpretation since the contributions from both the electrons and the holes must be taken into account. In spite of such complications, a clear trend has emerged in the results of interstitial electromigration as summarized in Table 2.2. First, except for a few exceptions, the transport direction of the interstitials is the same in the host metal. Second, the sign of Z^* is positive for elements in columns V, VI, and VIII, but negative for columns I, III, and IV. These results are interpreted as an indication that the wind force dominates, so the direction is the same for all interstitials in the same matrix. The direction of the wind force depends on the balance between the scattering from the electrons versus the holes. Thus, the sign of Z^* reflects the dominance of electrons in metals with less than half-filled d bands and holes in metals with more than half-filled d bands. Quantitatively, the magnitude of Z^* is less satisfactorily accounted for due to the difficulty in treating the scattering in metals with d bands, so their values are not shown in Table 2.2. In some cases, e.g., H in Fe and Ni [59], Z^* is small (about 0.5), indicating a delicate balance between electron and hole scatterings.

Interestingly, results on H and D showed that the sign and magnitude of Z^* in Ni are the reverse of those in Ag [60]. These results do not follow the theory developed by Flynn and Stoneham [53, 54]. These authors predicted a larger wind force for the proton than for the deuteron because the proton has a larger displacement field normal to the jumping direction due to its higher zero-point energy. In addition, the theory of Flynn and Stoneham cannot account for the magnitudes of Z^* in Ag, which are about an order of magnitude higher than those in Ni and Fe. Huntington [2] suggested that H behaves rather like a proton in Ag and thus experiences a strong scattering from the charge carriers.

For electromigration of interstitials, the controversy on the direct force about the screening of interstitial impurities is of basic interest. The problem was first investigated by Herold et al. [61] for H and D in V, Nb, and Ta. The experiment has been repeated by Erckmann and Wipf [56], who improved the accuracy of the measurement by using a steady-state resistometric technique and by extending the measurement over a range of temperatures. The steady-state approach provided an accurate measurement of Z^*, where the temperature dependence of Z^* allowed a direct determination of Z^e according to the following expression:

$$Z^* = Z^e + K/\rho(T), \tag{2.22}$$

where K is a parameter defined according to (2.18). The results obtained between 49°C and 245°C were analyzed according to (2.22), and Z^* was found to be close to unity. Recently the experiment was repeated by Verbruggen et al. [57] using a more sensitive capacitance technique to measure the volume change of the sample due to the steady-state distribution of H driven by electromigration over a temperature range from 275–525°K. The improved accuracy of this measurement allowed these authors

Table 2.2. Electromigration of interstitial impurities.

Solvent	Column in periodic table	Impurity + to cathode	Impurity − to anode	Reference*
Cu	IB		H	I
Ag	IB		H, D	II
			H	I
Y	IIIB		C, O, N, H	II
			H, C, N, O	I
Ti	IIIB	C	O	II
		H, C	N, O	I
Zr	IVB		O, N, C	II
		H	C, N, O	I
Hf	IVB		C, N, O	I
V	VB	C, O, N, H, D		II
		H, C, N, O		I
Nb	VB	C, O, H, D		II
		H, N		I
Ta	VB	C, H, D		II
		C, N, O		I
Mo	VI B	C, N, O		I
W	VI B	C		II
		C		I
Fe	VIII	H, C, N, O	N	I
Fe-α	VIII	H, D, C		II
Fe-γ	VIII	C, B	N	II
Co	VIII	C		II
		C		I
Ni	VIII	H, D, C		II
		H, C, N, O		I
Pd	VIII	H, D		II
		H		I
Ce	Lanthanum		O	I
Gd	Lanthanum		C, N, O	II
			C, N, O	I
Tb	Lanthanum		N, O	I
Lu	Lanthanum		C, N, O	II
			C, N, O	I
Th	Actinium		C, N	II
			C, N, O	I
U	Actinium		C, N, O	I

* Reference: I: [55]. II:[2].

to determine directly Z^e from the temperature dependence of Z^*, as shown in Figure 2.4, by plotting Z^* versus inverse resistivity. The experimental results show that Z^* follows closely a linear relationship that allows a quantitative determination of Z^e for the first time. This yields a value of 1.11, 1.23, and 0.44 for interstitial H in V,

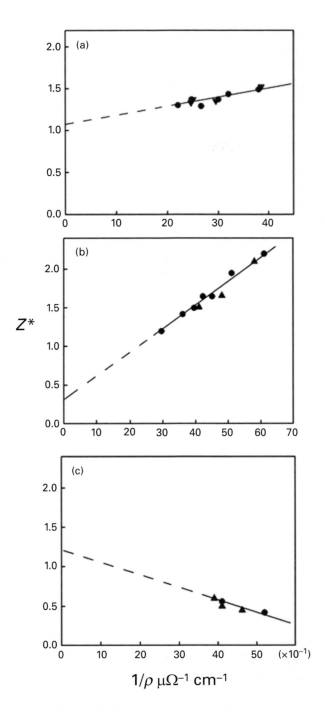

Figure 2.4 Effective valence Z^* vs. inverse resistivity for H in (a) V (● $VH_{0.0115}$, ▼ $VH_{0.0100}$); (b) Nb (● $NbH_{0.0110}$, ▲ $NbH_{0.0115}$, ■ $NbH_{0.0103}$), and (c) Ta (● $TaH_{0.0055}$, ▲ $TaH_{0.0108}$). The lines are least square fits to the expression $Z^* = Z^e + K/\rho$. From [57], © IOP Publishing. Reproduced with permission. All rights reserved.

Nb, and Ta respectively, with a measurement error of 10–20%. It seems that the results of these experiments have finally resolved the controversy of the direct force in electromigration, indicating that the impurity is not completely screened, but not equal to that of a bare ion neither. Interestingly, the value of the direct force Z^e was found to be able to deviate from unity, either positively (1.23 for Ta) or negatively (0.44 for Ta), for the column 5 transition metals. This was attributed to the dynamic nature of the scattering process in electromigration.

The study of fast interstitial diffusers in Pb alloys is of basic interest since those diffusers are metallic elements and their diffusion mechanism has been studied for a long time. This study has considerable technological interest as well, since those alloys are commonly used to form solder contacts in microelectronics, where electromigration is a key reliability factor. Results of electromigration studies in these alloys have been reviewed by Hu and Huntington [58]. The results of this study provided an interesting contrast to the gas interstitial systems discussed so far since those fast diffusants are metallic elements and their transport occurs via a more complex interstitial or interstitialcy mechanism.

2.5.3   Electromigration in Substitutional Systems

The study of substitutional electromigration can be traced to the early work of Seith and Wever [62] in multiphase binary alloys in the early 1950s. The interest in such studies started to flourish in the early 1960s, when the theoretical models were first formulated (see Section 2.4). These models proposed the origin of the wind force as being electron–ion collision, thus indicating the possibility of using electromigration to probe the nature of electron–ion interaction in metals. It became clear also that it is simpler to study electromigration in pure elements instead of alloys. This led to about a decade of systematic studies on noble metals and other monovalent and divalent metals with relatively simple electronic structures. The study was subsequently extended to systems with complex electronic structures, including polyvalent, transition, and refractory metals, as well as substitutional alloys. Substitutional electromigration in bulk metals has been reviewed by Huntington [2] and Wever [55]. Selected values of Z^*/f for pure metals summarized by Huntington [2], Verhoeven [63], and Pai and Marton [64] are listed in Table 2.3. Except for a few additional studies since then, these articles provide comprehensive summaries of the experimental results. Here only the salient features of these results are included.

The results in Table 2.3 for the noble metals, Au, Ag, and Cu, are in general agreement on the value of Z^*/f obtained in different laboratories except for the rather high value obtained by Doan [66] for Ag. Electromigration in these metals is clearly dominated by the electron wind force, about 5 to 10 times larger than the direct force. The behavior of the alkali metals, Li and Na, is similar as the noble metals with the direction of electromigration dominated by the wind force. In comparison, the magnitude of the wind force is generally smaller, giving an effective charge of only about −2. It is interesting that the relative magnitudes of Z^* for the noble metals and alkali metals seem to be consistent with their relative magnitudes of electrical

Table 2.3. Z^*/f for substitutional electromigration in metals.

Metal	Z^*/f	Temperature range (°C)	Reference
Noble metals			
Gold	−9.5−−7.5	850–1,000	Huntington and Grone (1961) [15]
	8.0		Gilder and Lazarus (1966) [65]
Silver	−21 ± 5	830–890	Doan (1971) [66]
	−8.3 ± 1.8	795–900	Patil and Huntington (1970) [67]
Copper	−5.5 ± 1.5	845–1,030	Sullivan (1967a) [68]
	−4.8 ± 1.5	870–1,005	Grimme (1971) [69]
Alkali metals			
Lithium	−2.5−−1.6	90–160	Themquist and Lodding (1968) [70]
Sodium	−0.3 ± 0.7	45–80	Sullivan (1967b) [71]
Divalent metals zinc			
// to c axis	−2.5 ± 0.2	366–400	Routbort (1968) [72]
− to c axis	−5.5 ± 0.6		
Cadmium			
// to c axis	−2.0 ± 0.2	215–290	Alexander (1971) [73]
− to c axis	−4.1 ± 0.4		
Magnesium	2.0 ± 0.3	500–580	Wohlgemuth (1975) [74]
Trivalent metals			
Aluminium	−30−−12	480–640	Penney (1964) [41]
	Comparable	450–610	Heumann and Meiners (1966) [75]
Indium	−11.5	115–150	Lodding (1965) [76]
Gallium (liquid)	−1.3	18–312	Lodding (1967) [77]
Thallium	−4.6 ± 0.5	233–303	Lundan et al. (1972) [78]
Transition metals			
Nickel	−3.5	1,000–1,400	Hering and Wever (1967) [79]
Iron	+2 ± 1	700–1,300	Hering and Wever (1967) [79]
Cobalt	+1.6 ± 0.3	1,260–1,360	Ho (1966) [80]
	Negative	540–640	Van Gurp (1976, 1977) [81, 82]
Platinum	+0.28 ± 0.04	1,480–1,670	Huntington and Ho (1963) [83]
Refractory metals			
Zirconium	+0.3	930–1,730	Campbell and Huntington (1969) [84]
Uranium	−1.6 ± 0.1	830–1,100	D'Amico and Huntington (1969) [85]
Quadrivalent metals			
Lead	−47	250	Kuz'menko (1962) [86]
Tin	−80	190	Khar'kov and Kuz'menko (1960) [87]
Tin	−18	180–213	Khosla (1973) [88]

conductivities. This suggests that the electrons, instead of the holes, dominate in the scattering processes underlying these two phenomena.

The studies on the divalent metals, Zn and Cd, were carried out on single-crystal specimens. The results showed clearly measurable anisotropy in the values of

Z^* along the a and c crystalline axes. The observed anisotropy in Z^* is inconsistent with that of the electrical conductivity, which is only slightly higher along the a axis. This reveals the different nature of the scattering processes responsible for electromigration and conductivity of these metals. This question has been resolved in a theoretical study by Genoni and Huntington [37]. Based on the band structure of Zn, these authors showed that the anisotropies in Z^* and conductivity are different primarily due to the different averages of the relaxation time of the electron–ion scattering for the two phenomena.

Results for transition and refractory metals show that the value of Z^* is generally small and the wind force can be positive or negative. Thus, for these metals with complex band structures, the wind force is a result of the balance of the scattering from electrons and holes, and the outcome is difficult to predict. This was first demonstrated for cobalt [80], where the positive Z^* was measured at about 1,300°C in bulk Co and found to correlate with the positive Hall coefficient. However, subsequent studies in Co films by van Gurp [81, 82] found that electromigration at a lower temperature of 530°C was directed toward the anode with a negative Z^*, which is correlated with the sign of the normal Hall coefficient R_0, being negative below about 400°C and positive at higher temperatures. Such a correlation does not seem very strong for other metals of this category, e.g., Pt has a small positive Z^* but a negative Hall coefficient [83]. It appears that the complex band structure of these metals is making it difficult to establish a simple correlation. For the same reason, the effective charges of these metals have not been as satisfactorily accounted for by theory as those of the monovalent or divalent metals.

For the trivalent and quadrivalent metals, electromigration is in general observed to be dominated by the electron wind force. The magnitude of Z^* appears to vary considerably and there seems to be no systematic pattern for these metals. It is interesting to note that Al, a commonly used metal in microelectronics, has a high value of Z^*. This high driving force presents a basic reliability problem due to electromigration in microelectronic devices.

Electromigration of impurities in bulk metals or alloys has not been as extensively studied as self-electromigration in pure metals. Nevertheless, there are some interesting results, notably the study by Doan [89] on the valence effect on electromigration in Ag using elements in the same row of the periodic table as Ag (Cd, In, Sn, and Sb). Results of this measurement showed a good correlation of Z^* to the parameter $z(z + 1)$, where z is the difference in nominal valence between the solute and Ag. This correlation was interpreted to show that Z^* originates from an average of the wind forces for the substitutional and saddle-point configurations. However, such a correlation was found to be less clear in a subsequent study of valence effects on electromigration in Cu and Ni [19]. Using another series of solute atoms, Mn, Fe, Co, and Ni, in Ag, Doan [7, 66] has investigated the thermodynamic factors discussed in Section 2.3 of the vacancy flow effect for electromigration in dilute alloy systems and showed that the valence effect is complicated by the kinetics of the correlated jumping processes in dilute alloy systems as discussed in Section 2.3.

2.6   Summary

In this chapter, we have reviewed the basic studies of electromigration aiming to understand the phenomenon as a force-driven diffusion process. On the experimental side, electromigration has been used effectively to probe the nature of electron–ion interaction in substitutional and interstitial diffusion systems. Substantial results have been accumulated, revealing a systematic trend in the effective charge for metals with electronic structures varying from the simple alkali elements to the complex refractory metals. On the theoretical side, thermodynamic formulation and kinetic analysis have provided a framework in which the macroscopic behavior of electromigration can be analyzed. An example using this approach to treat the solute effect on electromigration has been discussed in terms of atomic jumps in Section 2.3. The theory for the effective charge has been relatively well developed, including the use of many-body quantum mechanical formalism to treat the scattering between ions and charge carriers. The controversy regarding the extent of screening on the electrostatic force seems to have been resolved by experiments indicating incomplete screening due to the dynamical nature of the scattering process. All these efforts have provided a good level of understanding of electromigration in bulk metals, where the migration and the scattering processes can be examined in atomistic detail.

The direction of electromigration studies was significantly changed in the early 1970s, following the discovery of the cracked Al stripes in integrated circuits by Blech and Sello in 1966 [90]. This happened at the beginning of development of very-large-scale integration (VLSI) of integrated circuits (ICs) and has generated great interest to study electromigration in thin films and conductor lines. It soon became clear that the problem was quite different, dealing primarily with mass transport at grain boundaries and interfaces in the conductor lines under significantly larger current density but at moderate temperatures of about half of the absolute melting point T_m. Here the material of interest was largely confined to Al and Cu with high electrical conductivity because they are used as on-chip interconnects in ICs. This shifted the direction and broadened the scope of interests of the studies to focus on the effects of materials, microstructure, and line dimensions on mass transport and the development of reliable EM structures. Since then, the study of electromigration has been sustained by unprecedented advances of the IC miniaturization from the micro- to the nanoscale with continued downscaling of the Cu interconnects. The study has now been extended to Co and Ru as alternate interconnect materials. As electromigration has evolved from the bulk metals, the information in this chapter provides the basics for understanding more complex electromigration phenomena and reliability studies in conductor lines. Such studies will be discussed in subsequent chapters.

References

1. Y. Adda and J. Philibert, *La Difusion dans les Solides,* vol. 1 (Paris: Presses Universitaires de France, 1966).
2. H. B. Huntington, *Diffusion in Solids: Recent Developments,* ed. A. S. Nowick and J. J. Burton (New York: Academic Press, 1974), pp. 303–352.

3. R. W. Balluffi, S. M. Allen and W. C. Carter, *Kinetics of Materials* (Wiley Pub., 2005).
4. P. S. Ho and T. Kwok, Electromigration in metals, *Reports on Progress in Physics* **52** (1989), 301–348.
5. P. S. Ho, Solute effects on electromigration, *Physical Review* **B8** (1973), 4534: https://doi .org/10.1103/PhysRevB.8.4534.
6. F. M. d'Heurle and R. Rosenberg, *Physics of Thin Films,* vol. 7 (New York: Academic Press, 1973) p. 257.
7. N. V. Doan, Effet de valence en electromigration dans l'argent, *Journal of Physics and Chemistry of Solids* **31** (1970), 2079–2085.
8. J. R. Manning, *Diffusion Kinetics for Atoms in Crystals* (Princeton: Van Nostrand-Reinhold, 1968).
9. J. L. Bocquet, Correlation factor for diffusion in cubic crystals with solute–vacancy interactions of arbitrary range, *Philosophical Magazine* **94** (2014), 3603–3631.
10. R. E. Howard and J. R. Manning, Kinetics of solute-enhanced diffusion in dilute face-centered-cubic alloys, *Physical Review* **154** (1967): https://doi.org/10.1103/PhysRev.154.561.
11. N. V. Doan, A new method of determination of the vacancy jump frequency ratios by electromigration in dilute alloy, *Journal of Physics and Chemistry of Solids* **33** (1972), 2161–2166.
12. I. Ames, F. M. d'Heurle, and R. E. Horstmann, Reduction of electromigration in aluminum films by copper doping, *IBM Journal of Research and Development* **44** (1970): 89–91.
13. C. K. Hu, J. Ohm, L. M. Gignac et al., Electromigration in Cu(Al) and Cu(Mn) damascene lines, *Journal of Applied Physics* **111** (2012), 093722.
14. V. B. Fiks, Forces produced by conduction electrons in metals located in external fields, *Soviet Physics of the Solid State* **1** (1959), 14.
15. H. B. Huntington and A. R. Grone, Current-induced marker motion in gold wires, *Journal of Physics and Chemistry of Solids* **20** (1961), 76–81.
16. C. Bosvieux and J. Friedel, Sur l'electrolyse des alliages metalliques, *Journal of Physics and Chemistry of Solids* **23** (1962), 123–136.
17. R. S. Sorbello, *Electro- and Thermo-Transport in Metals and Alloys,* ed. R. E. Hummel and H. B. Huntington (New York: AIME, 1977), ch.1.
18. R. S. Sorbello, Theory of the direct force in electromigration, *Physical Review B* **31** (1985), 798. https://doi.org/10.1103/PhysRevB.31.798.
19. A. H. Verbruggen, Unpublished PhD Thesis, Free University of Amsterdam (1985).
20. P. Guilmin, L. Turban, and M. Gerl, Electrotransport d'impuretes dans Cu et Ni, *Journal of Physics and Chemistry of Solids* **34** (1973), 951–959.
21. P. S. Ho and H. B. Huntington, Electromigration and void observation in silver, *Journal of Physics and Chemistry of Solids* **27** (1966), 1319–1329.
22. A. K. Das and R. Peierls, The force of electromigration, *Journal of Physics C: Solid State Physics* **8** (1975), 3348.
23. R. S. Sorbello and B. Dasgupta, Local fields in electron transport: Application to electromigration, *Physical Review B* **16** (1977), 5193; https://doi.org/10.1103/PhysRev B16.5193.
24. W. L. Schaich, Theory of the driving force of electromigration: Weak-charge solutions, *Physical Review* **B 19** (1979), 620; doi.org/10.1103/PhysRevB.19.620.
25. P. Kumar and R. S. Sorbello, Linear response theory of the driving forces for electromigration, *Thin Solid Films* **25** (1975), 25–35.
26. L. J. Sham, Microscopic theory of the driving force in electromigration, *Physical Review B* **12** (1975), 3142; https:// doi.org/10.1103/PhysRevB.12.3142.

27. W. L. Schaich, Theory of the driving force for electromigration, *Physical Review* B **13** (1976), 3350: https://doi.org/10.1103/PhysRevB.13.3350.
28. P. R. Rimbey and R. S. Sorbello, Strong-coupling theory for the driving force in electro-migration, *Physical Review B* **21** (1980), 2150: https://doi.org/10.1103/PhysRevB.21.215.
29. R. Landauer and J. W. Woo, Driving force in electromigration, *Physical Review B* **10** (1974), 1266: https://doi.org/10.1103/PhysRevB.10.1266.
30. R. Landauer, Spatial carrier density modulation effects in metallic conductivity, *Physical Review B* **14** (1976),1474: https://doi.org/10.1103/PhysRevB.14.1474.
31. A. H. Verbruggen, R. Griessen, and D. G. de Groot, Electromigration of hydrogen in vanadium, niobium and tantalum, *Journal of Physics F: Metal Physics* **16** (1986), 557.
32. M. Y. Hsieh, H. B. Huntington and R. N. Jeffrey, Electromigration of Au and Ag in lead. *Crystal Lattice Defects* **7** (1977), 9–22.
33. A. Gangulee and F. M. d'Heurle, Electromigration and transport reversal in copper–silver thin films, *Journal of Physics and Chemistry of Solids* 35 (1974), 293–299.
34. R. S. Sorbello, A pseudopotential based theory of the driving forces for electromigration in metals, *Journal of Physics and Chemistry of Solids* **34** (1973), 937–950.
35. M. D. Feit and H. B. Huntington, Transport in nearly-free-electron-model metals. I. Point-defect scattering, *Physical Review* B5 (1972), 1416; https://doi.org/10.1103/PhysRevB.5.1416.
36. H. B. Huntington, W. B. Alexander, W. B. Feit, and J. L. Routbout, *Atomic Transport in Solids and Liquids,* ed. A. Lodding and T. Lagerwall (Tubingen: Z. Naturf: 1971), p. 91.
37. T. C. Genoni and H. B. Huntington, Transport in nearly-free-electron metals. IV. Electromigration in zinc, *Physical Review* 16 (1977), 1344. https://doi.org/10.1103/PhysRevB.16.
38. W. Jones and H. N. Dunleavy, The calculation of electromigration forces and resistivities for liquid binary alloys, *Journal of Physics F: Metal Physics* **9** (1979), 1541.
39. R. P. Gupta, Y. Serruys, G. Brebec and Y. Adda, Calculation of the effective valence for electromigration in niobium, *Physical Review B* 27 (1983), 672. https://doi.org/10.1103/PhysRevB.27.672.
40. F. M. d'Heurle and P. S. Ho, *Thin Films: Interdiffusion and Reactions,* ed. J M Poate et al. (New York: Wiley-Interscience, 1978), p. 183.
41. R. V. Penney, Current-induced mass transport in aluminum, *Journal of Physics and Chemistry of Solids* **25** (1964), 335–345.
42. P. S. Ho and J. K. Howard, Grain-boundary solute electromigration in polycrystalline films, *Journal of Applied Physics* 45 (1974), 3229. https://doi.org/10.1063/1.1663763.
43. K. L. Tai and M. Ohring, Grain-boundary electromigration in thin films. I. Low-temperature theory, *Journal of Applied Physics* 48 (1977), 28; https://doi.org/10.1063/1.323375.
44. K. L. Tai and M. Ohring, Grain-boundary electromigration in thin films. II. Tracer measurements in pure Au, *Journal of Applied Physics* 48 (1977), 28; https://doi.org/10.1063/1.323336.
45. P. S. Ho, Analysis of grain boundary electromigration, *Journal of Applied Physics* 49 (1978), 2735; https://doi.org/10.1063/1.325196.
46. P. S. Ho, J. E. Lewis, and J. K. Howard, Kirkendall study of electromigration in thin films, *Thin Solid Films* 25 (1975), 301–315.
47. I. A. Blech and E. Kinsbron, Electromigration in thin gold films on molybdenum surfaces *Thin Solid Films* **25** (1975), 327–334.

48. 1. Coehn, A. and W. Specht, Über die Beteiligung von Protonen an der Elektrizitätsleitung in Metallen. *Z. Physik* **62**, 1–31 (1930). https://doi.org/10.1007/BF01340398.

49. W. Seith and O. Kubaschewski, Die elektrolytische Überführung von Kohlenstoff in festen Stahl, *Zeitschrift für Elektrochemie* 41 (1935), 551. https://doi.org/10.1002/bbpc .19350410755.

50. O. N. Carson, F. A. Schmidt, and D. T. Pederson, Electrotransport of interstitial atoms in yttrium, *Journal of Less-Common Metals* **10** (1966), 1–11.

51. D. T. Pederson and F.A. Schmidt, Electrotransport of carbon, nitrogen and oxygen in lutetium, *Journal of Less-Common Metals* **18** (1969), 111–116.

52. D. T. Pederson and F. A. Schmidt, Electrotransport of carbon, nitrogen and oxygen in gadolinium, *Journal of Less-Common Metals* **29** (1972), 321–327.

53. C. P. Flynn and A. M. Stoneham, Quantum theory of diffusion with application to light interstitials in metals, *Physical Review B* **1** (1970), 3966. https://doi.org/10.1103/PhysRevB .1.3966.

54. A. M. Stoneham and C. P. Flynn, *Journal of Physics F: Metal Physics* **3** (1973), 503.

55. H. Wever, Elecktro und thermotransport in metals. *Electro- and Thermo-Transport in Metals and Alloys*, ed. R. E. Hummel and H. B. Huntington (New York: AIME, 1977) Ch. 8, 140–159.

56. V. Erckmann and H. Wipf, Electrotransport of Interstitial H and D in V, Nb, and Ta as Experimental Evidence for the Direct Field Force, *Physical Review Letters* **37** (1976), 341. https://doi.org/10.1103/PhysRevLett.37.341.

57. A. H. Verbruggen, R. Griessen, and D. G. de Groot, Electromigration of hydrogen in vanadium, niobium and tantalum, *Journal of Physics F: Metal Physics* 16 (1986), 557. https://doi.org/10.1088/0305-4608/16/5/006.

58. C. K. Hu and H. B. Huntington, Diffusion and electromigration of impurities in lead solders, *Diffusion Phenomena in Thin Films and Microelectronic Materials,* ed. D. Gupta and P. **S.** Ho (Park Ridge: Noyes Publications, 1988), 2546–581.

59. R. A. Oriani and O. D. Gonzales, Electromigration of hydrogen isotopes dissolved in alpha iron and in nickel, *Transactions of the Metallurgical Society of AIME* **239** (1967), 1041.

60. R. E. Einzinger and H. B. Huntington, Electromigration and permeation of hydrogen and deuterium in silver, *Journal of Physics and Chemistry of Solids* 35 (1974), 1563–1573.

61. A. Herold, J. F. Mareche, and J. C. Rat, Electromigration des isotopes de l'hydrogene dans le vanadium, le niobium et le tantale, *Academy of Science, Paris* **273** (1971), 1736–1739.

62. W. Seith and H. Wever, A new effect in the electrolytic transfer in solid alloys, *Z. Elektrochem* **57** 891 (1953), 61.

63. J. Verhoeven, Electrotransport in metals, *Metall. Rev.* **8** (1963), 311–368.

64. S. T. Pai and J. P. Marton, Electromigration in metals, *Canadian Journal of Physics* **55** (1977), 103. https://doi.org/10.1139/p77–013.

65. H. M. Gilder and D. Lazarus, Effect of high electronic current density on the motion of Au 195 and Sb 125 in gold, *Physical Review* 145 (1966), 507–511.

66. N. V. Doan, Vacancy flow effect on electromigration in silver, *Journal of Physics and Chemistry of Solids* **32** (1971), 2135–2143.

67. H. R. Patil and H. B. Huntington, Electromigration and associated void formation in silver, *Journal of Physics and Chemistry of Solids* 31 (1970), 463–474.

68. G. A. Sullivan, Search for reversal in copper electromigration, *Journal of Physics and Chemistry of Solids* 28 (1967a), 347–350.

69. D. Grimme, *Atomic Transport in Solids and Liquids*, ed. A. Lodding and T. Lagerwall (Tübingen: Z. Naturf. 1971), 178.

70. P. Thernquist and A. Lodding, Electrotransport of lattice defects in lithium metal, *Z. Naturforsch.* 23a (1968), 627–628.

71. G. A. Sullivan, Electromigration and thermal transport in sodium metal, *Physical Review* 154 (1967b), 605.

72. J. L. Routbort, Electromigration in zinc single crystals, *Physical Review* 176 (1968), 796.

73. W. B. Alexander, Electromigration in single crystal cadmium 1, *Zeitschrift für Naturforschung A* 26 (1971), 18–20.

74. J. Wohlgemuth, Electromigration in polycrystalline and single crystal magnesium. *Journal of Physics and Chemistry of Solids* 36 (1975), 1025–1031.

75. T. Heumann and H. Meiners, Electrotransport in aluminium, *Z. Phys.* 57 (1966), 571.

76. A. Lodding, Current induced motion of lattice defects in indium metal, *Journal of Physics and Chemistry of Solids* 26 (1965), 143–151.

77. A. Lodding, Electrotransport and effective self-diffusion in pure liquid gallium metal, Journal of Physics and Chemistry of Solids 28 (1967), 557–568.

78. A. Lundan, S. Christofferson, and A. Lodding, Electrotransport in molten indium-thallium alloys, *Z. Naturf A* 27 (1972), 156.

79. H. Hering and H. Wever, Electro- and thermotransport in nickel, *Z. Phys. Chem.*1 (1967), 310–325.

80. P. S. Ho, Electromigration and Soret effect in cobalt, *Journal of Physics and Chemistry of Solids* **27** (1966), 1331–1338.

81. G. J. van Gurp, Electromigration in cobalt films, *Thin Solid Films* **38** (1976), 295–311.

82. G. J. van Gurp, Electromigration and Hall effect in cobalt films, *Journal of Physics and Chemistry of Solids* 38 (1977), 627–633.

83. H. B. Huntington and P. S. Ho, *Journal of the Physical Society of Japan Supplement* I1 **18** (1963), 202.

84. D. R. Campbell and H. B. Huntington, Thermomigration and electromigration in zirconium, *Physical Review* 179 (1969), 601.

85. J. F. D'Amico and H. B. Huntington, Electromigration and thermomigration in gamma-uranium, *Journal of Physics and Chemistry of Solids* 30 (1969), 2607–2621.

86. P. P. Kuz'menko, *Ukrain. Fig. Zhur.* **7** (1962), 117.

87. E. I. Khar'kov and P. P. Kuz'menko, *Ukrain. Fig. Zhur.* 5 (1960), 428.

88. A. Khosla and H. B. Huntington, Electromigration in tin single crystals, *Journal of Physics and Chemistry of Solids* 36 (1975), 395–399.

89. N. V. Doan, Effet de valence en electromigration dans l'argent, *Journal of Physics and Chemistry of Solids* **31** (1970), 2079–2085.

90. I. A. Blech and H. Sello, The failure of thin aluminum current carrying strips on oxidized Silicon, *Fifth Annual Symposium on the Physics of Failure in Electronics*, (1966), 496–505.

3 Thermal Stress Characteristics and Stress-Induced Void Formation in Aluminum and Copper Interconnects

3.1   Technology Impact and Stress Effect on Electromigration

The study of electromigration in conductor lines can be traced to the report in 1966 by Blech and Sello [1], who first identified that the crack formation in aluminum conductor lines on silicon chips was induced by electromigration. This problem was found when the microelectronics industry was beginning to develop large-scale integrated circuits where the reliability of the on-chip interconnects would be critical, and thus had generated great interest in the industry. Compared with bulk metals as discussed in the previous chapter, EM in conductor lines has several distinguished characteristics due to the way that the conductor lines are integrated into the silicon chip and their functional and reliability requirements. As on-chip interconnects, the Al lines are formed by a subtractive etching process where a thin Al film is first deposited on a Si subtract containing transistor devices, then patterned by lithographic technique and integrated to form multilayered conductor structures as on-chip interconnects [2]. In this way, multiple levels of conductor lines form a hierarchical wiring structure connecting the transistors to form a dynamic memory or a microprocessor chip. The interconnect structure serves multiple functions, including to supply current and voltage to operate the transistors, to distribute power and electrical signals, and to synchronize the clock signals for the chip. Among the functional requirements, supplying a high current pulse, unipolar or bipolar, to drive the transistor devices to function reliably as designed is essential. As the technology advanced following Moore's law [3], the device density and chip performance have improved exponentially with time. This has led to the development of very-large-scale integration (VLSI) and subsequently the copper interconnects in the 1990s [4]. With the technology driving device scaling, the interconnects have evolved to become a complex multilevel wiring structure with decreasing dimensions and increasing current density. Interconnect failures due to formation of electrical opens or shorts induced by electromigration has become an important reliability problem. As the technology advances into nanoscale integration, there are fundamental changes in the perspective and focus of electromigration studies.

A schematic representation of a multilayered Cu wiring structure [5] is shown in Figure 3.1 for a first generation of Cu interconnects used in microprocessors containing six Cu wiring levels vertically stacked on top of the complementary metal–oxide semiconductor (CMOS) transistors on a Si substrate. In the structure as shown in

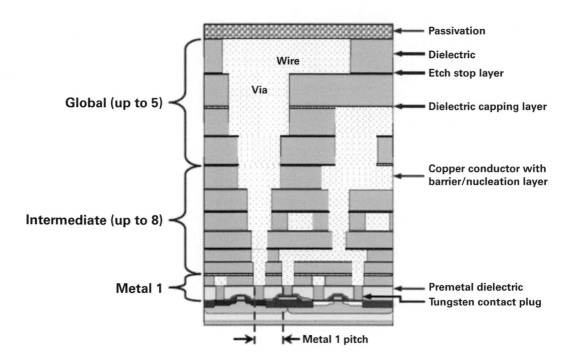

Figure 3.1 Schematic representation of a multilayered Cu interconnect structure (Semiconductor Industry Association. *The International Technology Roadmap for Semiconductors*, 2007 Edition).

Figure 3.1, each single wiring level is made up of two adjacent layers. The first wiring level is identified as the "metal 1" level, followed with the next three wiring levels as the "intermediate" levels, and the top two wiring levels as the "global" levels. In each wiring level, the upper layer contains a set of interconnect wires while the lower layer has a set of vias or plugs connecting to a lower wiring layer. The dimensions of the conductor line increase usually in ratios of $1\times$, $2\times$, and $4\times$ from the metal 1 level to the intermediate levels and the global levels, respectively, as more wiring structures are integrated on the chip. In the Cu wiring structure as illustrated, the conductor lines are cladded with thin but stiff dielectric layers, such as oxides or nitrides, which serve as diffusion barriers preventing Cu from diffusing into the dielectric to degrade the device performance. Constructed in this way, the conductor lines have distinct structural features that can affect the EM and thermal stress characteristics. First, the efficient cooling from the interfaces enables the conductor line to carry a much large current density, up to three or four orders of magnitude higher than an isolated line. While this serves an important function to provide the current density required to drive the transistors, it significantly increases the mass transport to induce damage and degrade the interconnect reliability. The device operating temperature is generally lower than 0.5 of the absolute melting temperature T_m, where mass transport is dominated by microstructure and defects, such as grain boundaries and interfaces, instead of bulk diffusion. During fabrication, the silicon chip is subjected to thermal processes where the annealing at an elevated temperature can induce grain growth and material reactions as well as thermal stresses due to the dielectric confinements. Such processes can change the characteristics of EM and thermal stresses, rendering microstructural effects from grain boundaries and interfaces to be important in controlling mass transport leading to the formation of voids and hillocks to fail the conductor line.

One distinct aspect in conductor lines that can affect EM arises from the thermal stresses generated by the confinement of the interlevel dielectrics and the barrier and cap layers used to contain the metal line. The thermal stresses constitute an additional driving force to supplement the electrical current for mass transport and can substantially affect the kinetics and mechanism for damage formation. The stress effect on EM was first observed by Blech in 1976 in a study of EM in Al conductor lines [6]. He found an interesting length dependence of the amount of mass depletion driven by EM from the cathode end, which was inversely proportional to the line length and vanished completely at a "critical length." This is shown in Figure 3.2, where a series of strips deposited on TiN were subjected to a current of 3.7×10^5 A/cm^2 for 15 hours at 350°C. The strip on the left, 10 μm long, drifted very little while the strip on the right, 90 μm long, drifted as much as 20 μm. The reduction in drift velocity was found to be inversely proportional to the strip length, which was attributed to the buildup of a "backflow" stress by mass accumulation at the anode end with its gradient acting against the EM driving force. The initial observation was followed up with a series of studies by Blech and coworkers [6–8] to clarify the nature of the backflow stress. This established the concept of a critical current density-length (jLc) product with a "critical length" Lc, where the stress gradient can completely counteract the current j driving force to stop EM [6]. This turned out to be important for design of on-chip interconnects in providing a criterion for specifying the current carrying capability of conductor lines below which EM damage would not occur.

As the industry embarked on the development of VLSI technology, the stress effect took a drastic turn in the 1980s to become an important reliability problem with the finding of stress-induced void formation in passivated Al lines, which can cause line failure without EM [9, 10]. This phenomenon has generated great interest at that time to investigate the thermal stress characteristics and the stress voiding phenomenon. Such studies revealed that voids formed only in passivated Al lines on a Si substrate, maximized at a moderate temperature of about $0.5T_m$ and with void configurations, obtuse or slit-like, depending on the local grain structure. These characteristics indicated that void formation is induced by the thermal stress due to the thermal mismatch between the oxide passivation and Al, where the grain structure is important in controlling the mass transport and void formation. It soon became clear that stress voiding is synergetic to electromigration, and together they led to a fundamental change in the EM studies from basic studies of bulk metals to the studies of the kinetics and failure mechanisms in passivated Al lines. As the technology advanced with the development of the Cu interconnects in the 1990s, the research focus shifted to Cu on-chip interconnects. The Cu interconnects are fabricated by a dual damascene

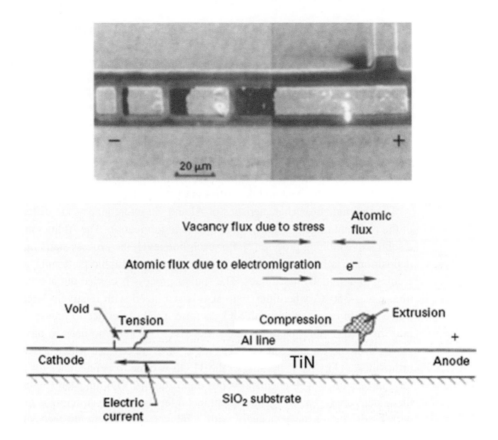

Figure 3.2 Scanning electron microscopy (SEM) images of electromigration in a set of short Al strips on TiN under current density of 3.7×10^5 A/cm^2 at 350°C for 15 h. The Al strips are 15 μm wide, 100 nm thick, and lengths of 10, 20, 30, and 90 μm. (b) Schematic diagram depicting void and hillock formation in the strip. The mass transport is along the electron flow: from left to right. Reprinted from [6], with the permission of AIP Publishing.

process that has incorporated new materials and structural elements with distinct stress and EM characteristics and mechanisms. As the Cu damascene structure evolved, results from EM and stress studies revealed that dimensional scaling, materials, and the Cu microstructure are important parameters in determining the EM and thermal stress characteristics.

In this chapter, we focus on the thermal stress characteristics and stress-induced void formation in passivated Al and Cu lines. We first discuss in Sections 3.2 and 3.3 the effect of dielectric confinement on the thermal stress characteristics and the X-ray diffraction measurements of thermal stresses for passivated Al and Cu lines. Then we discuss stress relaxation in passivated Al and Cu lines in Sections 3.5 and 3.6 and stress-induced void formation in Section 3.7, where the results are relevant to the study of EM reliability in Chapters 4 and 5.

3.2　Effect of Passivation on Thermal Stress Characteristics in Interconnect Lines

For a metallic film such as Al embedded on a silicon substrate, a biaxial stress is generated due to the thermal mismatch of Al and Si when subjected to a temperature change of ΔT. As the film is patterned into lines, the stress becomes biaxial with unequal components, depending on the line dimensions and the aspect ratio. When the metal line is passivated by a dielectric overlayer such as SiO_2, the stress state becomes triaxial with an additional component normal to the line surface, as shown in Figure 3.3c. The stress characteristics depend on the materials, structure, and dimensions of the interconnect and the stiffness of the dielectric overlayer.

The thermal stress characteristics of Al and Cu interconnects are different because of the different materials and processes used in fabrication. The Al lines are formed by a subtractive etching process [2]. For each line level, the process starts first by sputter deposition of a blanket Al(Cu) film, then photolithographic patterning and selective etching to define the metal lines. The sputter process is carried out at a high temperature, e.g., ~350°C, where the grain size is stabilized with relatively large grain size. Since Al can readily react with SiO_2 to form a tight interface to prevent Al diffusion into the surrounding interlevel dialectics (ILD), there is no need to have a sidewall barrier as in the Cu interconnects. The metal lines are covered with a layer of tetraethyl orthosilicate (TEOS). Patterning of the TEOS is done to define holes for the vias, and the vias are filled with chemical vapor deposition (CVD) tungsten (Figure 3.4a). All these steps are repeated for subsequent metal levels, and the structure is finally capped with a passivation layer, usually with a bilayer of plasma-enhanced silicon nitride (PEN) and borophosphosilicate glass (BPSG). As shown in Figure 3.4a, the metal layer in Al interconnects is Al(Cu) sandwiched by thin films of refractory metal layers of Ti/TiN, which serve a variety of functions, including as redundant metal to

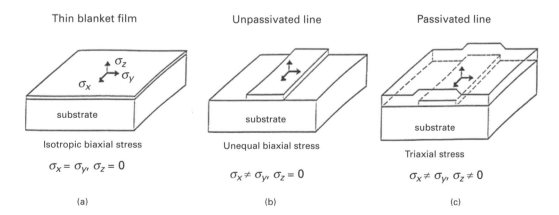

Thin blanket film

σ_z

σ_y

σ_x

substrate

Isotropic biaxial stress

$\sigma_x = \sigma_y, \sigma_z = 0$

(a)

Unpassivated line

substrate

Unequal biaxial stress

$\sigma_x \neq \sigma_y, \sigma_z = 0$

(b)

Passivated line

substrate

Triaxial stress

$\sigma_x \neq \sigma_y, \sigma_z \neq 0$

(c)

Figure 3.3 Effect of substrate confinement on the stress characteristics for (a) a blanket thin film, (b) a line without passivation, and (c) a line with passivation.

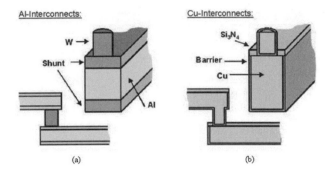

Figure 3.4 Schematics of interconnect structures: (a) Al interconnect, with W-filled vias and conducting Ti/TiN shunt layers at the top and bottom of the interconnect line; (b) dual-damascene Cu interconnect, with Cu-filled vias, thin refractory liners at the sides and bottom of the line, and a dielectric capping layer at the top of the line. Reprint from [11] with permission from Elsevier.

minimize electrical opens due to voids and as an antireflective coating (ARC) for the lithography process.

In comparison, Cu is difficult to etch chemically and can readily diffuse through the dielectric to cause device leakage, so Cu interconnects are fabricated using a damascene or inlaid process with a basic structure, as schematically shown in Figure 3.4b. The Cu damascene line is surrounded by a thin barrier and embedded in a dielectric layer, starting as Si dioxide and then changing to materials with a low dielectric constant (k) as the technology evolved. The damascene process is more complex than the subtractive etching process of Al interconnects and will be described in Section 3.5.

To discuss the passivation effect on thermal stresses of metal lines, we follow the analysis by Korhonen et al. [12–14]. This is based on an extension of the Eshelby's analysis of elastic inclusions for a passivated Al line embedded in SiO_2 on a thick Si substrate as shown in Figure 3.5a. Since SiO_2 and Si are quite similar in the coefficient of thermal expansion (CTE) and elastic modulus, the problem was simplified by assuming that the Al line is completely embedded in an Si matrix. Accordingly, the thermal strains induced by a temperature change from T_0 to T are as follows:

$$\varepsilon_1^T = \varepsilon_2^T = \varepsilon_3^T = \Delta\alpha(T - T_0). \tag{3.1}$$

Here ε_i^T are the principal thermal strains along the principal axes of the conductor line, and the CTE mismatch $\Delta\alpha$ between Al and Si is assumed to be isotropic. In general, the thermal stresses depend on the grain orientations in the line; for an Al line with width in the micrometer range, it usually exhibits a strong (111) fiber grain texture, so the σ_1 stress component along the line is placed along the (111) orientation and the σ_2 component in the width direction along the (101) orientation and the σ_3 component along the (121) orientation in the thickness direction. The stress components were deduced based on the Eshelby model of an elliptical cylinder inclusion embedded in a

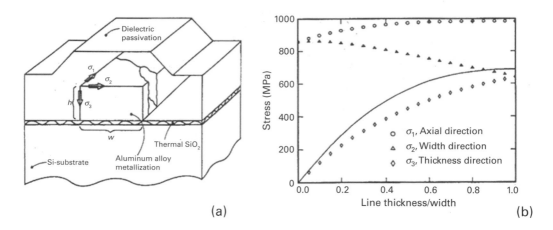

(a)   (b)

Figure 3.5 (a) Schematic of a passivated Al interconnect line on a silicon substrate with the principal stress components. (b) Thermal stresses for a passivated Al interconnect calculated based on Eshelby inclusion analysis after heat treatment from 400°C to room temperature. The solid line indicates the hydrostatic stress after complete relaxation of shear stresses. Reprinted from [12], with the permission of Springer Nature.

matrix with different elastic properties and then generalized to deduce the thermal stresses in a passivated line due to thermal mismatch and plastic deformation [14]. The results obtained from this model for Al lines subjected to a 400°C anneal then cooled down to room temperature are shown in Figure 3.5b as a function of the line aspect ratio (thickness/width h/w).

The results in Figure 3.5b show a strong confinement effect from the Si surrounding the Al line; for example, the biaxial stresses for a confined film ($h/w = 0$) reach a value as high as 850 MPa, far exceeding the 350 MPa reported for a 300 nm thick Al film bound on a ceramic substrate [15] and the biaxial stresses of 500 MPa reported for an Al film with Si nitride passivation on top [16]. While the calculation based on the Eshelby model tends to overestimate the confinement effect by assuming an infinite dielectric passivation and ignoring the polycrystalline structure of the Al line, the error was estimated to be only a few percent. This strengthening mechanism has been attributed to piling up of geometrically necessary dislocations at the interface to offset the interfacial strain due to thermal mismatch, which was originally proposed by Ashby [17] and extended by Nix to line structures [18]. With increasing aspect ratio, the extent of the dielectric confinement increases and thus the triaxial stress levels.

The increase in the thermal stresses found in passivated Al lines indicates that the yield strength of the Al lines is significantly increased by the dielectric confinement. However, such a stress state is not kinetically stable since the interconnect metallization will undergo additional heat treatments during device fabrication or operation, where stress relaxation could occur through plastic yield or void formation and raise reliability concern. This interesting and important problem will be discussed later in Section 3.6 following the discussion on thermal stresses and the passivation effects of Al and Cu lines.

3.3 Thermal Stress Measurements by X-Ray Diffraction for Passivated Metal Lines

The stress behaviors of Al and Cu thin films and lines have been extensively studied both experimentally and by modeling [12, 19–22]. Two methods are commonly used for stress measurements on thin films and line structures: the wafer curvature method and X-ray diffraction. In the wafer curvature method, the stress state of a uniform blanket film on a substrate of known thickness is derived from measurement of the curvature of the substrate, using the elastic constants of the substrate material [23, 24]. The measurement is usually carried out using a laser deflection technique to monitor the stress continuously as the sample is thermally cycled. This method can be applied to most materials, crystalline or noncrystalline, without knowing the properties of the materials. However, to deduce the triaxial stress state using the wafer curvature method is not straightforward and would require micromechanical analysis. In comparison, the X-ray diffraction method is more direct for measuring the stresses in thin films and interconnect lines. In this method, the change of lattice parameter for samples in a stressed state can be directly measured as a lattice strain and converted to the corresponding stress. This technique is applicable to the interconnect lines under passivation and can directly measure the triaxial stress components to investigate the passivation effect. Measurements for line structures such as metal interconnects of micrometer dimensions are quite challenging due to the small sampling volume of the line structures and would require a very intense X-ray source and specially designed diffractometers to delineate the individual stress components.

X-ray diffraction techniques have been well developed to measure thermal stresses in thin films [25, 26]. Here we discuss the results of X-ray diffraction measurements on thermal stresses that were measured in parallel arrays of Al and Cu line structures where the array configuration can increase the diffraction volume and enable measurements into submicron line dimensions. Experimental details are described in [20] and [27]. Essentially, the stress state is determined by measuring the strain components from the thermally induced changes of selected lattice spacings. For this purpose, a special X-ray diffractometer was developed, consisting of a four-circle diffractometer with an 18 kW high-intensity rotating anode source as schematically shown in Figure 3.6a. Custom-designed high-precision slits and sample position adjustments were incorporated, making it possible to use the low-index lattice parameters for strain measurement. This increased the X-ray intensity to facilitate in situ measurements during thermal cycling, although it required high-precision sample position adjustments and calibration. For the results reported here for Al and Cu line structures, a special sample heating stage was constructed to carry out in situ strain measurements as a function of temperature. All stress measurements were made in situ during thermal cycling at a constant ramp rate chosen between 2.4 and 2.8°C/min.

X-ray diffraction measures the strain of a crystal by measuring the change in plane spacing d based on the Bragg's law

$$n\lambda = 2d_{hkl} \sin \theta, \qquad (3.2)$$

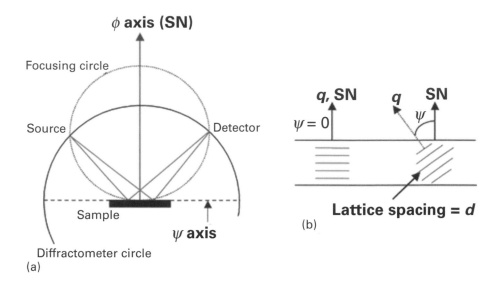

Figure 3.6 (a) Schematics of a four-circle X-ray diffractometer with high-intensity rotating anode X-ray source with the ψ-geometry for a flat sample and symmetrical diffraction, and (b) schematics of lattice planes used in X-ray diffraction measurements. The ϕ axis is normal to the sample surface, and ψ is an angle between the crystal plane and the sample surface. Reproduced from [27] with permission of S. H. Rhee.

where n is an integer, λ is the X-ray wavelength, d_{hkl} is the lattice spacing of the (hkl) diffraction plane, and θ is half of the diffraction angle. If the crystal is subject to a stress, then there will be a corresponding dilatation of the crystal lattice (i.e., strain) to balance this stress. The elastic strain is then given by

$$\varepsilon = (d_{hkl} - d_0)/d_0, \tag{3.3}$$

where d_0 is the spacing of the unstressed lattice plane. For thin films and line structures, the stress state σ can be deduced by measuring the strain state ε as

$$\sigma = C' \cdot \varepsilon, \tag{3.4}$$

where C' is the compliance matrix. To fully specify the strain in a solid, all six independent components of the strain tensor must be known. Fortunately, cubic metals such as Al and Cu are characterized by only three independent components in the compliance tensor, although Al or Cu films or line structures have grain textures with anisotropic elastic properties, which have to be taken into account in the strain measurements. For patterned and passivated lines, the thermal strain is triaxial (Figure 3.3c). Based on the symmetry of a parallel array of long interconnect lines, it is possible to choose a coordinate system a priori to be the principal coordinate system [25, 28]. The chosen coordinate system is defined by two Euler angles – the inclination angle ψ to the surface normal, and the rotation angle ϕ, as shown in

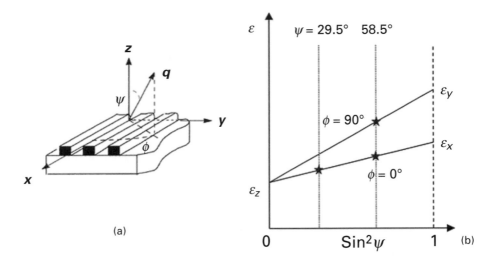

Figure 3.7 (a) The line coordinate system specified by Euler angles ϕ and ψ for strain measurements of Al and Cu lines; (b) appropriate (311) reflections at $\psi = 29.5°$ and $58.5°$ are chosen to determine the three strain components. Reproduced from [71] with permission of D. Gan.

Figure 3.7a, where all shear terms in this system are expected to be zero. In this system, the measured strains are related to the line strains as

$$\varepsilon_{\psi,0} = (\varepsilon_x - \varepsilon_z)\sin^2\psi + \varepsilon_z \text{ at } \phi = 0° \text{ (along the line length)}$$

and

$$\varepsilon_{\psi,90} = (\varepsilon_y - \varepsilon_z)\sin^2\psi + \varepsilon_z \text{ at } \phi = 90° \text{ (along the line width).} \quad (3.5)$$

Each of the preceding equations is linear with respect to $\sin^2\psi$ and even in ε_z. At $\psi = 0°$, the two functions intersect. Thus, with three measurements (minimum) of $\varepsilon_{\psi,\phi}$ in the $\phi = 0°$ and $\phi = 90°$ planes, all strain components can be determined. For Al line structures, appropriate (311) reflections at $\psi = 29.5°$ and $58.5°$ were chosen to optimize the X-ray intensity and minimize the experimental errors as shown in Figure 3.7b.

For stress measurements in Al and Cu line structures, the grain textures must be considered to account for the effect of elastic anisotropy. For the Al and Cu line structures reported here, we found a very strong (111) texture with more than 95% of the grains with a (111) pole approximately 5° of the film normal. Therefore, the stresses were determined from the measured strains elastic constants derived from the (111) grain texture, and the error was negligible by not considering the non-(111) textured grains. For (111) textured lines, it is convenient to orient the stress/strain state with the principal components along the length (σ_x), width (σ_y), and height (σ_z) of the line with corresponding principal strains. In this case, the stiffness coefficients in the line coordinate are as follows:

$$C'_{11} = 1/2\,(C_{11} + C_{12} + 2C_{44})$$
$$C'_{12} = 1/6\,(C_{11} + 5C_{12} + 2C_{44})$$
$$C'_{13} = 1/3\,(C_{11} + 2C_{12} + 2\,C_{44})$$
$$C'_{33} = 1/3\,(C_{11} + 2C_{12} + 4C_{44}) \tag{3.6}$$
$$C'_{44} = 1/3\,(C_{11} - C_{12} + C_{44})$$
$$C'_{66} = 1/6\,(C_{11} - C_{12} + 4C_{44}).$$

For Al, $C_{11} = 108.2$, $C_{12} = 61.3$, $C_{44} = 28.5$ and for Cu, $C_{11} = 168.4$, $C_{12} = 121.4$, and $C_{44} = 75.4$, all in units of GPa.

The samples used to measure thermal stresses in the Al lines were patterned as parallel line structures in a 0.6 µm thick Al(1.0 wt% Cu) film on a 0.7 µm thermal oxide on a Si wafer [20, 21]. The AlCu film was sputter deposited and then lithographically exposed and etched before annealing for 30 minutes at 400°C to produce parallel arrays of lines. The reticle was exposed to produce lines with widths of 6.0, 3.0, 1.0, 0.5, and 0.2 µm and corresponding aspect ratios of 0.1. 0.2, 0.6, 1.2, and 3.0. The patterned line samples and blanket wafer were passivated with 0.4 µm of BPSG and 0.7 µm of PEN at 300°C. The thermal cycles ran from 40–400°C with no interruptions, at a ramp rate chosen between 2.2–2.8°C/minute, and the measurements were performed in the second thermal cycle.

Results from X-ray diffraction measurements are presented in Figure 3.8. Overall, there is a systematic trend with a strong influence of decreasing linewidth on the thermal stress characteristics of the AlCu lines (the results from the 6 µm lines are very similar to the 3 µm lines and not shown). For the 3.0 µm wide lines, the stress state is almost equibiaxial with only small difference between the x (along the length) and y (along the width) stress components, and very little stress in the z (along the height) direction. This is very similar to a blanket film. In the 3.0 µm lines, the changes in the linear slope indicate yielding in compression and in tension induced by the thermal stresses. As the linewidth decreases to 1.0 µm, the stresses evolve into a triaxial state with increasing z-stress and decreasing y-stress components. In comparison, the narrower 1.0 µm lines show an almost perfectly elastic behavior with little stress hysteresis. The lack of hysteresis is accompanied by an increase in the magnitude of the stress excursion and the yield strength. This is an important feature in the stress behavior, indicating no yielding in the AlCu lines narrower than 1 µm. With the linewidth decreased to 0.5 µm, the increase in stress levels with increasing aspect ratio is due to increasing passivation confinement, rendering the z-component no longer small. For the 0.2 µm with the aspect ratio of 3, the z-stress becomes close to the y-stress, while the x-stress increases to 850 MPa. The extremely high stresses in the 0.2 µm lines are quite significant and had not been previously reported at that time.

In Figure 3.9, we summarize the stress characteristics of the AlCu lines as a function of the aspect ratio. The increase in residual stress with aspect ratio is consistent with the Korhonen's analysis in Figure 3.5. Starting with the 3 µm line with an aspect ratio of 0.2, the stress state is very close to biaxial, with σ_x and σ_y about equal at 350 MPa and σ_z about 0, as expected. With aspect ratio below unity, the

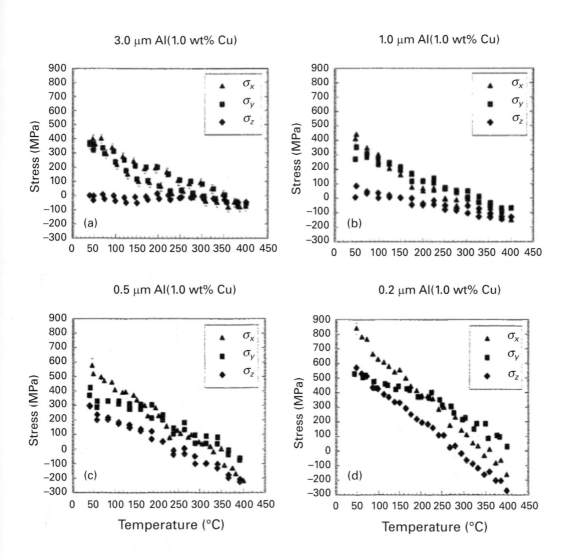

Figure 3.8 Thermal stress behavior of passivated Al(1.0 wt% Cu) lines, 0.6 μm in thickness and (a) 3.0 μm width, (b) 1.0 μm width, (c) 0.5 μm width, and (d) 0.2 μm width. Reproduced from [21] with permission of AIP Publishing.

stresses increase in the order of $\sigma_x > \sigma_y > \sigma_z$ as the stress state becomes triaxial. With aspect ratios above unity, σ_z continues to increase, changing the order to $\sigma_x > \sigma_z > \sigma_y$, indicating that as lines get narrower, there is an increased confinement from the sides and edges to change the stress characteristics. The magnitude of the stress components is generally smaller than those from the Korhonen model, which is to be expected since the latter has a higher confinement with an unlimited passivation of silicon.

The stress results presented here are consistent with the X-ray stress measurement by Yagi et al. [29] for Al(1.0% Si) lines as a function of aspect ratio. Their results show a monotonic increase in residual stress with aspect ratio, very similar to the

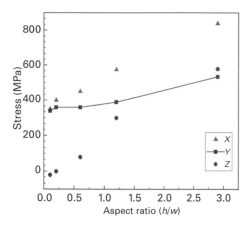

Figure 3.9 Principal stresses in passivated Al(1.0 wt% Cu) lines after cool down to 45°C, plotted as a function of line aspect ratio.

results reported here but with somewhat lower stress levels. The analytical results of Niwa et al. [30] and the finite element results of Sauter and Nix [31] are in general agreement with the results presented here for lines with aspect ratio less than 1. Above an aspect ratio of 1, Niwa et al. and Sauter and Nix showed a decrease in all the principal stress components. This is contrary to the X-ray results, showing that all the stress components continue to increase beyond an aspect ratio of 1. Yeo and his colleagues' bending beam experiments on Al(1.0%Cu) [32] showed the same trend, but do not include aspect ratios above 1.0.

The 850 MPa residual stress in the x-direction observed for the 0.2 μm lines is fundamentally significant. This corresponds to an elastic strain of about 1%, which is extremely high for a polycrystalline metal. The high yield strength of thin films and narrow lines is well known and is explained by the limited grain size and the restrained dislocation movement due to the increase in interface-to-volume ratio [17, 18]. Chaudhari's yield strain model [33] for blanket films was used to see if it can account for the high strains observed for the 0.2 μm lines. Consider a 0.2 μm thick Al film on Si and passivated with a 1.0 μm overlayer; the Chaudhari model yields a film strain of 0.8% for a grain size of 0.5 μm at room temperature, consistent with the X-ray result shown here. This highlights the significance of ultrafine structures where interfaces can play a crucial role in suppressing the motion of dislocations to relieve an elastic strain. As the line dimension shrinks down to submicron, the number of the dislocations becomes very small. As a result, dislocation glide and climb controlled deformation mechanisms will be less and less effective. From a stress-reliability standpoint, the submicron lines can bear very high elastic stresses and become more susceptible to nonvolume conserving relaxation mechanisms such as voiding, a topic that is discussed in the next section.

The effect of dielectric passivation on thermal stress of Al lines was measured using X-ray diffraction for a low-k dielectric material of polyarylene ether (PAE) with

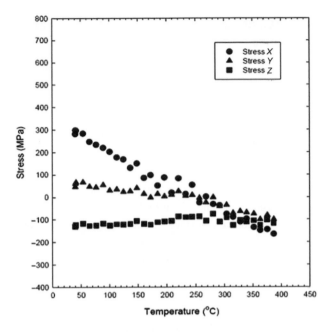

Figure 3.10 Thermal stress behavior of 0.6 μm wide, 0.6 μm thick, single-level AlCu lines passivated with PAE. Reproduced from [27] with permission of S. H. Rhee.

$k < 3.0$ [27, 34]. The results are shown in Figure 3.10 for AlCu lines of 0.6 μm wide and 0.6 μm high, revealing significantly decreased stress levels in PAE passivated lines due to reduced confinement effect. The stresses along and across the line directions (σ_x and σ_y) in the PAE passivated lines are 280 MPa and 70 MPa, respectively, and the stress normal to the surface (σ_z) becomes compressive at −120 MPa near room temperature. The stress values reflect the weak PAE passivation due to a lower modulus (2.2 vs. 71.4 GPa) and a higher CTE (75 vs. 0.51 ppm/°C) as compared with SiO$_2$. The stresses along and across the line directions in the PAE passivated lines are tensile near room temperature because the confinement from the thick, rigid Si substrate still dominates the in-plane stresses (σ_x and σ_y). However, the stresses are lower compared to the SiO$_2$ passivated lines (570 MPa and 400 MPa from Figure 3.8c, respectively), although they remain tensile near room temperature. In this case, the substrate confinement does not control the stress normal to the surface, which is primarily controlled by the CTE mismatch between Al and the dielectric layer and becomes compressive near room temperature due to the high CTE of PAE. The normal stress is relatively small due to the low modulus of PAE. The hydrostatic stress component of the PAE passivated lines near room temperature is found to be 76 MPa, which is significantly lower than 390 MPa of the SiO$_2$ passivated lines. This reduces substantially the driving force for stress-induced void formation in submicron AlCu/low-k interconnects.

3.4   Thermal Stress Characteristics and Effect of Dielectric Passivation on Cu Damascene Lines

The stress characteristics of blanket Cu films have been widely studied both experimentally and by modeling [19–22]. Quantitative information on the stress behavior of Cu lines was rather limited up to the late 1990s when Cu lines were introduced as on-chip interconnects for the 250 nm technology node [35, 36]. Compared to Al, the CTE of Cu is lower, by about 32%, while its elastic modulus is about twice larger. Therefore, the thermal strain of Cu is less but the thermal stress may be larger. Moreover, Cu is highly anisotropic with the $\langle 111 \rangle$ modulus almost three times that of $\langle 100 \rangle$, while Al is nearly isotropic. The AlCu interconnects are fabricated by a subtractive etching process, while the Cu interconnects by a damascene or inlaid process that has sidewall barriers and surface cap layers as structural elements. Driven by device scaling, the Cu interconnects have evolved with the implementation of the low-k and ultralow-k dielectrics, which are considerably weaker in thermal mechanical properties than SiO_2. This together with the scaling of the line dimensions has changed the thermal stress characteristics of the Cu interconnects. In this section, we discuss the results from X-ray measurements of the thermal stress characteristics of Cu damascene lines and the effect of the low-k dielectrics.

To discuss the stress characteristics, we first illustrate a "via first" dual damascene process for Cu interconnects [37]. As shown in Figure 3.11, the process starts with etching a via structure into a layer of SiCOH dielectric using a SiO_2 mask (step 1),

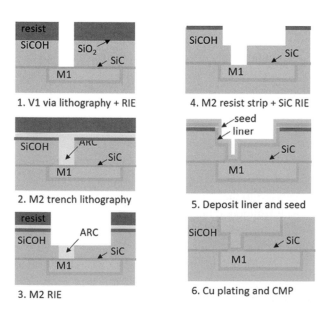

Figure 3.11 Schematics of via-first dual-damascene process for fabrication of Cu interconnect structures. © 2009 IEEE. Reprinted with permission from [37].

then follows by trench lithography and reactive ion etching (RIE) to pattern the M2 line (steps 2 to 3) and again RIE to form the line and via open structure (step 4). The process continues with filling the open structure with a Ta/TaN liner as an adhesion barrier to prevent Cu out-diffusion and coated with a seed layer for electroplating (step 5). Then the trench and via in one wiring level of the dual-damascene structure are formed by Cu electroplating, and the overburden layer is polished away by chemical-mechanical polishing (CMP) (step 6). After CMP, the one-level structure is completed by depositing a cap layer on top, usually made of Si_3N_4 or SiC_xNi_y, and ready for processing of subsequent wiring levels. In the damascene process, there are two thermal anneals, one after electroplating at about 250°C to desorb the plating additive and minimize the Cu resistivity, and another at about 400°C after the process is completed, to stabilize the line structure. The thermal stresses are primarily generated by the 400°C annealing, which also induces grain growth and void formation and can significantly affect the thermal stress and EM reliability, as discussed in this and subsequent chapters.

The Cu damascene structure gives rise to distinct thermal stress and stress relaxation characteristics, depending on the interlevel dielectrics and the barrier and cap layer materials. Here we discuss thermal stress characteristics of Cu damascene lines measured by X-ray diffraction, similar to that used for the AlCu lines [38]. Briefly, three sets of (311) reflections of inclination angle ψ and rotation angle ϕ (Figure 3.7a) were used to measure the lattice strains ε_x, ε_y, and ε_z and extract the stress components σ_x along the line, σ_y along the width, and σ_z along the height directions. The measurement was carried out for four sets of parallel Cu damascene lines with width/pitch of 1.0/2.0, 0.6/1.2, 0.4/0.7, and 0.25/0.55 µm as shown in Figure 3.12a. To study the passivation effect, tetraethyl orthosilicate (TEOS) and methyl silsesquioxane (MSQ) were used as interlevel dielectrics to fabricate the line structure. All measurements were performed in situ in two thermal cycles to 400°C at a ramp rate of 2°C/min, and only data from the second thermal cycle are presented.

The stress characteristics of the Cu lines are shown in Figure 3.13, indicating an almost elastic behavior with very high triaxial stress levels after cooling down from 400°C to room temperature. Overall, the characteristics are similar to that of the AlCu lines shown in Figure 3.8, with a smaller CTE mismatch between Cu and Si, but a higher modulus of Cu and the high yield strength of Cu lines can be attributed to the limited grain size and the restrained dislocation movement at interfaces [18, 33]. The stress characteristics of the 1 µm copper lines in Figure 3.13a show some small hysteresis along the length direction at intermediate temperatures. The amount is very small and can be estimated from the measured elastic strain during cooling from 400°C to 40°C, which is about 0.4%, while the thermal strain between the Cu line and the Si substrate is about 0.46%, so the plastic strain is only 0.06%. Near room temperature, the stress components follow the order of $\sigma_x > \sigma_y > \sigma_z$ reflecting the confinement effect of the line structure. With the TEOS passivation, the σ_x stress along the length direction is the highest because Cu is more rigid than TEOS, while the other stress components are affected by the Poisson effect. Figure 3.13b–d show the stress characteristics of the 0.6, 0.4, and 0.25 µm Cu lines, indicating increasing

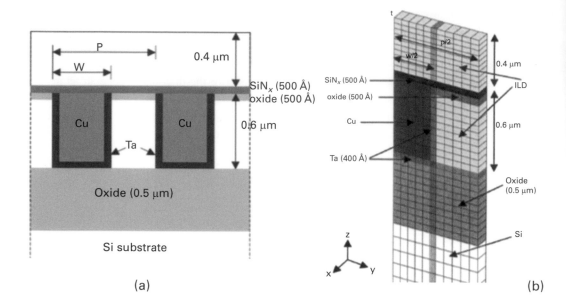

Figure 3.12 (a) Cross-sectional schematics of Cu line structures for X-ray stress measurement with TEOS or MSQ low-*k* dielectric as passivation dielectrics. The test structure contains four sets of parallel lines with width/pitch of 1.0/2.0, 0.6/1.2, 0.4/0.7, and 0.25/0.55 μm. (b) A unit segment of Cu interconnect structure used in the finite element model. Reprinted from [38] with the permission of AIP Publishing.

elastic behavior with decreasing linewidth. Since the height of the copper lines is a constant (0.6 μm), the decrease in the linewidth is equivalent to an increase in the aspect ratio. As the aspect ratio increases above unity starting with the 0.6 μm lines, the effect of the Ta barrier and the TEOS dielectric on the Cu lines along the height direction becomes clear. With decreasing linewidth or increasing aspect ratio, the stress normal to the surface σ_z increases, while the stress along the width σ_y decreases and the stress along the lines remain about the same. These characteristics can be attributed to the effect of the barrier layer and will be discussed later together with the stress behaviors of low-*k* passivated lines. The hydrostatic stress, calculated as $(\sigma_x + \sigma_y + \sigma_z)/3$, remains relatively constant in TEOS passivated Cu lines, ranging from 491 MPa for 1.0 μm lines to 548 MPa for 0.25 μm lines. Here the stress components are generally higher than those reported by Besser et al. [39], although the stress slopes of $d\sigma/dT$ are in good agreement. The different stress levels can be attributed to different processing conditions as observed by Besser et al. [40], so the stress slopes are analyzed in the finite element model.

The thermal stress characteristics of Cu lines passivated with MSQ low-*k* dielectric are shown in Figure 3.14a–d for 1.0, 0.6, 0.4, and 0.25 μm wide lines, respectively. The lines again behave elastically and exhibit triaxial tensile stress states after cooling down. The stress levels are generally lower than those observed for the TEOS passivated lines. Upon heating, the stresses decrease from zero at 200–250°C to become compressive

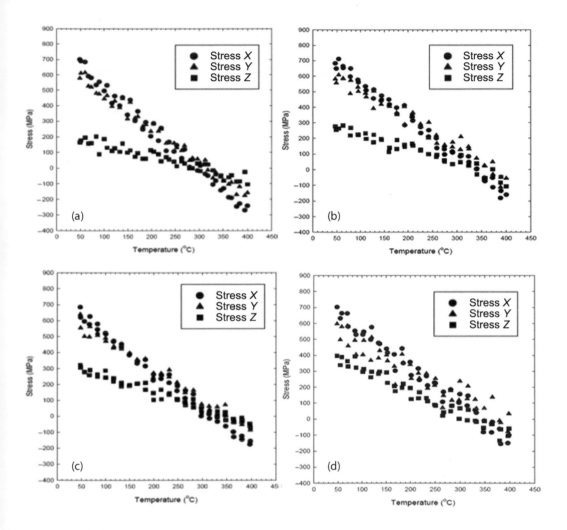

Figure 3.13 Thermal stress characteristics of Cu damascene lines with TEOS passivation. Lines are 0.6 μm in thickness and with widths of (a) 1.0 μm, (b) 0.6 μm, (c) 0.4 μm, and (d) 0.25 μm. Reprinted from [38] with the permission of AIP Publishing.

due to CTE mismatch between Cu and Si. The zero-stress level is about 100°C lower because the TEOS passivated lines were annealed at 300–350°C.

In the low-k structure, the in-plane stresses σ_x and σ_y are defined by the thermal mismatch between lines and substrate while the out-of-plane stress σ_z is defined by the mismatch between Cu and the dielectric. Thus, the tensile stresses near room temperature along and across the lines can be attributed to the substrate confinement. As shown in Table 3.1, the CTE of the MSQ low-k dielectric is comparable to that of Cu but more than an order of magnitude higher than TEOS, so a small amount of compressive stress normal to the sample surface is to be expected. However, the

Table 3.1. Properties of substrate, Cu, dielectric, and barrier material used in the finite element model (FEM).

	Si(100)	Cu	MSQ	TEOS	Ta
CTE (ppm/°C)	2.61	17.7	20.0	0.51	6.5
Modulus (GPa)	131.0	115.0	6.3	71.4	185.7

Source: Reprinted from [38], with the permission of AIP Publishing.

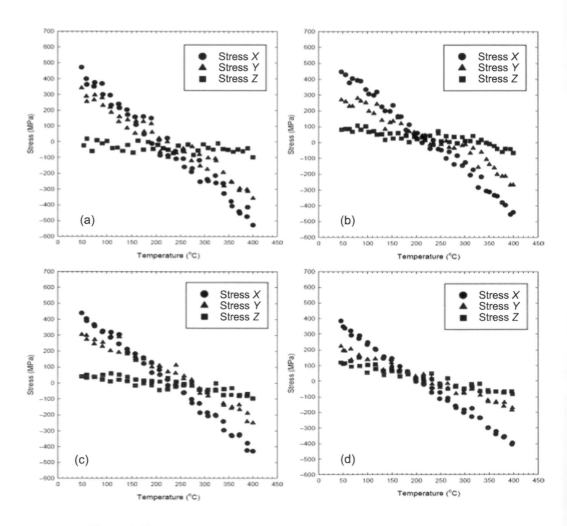

Figure 3.14 Thermal stress characteristics of Cu damascene lines with MSQ low-k passivation. Lines are 0.6 μm in thickness and with widths of (a) 1.0 μm, (b) 0.6 μm, (c) 0.4 μm, and (d) 0.25 μm. Reprinted from [38] with the permission of AIP Publishing.

X-ray data show tensile stresses in all directions near room temperature. The stress normal to the surface of the low-k passivated lines is similar as that of the TEOS passivated lines, increasing with decreasing linewidth. To explain this behavior, the effect of the Ta barrier layer must be considered, which has a CTE lower than MSQ and Cu but a significantly higher elastic modulus. Even with a thickness of 40 nm, the Ta barrier is sufficiently strong to confine the Cu lines to yield a tensile stress in the normal direction. As the barrier thickness remains constant for all linewidths, it increases relatively with decreasing linewidth and induces a higher tensile stress normal to the surface. With the Ta barrier deposited on both trench sidewalls, the thickness of the barrier layer accounts for a certain fraction of the total linewidth, especially for the narrow lines. Its effect is clearly not negligible, particularly for low-k structures. This reduces the shear stress in the Cu interconnects by introducing a tensile stress normal to the surface σ_z and effectively increases the hydrostatic stress in the lines to balance the effect of less confinement by the low-k ILD. As a result, the overall thermal stress is reduced for low-k passivated lines compared with that of TEOS passivated Cu lines. While this reduces the concern for stress voiding in low-k passivated lines, the shear stress remains a reliability concern for interfacial fracture in low-k passivated lines, which generally have a low fracture toughness to resist interfacial delamination.

A 3D finite element analysis was carried out to investigate the passivation effect on the stress characteristics of Cu damascene lines. To model the test structure with periodic parallel lines, a simplified model was set up to evaluate the effect of scaling, barrier thickness, and low-k passivation. The model is shown in Figure 3.12b, which contains only one unit segment of half a line with all structural layers, including the 400 Å Ta diffusion barrier and the SiNx cap layer. Periodic boundary conditions are used in the half-line model with the thicknesses of the metal lines and the ILD layer set as 0.6 and 0.4 μm respectively. All interfaces in the structures are assumed to be perfect and the metal lines to behave elastically within the temperature range of X-ray measurements. The interconnect structure is assumed to be stress free at 400°C, and then calculated during cooling from 400°C to room temperature. The simplified model was verified, and it was found that the thermal stresses deduced were consistent with a multiline model [27].

To analyze the stress behavior, the stress slope $d\sigma/dT$ is used to compare the stress characteristics since it is independent of the processing condition and the thermal history. The slope can be readily converted to $\Delta\sigma$ by multiplying with ΔT to compare with the stress data from X-ray measurements. In Figure 3.15a and b, we compare the stress slopes $-d\sigma/dT$ obtained from the simplified model with the X-ray results for TEOS and MSQ low-k ILD, respectively. Overall, the stress slopes are generally higher for the Cu/TEOS lines, reflecting more confinement by the stronger TEOS than the MSQ low-k dielectric. While the stress slopes along the line (x) and across the line (y) decrease, the slope normal to the surface (z) increases with decreasing linewidth or increasing aspect ratio. The results are in reasonable agreement with the X-ray data for the stress slopes along the lines (x) and normal to the line (z) directions but lower across the line (y) direction. This may be related to the grain structure in the Cu line,

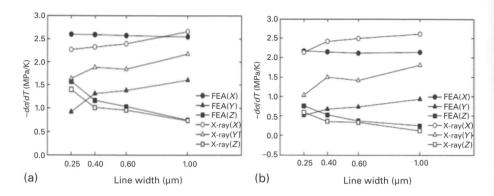

Figure 3.15 Comparison between measured X-ray data and FEM results of (a) TEOS and (b) MSQ low-k passivated Cu lines. Reprinted from [38] with the permission of AIP Publishing.

which is not considered in the stress model. In Cu damascene lines with the linewidths in this study, the microstructure is dominated by the (111) grains, which have an elastic modulus 3× higher than the (100) grains and about 50% higher than Young's modulus used in the model. This will increase the stress components σ_x along the line and σ_y across the line, improving the modeling results to agree with the X-ray measurements.

Synchrotron X-ray sources have been used to measure thermal strains in Al and Cu lines [41–43]. In one study, X-ray diffraction measurements were carried out using a synchrotron source on an individual Al line of 2.6 μm wide and Cu line of 2.0 μm wide with relatively large width-to-thickness ratios of 3.5 and 4.4 respectively [41, 42]. The thermal stresses were found far from the equiaxial, as to be expected; instead, the changes in linewidth and line thickness strains were found to relate to changes in line length strains by the uniaxial Poisson's ratio. The results showed that stress induced by confinement from the substrate and passivation to those line samples along the linewidths and line thicknesses is negligible compared with that along the line lengths, which suggested weak bonding between the conductor lines and adjacent substrate and passivation.

In another study [43], thermal stresses of Cu interconnects with organo-silicate glass (OSG) low-k dielectrics and linewidths scaled to 50 nm were determined using precision lattice parameter measurement at an advanced light facility. Grazing incidence and θ–2θ diffraction geometries were used to directly measure the strain tensor. The samples were prepared as periodic parallel line structures using a damascene fabrication process with anneal conditions varying from 30–240 sec. at 180°C and 250°C prior to chemical mechanical polishing. The results of the triaxial stress components and the hydrostatic stress are shown in Figure 3.16a and b, respectively. A substantial increase of about 3× in the stress components was observed as the linewidth reduced from 500 to 50 nm with corresponding aspect ratio increased from 0.35 to 3. The reductions observed in the triaxial stress components with increasing

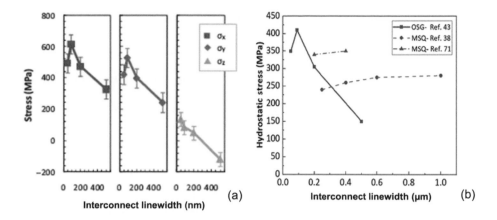

Figure 3.16 (a) Triaxial stress components in Cu damascene interconnects with porous OSG dielectric as a function of linewidth. (b) Measured hydrostatic stress as a function of interconnect linewidth and compared with results from two other studies with MSQ low-*k* dielectrics [38, 71]. Reprinted from [43] with the permission of AIP Publishing.

linewidth are not consistent with results from other X-ray measurements for Cu damascene lines with MSQ low-*k* dielectrics [38, 71]. In Figure 3.16b, we compare the hydrostatic stresses obtained from the synchrotron measurements with two other studies [38, 71] with the latter showing only a slight decrease in the hydrostatic stress with linewidths decreasing from 500 nm to 200 nm. The authors attributed the results to a bamboo-like grain structure with dominant (111) grain texture in the narrow lines, which has a high elastic modulus due to Cu elastic anisotropy. However, such a structure trend was not observed by transmission electron microscopy (TEM) in Cu damascene lines as reported in Chapter 6.

3.5 Stress Relaxation and Stress-Induced Void Formation in Passivated Al Lines

So far, we have demonstrated that dielectric confinement can alter the stress state of a conductor line to increase the hydrostatic stress, an effect that depends on the line aspect ratio and the properties of the passivation. Such a stress state, however, is not kinetically stable when the interconnect structure undergoes additional heat treatments during device fabrication or operation, where stress relaxation can occur via plastic yield and void formation. Stress-induced void (SIV) formation, stress voiding (SV), or stress migration (SM) was first reported in Al interconnects in 1984 [9, 10]. Since then, the problem has been extensively investigated in Al as well as in Cu interconnects, particularly when it is combined with EM (see, for example, [12, 44–50]). Generally, stress relaxation occurs via plastic deformation and void formation, depending on the stress characteristics. Plastic yield usually starts first to relieve the thermal stresses via dislocation and grain boundary sliding, although voids' nucleation and growth can also occur but with slower kinetics [44]. The onset of plastic yield can be evaluated based on the von Mises yield criterion:

$$2\sigma_{yield}^2 = (\sigma_1 - \sigma_2)^2 + (\sigma_2 - \sigma_3)^2 + (\sigma_3 - \sigma_1)^2, \tag{3.7}$$

where σ_i are the principal stresses and σ_{yield} the tensile yield strength. In general, plastic flow depends only on the deviatoric components of the stress state, so only the shear stresses are relieved, and since the process is volume conserved, it does not affect the hydrostatic stress generated by dielectric confinement. This can readily be illustrated by evaluating the hydrostatic stress in the line, assuming all the shear stress components are completely relaxed. As shown in Figure 3.5b, this yields a hydrostatic stress as high as 680 MPa in a 300 nm thick line with an aspect ratio of 1, while the von Mises stress is reduced to 350 MPa. For other aspect ratios, the hydrostatic stress is lower with corresponding higher von Mises stresses to allow some plastic yield. This gives rise to a high elastic energy to drive void growth, which can be evaluated as $(\sigma_1^2 + \sigma_2^2 + \sigma_3^2)/2E$, with E being the elastic modulus of the conductor line. For the 300 nm thick line, the elastic energy increases by 7×. Such a large increase in the elastic energy provides a substantial drive force for void growth.

The stress voiding problem has generated great interest to develop kinetic models to analyze stress relaxation and void formation, first for Al interconnects and subsequently for Cu interconnects [12, 13, 44]. Stress voiding characteristics in Al and Cu lines are quite different because of the different materials and interconnect structures. For Al interconnects, voids are commonly found at structural singularities such as triple points, where grain boundaries intercept the line edge [45–47]. For Cu interconnects, stress voiding is found to occur instead under or near a via contact, so different questions were raised regarding the vacancy source and the mechanism leading to void formation and growth [48–50]. Here we first discuss stress voiding in Al interconnects, then follow with discussion on Cu interconnects in the next section.

An early kinetic model was developed by Rosenberg and Ohring [51] that examined void formation and growth induced by vacancy flux divergences at grain boundaries in Al interconnects. This model was aimed to project the vacancy accumulation rate and the maximum supersaturation that can be achieved at grain boundary divergence sites for void formation. A mass balance equation was set up to describe the rate of vacancy accumulation/depletion as

$$\frac{\partial c(x,t)}{\partial t} = -\nabla J + \frac{c_0 - c(x,t)}{\tau_0}. \tag{3.8}$$

Here $c(x,t)$ is the vacancy concentration at position x and time t, and $c_0 = \Omega^{-1} \exp\{-E_V/kT\}$ is the equilibrium vacancy concentration, where Ω is the atomic volume, E_V the energy of vacancy formation, and k the Boltzmann constant, T the absolute temperature. In (3.8), J is the vacancy flux, $(c_0 - c)/\tau_0$ is the vacancy generation/annihilation rate, and τ_0 is the lifetime of a vacancy in the presence of sinks and derived from an average distance to sinks $\bar{x} = \sqrt{2D\tau_0}$, where vacancy generation/annihilation occurs. The vacancy flux can be written as

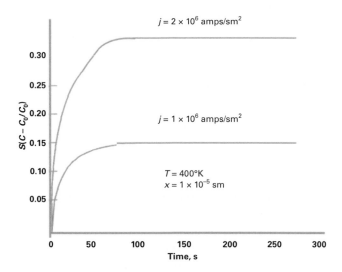

Figure 3.17 Buildup of a vacancy supersaturation in an Al lines at $400°$K for an average sink distance of 10^{-5} cm and a current density between 1 and 2×10^6 A/cm^2. Reprinted from [51] with the permission of AIP Publishing.

$$J = -D\nabla c + c\frac{D\Omega}{kT}\nabla\sigma + Dc\frac{eZ^*\rho j}{kT}. \tag{3.9}$$

Here three driving forces for vacancy migration are included: vacancy concentration gradient, the EM driving force, and the stress gradient, where D is the vacancy diffusivity, σ is the stress, Ω is the atomic volume, eZ^* is the effective charge, ρ is the metal resistivity, and j is the current density.

A key question in this model was to clarify whether the time required to reach a maximum vacancy concentration for void formation is consistent with a nucleation or a growth-limited process as observed in experiments. In Figure 3.17, the model showed that the buildup of a vacancy supersaturation $S = (c - c_0)/c_0$ at a blocking grain boundary in the range of $0.1 < S < 1$ at $400°$K for void nucleation at an average sink distance of 10^{-5} cm and a current density between 1 and 2×10^6A/cm^2 is in the range of seconds to minutes. Since the experiments conducted under a similar stressing condition showed that the onset of void formation required days to occur, it was concluded that the void formation observed was most probably growth instead of nucleation controlled. Another important conclusion was reached regarding whether a large vacancy supersaturation ($S \gg 1$) is required for the spontaneous void formation by vacancy condensation at a triple point. This condition was found impossible to reach under the test condition; instead, the presence of heterogeneous nucleating sites of high surface energy was found for void stabilization. This suggested that in the case of Al lines, a trace amount of aluminum oxide particles or some other processing defects is required at grain boundary sites to initiate void nucleation, and once a void is nucleated, its growth to a macroscopic size can occur by further accumulation of vacancies. Two most probable mechanisms for damage formation were proposed: one due to the growth of submicroscopic voids at a grain boundary triple-point junction, and the other attributed the formation of a pinhole induced by a grain boundary grooving process.

The conclusion regarding the heterogeneous void formation was supported by later studies, for example [52, 53], where the void was found to start by growth of preexisting flaws at the metal–passivation interface. Such flaws were attributed to etch residues from the fabrication process, where the metal is weakly bonded to the barriers or dielectric. However, other relevant questions remain regarding the growth-controlled mode of damage formation under EM and the effect of the EM-induced mechanical stress on void formation in passivated metal lines. These topics will be discussed in Chapter 4.

To investigate stress relaxation and void formation, measurements must be performed over a long period of time to quantify the kinetics of void growth. Such studies were carried out by Yeo et al. [32, 54] using a bending beam technique to investigate isothermal stress relaxation in passivated AlCu line structures. In one study, in situ measurements were carried out simultaneously on two types of samples, one with metal lines parallel to the beam and the other perpendicular to the beam over a period of 20 days so the stress state and the relaxation characteristics can be determined. Additionally, identical AlCu line structures were used for X-ray measurements to define the stress state of the bending beam samples.

Stress relaxation measurements were carried out for samples with linewidths of 3.0, 1.0, and 0.5 μm over a period of 15 hours and 20 days. The results for the 1 μm wide lines observed over a period of 20 days at 150, 200, and 250°C are shown in Figure 3.18a–c. Here a transition in the kinetics was clearly observed after annealing for 100–1,000 minutes, which delineated two kinetic stages. In the first stage, thermal stresses relax via plastic flow, which is driven by the shear stress arising from the

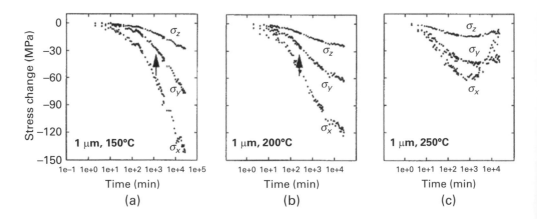

Figure 3.18 Stress relaxation in 1 μm wide AlCu lines tested for 20 days at (a) 150°C, (b) 200°C, and (c) 250°C. The arrows indicate the transition time from short-term to long-term stress relaxation. Reprinted from [54], with the permission of AIP Publishing.

difference of the principal stress components. In this stage, the highest relaxation rate occurs for the largest principal stress along the line direction and decreases in the order of across the line and normal to the line directions. This leads to a global relaxation of the stresses to a nearly hydrostatic state for subsequent stress relaxation. In the second stage, diffusion flow occurs under a nearly hydrostatic stress state, which leads to void formation and local stress relaxation adjacent to voids. During annealing, both contribute to stress relaxation, but because of the different kinetics, stress relaxation via plastic flow dominates initially, and then at longer times, diffusional flow dominates the relaxation process, leading to void formation.

In Figure 3.18, the transition in the stress relaxation kinetics for the 1 μm wide lines was observed at anneal temperatures of 150°C and 200°C but not at 250°C. Annealing at 250°C, stress relaxation was observed for an initial period up to about 1,000 minutes, with no further relaxation afterward. This can be attributed to low thermal stresses at 250°C as seen from the X-ray results in Figure 3.8b, so after the initial period, there was little stress left to drive further relaxation. In comparison, the transition point in relaxation kinetics seems to occur earlier at 200°C than at 150°C. An early transition in the kinetics of stress relaxation implies that void growth starts earlier in samples annealed at 200°C. Although stress relaxation occurs later at 150°C, its rate as measured from the stress-log(time) slope is higher. Considering that the rate of diffusional flow should increase with temperature, the opposite result observed here must be attributed to a larger stress driving force at 150°C. This result suggests that once voids start to grow at 150°C, they will grow at a faster rate relatively to that of the 200°C anneal. Overall, the stress relaxation rate found for the 1.0 μm wide lines is in good agreement with the model prediction and maximized between 150°C and 200°C. A similar stress relaxation behavior was observed for the 0.5 μm wide lines, but no clear transition in kinetics was found for the 3 μm wide lines because the latter had a near biaxial stress state as discussed later in conjunction with short-term stress relaxation.

The short-term stress relaxation kinetics are shown in Figure 3.19 as a function of linewidth and temperature. The stress relaxation of the 3 μm wide lines does not show significant changes in kinetics, although at 150°C anneal, there is a slight change in the stress-log(time) slope along the line direction. This indicates that, for the 3 μm wide lines, the kinetics of the entire stress relaxation process is dominated by plastic flow that is controlled by dislocation glide rather than due to diffusional flow and void growth. This is to be expected from the X-ray stress data in Figure 3.8a showing a near biaxial stress state for the 3 μm AlCu lines with only a small stress component normal to the line. With such a near biaxial stress state, stress relaxation is mainly driven by the shear stress as plastic flow, making it difficult to observe the kinetic stage of diffusional flow. The stress relaxation data also explain why stress-induced voiding phenomena have been observed only for interconnects with a certain minimum linewidth.

Results of the long-term relaxation rates obtained up to 20 days are summarized in Figure 3.20. The 0.5 μm wide lines have long-term relaxation rates for all three stress components higher than the 1 μm wide lines, although the relaxation generally starts

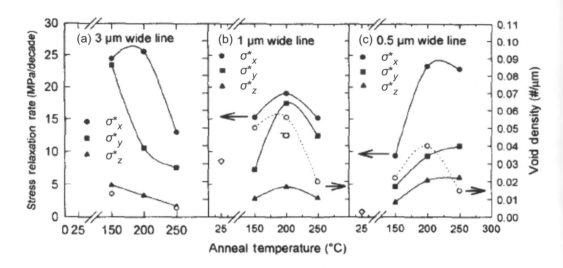

Figure 3.19 Comparison of short-term stress relaxation and void density for (a) 3 μm, (b) 1 μm, and (c) 0.5 μm wide lines. Void density is plotted as the dash line and with scale on the right. Reprinted from [54], with the permission of AIP Publishing.

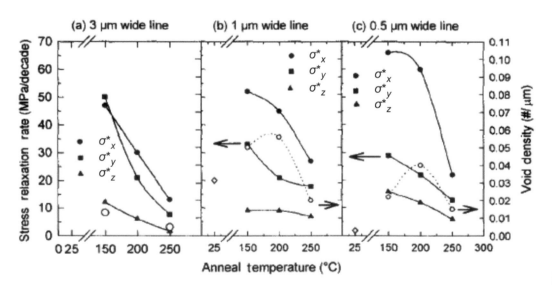

Figure 3.20 Comparison of the rate of stress relaxation during the second stage and void density for (a) 3 μm, (b) 1 μm, and (c) 0.5 μm wide lines. Void density is plotted as the dash line and with scale on the right. Reprinted from [54], with the permission of AIP Publishing.

later. The average grain sizes for these lines have been measured by transmission electron microscopy and found to be 1.69, 0.88, and 0.83 μm for the 3, 1, and 0.5 μm lines respectively. This indicates that the 1 μm line has a near-bamboo and the 0.5 μm line a bamboo grain structure, and they should have comparable rates of grain boundary diffusion. Thus, the higher relaxation rate for the 0.5 µm line can be attributed to a higher stress driving force, which is consistent with the temperature dependence of the relaxation rate. For the 3 µm wide lines, the kinetics of the long-term stress relaxation process continues to be dominated by plastic flow rather than diffusional flow and void growth.

The rate of stress relaxation depends on the driving force (thermal stress) and mass transport. While the rate of mass transport increases with increasing temperature, the driving force from thermal stresses decreases with increasing temperature. The combination of these two factors yields a finite temperature range, which is approximately from 150 to 250°C for AlCu lines, where the stress relaxation rate reaches a maximum and was observed. Stress relaxation due to void growth has been analyzed for Al line structures, based on atomic diffusion to the grain boundaries. An exponential time decay was predicted with a rate depending on the balance between the stress driving force and the grain boundary diffusivity, which is maximized at some intermediate temperature [55, 56].

To correlate void formation to the observed stress relaxation kinetics, measurements on void density were carried out using SEM and TEM. Voids were found located at grain boundary and line edge intersections and predominantly wedge shaped. Void density depended on both the anneal temperature and linewidth, which was consistent with the stress relaxation rate. The typical size of the voids after a short-term anneal at 250°C for one day is in the range of 0.1–0.2 µm and increases with increasing anneal time. A high number of voids were found in the 1 µm wide lines, although fewer voids were observed in the 3 and 0.5 µm wide lines. Yue et al. [57] also observed a similar linewidth dependence of void formation where void density showed a maximum for 2 to 3 µm wide lines. Considering that the 3 µm has a polycrystalline grain structure while the 1 and 0.5 µm lines have a near-bamboo and bamboo grain structures respectively, the results observed indicate that void formation is a general phenomenon and does not depend on grain structure.

Void density has been obtained for each linewidth, and the void densities after the short-term annealing for five days are compared with their respective stress relaxation rates in Figure 3.19. The overall trend of decreasing void density with increasing anneal time indicates that voids probably grow by coalescence of preexisting small voids during isothermal annealing. Void densities for 1 µm and 0.5 µm wide lines show good correlation with the relaxation rate at anneal temperatures of 150 and 200°C, although the 250°C anneal data for the 0.5 µm lines do not correlate well with the short-term stress relaxation data. This indicates that the correlation of the short-term relaxation rates with void density is not good enough for all test temperatures. The void densities are compared with the long-term stress relaxation rate in Figure 3.20, where the relaxation rate correlates with a decrease in void density only above 200°C. Thus, neither the short-term nor the long-term relaxation behavior can explain the observed behavior of void density over the entire temperature range. Compared with the narrower 1 and 0.5 µm lines, the 3 µm lines have a relatively low void density and do not correlate well with the stress relaxation rate. Because the 3 µm lines exhibit a nonbamboo structure that offers many pathways for grain boundary diffusion, mass transport cannot be the reason for the observed low void density. Instead, this suggests that stress relaxation in the 3 μm lines is dominated by plastic yield, which leaves a low stress driving force for void formation, as evidenced by comparing the stress levels for the 3 μm lines with the 1 and 0.5 μm lines in Figure 3.8. Overall, the void density results indicate that the short-term stress relaxation rate is correlated with the void density at low temperature while the long-term stress relaxation is correlated at high temperature. This can be attributed to the nucleation and growth of voids. It is reasonable that the long-term stress relaxation is due to void growth, thus it dominates at high temperature. The correlation shown in Figure 3.19 suggests that the short-term stress relaxation may be related with void nucleation. As a result of the combination of void nucleation and growth, the void density is maximized at 200°C for 1 and 0.5 μm wide lines and thus confirms that void formation due to stress relaxation reaches a maximum depending on the balance of the thermal stress driving force and the mass transport.

The results presented here are consistent with the SIV model proposed by McPherson and Dunn [56] based on the two factors controlling the rate of void formation, one due to the mass transport that increases with increasing temperature, and the other from the thermal stresses that decrease with increasing temperature. The combination of these two factors yields a finite temperature range, which is approximately from 150 to 250°C for AlCu lines, where the void formation rate reaches a maximum. Based on a power-law creep, the rate of void formation is presented as

$$R = C(T_0 - T)^N \exp\left(\frac{-Q}{kT}\right), \tag{3.10}$$

where T_0 is the stress-free temperature where the thermomechanical stress changes from tensile to compressive, T is the temperature, N is the "creep exponent," Q is the diffusional activation energy, k is the Boltzman constant, and C is a proportionality constant. For a 3 um wide Al(1%Si) line, the values for Al-based interconnects determined are $T_{0,Al} = 232°C$, $N_{Al} = 2.33$, and $Q_{Al} = 0.58$ eV [56]. This SIV model has been generally accepted for Al-based interconnect, although the stress relaxation leading to void formation is considerably more complex as seen from the results presented here. It is worth noting that the stress-voiding problem has been greatly mitigated by using redundant metal layers (e.g., TiW, TiN, W) under the Al layer in the form of a bilayer composite.

3.6   Stress Relaxation and Passivation Effects in Cu Damascene Films and Lines

To discuss stress relaxation in Cu interconnects, it is relevant to consider the damascene structure, which incorporates structural elements and fabrication steps with distinct mass transport paths and failure mechanisms for stress-induced void formation and EM [36, 58, 59]. The dual damascene structure has introduced a new stress voiding mechanism due to void formation under the via [48]. This problem is discussed in the next section. In addition, mass transport under EM in Cu damascene lines was found to be dominated by diffusion at the cap layer interface due to the presence of defects induced by CMP (step 6 in Figure 3.11), which can seriously degrade the EM lifetime [60, 61]. Lane et al. [62] showed that the EM lifetime of Cu interconnects is correlated to the chemical bonds at the Cu–cap layer interface and can be improved by optimization. This was corroborated by Hu et al. [63], who demonstrated a significant improvement in EM lifetime by coating the Cu interface with a thin metal layer. As interconnect scaling continues, the interface-to-volume ratio increases with decreasing linewidth, making interfacial diffusion increasingly important in controlling the EM reliability, a topic that is discussed in Chapter 5. It is difficult, however, to delineate the kinetics of mass transport from the EM lifetime because EM failure is a complex process, depending also on the defect and microstructure characteristics. In this section, we present stress relaxation studies to investigate the effect of the cap layer on mass transport, first in Cu damascene films and then in line structures.

Stress relaxation processes and deformation mechanisms have been extensively studied for Cu films, including those deposited by e-beam and sputtering [64–68]. Several authors have shown that, at moderate temperatures, stress in polycrystalline Cu films is relaxed by diffusion flows at the surface and grain boundaries [65–68]. Thouless [69] and Gao et al. [70] have developed diffusional creep models based on coupling of grain boundary and surface diffusion to account for thermal stress behavior under thermal cycling in unpassivated Cu films. However, the kinetics under thermal cycling is usually dominated by fast processes such as plastic yield, so except for very slow ramping rates, it would be difficult to extract information for a slow process such as diffusional creep and the associated mass transport. The stress relaxation studies discussed here were carried out using a bending beam method under isothermal conditions similar to that used for AlCu lines in Section 3.5 [71, 72]. A kinetic model was formulated to analyze stress relaxation by coupling interfacial and grain boundary mass transports in thin films. The interface and grain boundary diffusivities can be deduced using this model to investigate the effect of passivation on interfacial mass transport in Cu damascene structures. Following the damascene film studies, bending beam experiments were carried out on Cu damascene lines to investigate stress relaxation and the effect of interfacial passivation in line structures.

For these studies, electroplated Cu films were prepared following a damascene process with structural elements as shown in Figure 3.21. A thin layer of silicon oxide (100 nm) was deposited on the wafer surface, followed by depositing an etch-stop layer and a 450 nm layer of carbon-doped oxide (CDO), a low-k dielectric material. A TaN/Ta diffusion barrier was deposited before physical vapor deposition of a Cu seed layer. The rest of the Cu film was electroplated, followed by CMP, to yield a thickness of 0.8 μm, and then passivated with a cap layer. Four different cap layers were investigated, including SiN, SiC, SiCN, and a metal cap. A thin Co cap layer (about 10–20 nm) was used as the metal cap at the interface between the Cu film and a SiN passivation. The thickness of the other cap layers was 100 nm.

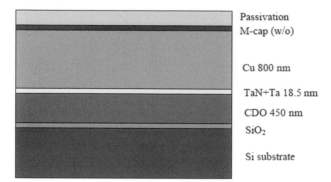

Passivation
M-cap (w/o)

Cu 800 nm

TaN+Ta 18.5 nm

CDO 450 nm

SiO₂

Si substrate

Figure 3.21 Schematic of the film stack of the Cu samples used for stress relaxation studies. Reprinted from [71], with the permission of D. Gan.

In this study, all samples were first annealed in a thermal cycle with a peak temperature of 450°C, then followed with a second thermal cycle where the peak temperatures were adjusted to normalize the starting stress of the samples for the isothermal stress measurements. These thermal cycling conditions were chosen to yield similar microstructure and stress state in the Cu films after the 450°C annealing in the first thermal cycle and to complete the short-term stress relaxation in the second thermal cycle at a lower temperature of about 350°C before the long-term stress relaxation. In this way, the isothermal measurements can be set up with similar initial tensile stresses at different temperatures. Following this procedure, stress relaxations were carried out for the four cap layers, and the results at 210°C for 30 hours with a starting stress of 190 MPa are shown in Figure 3.22. The results show that the stress relaxation rate with a Co metal cap layer is the slowest and thus the most effective in reducing the interfacial mass transport, followed in the order of SiN, SiCN, and SiC cap layers. The order of the stress relaxation rates for the four cap layers correlates very well with the EM results of Cu interconnects passivated with those cap layers [73].

For data analysis, a kinetic model for isothermal stress relaxation in polycrystalline thin films with and without passivation was developed based on the coupling of grain boundary and interface diffusion [74]. In this model, the film consists of a single-layer periodic array of grains with grain boundaries perpendicular to the interface as shown in Figure 3.23.

For the unpassivated film in Figure 3.23a, the model considers mass transport by diffusion along the free surface and the grain boundary, but neglects lattice diffusion and diffusion along the film/substrate interface. The chemical potential is defined by the local curvature for the surface and the normal stress for the grain boundaries. The chemical potential gradient drives atoms to diffuse into or out of grain boundaries, relaxing the average stress in the film. At a grain boundary, the chemical potential is proportional to the normal stress and evolves with time as atoms diffuse along the grain boundary. The divergence of mass transport at the grain boundary induces an

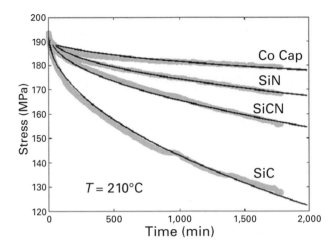

Figure 3.22 Isothermal stress relaxation in Cu damascene films passivated with four different cap layers of Co metal cap, SiN, SiCN, and SiC measured at 210°C. The thin solid lines are stress relaxation curves calculated from the kinetic coupling model. Reprinted from [72] with permission from Springer Nature.

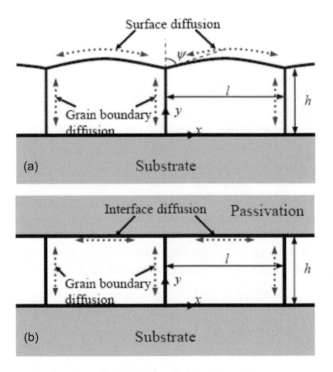

Figure 3.23 Schematic illustration of the grain structure and diffusion paths in the kinetic model for (a) unpassivated Cu films and (b) passivated Cu films. Reprinted from [74], with the permission of AIP Publishing.

inelastic strain (or creep strain), which is related to the curvature of the grain boundary stress. In terms of the normal stress, the governing equation for the grain boundary diffusion is

$$\frac{\partial \sigma_B}{\partial t} = \frac{M\Omega\delta_B D_B}{\ell k T} \frac{\partial^2 \sigma_B}{\partial y^2},$$

(3.11)

where σ_B is the normal stress at the grain boundary, M the biaxial modulus of the film, Ω the atomic volume, δ_B the grain boundary width, D_B the grain boundary diffusivity, l the grain size, k the Boltzmann's constant, T the absolute temperature, and y the coordinate along the grain boundary.

For the film with a passivation layer in Figure 3.23b, the biaxial in-plane stress is relaxed by diffusional flow of atoms via the grain boundaries and the interface with a cap layer. In contrast to the case of the free surface for unpassivated films, the interface is assumed to remain flat due to the constraint of the cap layer. Mass transport along the interface induces nonuniform local normal stresses, which define the chemical potential and drive interface diffusion, and we have

$$\frac{\partial \sigma_I}{\partial t} = \frac{M\Omega\delta_I D_I}{h k T} \frac{\partial^2 \sigma_I}{\partial x^2} + \frac{2M\Omega}{h l} J_0(t),$$

(3.12)

where $\sigma_I(x,t)$ is the locally normal stress at the interface, x is the coordinate along the interface, and $J_0(t)$ is the atomic flux rate at the junction between the interface and a grain boundary.

The mass transport processes are coupled since the chemical potential and the atomic flux are continuous at the junction between a grain boundary and the surface or interface. Additional boundary conditions are specified by assuming no flux at the film–substrate interface and a periodic grain structure. The coupled problem can be solved numerically, and the details are presented in [71]. In general, the diffusivities follow the trend of $\delta_s D_s > \delta_B D_B > \delta_I D_I$ (the indices S, B, and I refer to surface, grain boundary, and interface, respectively); in this case, the stress relaxation is much more sensitive to the interface diffusivity for passivated films than it is to the grain boundary and the surface diffusivities. This allows quantitative characterization of the kinetics of mass transport along the interface, which can then be used to evaluate the effect of interfacial passivation layer. In the limiting case when the grain boundary diffusion is about 10× faster than at the interface, stress relaxation is fully controlled by interface diffusion. Then an analytical solution is obtained for this limiting case as

$$\sigma(t) = \sigma_0 \frac{8}{\pi^2} \sum_{n=0}^{\infty} \frac{\exp\left[-(2n+1)^2 t/\tau_I\right]}{(2n+1)^2},$$

(3.13)

where $\sigma(t)$ is the locally normal stress at the interface, σ_0 is the initial stress and

$$\tau_I = \frac{k T h l^2}{\pi^2 M\Omega\delta_I D_I}$$

(3.14)

By fitting the solution with experimental data from isothermal stress relaxation measurements for unpassivated and passivated films, the grain boundary and the interface diffusivities were deduced with the grain boundary diffusivity more than $100\times$ larger. The fitting parameters used are grain size, $l = 1.04\,\mu m$; modulus, $M = 155$ MPa; and atomic volume, $\Omega = 1.18 \times 10^{-29}\,m^3$. The stress relaxation curves calculated from the kinetic coupling model are plotted as the thin solid lines in Figure 3.22, in good agreement with the experiments.

To measure the temperature dependence of stress relaxation, the peak temperatures in the second thermal cycle were adjusted to normalize the starting stress of the samples for the isothermal stress measurements. In this way, the isothermal measurements were set up with similar initial stresses at different temperatures to minimize the effect of initial stress. The stress relaxation curves of the Cu damascene films passivated with Co metal cap, SiN, SiCN, and SiC were measured, each at three temperatures, 179°C, 210°C, and 247°C. The samples were cooled in the second thermal cycle from 300°C, 350°C, and 400°C, respectively, which adjusted the initial stresses to about 190 MPa to start the stress relaxation measurements. The stress relaxation curves are shown in Figure 3.24a and b for SiN cap and Co metal cap, respectively, and with a much slower kinetics for the Co metal cap. The kinetic model was used to deduce the interfacial diffusivities by fitting to the measured stress relaxation curves and the results are plotted in Figure 3.24 as the thin solid lines.

The deduced interface diffusivities for different cap layers are summarized in the Arrhenius plots in Figure 3.25, where the previously deduced grain boundary diffusivities are also plotted for comparison. Using linear regression, the interfacial diffusivities as functions of temperature were deduced as

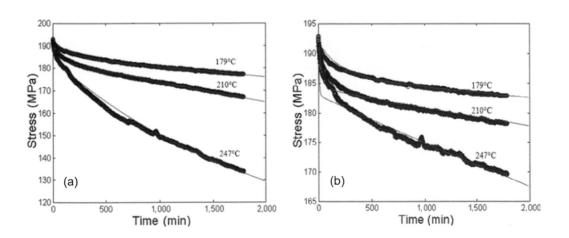

Figure 3.24 Isothermal stress relaxation of electroplated Cu films passivated with (a) SiN, and (b) Co metal cap layers. Note the difference in the stress scale. The thin solid lines are stress relaxation curves calculated from the kinetic coupling model. Reprinted from [72] with permission from Springer Nature.

$$\delta_I D_I = 3.18 \times 10^{-23} \exp\left(-0.51\ \text{eV}/kT\right)\ \text{m}^3/\text{sec} \quad \text{for Co metal cap}$$
$$\delta_I D_I = 1.25 \times 10^{-20} \exp\left(-0.70\ \text{eV}/kT\right)\ \text{m}^3/\text{sec} \quad \text{for SiN cap}$$
$$\delta_I D_I = 2.79 \times 10^{-21} \exp\left(-0.61\ \text{eV}/kT\right)\ \text{m}^3/\text{sec} \quad \text{for SiCN cap} \qquad (3.15)$$
$$\delta_I D_I = 1.91 \times 10^{-21} \exp\left(-0.56\ \text{eV}/kT\right)\ \text{m}^3/\text{sec} \quad \text{for SiC cap}$$
$$\delta_B D_B = 1.1 \times 10^{-14} \exp\left(-1.07\ \text{eV}/kT\right)\ \text{m}^3/\text{sec} \quad \text{for grain boundaries}$$

Results from the isothermal stress relaxation measurements in Figures 3.22 and 3.24 clearly show a significant effect of the passivation layer on the kinetics of the interfacial mass transport. Results from this study directly correlate the cap layer effect on the interfacial mass transport, which is a key factor to control EM reliability. Among the four cap layers used in this study, the Co cap layer provides the most significant reduction in the stress relaxation rate by suppressing the mass transport at the interface between Cu and the cap layer. Quantitatively, the interface diffusivity decreases by almost two orders of magnitude from SiC to the Co metal cap. The order of decreasing relaxation rates in the Cu films passivated with SiC, SiCN, SiN, and metal cap agrees well with other studies correlating interfacial adhesion to electromigration [62, 73], which was attributed to increasingly enhanced chemical bonding at the interfaces. Figure 3.25 shows that the interface diffusion is substantially slower than the grain boundary diffusion in the temperature range of this study. This agrees with the general understanding about the passivation effect on the thermomechanical behavior in thin films subjected to thermal cycling [69, 70]. The grain boundary

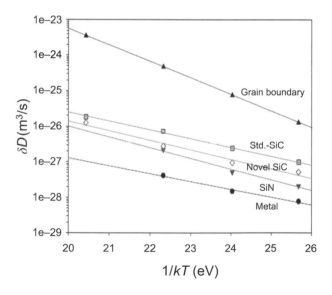

Figure 3.25 Temperature dependence of the interface diffusivities in electroplated Cu films passivated with four different cap layers: SiC, SiCN (novel SiC), SiN, and Co metal cap. The grain boundary diffusivities are plotted for comparison. Reproduced from [71] with permission of D. Gan.

diffusivity was found to have an activation energy of 1.07 eV, which is in agreement with the 0.95 eV reported by Gupta et al. for Cu thin films [75]. The activation energies, however, are systematically lower than the activation energy (0.8–1.0 eV) deduced from EM lifetimes of Cu interconnects [63] and coupled with an unusually large preexponential factor. The difficulty may rest with the stress relaxation measurement, which requires a reasonable starting stress value, thus limiting the experiments to a relatively narrow temperature range around 180°C to 250°C and increased the error in determining the activation energy and the preexponential factor.

The effect of the passivation layer on stress relaxation was measured for Cu lines using the bending beam technique [76]. The Cu lines were fabricated by a damascene process with a CDO low-k dielectric with two line configurations, one oriented parallel and one perpendicular to the length direction of the sample beam. The curvatures of the parallel and the perpendicular samples were measured to determine the stress along (parallel to) the line and across (perpendicular to) the line directions. Isothermal stress relaxation measurements were carried out following the procedure used for Cu films, starting with thermal cycling to stabilize the stress state and the Cu microstructure, then setting the starting stress and temperature for isothermal stress relaxation measurements. In this way, stress relaxations were measured at 160°C, 200°C, and 240°C, and the relaxation behaviors of σ_x along the line and σ_y across the line directions at 240°C are shown in Figure 3.26 as a function of time. Overall, the stress relaxation was found to be faster along the line x-direction than across the line y-direction, reflecting the directionality of the passivation effect on stress relaxation of line structure. The observed results indicate that stress relaxation rates in damascene Cu line structures strongly depend on the passivation in the order of SiC > SiN > Co metal cap, showing that the stress relaxation is significantly suppressed by the Co

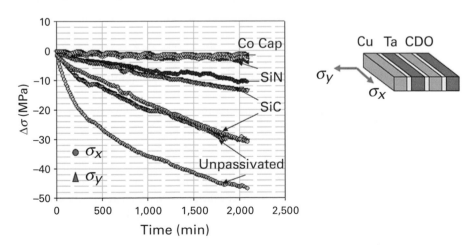

Figure 3.26 Stress relaxation of Cu damascene line structures with different passivation; for each passivation, σ_x is the lower curve and σ_y the upper curve. The insert on the right shows the schematic configuration of the Cu damascene line structure used in the measurement. Reproduced from [71] with permission of D. Gan.

Table 3.2. Correlation of stress relaxation results with adhesion energy and electromigration lifetime for various passivation layers in Cu/CDO structures.

Passivation layer	Interfacial adhesion (J/m^2)	Electromigration MTTF, hours (normalized)	$\Delta\sigma_x$ (MPa)	$\Delta\sigma_y$ (MPa)
SiC	4	1	31	13.5
SiN	10	2.0	10	3.1
Co metal cap	12	3.1	3.3	2.0

Source: Reproduced from [71] with permission of D. Gan.

metal cap. The results are consistent with the passivation effect on stress relaxation observed for the Cu damascene films.

A 3D finite element model was set up to analyze the passivation effect on the stress relaxation rate observed in the Cu damascene line structures [77]. In this model, the passivation effect was attributed to two factors, one from changes in the interfacial diffusivity and the other from the confinement of the passivation layer and the line structure. The confinement effect was evaluated using an effective bulk modulus B, which is related to the stiffness of the line structure responding to the stress generated by mass transport under confinement [78]. Assuming mass transport only at the cap layer interface, B was calculated for the line structures, and it was found that the effect due to different cap layers is relatively small and cannot account for the passivation effect observed, particularly for the Co metal cap to have the slowest stress relaxation. Therefore, the results as observed indicate an effect primarily due to the interfacial chemistry of the passivation layer on the mass transport. For the Cu/CDO line structures in this study, the EM lifetime and the interfacial adhesion have been measured, and the results showed a good correlation with the stress relaxation rate as shown in Table 3.2.

3.7   Stress-Induced Void Formation in Cu Damascene Line Structures

When Cu was introduced in the 1990s to replace Al as on-chip interconnects, stress voiding in Cu interconnects was generally considered less a reliability problem. This is because the stress level in Cu lines is about the same (compare the results in Sections 3.4 and 3.5), but the atomic mobility in Cu is lower due to a higher activation energy. However, the Cu damascene line structure has introduced new structural elements and processes, leading to different stress characteristics and voiding mechanisms. In 2002, Ogawa et al. reported stress-induced void formation under the Cu via that was different from void formation found mostly at the line edges in Al interconnects [48]. This was attributed to condensation of supersaturated vacancy concentration driven by thermal stresses and has generated considerable interest subsequently to investigate stress voiding in Cu interconnects. Subsequently, with the introduction of low-*k* dielectrics, there are additional questions concerning the effect of dielectric passivation on the thermal stress and stress voiding.

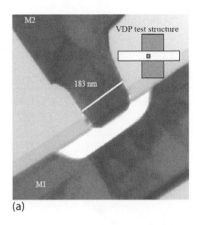
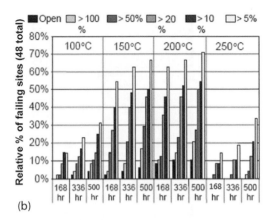

(a)   (b)

Figure 3.27 (a) Void formation under a small via placed over a wide metal line. The M1 line is about 3 μm wide where the void was formed after baking for over 100 hours at 150°C. The inset graphic shows the test structure. (b) Temperature-dependent bake data on void density population between 100°C to 250°C. © 2002 IEEE. Reprinted with permission, from [48].

An example of void formation in Cu interconnects is shown in Figure 3.27a in a single 0.18 μm via structure in contact to a wide (3 μm) Cu stripe after baking for over 100 hours at 150°C. A detailed study was carried out using dual-damascene Van der Pauw (VDP) via test structures embedded in fluorinated silicate glass (SiOF) to measure the void population as a function of temperature between 100°C and 250°C. A high void formation rate was found between 150°C and 200°C, which reduced sharply above 200°C, as shown in Figure 3.27b.

The void formation observed was attributed to the condensation of supersaturated vacancies driven by stress gradients under the via in a fully constrained damascene line. In the damascene structure, Cu is deposited by electrochemical deposition (ECD) at a fast deposition rate where grain growth readily occurs during subsequent thermal annealing. This generates supersaturated vacancies at grain boundaries and interfaces and that are then swept by the local stress gradient toward the via to induce void formation. This stress voiding mechanism is schematically shown in Figure 3.28a.

The SIV data shown in Figure 3.27b were analyzed using the stress migration model developed by McPherson and Dunn [56]. This yielded a maximum void formation rate at $T_{crit} \sim 190°C$ and a stress-free temperature T_0 about 270°C. An "effective" activation energy was found to be about 0.74 eV, which is lower than Cu grain boundary diffusion (~1 eV [75]) and suggests that interfacial diffusion may be important for void formation under the via. The thermomechanical stress exponent of $N = 3.2$ is in the generally accepted range of 2–4 for metals and is consistent with the exponents obtained from power-law creep ($3 < N < 8$) [46].

Following the Ogawa study, several groups have extended stress voiding studies in Cu damascene interconnects to investigate the effects of materials and structures, including Cu line structures integrated with low-k dielectrics. Results from these

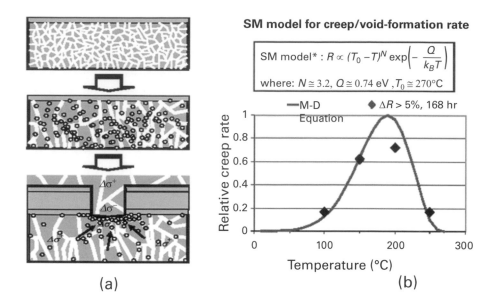

(a)   (b)

Figure 3.28 (a) Vacancy generation through grain growth in constrained metal leads. Small grains that are constrained by barrier and capping layers that undergo grain growth will be unable to eliminate trapped vacancies at a free surface. The stress gradient developed underneath the via will then attract vacancies to coalesce into voids. (b) McPherson and Dunn creep/voiding rate model. © 2002 IEEE. Reprinted with permission, from [48].

studies with the TEOS passivation have largely confirmed the basic voiding mechanism with higher failure rates occurring in wider M1 lines and maximized at an intermediate temperature around 200°C to 240°C [79–81]. Several subsequent studies [82 ,83] reported also stress voiding induced by processing defects near the line edges in Cu/low-k interconnects. Kawano et al. [79] showed that stress-induced voiding can occur in the via as well as under the via. In this study, they observed that a via connected to a wide M2 line above it was more susceptible to stress-induced voiding (SIV). On the other hand, a via connected to a wide M1 line was vulnerable to SIV under the via bottom. In these cases, wide Cu lines can provide more supersaturated vacancies, while narrow Cu lines can provide only a limited number of supersaturated vacancies. In a study of the line dependence of void formation, Shao et al. [80] found that the void formation rate depended not only on the stress but also on the stress gradient and linewidth, particularly across the line direction, which can drive vacancy migration to accumulate near the via bottom leading to a much higher resistance change at wider M1 lines. This was confirmed by Paik et al. [81] reporting that the stress gradient can be more important than stress itself for stress voiding, particularly for line structures with low-k dielectric passivation. They showed by modeling that the stress gradient is larger in Cu/low-k line structures and provides a larger driving force for vacancy migration and void formation. These studies showed that stress gradient can be effective in driving vacancy migration, leading to void formation as described by the vacancy flux (3.9) in Section 3.6. This can account for void formation in addition to the vacancy concentration gradient used in the Ogawa model, where their relative contributions depend on the dielectric passivation and the line/via dimensions. For Cu interconnects, grain growth in the electroplated Cu line can readily supply vacancies for void formation, a process not readily available in Al interconnects.

In two other studies [82, 83], stress voiding in Cu/low-k structures was found to be enhanced by processing defects with significantly reduced median time to failure (MTF) under accelerated testing. The linewidth was also found to have a much more profound effect on stress voiding in Cu/low-k than in Cu/FTEOS with higher failure rates for wider lines, which was attributed to higher stresses in Cu and larger diffusion volumes. Void formation has been found to have an Arrhenius temperature dependence instead of being maximized at an intermediate temperature in the Ogawa model, indicating that it is driven primarily by mass transport with the stress playing only a minor role [84].

It is interesting that void formation has been observed in passivated Cu films and wide lines under a biaxial stress state, which is unexpected and seldom observed in Al films or wide lines. Void formation was first reported by Shaw et al. [85] in passivated Cu lines of 10–30 microns in linewidths where the stress state is predominantly biaxial. Void formation was subsequently observed in passivated Cu films under isothermal annealing [86], which was attributed to localized triaxial stress states at the $\langle 111 \rangle$/$\langle 200 \rangle$ type grain boundary junctions due to the elastic anisotropy of $\langle 111 \rangle$ and $\langle 200 \rangle$ Cu grains. To account for void density, X-ray analysis was performed and found in good agreement with the population of the $\langle 111 \rangle$/$\langle 200 \rangle$ boundary junctions in the Cu films. The kinetics of void formation were found to be consistent with the McPherson and Dunn model, with a peak growth rate at about 250°C and an activation energy of 0.75 eV, but followed the kinetics of a linear diffusional creep instead of a power-law creep to reflect a different stress relaxation mechanism in Cu films.

Stress voiding reliability has been extensively evaluated by Li and Badami for Cu interconnects [87]. In this study, a broad range of SIV structures were used and tested at multiple temperatures for up to 20,900 hours to probe the voiding statistics and mechanisms. Four types of SIV characteristics were observed after long-term stressing with a wide range of test structures: (I) the voiding rate is predominately driven by chemical potential gradient such as the vacancy concentration gradient, and increases with temperature; (II) the voiding rate is dominated by the combination of stress and chemical potential gradients, reaching a peak at an intermediate temperature; (III) voiding behavior shows defect-like characteristics, with a portion of the samples showing open circuit failures; and (IV) the voiding rate is very low, and virtually no resistance change is observed at all stress temperatures. The overall stress voiding statistics exhibited a bimodal behavior for all linewidths as a function of temperature with a type I strong mode and a type II weak mode as shown in Figure 3.29. The type I strong mode showed a low resistance increase with increasing stress temperature. In contrast, the type II weak mode showed larger resistance increases saturated at an intermediate temperature of about 225°C. Although the two failure modes exhibited

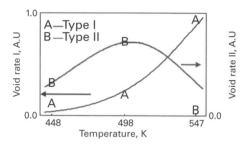

Figure 3.29 Examples of void growth rate vs. stress temperature for Group A with a strong type I and group B with a weak type II stress voiding kinetics. © 2012 IEEE. Reprinted with permission, from [87].

different resistance growth characteristics, both can be fit with (3.10) of the McPherson and Dunn model with a similar activation energy Q of 0.9–1.0 eV and a stress-free temperature T_0 of 340°C but different N (~1 for type I and ~6 for type II). It was concluded that, only for the type II weak mode, the characteristics were associated with stress-induced voiding, and only this group could pose SM concerns and should be taken into account for circuit and interconnect design.

3.8   Summary

In this chapter, we trace the EM study of metal lines to the first observation of EM-induced failure in Al interconnects. Since then, the study of EM has made a significant and basic change from bulk metals to metal lines used as on-chip interconnects. Driven by the rapid development of semiconductor technology, EM soon became an important reliability problem for interconnects, and its study has evolved with new perspectives and focus. Such studies are defined on one hand by the material, structure, and configurations of the metal lines, and on the other hand by the functionality and reliability requirement of the on-chip interconnects. As the technology advances, the interconnect technology continues to evolve, migrating from Al to Cu interconnects, and with line dimensions decreasing from microns to nanometers, leading to fast but significant changes in EM studies.

The focus of this chapter is on thermal stresses and stress voiding of Al and Cu line structures. For line structures, the passivation from interlevel dielectrics induces a triaxial stress state where the stress can be very large, reaching about 0.8% strain for Al on Si, and with it a substantial increase in the elastic energy. This sets up a reference frame to study EM where thermal stress is as important as the electrical current in contributing to the mass transport and damage formation in metal line structures. In this regard, Cu interconnects are quite different from the Al interconnects because of the damascene structure and the fabrication process, which incorporate distinct materials and structures that can affect the mass transport and damage mechanisms. Driven by device scaling, the dimensions of the Cu lines continue to decrease to the nanoscale, which has caused the Cu microstructure to evolve and, together with Cu elastic anisotropy, change the thermal stress and EM characteristics for damage formation. At the same time, the demand in performance continues to increase the current density and the power, accelerating the rate of damage formation in the wiring structure.

The discussion in this chapter is focused on the topics of thermal stresses and stress-induced void formation, primarily on the methodology and early results of Al and Cu line structures. Such studies serve as a basic framework for discussing EM and stress voiding in subsequent chapters. The discussion will continue with the formulation of the Korhonen equation and kinetic analysis in Chapter 4 and a more general 3D formulation and analyses in Chapter 7, EM reliability in Chapter 5 and microstructure evolution in Chapter 6.

References

1. A. Blech and H. Sello, The failure of thin aluminum current carrying strips on oxidized silicon, *Fifth Annual Symposium on the Physics of Failure in Electronics* (1966), 496–505.
2. G. C. Schwartz and K. V. Srikrishnan, Metallization. *Handbook of Semiconductor Interconnection Technology*, 2nd ed., ed. G. C. Schwartz and K. V. Srikrishnan (London: Taylor & Francis Group, 2006), 311–384.
3. G. E. Moore, Cramming more components onto integrated circuits, *Electronics* **38** (8) (1965), 8–11.
4. D. Edelstein, J Heidenreich, R Goldblatt, et al. Full copper wiring in a sub-0.25 micron CMOS ULSI technology, *Tech. Dig. IEEE International Electron Devices Conference* (1997), 773–776.
5. International Technology Roadmap for Semiconductors, Interconnect Section, 2007 Edition. Semiconductor Industry Association (2007).
6. I. A. Blech, Electromigration in thin aluminum films on titanium nitride. *Journal of Applied Physics* **47** (1976) 1203–1208.
7. I. A. Blech and C. Herring, Stress generation by electromigration. *Applied Physics Letters* **29** (1976) 131–133.
8. I. A. Blech and K. L. Tai, Measurement of stress gradients generated by electromigration, *Applied Physics Letters* **30** (1977), 387.
9. J. Klema, R. Pyle, and E. Domangue, Reliability implications of nitrogen contamination during deposition of sputtered aluminum/silicon metal films, *IEEE Proceedings International. Reliability Physics Symposium* **22** (1984), 1–5.
10. J. Curry, G. Fitzgibbon, Y. Guan, R. Muollo, G. Nelson, and A. Thomas, New failure mechanisms in sputtered aluminum-silicon films, *IEEE Proceedings International. Reliability Physics Symposium* **22** (1984), 6–10.
11. S. M. Alam, C. L. Gan, C. V. Thompson, and D. E. Troxel, Reliability computer-aided design tool for full-chip electromigration analysis and comparison with different interconnect metallizations. *Microelectronics Journal* **38** (2007), 463–473.
12. M. A. Korhonen, P. Borgesen, and Che-Yu Li, Mechanisms of stress-induced and electromigration-induced damage in passivated narrow metallizations on rigid substrates, *Materials Research Society Bulletin* **17** (1992), 61–69.

13. M. A. Korhonen, R. D. Black, and C-Y. Li, Stress relaxation of passivated aluminum line metallizations on silicon substrates, *Journal of Applied Physics* **69** (1991), 1748.

14. M. A. Korhonen, P. Borgesen, and C-Y. Li, Thermal stress and stress induced voiding in passivated narrow line metallizations. *Thermal Stress and Strain in Microelectronics Packaging*, ed. J. H. Lau (New York: Van Nostrand Reinhold, 1992), 133.

15. J. E. Steinwall and H. H. Johnson, Mechanical properties of thin film aluminum fibers: grain size effects. Thin films: stresses and mechanical properties II. Materials Research Society Symposium Proceedings 188, ed. M. Doerner, W. C. Oliver, G. M. Pharr, and F. R. Brotzen (1990), 177.

16. C. A. Paszkiet, M. A. Korhonen, and C-Y. Li, The effect of a passivation over-layer on the mechanisms of stress relaxation in continuous films and narrow lines of aluminum, *Materials Reliability Issues in Microelectronics*, ed. J.R. Lloyd, P.S. Ho, C.T. Sah, and F. Yost (Materials Resource Society Symposium Proceedings **225** (1991), 161.

17. M. F. Ashby, The deformation of plastically non-homogeneous materials, *Philosophical Magazine* **21,** (1970), 299–424.

18. W. D. Nix, Mechanical properties of thin films. *Metallurgical and Materials Transactions A,* **20** (1989), 2217–2245.

19. R. P. Vinci, E. M. Zielinski, and J. C. Bravman, Thermal strain and stress in copper thin films, *Thin Solid Films* 262 (1995), 142–153.

20. J. Kasthurirangan, Thermal stress and microstructure in Al(Cu) and Cu interconnects for advanced ULSI applications, unpublished Ph.D. dissertation, Materials Science and Engineering, University of Texas at Austin, 1998.

21. J. Kasthurirangan, Y. Du and P. S. Ho et al., Thermal stresses in Cu damascene submicron line structures, *AIP Conference Proceedings* **491** (1999), 304–314.

22. D. S. Gardner, J. Onuki, K. Kudoo, Y. Misawa, and Q. T. Vu, Encapsulated copper interconnection devices using sidewall barriers, *Thin Solid Films* **262** (1995), 104–119.

23. P. A. Flinn, D. S. Gardner and W. D. Nix, Measurement and interpretation of stress in aluminum-based metallization as a function of thermal history, *IEEE Transactions on Electron Devices* **34** (1987), 689–698.

24. M. F. Doerner, D. S. Gardner, and W. D. Nix, Plastic properties of thin films on substrates as measured by submicron indentation hardness and substrate curvature techniques, *Journal of Materials Research* **1** (1986) 845–851.

25. I. C. Noyan and J. B. Cohen, *Residual Stress Measurement by Diffraction and Interpretation* (New York: Springer, 1987).

26. A. S. Segmuller and M. Murakami, *Analytical Techniques for Thin Films, Treatise on Materials Science and Technology*, Vol. 27 (San Diego: Academic Press, 1988), 143–200.

27. S. H. Rhee, Thermal stress behaviors of Al(Cu)/low-*k* and Cu/low-*k* submicron interconnect structures, unpublished PhD dissertation, Materials Science and Engineering, University of Texas at Austin, 2001.

28. A. P. Clarke, High intensity X-ray instrumentation for residual strain measurements in composites and textured materials, unpublished PhD dissertation, Queen's University at Kingston, 1993, 174.

29. H. Yagi, H. Niwa, T. Hosoda, M. Inoue, and H. Tsuchikawa, Analytical calculation and direct measurement of stress in an aluminum interconnect of very large scale integration, *American Institute for Physics Conference Proceedings* 263 (1992), 44.

30. H. Niwa, H. Yagi, H. Tsuchikawa, and M. Kato, Stress distribution in an aluminum interconnect of very large scale integration, *Journal of Applied Physics* **68** (1990), 328.

31. A. I. Sauter and W. D. Nix, Finite element calculations of thermal stresses in passivated and unpassivated lines bonded to substrates, *Thin Films: Stresses and Mechanical Properties II*, ed. M. Doerner, W. C. Oliver, G. M. Pharr, and F. R. Brotzen (Pittsburgh: Materials Research Society Symposium Proceedings **188** (1990), 15.

32. I.-S. Yeo, S. G. H. Anderson, P. S. Ho, and C. K. Hu, Characteristics of thermal stresses in Al (Cu) fine lines. II. Passivated line structures, *Journal of Applied Physics* **78**, (1995), 953–964.

33. P. Chaudhari, Plastic properties of polycrystalline thin films on a substrate, *Philosophical Magazine A* **39** (1979), 507–516.

34. P. H. Wang, Thermal stresses and electromigration behaviors of Al/low *k* polymer interconnect structures, unpublished PhD dissertation, University of Texas at Austin (1997).

35. D. Edelstein, J. Heidenreich, R. Goldblatt, et al. Full copper wiring in a sub-0.25 micron CMOS ULSI technology, *Tech. Dig. IEEE International Electron Devices Conference* (1997), 773–776.

36. R. Rosenberg, D. C. Edelstein, C.-K. Hu, and K. P. Rodbell, Copper metallization for high performance silicon technology, *Annual Review of+ Materials Science* **30** (2000), 229–262.

37. J. Gambino, F. Chen, and J. He, Copper interconnect technology for the 32 nm node and beyond, *2009 IEEE Custom Integrated Circuits Conference, Rome* (2009), 141–148, doi: 10.1109/CICC.2009.5280904.

38. S. H. Rhee, Y. Du, and P. S. Ho, Thermal stress characteristics of Cu/oxide and Cu/low-submicron interconnect structures, *Journal of Applied Physics* **93** (2003), 3926–3933.

39. P. R. Besser, Y.-C. Joo, D. Winter, M. Van Ngo, and R. Ortega, Mechanical stresses in aluminum and copper interconnect lines for 0.18 μm logic technologies, *Materials Research Society Symposium Proceedings* **563**, (1999), 189–200.

40. P. R. Besser, E. Zschech, W. Blum, et al. Microstructural characterization of inlaid copper interconnect lines. *Journal of Electronics Materials* **30** (2001) 320–330.

41. H. Zhang, G. S. Cargill, and A. M. Maniatty, Thermal strains in passivated aluminum and copper conductor lines, *Journal of Materials Research* **26** (2011), 633–639.

42. P.-C. Wang, G. S. Cargill, and C.-K. Hu, Local strain measurements during electromigration, *American Institute for Physics Conference Proceedings* **491** (1999), 112–122.

43. C. J. Wilson, K. Croes, C. Zhao, et al., Synchrotron measurement of the effect of linewidth scaling on stress in advanced Cu/low *k* interconnects, *Journal of Applied Physics* **106** (2009), 053524.

44. M. A. Korhonen, C. A. Paszkiet, and C-Y. Li, Mechanisms of thermal stress relaxation and stress-induced voiding in narrow aluminum-based metallizations, *Journal of Applied Physics* **69** (1991), 8083–8091.

45. H. Okabayashi, Stress-induced void formation in metallization for integrated circuits, *Materials and Science Engineering R* **11** (1993), 191–241.

46. T. D. Sullivan, Stress-induced voiding in microelectronic metallization: void growth models and refinements, *Annual Reviews in Materials Science* **26** (1996), 333–364.

47. A. H. Fischer and A. E. Zitzelsberger, The quantitative assessment of stress-induced voiding in process qualification, *2001 IEEE International Reliability Physics Symposium Proceedings 39th Annual* (2001), 334–340.

48. E. T. Ogawa, J. W. McPherson, J. A. Rosal, et al., Stress-induced voiding under vias connected to wide Cu metal leads, *40th Annual International Reliability Physics Symposium (IRPS)* (2002), 312–331.

49. H. Y. Lin, S. C. Lee, and A. S. Oates, Characterization of stress-voiding of Cu/low k narrow lines, *46th International Reliability Physics Symposium* (2008), 687–688.

50. Z. Tőkei, K. Croes, and G. P. Beyer, Reliability of copper low-k interconnects, *Microelectronic Engineering* **87**, (2010), 348–354.

51. R. Rosenberg, and N. Ohring, Void formation and growth during electromigration in thin films, *Journal of Applied Physics* **42** (1971) 5671–5679.

52. W. D. Nix and E. Arzt, On void nucleation and growth in metal interconnect lines and electromigration conditions, *Metallurgical and Materials Transactions A* **23** (1992), 2007–2013.

53. Z. Suo, Reliability of interconnect structures. *Volume 8: Interfacial and Nanoscale Failure. Comprehensive Structural Integrity*, ed. W. Gerberich and W. Yang (Amsterdam: Elsevier, 2003), 265–324.

54. I.-S. Yeo, P. S. Ho, S. G. H. Anderson, and H. Kawasaki, Stress relaxation and void formation in passivated Al(Cu) line structures, *AIP Conference Proceedings* **418** (1998), 262–276.

55. A. Tezaki, T. Mineta, and H. Egawa, Measurement of three dimensional stress and modeling of stress induced migration failure in aluminum interconnects, *1990 IEEE International Reliability Physics Symposium Proceedings 28th Annual* (1990), 221–229.

56. J. W. McPherson and C. F. Dunn, A model for stress-induced metal notching and voiding in very large-scale-integrated Al-Si(1%) metallization, *Journal of Vacuum Science and Technology B* **5** (1987), 1321–1325.

57. J. T. Yue, W. P. Funsten, and R. V. Taylor, Stress induced voids in aluminium interconnects during IC processing, *IEEE International Reliability Physics Symposium Proceedings* **23** (1985), 126.

58. C. K. Hu and J. M. E. Harper, Copper interconnects and reliability, *Materials Chemistry and Physics* **52** (1998), 5–16.

59. E. Ogawa, K. D. Lee, and P. S. Ho, Electromigration reliability issues in dual-damascene Cu interconnections, *IEEE Transactions on Reliability* **51** (2002), 403–419.

60. C. K. Hu, R. Rosenberg, and K. Y. Lee, Electromigration path in Cu thin film lines, *Applied Physics Letters* **74** (1999), 2945.

61. P. Besser, A. Marathe, L. Zhao, M. Herrick, C. Capasso, and H. Kawasaki, Optimizing the electromigration performance of copper interconnects, *IEDM Technical Digest 2000* (Piscataway: IEEE, 2000), 119.

62. M. W. Lane, E. G. Liniger, and J. R. Lloyd, Relationship between interfacial adhesion and electromigration in Cu metallization, *Journal of Applied Physics* **93,** 1417 (2003), 1417.

63. C. K. Hu, L. Gignac, R. Rosenberg, et al., Reduced electromigration of Cu wires by surface coating, *Applied Physics Letters* **81,** (2002), 1782.

64. P. A. Flinn, Measurement and interpretation of stress in copper films as a function of thermal history, *Journal of Materials Research* **6** (1991), 1498–1501.

65. M. D. Thouless, J. Gupta, and J. M. E. Harper, Stress development and relaxation in copper films during thermal cycling, *Journal of Materials Research* **8** (1993), 1845–1852.

66. R. P. Vinci, E. M. Zielinski, and J. C. Bravman, Thermal strain and stress in copper thin films, *Thin Solid Films* **262** (1995), 142–151.

67. Y. L. Shen, S. Suresh, M. Y. He, et al., Stress evolution in passivated thin films of Cu on silica substrates, *Journal of Materials Research* **13**, (1998), 1928–1937.

68. R. M. Keller, S. P. Baker, and E. Arzt, Quantitative analysis of strengthening mechanisms in thin Cu films: effects of film thickness, grain size, and passivation, *Acta Materialia* **47** (1999), 1307–1317.

69. M. D. Thouless, Effect of surface diffusion on the creep of thin films and sintered arrays of particles, *Acta Metallurgica et Materialia* **41** (1993), 1057–1064.

70. H. Gao, L. Zhang, W. D. Nix, C. V. Thompson, and E. Arzt, Crack-like grain-boundary diffusion wedges in thin metal films, *Acta Materialia* **47** (1999), 2865–2878.

71. D. Gan, Thermal stresses and stress relaxation in copper metallization for ULSI interconnects, unpublished PhD dissertation, Materials Science and Engineering, University of Texas at Austin, 2005.

72. D. Gan et al. Effect of passivation on stress relaxation in electroplated copper films, *Journal of Materials Research* 21, 6 (June 2006), doi: 10.1557/JMR.2006.0196.

73. J. R. Lloyd, M. W. Lane, E. G. Liniger, C-K. Hu, T. M. Shaw, and R. Rosenberg, Electromigration and adhesion, *IEEE Transactions on Device Materials Reliability* **50,** (2005), 113.

74. R. Huang, D. Gan and P. S. Ho, Isothermal stress relaxation in electroplated Cu films. II. Kinetic modeling, *Journal of Applied Physics* **97** (2005), 103532.

75. D. Gupta, C. K. Hu, and K. L. Lee, Grain boundary diffusion and electromigration in Cu-Sn alloy thin films and their VLSI interconnects, *Defect and Diffusion Forum* **143–147** (1997), 1397–1406.

76. D. Gan, S. Yoon, P. S. Ho, et al., Effects of passivation layer on stress relaxation in Cu line structures, *Proceedings of the IEEE 2003 International Interconnect Technology Conference* (2003), 180–182.

77. N. Singh, A. F. Bower, D. Gan, S. Yoon, and P. S. Ho, Numerical simulations and experimental measurements of stress relaxation by interface diffusion in a patterned copper interconnect structure, *Journal of Applied Physics* **97** (2005)**,** 013539.

78. M. A. Korhonen, P. Borgesen, K. N. Tu, and C. Y. Li, Stress evolution due to electromigration in confined metal lines, *Journal of Applied Physics* **73** (1993), 3790–3796.

79. M. Kawano et al., Stress relaxation in dual-damascene Cu interconnects to suppress stress-induced voiding, *Proceedings of the IEEE 2003 International Interconnect Technology Conference* (2003), 210–212. 10.1109/IITC.2003.1219756.

80. W. Shao, Z. H. Gan, S. G. Mhaisalkar, et al., The effect of line width on stress-induced voiding in Cu dual damascene interconnects, *Thin Solid Films* **504** (2006), 298–301.

81. J. M. Paik, I. M. Park, and Y. C. Joo, Effect of grain growth stress and stress gradient on stress-induced voiding in damascene Cu/low-*k* interconnects for ULSI, *Thin Solid Films* **504** (2006), 284–287.

82. C. J. Zhai, H. W. Yao, A. P. Marathe, P. R. Besser and R. C. Blish, Simulation and experiments of stress migration for Cu/low-*k* BEOL, *IEEE Transactions on Device and Materials Reliability* **4** (2004), 523–529.

83. W. C. Baek et al. Stress migration studies on dual damascene Cu/oxide and Cu/low k interconnects, *American Institute of Physicists Conference Proceedings* **741** (2004), 249–255.

84. A. Glasow and A. H. Fischer, New approaches for the assessment of stress-induced voiding in Cu interconnects, *Proceedings of the IEEE International Interconnect Technology Conference* (2002), 274–276.

85. T. M. Shaw, L. Gignac, X-H. Liu, and R. Rosenberg, Stress voiding in wide copper lines, *Sixth International Workshop on Stress induced Phenomena in Metallization* (2001), 25–27.

86. D. Gan, B. Li, and P. S. Ho, Stress-induced void formation in passivated Cu films, *Materials Research Society Symposium Proceedings* **863** (2005), 259–266.

87. B. Li and D. Badami, Stress voiding characteristics of Cu/low k interconnects under long term stresses, *2012 IEEE International Reliability Physics Symposium (IRPS), Anaheim, CA* (2012), 5E.2.1–5E.2.6, doi: 10.1109/IRPS.2012.6241857.

4 Stress Evolution and Damage Formation in Confined Metal Lines under Electric Stressing
1D Analysis

4.1 Introduction

The experimental breakthrough in the early 1970s by Blech et al. [1, 2], as discussed in Chapter 3, demonstrated that mechanical stresses play an important role as a driving force to counter the EM mass transport. It took quite a long time, however, to incorporate the stress effect into the kinetic analysis of EM damage formation in a metal line taking into account local microstructure variations. The first set of papers dealing with the stress effect on EM-induced damage was published in the early 1990s from three independent groups of Nix and Arzt [3, 4], Kirchheim et al. [5, 6], and Korhonen et al. [7, 8]. The main question addressed by these groups concerned the nature of the stress developed by EM mass transport and the kinetics of stress evolution in a conductor line. The formulation in these studies has incorporated the key features to analyze the effect on mass transport and stress evolution due to local microstructure variations, such as grain boundaries and passivation barriers. A consensus was reached concerning the deformation caused by local variations in atomic density and configuration subjected to the passivation constraint as a source of the elastic stresses. The results showed that microstructure and passivation effects are important in controlling stress evolution and damage formation for Cu interconnects fabricated using a damascene process with specific barriers and microstructures.

In this chapter, we analyze the mass transport and stress evolution in confined lines with structural elements pertaining to Cu damascene lines. We will discuss the effect of the structural elements on mass transport and void formation in a damascene line. The overall analysis is 3D in nature and quite complex; therefore, the discussion in this chapter is limited to a simplified one-dimensional analysis based on the Korhonen equation, aiming to elucidate the basic analysis for stress evolution and void formation in damascene lines. Extensive experimental studies have been carried out on EM reliability and microstructure characteristics for Cu damascene lines. In Chapter 5, we will first present results from high-resolution microscopy studies on the grain structure due to scaling of the Cu line dimensions and the implication on electrical resistivity, which plays an important role in controlling the mass transport and void formation as the technology advances to the nanoscale interconnects. This is followed by Chapter 6, where the results from the experimental studies of EM characteristics for Cu damascene interconnects will be reviewed, emphasizing the scaling effects and the approaches developed to improve EM reliability. Supplemented by the detailed experimental results, a general 3D analysis for EM-induced stress evolution and void formation will be presented in Chapter 7.

4.2 Kinetics of EM-Induced Mass Transport and Stress Evolution in Confined Metal Lines

To analyze EM-induced stress evolution, we consider a typical Cu damascene line as shown in Figure 4.1, embedded in dielectric materials and surrounded by thin sidewall barriers and top interface layers. In general, a damascene line contains multilinks of line segments represented by bamboo-like grains and mixing with polycrystalline grains in between, as shown in Figure 4.1b. The sidewall barrier layers are formed by refractory metals (Ti, Ta, W) serving as barriers to prevent Cu out-diffusion into the surrounding dielectrics. The top interface dielectric barrier is deposited on the line after chemical mechanical polishing (CMP), which generates high defect density at the interface, contributing to mass transport at the interface. The Cu grains, the barriers, and the vias and pads are important structural elements to control the EM mass transport, stress evolution, and reliability of the Cu damascene line.

Under EM, the mass transport will lead to stress generation, depending on the diffusion mechanism and the structural elements of the metal line. For Al and Cu interconnects, the vacancy mechanism is generally accepted for diffusion, which consists of two steps: vacancy formation and migration. A vacancy can be generated at a lattice site by removing a lattice atom to another site where the minimum energy is required, as shown in Figure 4.2, which would correspond to a free surface if available, or in case of a confined line, to an interface, a grain boundary, or at a dislocation edge, etc. In this process, a vacancy is formed in pair with a displaced atom, which is to be added, or plated, to a free or a confined surface, depending on the line structure. There are localized atomic relaxations around the vacancy and the plated atom to induce stress, leading to a new stress state. For a damascene line as illustrated in Figure 4.1, the displaced atom is placed on the edge of the interface or at a grain boundary that requires the minimum energy, as illustrated in Figures 4.3 and 4.4 respectively.

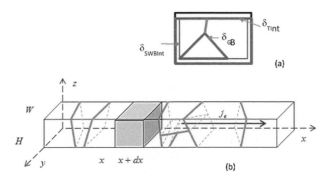

Figure 4.1 Schematics of interconnect segments of a damascene line with thin sidewall barriers and top interface layer: (a) a cross section showing typical grain structures, and (b) a line segment consisting of polygrain clusters separated by bamboo-like grains.

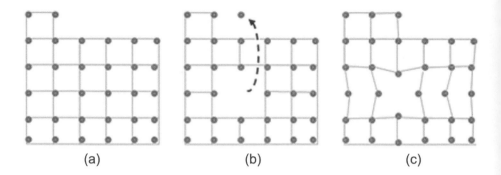

Figure 4.2 Formation of a vacancy near the crystal surface: (a) an initial lattice with the surface step; (b) transfer of a subsurface atom to the surface site; (c) a volume relaxation around a vacant site in the bulk.

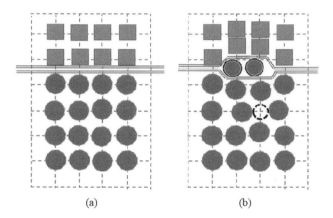

Figure 4.3 Formation of a vacancy near the interface: (a) initial lattice with the interface; (b) transfer of subsurface atoms to the interface (plating), and volume relaxation around a vacant site in the grain interior.

An infinite crystalline solid in the state of zero stress is characterized by an equilibrium concentration of vacancies as $c_0^{ZS} \equiv c_0(T) = \Omega^{-1} \exp\{-E_V/k_B T_{ZS}\}$, where E_V is the energy of vacancy formation, T_{ZS} is the zero-stress temperature, and k_B is the Boltzmann constant. Stressing the solid with different types of loads (electrical, mechanical, thermal) will disturb the state of thermochemical equilibrium due to vacancy generation and migration. Herring [9] and Larche and Cahn [10] showed that the new equilibrium concentration of vacancies in the presence of stresses equals the following:

$$c_{Eq} = \Omega^{-1} \exp\left\{-\frac{E_V - f\Omega\sigma_{Hyd}}{k_B T}\right\} = c_0(T) \exp\left\{\frac{f\Omega\sigma_{Hyd}}{k_B T}\right\}. \qquad (4.1)$$

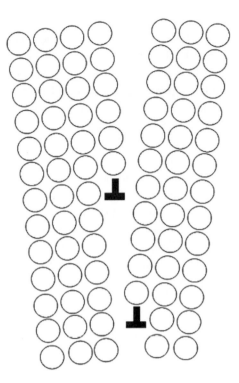

Figure 4.4 Schematics of a low-angle grain boundary showing interfacial edge dislocations, following the classical Fisher model.

Here, $f = \Omega_V/\Omega$ is a ratio of the lattice volume occupied by a vacancy Ω_V to the volume occupied by an atom Ω; $f\Omega = \Omega + (\Omega_V - \Omega)$ is the formation volume, which is the crystal volume change upon formation of a vacancy in its standard state; and $f\Omega\sigma_{Hyd}$ represents the work required against the hydrostatic stress for vacancy formation, which is simplified by replacing the normal stress component used by Herring [9] with the hydrostatic pressure σ_{Hyd} to calculate the work for grain boundary or interface deformation. This includes the work done against pressure in transferring a lattice atom to the interface or grain boundary: $\Omega\sigma_{Hyd}$, and the work performed to relax the volume around the newly formed vacancy: $-(1-f)\Omega\sigma_{Hyd}$, as illustrated in Figures 4.2 and 4.3.

Under EM, the vacancy and plated atom are driven by the current to migrate at the grain boundaries (GBs) or interfaces, where the migration of the vacancy and the plated atom will involve local atomic relaxation and lead to a new stress state in the line, as discussed later. The vacancy formation and migration through the grain boundaries strongly depends on the GB structure and thus the crystallography of the neighboring grains and their boundaries; the chemical composition of the GB; lattice relaxation around the GB; and so on. All these parameters can affect the diffusivity of migrating species. The fundamentals of GB diffusion can be found in a number of review papers; see, for example, [11, 12]. While a majority of experimental data are analyzed based on the classical Fisher model [13], new computational techniques have been using GB diffusion modeling. These are molecular dynamics, molecular static, and kinetic Monte Carlo modeling [14–16]. Molecular dynamic simulations were employed to analyze the effect of stress on GB diffusion [17]. Sorensen et al. [15] established that symmetric tilt GBs support both vacancy and interstitial-mediated diffusion at low temperatures, while at elevated temperatures, more complex defect interactions such as Frenkel pairs and even ring mechanisms contribute to GB diffusion. Suzuki and Mishin have demonstrated that vacancy and interstitial mechanisms can have comparable formation energies in symmetric tilt boundaries in copper [18]. Nomura and Adams have extensively studied diffusion mechanisms in and near both symmetric tilt and pure twist GBs in copper [14]. Assuming only a vacancy mechanism for diffusion in twist boundaries, they found that diffusion at low temperatures is dominated by migration along the screw dislocations comprising high angle twist boundaries, while high-temperature diffusion occurs primarily through the bulk. Nomura has also established that vacancies are strongly bound to the GB itself so that vacancy diffusion is primarily restricted to the plane defined by the GB.

The interfaces in damascene lines and components, such as vias and pads. are covered by special liners with very low diffusivity to prevent Cu atoms from diffusing into the surrounded dielectric. These diffusion barrier layers are formed using either refractory metals (Ti, Ta, W) or dielectrics (SiN), and depending on the fabrication process, the interfaces generally contain structural defects and impurities to yield high atom diffusivity and high vacancy concentration.

Under EM, the atoms accumulation (by plating) occurs due to the work done by the current-induced driving force. The plating or dissolving of atomic layers is accompanied by the volumetric changes that are responsible for the additional stress generation through the interaction with the rigid confinement. In a damascene line, the atomic flux driven by the electric current will deplete the atoms at the cathode area and accumulate at the anode area, in the GBs near the diffusion barriers of the line ends, or in a line segment with blocking bamboo-like grains, as illustrated in Figure 4.5.

For vacancy migration, we can approximate the multiparticle problem as a quasispecies migrating through a crystal lattice [19]. As illustrated in Figure 4.6, the vacancy mechanism generates equal fluxes of atoms and vacancies, where the transfer of a vacancy (the clear circle) from the right end to the left end of the chain of nine lattice atoms is equivalent to the transfer of each of nine atoms on one lattice distance.

Thus, we can write the atomic flux as follows:

$$N_{LC} D_a = cD, \tag{4.2}$$

where $N_{LC} = \Omega^{-1}$ is the concentration of the lattice cites, $c(x, t)$ is the instantaneous vacancy concentration at position x and time t, D_a the atomic diffusivity, and D the vacancy diffusivity, which is described as

$$D = D_0 \exp\left\{ -\frac{E_{VD} - \Omega^m \sigma_{Hyd}}{k_B T} \right\}. \tag{4.3}$$

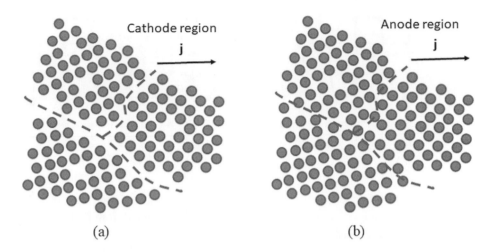

Figure 4.5 Schematic of the dissolution of portions of crystallographic planes of individual grains in (a) and plating of new planes in (b) at the line cathode and anode regions.

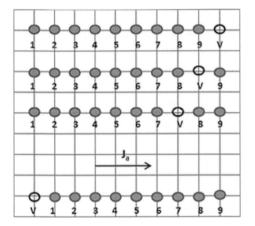

Figure 4.6 Schematics of the matter transfer process for vacancy diffusion mechanism.

Here E_{VD} is the effective activation energy of vacancy diffusion and Ω^m is the migration volume, which represents the volume change that occurs when the defect reaches the saddle point in its migration path, as illustrated in Figure 4.7 [20, 21]. Combining (4.2) and (4.3), we have the effective atom diffusivity as

$$D_a = D_0 \exp\left\{-\frac{E_{aD} - \Omega^* \sigma_{Hyd}}{k_B T}\right\}, \tag{4.4}$$

where the activation energy $E_{aD} = E_V + E_{VD}$ is the sum of the energies of vacancy formation and migration, and the activation volume $\Omega^* \approx 0.95\Omega$ is the combined

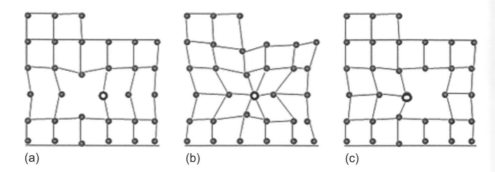

(a)   (b)   (c)

Figure 4.7 Schematics of the migration volume: (a) an initial state, (b) the sandle point in the migration path, and (c) the final state. Reprinted from [20], with the permission of AIP Publishing.

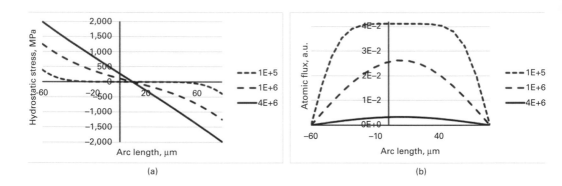

(a)   (b)

Figure 4.8 Evolution of stress (a) and atomic flux (b) distributions along the line.

volumes of vacancy formation and migration [20]. Detailed discussion of the effect of hydrostatic pressure on atomic diffusion can be found in [21].

The depletion and accumulation of atoms and vacancies occur in a damascene line will generate the corresponding tension and compression at the cathode and anode ends, or at the ends of the polygrain clusters, respectively. The stress gradient will in turn induce an atomic flux in a direction opposing the electron flow as a back-stress flux [1]. Under a unipolar DC current, the EM flux is constant in time, in contrast to that induced by the stress gradient. Acting together, this will change the overall stress distribution with time with the stress gradients initially located near the ends of the line, then expand with time as shown in Figure 4.8a. There is a corresponding change in the atomic flux distribution along the line as a function of time as shown in Figure 4.8b. Excessive or depleted, as shown in Figure 4.5, atom density, which is generated by the atomic flux divergence, results in a plating or a dissolution of the atomic layers in the GB and interfaces. These volumetric transformations generate the compressive or tensile stresses. Thus, we observe a dual character of stress evolution: the atomic flux divergence generates a stress, which in turn initiates the back-flux of

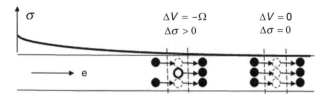

Figure 4.9 Volumetric strain evolution caused by a divergence of the atomic back-flux.

atoms resulting in an expansion of the portion of the line where the flux divergence occurs, as shown in Figure 4.9.

4.3   Kinetics of EM-Induced Evolution of Vacancy Distribution in Interconnect Segments

Following the discussion on stress evolution, we proceed to analyze the kinetics of stress evoluation induced by an electric current in a metal line segment embedded into a rigid confinement. Here the distributions of vacancies and plated atoms across the polycrystalline segment are important in determining the stress components in the line segment. At the same time, these species distributions are in turn functions of stress. Thus, we have a self-consistent problem of a simultaneous determination of the kinetics of stress evolution together with the migration of vacancies and plated atoms in a metal line segment induced by stresses under electromigration. As already mentioned, the overall analysis is 3D in nature and quite complex, so the discussion in this chapter is limited to a simplified one-dimensional analysis.

Considering a damascene line segment containing some general grain structure as shown in Figure 4.1, the vacancy accumulation or depletion occurring at any location can be related to the vacancy flux divergence. Let us assume that before applying the electric current to induce mass transport, the line was in a stress-free condition, i.e., $c = c_0$. Another case where the initial stress is generated by cooling down from the anneal temperature (a zero-stress condition) to the test temperature will be discussed in Chapter 7. Referring to Figure 4.1b, the evolution of the vacancy concentration in a volume dV of the line segment with coordinates x and $x + dx$ with a line thickness h and a width w can be derived using the continuity equation:

$$\frac{\partial c(x,t)}{\partial t} = -\nabla J + R(x,t). \tag{4.5}$$

This problem was first formulated by Rosenberg and Ohring [22] as discussed in Section 3.5 but with a more specific form of the generation/annihilation term $R(x,t)$; see (3.8). As we already discussed, a key input in the EM stress analysis deals with the physics of generation/annihilation of the vacancy and atom pairs at structural elements, such as grain boundaries and interfaces, which are responsible for stress generation. Indeed, since the temperature is constant and uniform, the vacancy excess or depletion is due to divergence of the vacancy flux and vacancy generation or annihilation at grain boundaries and interfaces. In this way, any forms of vacancy excess or depletion can be accounted for by creating or eliminating a local volume, which under a rigid passivation will result in an elastic stress. Indeed, since a vacancy is characterized by a volume $\Omega_V = f\Omega$, where $f = \Omega_v/\Omega \approx 0.6$ is the ratio of vacancy and atom volumes, a volume reduction per vacancy is $\Omega_v - \Omega = f\Omega - \Omega = -(1 - f)\Omega$. With excessive vacancies of concentration c accumulated at grain boundaries, the volume change of the line segment can be written as follows:

$$\delta V \approx -\frac{\delta}{d}(1 - f)\Omega(c - c_0)whdx. \tag{4.6}$$

Here δ is the grain boundary thickness, and d the average grain size. Thus, for the volumetric strain caused by vacancy migration, we have

$$\theta = \frac{\delta V}{V} \approx -\frac{\delta}{d}(1 - f)\Omega(c - c_0). \tag{4.7}$$

Without a passivation confinement, the metal volume will certainly change. Since the confinement does not allow the metal volume to change, an elastic stress would be generated, opposite in sign to the inelastic strain. Hence, when $c > c_0$, there is an inelastic contraction $\delta V < 0$, and a tensile elastic stress $\sigma > 0$ is generated. In general, the stress field generated in the damascene line encapsulated by diffusion barriers and dielectrics is quite complicated, as it is nonuniform along the line and contains various stress components. This complexity can be simplified if we take into account the long time scales associated with EM-induced growth of large voids or establishing a steady-state stress distribution along the line. In such cases, the vacancies should migrate over a distance much longer than the line cross-sectional dimensions (width or height), so a vacancy equilibrium can be established at grain boundaries. This will allow enough time for stress-induced creep to occur to reach a hydrostatic stress state at each point of the line [8, 23]. This greatly simplifies the analysis, since one has to consider only the effect of the hydrostatic stress on mass migration along the line. We can then relate the hydrostatic stress σ with the volumetric mismatch strain between the metal and the passivation as

$$\sigma = -B\theta, \tag{4.8}$$

where B is the effective modulus, which depends on the shape of the line cross section, the elastic constants of the surrounding dielectric and barriers, and the silicon substrate, as well as the bulk modulus of the metal [23, 24]. Thus, combining (4.5)–(4.8) allows us to calculate the stress evolution at flux divergence sites.

4.4   EM-Induced Stress Evolution in a Confined Metal Line: Korhonen's Equation

In this section, we formulate the kinetics of stress evolution induced by an electric current in a confined metal line. As already demonstrated, the stress state depends on the formation and migration of vacancies and plated atoms in the polycrystalline line, which in turn is driven by the stress components. So far, the discussion reflects an important atomistic detail for vacancy formation or annihilation in that the vacancy cannot be formed alone but has to be paired with another atomic species. Indeed, vacancy formation is accompanied by plating the atom occupying the vacant site to another place in the line, such as a confined interface, grain boundary, or dislocation edge. The same processes but in reverse order are involved in the case of vacancy annihilation. Thus, the energy of vacancy formation should be calculated by considering the formation of the paired species. This approach was adopted by Kirchheim [5] and Korhonen et al. [8], and later by Sarychev et al. [25] and Sukharev et al. [26] to formulate the stress effect on EM damage formation in confined metal lines.

The mass transport along the line, which leads to stress generation, depends on the applicable diffusion mechanism. For Al and Cu interconnects, *vacancy mechanism of diffusion* is generally accepted as the underlying mechanism for atomic diffusion: atoms propagate along the line by jumping into vacant nearby lattice sites, i.e., the sites occupied by vacancies. Thus, the mass transport depends on *vacancy concentration*, which is extremely small at typical chip operating temperatures, and can be described by vacancy diffusion. Furthermore, the nonuniform distribution of stress along the line leads to an additional driving force that drives vacancy migration, which is due to the inhomogeneous elastic energy associated with the hydrostatic stress. Indeed, when an atom jumps into a vacancy site, the site volume will increase by the difference of the vacancy volume $v_v = f\Omega$ and the atomic volume $v_a = \Omega$. This requires work to push apart the atoms neighboring the vacant site. When under tension, the existing hydrostatic stress can help this work to be done or resist it in the case of compression. Thus, the change in the free energy of the migrating atom under a hydrostatic stress $\sigma\left(\overrightarrow{r}\right)$ is $\Delta U = -(1-f)\Omega\sigma$.

Assuming that mass transport along the interconnect line is determined by grain boundary diffusion alone with the vacancy diffusion coefficients D_v^{GB}, the effective vacancy transport corresponding to the entire cross section of the line is characterized by the effective diffusion coefficients $D = D_v^{GB}\delta/d$. Here δ, as before, is the grain boundary thickness and d the average grain size. Thus, as a 1D approximation, total vacancy flux (assumed to have a reference direction in the negative x-direction) due to electric current and stress is given by

$$J = \frac{cD}{k_B T}\left[(1-f)\Omega\frac{\partial\sigma}{\partial x} - q^* \rho j\right]. \tag{4.9}$$

The corresponding *atomic flux*, denoted J_a, and assumed to have a reference direction in the positive x-direction, is equal in value to the vacancy flux, i.e.,

$$J_a = -\frac{ND_a}{k_B T}\left[(1-f)\Omega\frac{\partial\sigma}{\partial x} - q^* \rho j\right], \tag{4.10}$$

where N is the atomic concentration; D_a is the coefficient of atomic diffusion corresponding to the entire cross section of the line, which is related to the grain boundary atomic diffusivity as $D_a = D_a^{GB}\delta/d$; $q^* = eZ$, as it was discussed already in Chapter 2, is the effective charge of the migrating atom; e is the elemental charge; and Z is the effective valence. It should be noted that similar relations exist between a local grain boundary concentration of vacancies c_v and their concentration normalized to the entire cross section of the line c: $c = c_v \delta/d$.

In general, the new vacancy concentration resulting from the flux divergence is not in equilibrium with the stress generated by these vacancies. Indeed, when a vacancy flux divergence occurs at an arbitrary cross section of a conductor line, this changes the initial stress-free vacancy concentration c_0 to a concentration c^* and generates a volumetric strain of $\theta = -\Omega(\delta/d)(1-f)(c^* - c_0)$. The strain will interact with the rigid confinement to yield an elastic stress of $\sigma = B\Omega(\delta/d)(1-f)(c^* - c_0)$ and a corresponding local equilibrium vacancy concentration of $\tilde{c}_{Eq} = c_0 \exp\{(\delta/d)B\Omega f(1-f)\Omega(c^* - c_0)/k_B T\}$. It can be shown that the new equilibrium vacancy $\tilde{c}_{Eq}$ is smaller than c^* in the case that excessive vacancy concentration is generated, $c^* > c_0$, and larger than c^* in the opposite case of $c^* < c_0$. This indicates that some number of vacancies will disappear or generate in order to reach a new state of equilibrium. It should be noted that the kinetics of bringing the vacancy concentration to equilibrium under a hydrostatic stress is governed by the generation/annihilation rate $R(x,t)$ for a vacancy-plated atom pair, which depends on the diffusion mechanism. In the case that excessive or deficient vacancies are located only in grain boundaries and interfaces, and with little vacancy exchange with the grain interiors, the rate of pair generation or annihilation will be quite high, approaching the rate of relaxation of the hydrostatic stress. Thus, the concentration of vacancies in the grain boundaries at any time will be close to the equilibrium concentration, c_{Eq}, as described by (4.1), and plated atoms of the amount $c_{PL} = c_{Eq} - c^*$ will be deposited at or removed from the grain boundaries. This yields a total volumetric strain as follows:

$$\theta \approx -\frac{\delta}{d}\left[(1-f)\Omega(c_{Eq} - c_0) - \Omega c_{PL}\right], \tag{4.11}$$

with a corresponding hydrostatic stress:

$$\sigma \approx \frac{\delta}{d}B\left[(1-f)\Omega(c_{Eq} - c_0) - \Omega c_{PL}\right]. \tag{4.12}$$

The ratio δ/d is used to get the volumetric strain of the line unit volume when excessive or deficient vacancies and plated atoms are generated at the grain boundaries.

The preceding results allow us to estimate the number of vacancies used to adjust for the hydrostatic stress. Combining (4.1) and (4.12), we have the following:

$$c_{Eq} = c_0 \exp\left\{\frac{f\Omega\frac{\delta}{d}B\left[(1-f)\Omega(c_{Eq} - c_0) - \Omega c_{PL}\right]}{k_B T}\right\}.$$

Expanding the exponential function in series and keeping the first two terms, which is a reasonable approximation since $f\Omega\sigma/k_B T \ll 1$, we get

$$c_{PL} \approx -\frac{d}{\delta}\frac{k_B T}{f\Omega^2 B}\frac{1}{c_0}(c_{Eq} - c_0). \tag{4.13}$$

For typical conditions of $d/\delta \sim 10^3$, $k_B T/f\Omega B \sim 10^{-2}$ and $(\Omega c_0)^{-1} = \exp\{E_V/k_B T\} \sim 10^{12}$, we have $c_{PL} \gg (c_{Eq} - c_0)$. This result was mentioned in several previous publications [3–8, 27], indicating that most of the vacancies induced by flux divergence are used to equilibrate with the stress. In this way, most of the vacancies are removed from the pool of free migrating vacancies.

From (4.12), we have a simple relation between the generation rate of the vacancy-plated atom pairs and the rate of hydrostatic stress evolution:

$$R = -\frac{\partial c_{PL}}{\partial t} = \left(\frac{1}{\Omega B} \frac{d}{\delta} - (1-f) \frac{\Omega}{k_B T} c_{Eq} \right) \frac{\partial \sigma}{\partial t} \approx \frac{1}{\Omega B} \left(\frac{d}{\delta} \right) \frac{\partial \sigma}{\partial t}, \tag{4.14}$$

which is valid since $(k_B T/\Omega B)(d/\delta) \sim 1$ and $(1-f)f\Omega c_{Eq} \ll 1$.

So far, we have shown that if EM-induced matter transport occurs only through the grain boundaries, the vacancy concentration everywhere is in equilibrium with the stress. In this case, the vacancy evolution rate, which is governed by the vacancy flux divergence and the generation/annihilation of the vacancy-plated atom pairs can be be described by (4.5), where $c(x, t)$ is substituted with $c_{Eq}(x, t)$ and $R(x, t)$ is given by (4.14). Thus,

$$-\nabla J_v = \frac{\partial}{\partial t} \left(c_{Eq} + c_{PL} \right) \left(\frac{\delta}{d} \right) = \frac{\partial \sigma}{\partial t} \left[\left(\frac{\delta}{d} \right) \frac{f\Omega}{k_B T} c_{Eq} + \frac{1}{\Omega B} \right] \approx \frac{1}{\Omega B} \frac{\partial \sigma}{\partial t}. \tag{4.15}$$

The preceding derivation was first done by Arzt and Nix in [4] to provide a physics-based description of the EM-induced vacancy generation/annihilation mechanism.

In a later paper [8], Korhonen et al. followed this result closely to derive an equation for the kinetics of 1D stress evolution in a metal line embedded in a rigid confinement driven by a constant current density j. The analysis has also been performed by Clement [27]. For a 1D conductor line, (4.15) is simplified as follows:

$$-\frac{1}{\Omega B} \frac{\partial \sigma}{\partial t} = \frac{\partial J}{\partial x}, \tag{4.16}$$

which after replacing the vacancy flux J with the atomic flux $J_a(J_a = -J)$ becomes

$$\frac{\partial \sigma}{\partial t} = \frac{\partial}{\partial x} \left[\frac{\delta}{d} D_a^{GB} \frac{B\Omega}{k_B T} \left((1-f) \frac{\partial \sigma}{\partial x} - \frac{eZ\rho j}{\Omega} \right) \right]. \tag{4.17}$$

Following Korhonen et al. [8], for the sake of simplicity, we omit the vacancy volume relaxation, because this will not affect our results, and employ the previously introduced effective atomic diffusivity $D_a = D_a^{GB}\delta/d$ corresponding to the entire cross section of the line. Then (4.17) takes the form of the well-known 1D Korhonen's equation describing the stress evolution in a metal line embedded in a rigid confinement under a current density j:

$$\frac{\partial \sigma}{\partial t} = \frac{\partial}{\partial x} \left[\frac{D_a B\Omega}{k_B T} \left(\frac{\partial \sigma}{\partial x} - \frac{eZ\rho j}{\Omega} \right) \right]. \tag{4.18}$$

The most important difference between this equation and that employed in the early analyses [22, 28] is that the rate of stress evolution is now characterized by the quasidiffusion coefficient $\kappa^2 = D_a B \Omega / k_B T$ while the evolution rate of vacancy concentration is determined by the grain boundary diffusivity D. It is easy to see that

$$\kappa^2 = D_a B \Omega / k_B T = (\delta/d)(\Omega B / k_B T) \Omega c_{Eq} D \ll D. \tag{4.19}$$

To account for the retardant kinetics of the stress propagation, we need to come back to the derivation by Nix and Artz [3]. The mass balance (4.5) for the equilibrium vacancy concentration with the vacancy generation/annihilation term from (4.14) takes the form

$$\frac{\partial c_{Eq}}{\partial t} = -\frac{\partial}{\partial x}\left(-D\frac{\partial c_{Eq}}{\partial x} + c_{Eq}\frac{DeZ\rho j}{k_B T} \right) + \frac{1}{f\Omega B}\frac{d}{\delta}\frac{\partial \sigma}{\partial t}, \tag{4.20}$$

which after substituting $\sigma = (k_B T / f \Omega) \ln \{c_{Eq}/c_0\}$ can be rewritten in the form exactly matching the equation derived in [3]:

$$\frac{\partial c_{Eq}}{\partial t} + \frac{k_B T}{\Omega^2 B}\frac{d}{\delta}\frac{\partial \ln c_{Eq}}{\partial t} = \frac{\partial}{\partial x}\left(D\frac{\partial c_{Eq}}{\partial x} - c_{Eq}\frac{DeZ\rho j}{k_B T} \right). \tag{4.21}$$

Using again the relation $(d/\delta)(k_B T / f\Omega B)(\Omega c_{Eq})^{-1} \gg 1$, (4.21) can be written in the final form:

$$\frac{\partial c_{Eq}}{\partial t} = c_{Eq}\frac{\partial}{\partial x}\left[\frac{D_a B \Omega^2}{k_B T}\left(\frac{\partial c_{Eq}}{\partial x} - c_{Eq}\frac{eZ\rho j}{k_B T} \right) \right]. \tag{4.22}$$

Numerical solution to (4.22), which was done by Clement in [29], has demonstrated that the rate of EM-induced evolution of the equilibrium vacancy concentration is characterized by the small, if compared with the vacancy coefficient of diffusion D, quasidiffusion coefficient $\kappa^2 = D_a B \Omega / k_B T$, same as for the rate of stress evolution, (4.18). Thus, the EM-induced evolution of vacancy concentration is a very slow process characterizing by the same slow kinetics as the stress evolution. This means that the required vacancy supersaturation is achieved in years, not in seconds as it was predicted when the stress effect was not included. A very slow evolution of the vacancy concentration can be explained by a consumption of the majority of vacancies by annihilation with the plated atoms leading to the stress changing, which is necessary for achieving the new state of equilibrium. The time scale for building up the supersaturation with a small fraction of the remaining vacancies is much longer than in the case with a large vacancy concentration if the stress effect is not considered. An analogy for this case would be diffusion along a line with distributed traps for the diffusing species. We should note that due to the obvious dual representation of the flux of equilibrated vacancies, we have

$$J = -D\frac{\partial c_{Eq}}{\partial x} + c_{Eq}\frac{DeZ\rho j}{k_B T} \equiv -c_{Eq}\frac{D\Omega}{k_B T}\frac{\partial \sigma}{\partial x} + c_{Eq}\frac{DeZ\rho j}{k_B T}. \tag{4.23}$$

Then (4.21) immediately yields Korhonen's equation (4.18) if the concentration gradient term is replaced by the corresponding stress gradient term. It should be noted that the original continuity equation in [3] contains only the concentration gradient and electromigration induced fluxes but without the back-flux of vacancies arisen from the stress gradient. As we just demonstrated, this would not cause a problem if the vacancy concentration is almost at equilibium with the stress. In a general case, where the equilibration of vacancy concentration with stress occurs not instantly, ignoring the stress gradient-induced back-flux of vacancies can give incorrect stress evolution kinetics. Analysis of the general case will be given in Sections 7.3–7.5. Later, Nix and co-workers [30–32] improved the model by including the stress gradient flux of vacancies to the general formulation. They also elaborated the physical meaning of the critical stress, which was proposed in [3], as a stress initiating a growth of the preexisting flaws at the metal/passivation interface. It was suggested that such flaws may be formed by etch residues from the fabrication process and act as regions where the metal is weakly bonded to the barriers or dielectric. We will discuss a critical stress phenomenon in Section 4.6.

4.5 Analytical Solutions for the 1D Korhonen Equation

Various solutions to Korhonen's equation with different boundary conditions have been reported in several papers; see, for example, [8, 23, 27, 33]. All such analyses have adopted the condition assumed by Korhonen et al. in [8] that the effective diffusion coefficient $D_a = D_a^{GB}\delta/d$ does not depend on stress. In the following, we will accept this approximation for the sake of simplicity but will return later in Chapter 7 to discuss the error by ignoring the stress dependency of D_{eff}.

We define this to be an initial boundary value problem (IBVP) for a finite line with diffusion being blocked at $x = 0$ and $x = L$, where L is the length of a line driven by a current density j. The solution can be obtained by the standard method of separation of variables, [34] for the problem specified as follows:

$$
\begin{aligned}
PDE &: \quad \sigma_t = \kappa^2 \sigma_{xx}, 0 < x < L, 0 < t < \infty \\
BC &: \quad G + \sigma_x(0, t) = 0, 0 < t < \infty \\
BC &: \quad G + \sigma_x(L, t) = 0, 0 < t < \infty \\
IC &: \quad \sigma(x, 0) = \sigma_T, 0 < x < L.
\end{aligned}
\tag{4.24}
$$

Here, PDE is for the partial differential equation, BC is for the boundary conditions, and IC is for the initial conditions. Thereafter, we use the notations $G = eZ\rho j/\Omega$, $\kappa^2 = D_a B\Omega/k_B T$ and σ_T as the residual stress preexisting in the line. The other standard notations are $\partial\sigma/\partial x = \sigma_x$; $\partial\sigma/\partial t = \sigma_t$; $\partial^2\sigma/\partial x^2 = \sigma_{xx}$. The compact representation (4.24) reads as follows: we are solving a given partial differential equation (PDE) to get a time-dependent stress distribution along the line $[0, L]$, satisfying the zero-flux BC at both ends of the line with uniformly distributed stress σ_T at the initial time.

It is well known that the method of separation of variables requires both PDEs and BCs being homogeneous, which is not our case. For the problem stated in (4.24), the BCs are not homogeneous. The standard solution of this kind of problem is to represent an original variable by a combination of two: $\sigma(x,t) = S(x,t) + u(x,t)$ with one $u(x,t)$ to satisfy the existing BCs. This ensures that new BCs for the variable $S(x,t)$ are homogeneous. Let us verify it for the first BC from (4.24): $-G = \sigma_x(0,t) = S_x(0,t) + u(0,t) = S_x(0,t) - G$.

This gives $S_x(0,t) = 0$, and the same is true for the second BC. Solving the PDE this way, we represent the solution as a sum of two parts: a steady-state term and a transient term. In this way, $u(x,t)$ is taken as a steady-state solution, which remains at $t \to \infty$, while $S(x,t)$ depending on the IC should vanish after a long time. A steady-state solution of the EM problem describes the stress distribution, which is achieved when the back-stress gradient-induced flow compensates the current density induced atomic flow. It is represented by a straight line when $j = const$:

$$u(x,t) = A\left(1 - \frac{x}{L}\right) + B\frac{x}{L}. \tag{4.25}$$

Since both BCs contain derivatives, we can determine just one of the unknown A and B, which provides: $u(x,t) = A - Gx$, and thus

$$\sigma(x,t) = A - Gx + S(x,t). \tag{4.26}$$

The IBVP (4.24) can be rewritten as

$$\begin{aligned}
PDE &: \quad S_t = \kappa^2 S_{xx}, \quad 0 < x < L, 0 < t < \infty \\
BC &: \quad S_x(0,t) = 0, \quad 0 < t < \infty \\
BC &: \quad S_x(L,t) = 0, \quad 0 < t < \infty \\
IC &: \quad S(x,0) = \sigma_T - A + Gx, \quad 0 < x < L.
\end{aligned} \tag{4.27}$$

We look for a solution of the form $S(x,t) = X(x)T(t)$, which provides $S_t = X(x)T'(t)$ and $S_{xx} = X''(x)T(t)$. Substituting these in the PDE generates the following equation:

$$X(x)T'(t) = \kappa^2 X''(x)T(t). \tag{4.28}$$

Dividing both parts by $\kappa^2 X(x)T(t)$ gives

$$\frac{T'(t)}{\kappa^2 T(t)} = \frac{X''(x)}{X(x)}. \tag{4.29}$$

Since the two sides of this equation are functions of different variables, they must be equal to a constant ψ. The constant $\psi < 0$ otherwise T would go to infinity at $t \to \infty$, which makes σ go to infinity. Since this does not happen due to the backflow stress, we take $\psi = -\lambda^2$ and deduce two ordinary differential equations (ODE):

$$\begin{aligned}
T' &= -\lambda^2 \kappa^2 T \\
X'' &= -\lambda^2 X.
\end{aligned} \tag{4.30}$$

The known solutions to these ODEs are the following:

$$T(t) = C_1 \exp\left\{-(\lambda\kappa)^2 t\right\}$$
$$X(x) = C_2 \sin \lambda x + C_3 \cos \lambda x,$$

where C_1, C_2, and C_3 are constants. Therefore,

$$S(x,t) = \exp\left\{-(\lambda\kappa)^2 t\right\}(C\sin\lambda x + D\cos\lambda x). \tag{4.31}$$

Since C, D, and λ are arbitrary constants, (4.31) represents an infinite number of solutions to the PDE. We should seek those solutions subjected to BCs and ICs of the IBVP (4.27). The first BC of (4.27) gives $S_x(0,t) = C\lambda e^{-(\lambda\kappa)^2 t} = 0$, so $C = 0$ and the second BC gives $S_x(L,t) = -D\lambda e^{-(\lambda\kappa)^2 t}\sin\lambda L = 0$, which for nonzero $S(x,t)$ leads to an equation $\sin\lambda L = 0$. The solutions to this equation, which are called the eigenvalues of the boundary value problem (BVP) for the first three equations of (4.27), are $\lambda_n = \pm\frac{n\pi}{L}$. Thus, we have an infinite number of functions satisfying both the PDE and BCs: $S_n(x,t) = D_n\exp\left\{-\left(\frac{\pi n\kappa}{L}\right)^2 t\right\}\cos\left(\frac{\pi n x}{L}\right)$. Their linear combination represents a general solution to the BVP

$$S(x,t) = \sum_{n=1}^{\infty} D_n\exp\left\{-\left(\frac{\pi n\kappa}{L}\right)^2 t\right\}\cos\left(\frac{\pi n x}{L}\right). \tag{4.32}$$

The unknown coefficient D_n and A can be found from the IC of (4.27): $S(x,0) = \sigma_T - A + Gx, 0 < x < L$. It yields

$$\sigma_T - A + Gx = \sum_{n=1}^{\infty} D_n\cos\left(\frac{\pi n x}{L}\right). \tag{4.33}$$

Using the orthogonality of $\cos\left(\frac{\pi n x}{L}\right)$, which is $\int_0^L \cos\left(\frac{\pi n x}{L}\right)\cos\left(\frac{\pi m x}{L}\right)dx = \begin{cases} 0, m \neq n \\ \frac{L}{2}, m = n \end{cases}$, we get from (4.33)

$$\int_0^L (\sigma_T - A + Gx)\cos\left(\frac{\pi m x}{L}\right)dx = \frac{L}{2}B_m,$$

which after integration provides $B_{2k+1} = -\frac{4GL}{\pi^2(2k+1)^2}$; $B_{2k} = 0$. When these are put into (4.32) and (4.26), we have

$$\sigma(x,t) = A - Gx + 4GL\sum_{k=0}^{\infty} \frac{\cos\frac{(2k+1)\pi x}{L}}{\pi^2(2k+1)^2}\exp\left\{-\left(\kappa\frac{(2k+1)\pi}{L}\right)^2 t\right\}.$$

Using the IC of $\sigma(x,0) = \sigma_T$ and the identity $\sum_{k=0}^{\infty} \frac{\cos\frac{(2k+1)\pi x}{L}}{\pi^2(2k+1)^2} = \frac{1}{4}\left(\frac{\pi}{2} - \frac{\pi x}{L}\right)$, we get

$$\sigma_T = A - Gx + 4GL\sum_{k=0}^{\infty} \frac{\cos\frac{(2k+1)\pi x}{L}}{\pi^2(2k+1)^2} = A - \frac{GL}{2},$$

which yields $A = \sigma_T + \frac{GL}{2}$. Finally, we obtain the evolution of the hydrostatic stress along the finite $(0, L)$ metal line subjected to a rigid confinement under a unidirectional electric current as follows:

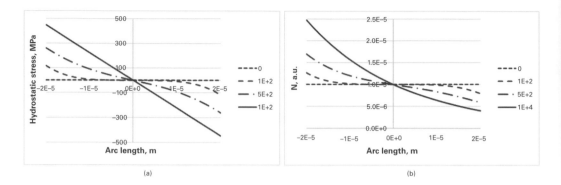

(a)   (b)

Figure 4.10 Evolution of the hydrostatic stress (a) and vacancy concentration (b) along the metal line driven by an electrical current j. Ends of the line block diffusion of atoms.

$$\sigma(x,t) = \sigma_T - Gx + \frac{GL}{2} - 4GL \sum_{k=0}^{\infty} \frac{\cos \frac{(2k+1)\pi x}{L}}{\pi^2(2k+1)^2} \exp\left\{-\left(\kappa\frac{(2k+1)\pi}{L}\right)^2 t\right\}. \quad (4.34)$$

Figure 4.10a shows the evolution of the hydrostatic stress $\sigma(x,t)$ along the line for the case of zero initial stress.

It can be seen from Figure 4.10a that the steady-state distribution of the hydrostatic stress developed at long times is linear. Indeed, (4.34) yields at $t \to \infty$

$$\sigma(x,\infty) = \sigma_T - \frac{eZ\rho jx}{\Omega}\left(x - \frac{L}{2}\right), \quad (4.35)$$

which is the same as originally derived by Blech and Herring [2] by assuming $\sigma_T = 0$. Figure 4.10b shows the kinetics of the vacancy concentration described by (4.1).

A different solution to the Korhonen's equation has been derived in [8] for a semi-infinite metal line with a diffusion blocking BC at the cathode end of the line $(x = 0)$ using the technique of Laplace transformations:

$$\sigma_{Hyd}(x,t) = -\frac{eZ\rho j}{\Omega}\left[\sqrt{\frac{4\kappa t}{\pi}}\exp\left(-\frac{x^2}{4\kappa t}\right) - x \cdot erfc\left(\frac{x}{\sqrt{4\kappa t}}\right)\right]. \quad (4.36)$$

This solution provides a square root time increase of stress at the blocking boundary of a semi-infinite line. In this case, a steady state cannot be achieved due to the absence of a second diffusion blocking boundary, where the total number of atoms in the metal line is not conserved. It is woth noting that the solutions to Korhenen's equation (4.34) and (4.36) describe a diffusion-driven evolution of stress corresponding to the stress propagating through the conductor line, not the migrating atoms.

It should be noted that both solutions (4.34) and (4.36) cannot be directly applied to the interconnect segment in a multibranched structure, such as an interconnect tree with multiply current ports (inlets and outlets), as shown in Figure 4.11. EM in such an on-chip interconnect in a power-ground network will be discussed later in

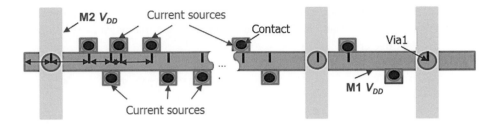

Figure 4.11 Interconnect segment confined by diffusion barriers/liners.

Chapter 9 where the solutions for the Korhonen's equation will be derived for each tree branch subjected to either a blocking BC at the ends terminated by the current inlet/outlet ports (vias) or internal BCs where the branches are joined with each other. In the latter case, the stress and atomic flux are continuous at each junction point where the branches are connected.

The stress evolution kinetics deduced in the preceding discussion allows us to estimate the time when the tensile stress developed at the cathode end of line will reach a critical stress (σ_{crit}) required for void formation [8, 23]. A similar criterion can be applied for hillock formation at the anode end of the line induced by a critical compressive stress to extrude metal to the dielectric surrounding. However, there are other more complicated interconnect structures, such as a power-ground grid as shown in Figure 4.11, where the open circuit or the resistance increase induced by EM leading to electrical failures would have characteristics quite different from the EM-induced void and hillock formation as discussed in this chapter. The topics of EM failures in power grids together with EM driven by an alternate current will be discussed later in Chapter 9. For the rest of this chapter, we focus on the discussion of void formation and movement under EM in confined metal lines.

4.6 Critical Stress for Void Formation

The physical meaning of the critical stress responsible for void nucleation has been discussed extensively; see, for example, [3, 6, 8, 22, 30–32, 35, 36]. A consensus seemed to have emerged that stress components can play a crucial role in void nucleation. As pointed out by Suo [23], the void nucleation condition based on the hydrostatic critical stress is overly simplistic. He has elaborated that the elastic anisotropy between metal grains and the elastic inhomogeneity between metal and the surrounding material can generate high stresses at various junctions under mechanical, thermal, or electrical current stressing. Such stresses can cause sliding of grain boundary (GB) [3], sliding at metal/liner interfaces [37], acceleration of GB diffusion [38], activation of the dislocation glide, [39], or relaxation by other mechanisms. Nevertheless, tracking the hydrostatic stress evolution can provide useful information about the potential locations for void nucleation in the interconnect segment.

An estimation of the σ_{crit} based on the classical model of homogeneous nucleation [40] can be found in [23]. For a spherical void of radius r in a solid under a tension σ, its free energy F is

$$\Delta F(r) = 4\pi r^2 \gamma - \frac{4}{3}\pi r^3 \sigma. \tag{4.37}$$

It is clear from (4.37) that the free energy will be reduced by growth of voids with sizes larger than $r_{crit} = 2\gamma/\sigma$ and increased with radius smaller than r_{crit}. Here γ is the surface energy per unit area, $4\pi r^2 \gamma$ is the increase of the free energy by void insertion, and $4\pi r^3 \sigma/3$ is the work done by the existing stress σ to relocate matter for void formation. Based on typical values of $\gamma = 1 \text{ J/m}^2$ and $\sigma = 200 \text{ MPa}$, Suo [23] deduced a value of $r_{crit} = 10 \text{ nm}$. Assume that a void can be nucleated only when a preexisting flaw exists; a flaw with the initial size r_f will start growing if the stress exceeds the following value:

$$\sigma_{crit} = \frac{2\gamma}{r_f}. \tag{4.38}$$

The assumption that a preexisting flaw is needed for void nucleation is supported by calculations done by Nix and Arzt in [3]. They showed that a large amount of energy is required for homogeneous nucleation of the critical void by agglomeration of vacancies, as verified later by an estimate of the critical stress by Gleixner et al. [30–32]. Nucleation rates were estimated at various locations in an interconnect line; for a copper dual-damascene interconnect, the free energy change upon the creation of a void precursor of volume V_{PR} can be expressed as follows:

$$\Delta F = -\sigma V_{PR} + \gamma_s A_s + \left(\gamma_{cap} - \gamma_{int}\right)A_{int} - \gamma_{GB}A_{GB}, \tag{4.39}$$

where γ_s, γ_{cap}, γ_{int}, and γ_{GB} are the surface energies of the metal free surface, capping layer surface, Cu/capping layer interface, and grain boundary, respectively, and A_j are the areas of the surfaces created or destroyed upon formation of the void. The energy barrier for void nucleation, ΔF_{crit}, is given by the condition

$$\frac{\partial \Delta F_{crit}}{\partial r} = 0, \tag{4.40}$$

which determines a critical void precursor volume. For homogeneous nucleation, the energy barrier is

$$\Delta F_{crit} = \frac{16\pi}{3}\frac{\gamma^3}{\sigma^2}. \tag{4.41}$$

Regarding the nature of the stress driving void growth, we can cite as an example the thermal stress in a metal line embedded in a rigid confinement of refractive metal, dielectric diffusion barriers, and interlayer dielectrics (ILD)/intermetal dielectrics (IMD). A thermal stress is generated during cooling from a stress-free anneal temperature T_{ZS} to the test or use temperature T_{test} due to the difference in the coefficients of thermal expansion (CTE) of metal α_{Me} and confinement α_{conf}, which is mainly determined by the silicon substrate as a primary source of the thermal stress. With a uniform distribution of temperature through the sample, the thermal stress can be approximated as

$$\sigma_T = B(\alpha_{Me} - \alpha_{conf})(T_{ZS} - T_{test}), \tag{4.42}$$

where B is the effective bulk modulus as defined in [24].

The migration of metal atoms through GBs and interfaces removes the nonuniformities existing in the sample. Such nonuniformities are developed in the metal line during thermal processing, such as material deposition, anneal, chemical-mechanical polishing, etc., due to variations in mechanical properties such as the elasticity modules caused by variations in grain size and orientation. A final state should be characterized by a uniform hydrostatic stress across the metal. However, such an equilibrium state, as stressed by Suo in [23], is not stable and can be destroyed by the presence of various flaws in the interconnect structure. For example, different cohesive or interfacial microcracks or process-induced microcaverns with a certain void size r_f can start growing when the stress exceeds $\sigma \geq 2\gamma/r_f$, where surface atoms will migrate from the stress-free void surface into the metal as driven by the tensile stress along the stress gradient. Due to the large difference in the diffusivities along GBs/interfaces and in the grain interior, the atoms originally in the void will redistribute through the network of GBs and interfaces to reduce the initial tensile stress. Such stress-driven GB diffusion process driving the void growth has been described by Yost [41], who showed that diffusion-induced void growth will persist until all initial deformation is accommodated by the void volume. Under that condition, a new equilibrium will be reached, which is characterized by the presence of one large "saturated" void and with zero stress everywhere in the metal. It is obvious, then, that the final zero stress state can be achieved with an infinitely rigid confinement. In reality, however, when the confined deformation is taken into account, the growing void will accommodate some confinement strain as well [23]. Therefore, in many practical cases the thermal stress generated in the metal line will exceed the yield strength. A description of the differences in stress developed in pure elastic and elastic-plastic materials can be found in [23], where the role of material plasticity in controlling the stress relaxation is described in a simple but elegant analysis.

4.7 An Approximate Derivation of the Void Nucleation Time

As discussed in Section 4.5, the void nucleation time t_{nuc} in a confined metal line under the stressing of the direct current (DC) of density j can be derived from (4.34). An approximate closed-form expression for the stress at the line cathode end $t = t_{nuc}$ can be deduced from (4.34) by keeping the slowest decaying term in the infinite series as

$$\sigma_{crit} = \sigma_T + \frac{GL}{2} - \frac{4GL}{\pi^2} \exp\left\{-\left(\kappa\frac{\pi}{L}\right)^2 t_{nuc}\right\}. \tag{4.43}$$

This gives an approximate t_{nuc} as follows:

$$t_{nuc} \approx \frac{L^2 k_B T}{D_a B \Omega} \ln \left\{ \frac{\frac{eZ\rho jL}{2\Omega}}{\sigma_T + \frac{eZ\rho jL}{2\Omega} - \sigma_{crit}} \right\}. \tag{4.44}$$

This analytical solution was derived previously in [42] by solving a full set of continuity equations for vacancies and plated atoms coupled with the stress equation, which will be discussed in Section 7.5. Equation (4.44) demonstrates that the relations between the stresses σ_T, σ_{crit}, and $\sigma_{EM} = eZ\rho Lj/2\Omega$ will yield different voiding conditions. For example, when $\sigma_{crit} > \sigma_T + \sigma_{EM}$ subjected to the condition $j \times L < 2\Omega(\sigma_{crit} - \sigma_T)/eZ\rho$, (4.44) yields a divergence of t_{nuc}, indicating that the line is immortal according to Blech's condition [1]. Another interesting case is when $\sigma_T > \sigma_{crit}$; then (4.40) yields a negative void nucleation time and shows that $\sigma_T > \sigma_{crit}$ is the proper condition for stress voiding. Indeed, if the residual stress exceeds the critical stress needed for void nucleation, voids will be nucleated before any electrical stressing is applied. The universal character of the void nucleation time can be further clarified by calculating t_{nuc} for different current densities and test temperatures, which is shown in Table 4.1. The values of parameters used in the calculation are $B = 3 \cdot 10^{10}$ Pa, $\Omega = 1.66 \cdot 10^{-29}$ m³, $k_B = 1.38 \cdot 10^{-23}$ J/K, $L = 1 \cdot 10^{-4}$ m, $Z = 10$, $e = 1.6 \cdot 10^{-19}$ q, $\rho = 3 \cdot 10^{-8}$ Ohm m, $D_0 = 7.56 \cdot 10^{-5}$ m²/s, $E_a = 1.6 \cdot 10^{-19}$ J, $D_a(T = 400$ K$) = 1.58 \cdot 10^{-17}$ m²/s, $\sigma_{crit} = 500$ MPa, and $\delta = 1 \cdot 10^{-9}$ m. The test current density is taken from the interval $5 \cdot 10^9 \div 3 \cdot 10^{10}$ A/m², with the test temperature in the range of 373 K $\div$ 573 K. To calculate the thermal stress, we use (4.44) with $\Delta \alpha = \alpha_{Me} - \alpha_{conf} = 1.7 \cdot 10^{-5}$ K^{-1}.

Table 4.1. Void nucleation time as a function of the test current density and temperature.

J, A/m²/T_{test}, K	323	373	423	473	523	573	623
1.00E+09	T-void	4.58E+05	IMMORT	IMMORT	IMMORT	IMMORT	IMMORT
2.00E+09	T-void	2.25E+05	1.79E+05	IMMORT	IMMORT	IMMORT	IMMORT
3.00E+09	T-void	1.49E+05	9.86E+04	1.88E+04	IMMORT	IMMORT	IMMORT
4.00E+09	T-void	1.11E+05	6.85E+04	1.06E+04	2.79E+03	IMMORT	IMMORT
5.00E+09	T-void	8.89E+04	5.26E+04	7.51E+03	1.54E+03	6.03E+02	IMMORT
6.00E+09	T-void	7.40E+04	4.27E+04	5.86E+03	1.10E+03	3.04E+02	2.19E+02
7.00E+09	T-void	6.34E+04	3.60E+04	4.81E+03	8.69E+02	2.19E+02	7.80E+01
8.00E+09	T-void	5.54E+04	3.11E+04	4.09E+03	7.19E+02	1.73E+02	5.60E+01
9.00E+09	T-void	4.22E+04	2.74E+04	3.55E+03	6.14E+02	1.44E+02	4.40E+01
1.00E+10	T-void	4.43E+04	2.44E+04	3.14E+03	5.37E+02	1.24E+02	3.70E+01
1.10E+10	T-void	4.03E+04	2.21E+04	2.82E+03	4.77E+02	1.08E+02	3.20E+01
1.20E+10	T-void	3.69E+04	2.01E+04	2.55E+03	4.29E+02	9.70E+01	2.80E+01
3.00E+10	T-void	1.47E+04	7.79E+03	9.54E+02	1.54E+02	3.30E+01	9.00E+00
5.00E+10	T-void	8.84E+03	4.63E+03	5.62E+02	9.00E+01	1.90E+01	5.00E+00
7.00E+10	T-void	6.31E+03	3.30E+03	3.99E+02	6.30E+01	1.30E+01	4.00E+00
9.00E+10	T-void	4.91E+03	2.56E+03	3.09E+02	4.90E+01	1.00E+01	3.00E+00

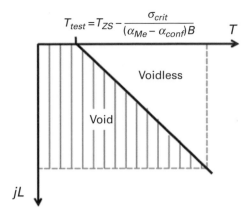

Figure 4.12 Critical product $(jL)_{crit}$ as a function of temperature, T_{ZS} is the temperature corresponding to the zero-stress condition.

It can be seen from Table 4.1 that the metal line is immortal at low current densities and high temperatures. This can be accounted for by the small thermal stress generated during cooling from the zero-stress temperature down to the higher test temperature, i.e., by reducing the thermal gap. Larger current densities are required in this case to compensate for the lower thermal stress. The leftmost column of the table shows the test temperature when the thermal stress exceeds the critical stress, so stress-induced voids can be nucleated. The mortal and immortal regions shown in Table 4.1 can be mapped into two regions in the "temperature–current density" plot as shown in Figure 4.12, which corresponds well to the traditional voiding conditions on a $T - j \times L$ plot.

The derived t_{nuc} in (4.44) can be represented in the form of the Black's equation [43] $t_{nuc} = Aj^{-n} \exp\{-E_a/k_B T\}$ to extract the current density exponent n and apparent activation energy E_a. The result confirms the published values of n and E_a as a function of T_{test} and j [44–47]. The predicted dependency $n(T_{test})$ is in good agreement with the measurement results [46, 47] as shown in Figure 4.13. In general, the value of $n(T_{test})$ depends on σ_T as a function of T_{test}. Neglecting this temperature dependency would result in a constant $n = 1.346$ for the chosen set of parameters, and the apparent activation energy E_a [42] would increase with the current density, which is consistent with the experimental observations [47].

Equation (4.44) provides an analytical form for the void nucleation time, which can be further simplified by expanding the logarithmic function in the series of $2\Omega(\sigma_{crit} - \sigma_T)/eZ\rho jL < 1$. This yields the following expression by keeping only the slowest decaying term:

$$t_{nuc} = \frac{2k_B T(\sigma_{crit} - \sigma_T)L}{D_a B e Z \rho j}. \tag{4.45}$$

The approximation of (4.45) clearly demonstrates how the void nucleation time depends on the experimental parameters of the test. It shows that a low residual

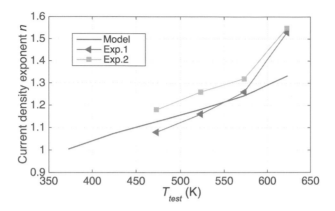

Figure 4.13 Extracted dependency of n on T_{test} for $T_{ZS} = 723$ K, $\sigma_{crit} = 600$ MPa vs. experimental data: Exp. 1 data are from [46], and Exp. 2 data are from [47]. © 2016 IEEE. Reprinted, with permission, from [48].

tension and long line length will increase t_{nuc} and, vice versa, a high atomic diffusivity and high test/use temperatures together with a large effective bulk modulus and a high current density will decrease t_{nuc}. The results clearly confirm the benefit from recent efforts of using interface strengthening to miminize void nucleation rate by reducing the interfacial atomic diffusivity. This is accomplished by introducing a metal cap between the top dielectric barrier and copper to optimize the Cu grain texture with better metal liner to improve the orientation and size of the Cu grains.

Such approaches to reduce void formation rate can be found in a recent review [49]. Qualitatively, the basic approach is to minimize the product $Z \cdot D_a$, i.e., the EM driving force in the metal line with Z being the effective charge and D_a the effective atomic diffusivity from the interface and GB mass transport. The product $Z \cdot D_a$ for a metal line with a near-bamboo grain structure and a mean grain size d can be approximated as follows [23]:

$$Z \cdot D_a = Z_{GR} \cdot D_{GR} + Z_{GB} \cdot D_{GB} \delta_{GB} \frac{1}{d} \left(1 - \frac{d}{W} \right) + 2 Z_{Int} \cdot D_{Int} \delta_{Int} \left(\frac{1}{W} + \frac{1}{H} \right). \quad (4.46)$$

Here the subscripts GR, GB, and Int refer to the grain interior, grain boundary, and interface, respectively, and δ_{GB} and δ_{Int} are the width of the GB and interface, respectively. This expression can be modified to account for the different atomic diffusivities between the top interface with the dielectric barrier and the sidewall/ bottom interfaces with the metal barriers [50]. In general, the metal and alloy caps are chosen for their capability to reduce significantly the interfacial diffusivity. The impurities segregating to the GBs can also reduce the GB diffusivity by reducing the fraction of the multigrain clusters to further minimize the GB contribution to the overall diffusivity. Reduction of the residual stresses can also be beneficial for increasing the void nucleation time (4.45). Finally, an improvement of the interface quality can reduce the sizes of process-induced flaws and increase the critical stress and the void nucleation time to improve the overall void formation characteristics.

4.8   Postvoiding Stress and Void Size Evolutions in Confined Metal Lines

Generally, the kinetics of the EM-induced failure can be controlled by a void nucleation mode or a void growth mode or by a mix of these two modes; see, for example, [8, 32, 47, 51–53]. Void-induced failure usually refers to a resistance increase of an individual line above a certain threshold under testing with controlled current densities and temperatures. It is clear, however, that only the void growth mode can be responsible for the line resistance degradation. Indeed, continuous reduction of the metal line cross section around a growing void can force the current to flow through the highly resistive metal liners/diffusion barriers when the void reaches the entire metal cross section to cause the line resistance increase. A void nucleation by transforming process-induced flaws located at the metal–passivation interface into a stable growing void [23, 31, 32] in general does not cause any noticeable change in the line resistance. Hence, the kinetics of the EM-induced degradation of the line resistance should be characterized by the resistance increase caused by void growth after an initial incubation time for void nucleation. This type of kinetics is usually considered an EM-induced failure after the resistance increases about 10% of the initial line resistance [54]. A different criterion for the EM-induced failure based on EM-induced voltage-drop degradation has been applied for on-chip interconnects, which will be discussed in Chapter 9. A model based on the kinetics of EM-induced resistance increase for interconnect trees has also been proposed by Hau-Riege and Thompson [54]. A similar approach based on further development of the 1D EM models [56] has been used to analyze EM failure in large interconnect nets. Such an analysis can become very important for modern on-chip power/ground grid designs, which can be enormous in scale to reach several billion nodes in. This topic will be discussed later in Chapter 9.

Standard industrial practice employed for estimating the EM-induced degradation of VLSI power grids consists of applying Black's model to individual lines to evaluate reliability and using the earliest line failure time as the failure time of the whole grid. The mean time to failure (MTTF), or t_{50}, according to Black [43, 57] for individual line is as follows:

$$MTTF = Aj^{-n}\exp\{E_a/k_BT\}. \tag{4.47}$$

Here T is the absolute temperature, k_B is the Boltzmann constant, j is the current density, n is the current density exponent (which Black found to be approximately equal to 2 [57]), E_a is the activation energy of the failure process, and A is a constant. Parameters n and E_a are extracted from measurements. The majority of measurements show that n varies over a range of 1–2, where two limiting cases of $n = 1$ and $n = 2$ are traditionally attributed to the void growth and void nucleation modes of failure. All other cases characterized by $1 < n < 2$ are considered a mix of these two failure modes and consistent with results of physics-based modeling of the void growth and the void nucleation time, [8, 23, 28–30, 52, 58]. The current density exponent $n = 2$ reported in the papers [8, 28, 29] was derived by considering infinitely long half-lines with the cathode edges blocking the atomic flux. This case is characterized by an absence of the steady state, i.e., the stress at the cathode edge grows infinitely with time. The stress at time t is a product of the atomic flux linearly dependent on the current density and the distance traveled by diffusing vacancies, which is proportional to the square root of time. It provides a reverse quadratic dependence of the void nucleation time on the current density. This type of dependence has not been derived in the case of a finite line. The commonly accepted void growth kinetics provides a well-known inverse dependence of the void volume on the current density; see, for example, [58]. This type of kinetics, however, is valid only for the cases of an unconfined line edge drift [48] and the initial growth of a saturated void preexisting in the confined metal line [33, 58], where the stress in the line has a negligible effect on the movement of the free surface.

Quite different void volume kinetics occur when an initially voidless metal line embedded into a rigid confinement is stressed by an electric current. In this case, we have a failure process consisting of two steps: void nucleation and following void growth to reach the critical volume to induce a specific, e.g., 10%, increase of line resistance. As an example, we can consider a void nucleation occurring near the cathode edge of the line when increase in the tensile stress creates a condition for a stable growth of preexisting process-induced flaws located at the metal/passivation interface [23, 32]. Similar voiding can occur in a microgranular cluster near the cathode end adjacent to a bamboo portion in the line. A flux divergence generated by different vacancy diffusivities in such two-line segments can lead to a large enough stress for void formation. In this case, the void growth can be accompanied by its drift toward the line cathode end. This type of the damage evolution will be analyzed later in Sections 4.10 and 7.8. Gleixner and Nix [32] estimated that a flaw of 4 nm long will start growing when the hydrostatic stress reaches a level of 500 MPa. However, Suo has shown that such a void initiation condition is overly simplified because elastic anisotropy of the polycrystalline metal lines, as well as misfit in the mechanical properties of metal and encapsulating materials, can generate highly different components of the stress tensor at grain boundaries and interfaces, which can cause growth of quite different flaws. In view of this limitation and by assuming a rapid development of a pure hydrostatic stress state, we will continue with the critical hydrostatic stress, σ_{crit}, as a void nucleation criterion.

In previous sections, the solution (4.34) to Korhonen's equation (4.18) provides a stress distribution along the line length at time t_{nuc} with a critical stress σ_{crit} at the cathode end. This stress distribution can be used as an initial condition to derive the postvoiding stress evolution kinetics. This condition shows that the metal near the flaw surface is under tension with a large stress gradient pulling atoms away from the surface of the growing void toward the bulk. It has been shown in [59] that the presence of a large stress gradient is the major difference between void evolution in an initially voidless line and the growth of a large saturated void preexisting in the line as analyzed by He and Suo [33, 57]. In the latter case, the initial stress is zero everywhere in the line outside the void, and initial void growth is controlled by the electric current [58].

The same approach, when applied to an interconnect tree, allows one to estimate t_{nuc} for different branches of the tree for different current densities and geometries. Employment of the proper boundary conditions at the segment junctions corresponding to the continuity of stress and atomic fluxes can yield an accurate deduction of the stress evolution in a multisegment tree [60]. Such kinetic analyses when applied to the interconnect trees can be extended to evaluate the voltage degradation in a power grid based on solution of the linear system [56, 61, 62]. This requires an extension of the analysis of the postvoiding evolution of the resistances of individual lines to describe the kinetics of voltage degradation in an on-chip power-ground grid. An implementation of this approach for a power-ground grid of an on-chip interconnect at a global scale will be discussed later in Chapter 9.

In the following, we continue to analyze the postvoiding stress evolution kinetics based on the Korhonen equation. We assume that the void is nucleated at the cathode edge of the line: $x = 0$ at t_{nuc}. We introduce an effective thickness of the void interface δ, which is infinitely small in comparison with all other lengths. This allows us to introduce a stress gradient between the zero-stress void surface and the surrounding metal as $\nabla\sigma = \sigma(\delta, t)/\delta$, where $\sigma(\delta, t) \approx \sigma(0, t)$ in the metal near the void surface. Thus, the IBVP based on Korhonen's equation can be expressed as follows:

$$PDE: \quad \sigma_t = \kappa^2 \sigma_{xx}; \quad 0 < x < L, \ 0 < t < \infty$$

$$BC: \quad \frac{\partial\sigma}{\partial x}(0, t) = \frac{\sigma(0, t)}{\delta}; \quad 0 < t < \infty$$

$$BC: \quad \frac{\partial\sigma}{\partial x}(L, t) = -G; \quad 0 < t < \infty$$

$$IC: \quad \sigma(x, 0) = \sigma_T - Gx + \frac{GL}{2}$$

$$-4GL \sum_{n=0}^{\infty} \frac{\cos \frac{(2n+1)\pi x}{L}}{\pi^2 (2n+1)^2} \exp\left\{ -\left(\kappa \frac{(2n+1)\pi}{L} \right)^2 t_{nuc} \right\}.$$

(4.48)

Here as before, $G = eZ\rho j/\Omega$, $\kappa^2 = D_{eff} B\Omega/kT$. The solution to this problem can be obtained using the standard method of separation of the variables, as follows:

$$\sigma(x, t) = -Gx + \sum_{m=0}^{\infty} \frac{A_m}{\sin(\lambda_m L)} \cos[\lambda_m(L - x)] \exp\left\{ -(\kappa\lambda_m)^2 t \right\}.$$

(4.49)

Here the eigenvalues λ_m can be obtained by solving the equation:

$$\lambda_m L = \frac{L}{\delta} \cot \lambda_m L,$$

(4.50)

which is derived from the BCs, all coefficients A_m take the following form:

$$\frac{A_m}{\sin(\lambda_m L)} = \frac{4}{\pi}\left(\sigma_T + \frac{GL}{2}\right)\frac{\sin(\lambda_m L)}{2m+1} - \frac{16}{\pi}GL$$

$$\times \sum_{n=0}^{\infty} \frac{\exp\left\{-(\kappa(2n+1)\pi/L)^2 t_{nuc}\right\}}{\pi^2(2n+1)^2} \frac{(2m+1)\sin(\lambda_m L)}{(2m+1)^2 - 4(2n+1)^2}. \quad (4.51)$$

It is easy to show that by employing the following trigonometric identities [59]:

$$\sum_{m=0}^{\infty}\frac{(2m+1)\sin[(2m+1)\xi]}{(2m+1)^2 - 4(2n+1)^2} = \frac{\pi}{4}\cos[2(2n+1)\xi] \text{ and } \sum_{m=0}^{\infty}\frac{\sin[(2m+1)\xi]}{(2m+1)} = \frac{\pi}{4},$$

$\sigma(x, t)$ given by (4.49) with the derived A_m satisfies the initial condition (4.48): $\sigma(x = 0, t = 0) = \sigma_{crit}$. The stress evolution starting from the void nucleation time $t = 0$ until the steady state is achieved is shown in Figure 4.14a, where the stress is calculated using the same parameters as in Section 4.7 except that the thermal stress is neglected, i.e., $\sigma_T = 0$.

Figure 4.14b shows the evolution of stress distribution along a line with a preexisting saturated void as defined in [58]. Comparison of these two figures side by side shows that stress evolution is different mainly at the initial time, then becomes the same after a long time at the steady state. The different initial stress evolution comes mainly from the different void growth kinetics in these two cases. For the initially voidless line, the void volume evolves as follows:

$$V_{void}(t) = \frac{V_{line}}{L}\int_0^L \frac{\sigma(x, t)}{B}dx. \quad (4.52)$$

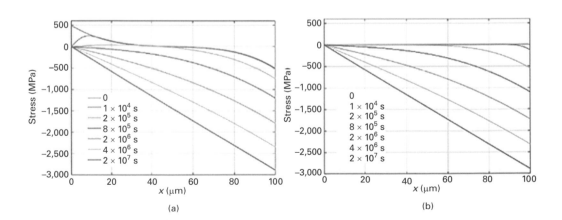

(a)   (b)

Figure 4.14 Evolution of the distribution of hydrostatic stress along the metal line loaded with the DC current of 1×10^{10} A/m^2 at $T = 400$ K at two different IC for (a) a voidless metal line and (b) a line with a preexisting saturated void at the cathode end. © 2016 IEEE. Reprinted, with permission, from [59].

Thus,

$$
\frac{V_{void}(t)}{V_{line}} = \left(\frac{\sigma_T}{B} + \frac{GL}{2B}\right)\left(1 - \frac{8}{\pi^2}\sum_{m=0}^{\infty}\frac{e^{-(\pi\kappa(2m+1)/2L)^2 t}}{(2m+1)^2}\right)
$$

$$
+ \frac{32}{\pi^2}\frac{GL}{B}\sum_{n=0}^{\infty}\frac{e^{-(\pi\kappa(2n+1)/L)^2 t_{nuc}}}{\pi^2(2n+1)^2}\sum_{m=0}^{\infty}\frac{e^{-(\pi\kappa(2m+1)/2L)^2 t}}{(2m+1)^2 - 4(2n+1)^2}
$$

$$
= \left(\frac{\sigma_T}{B} + \frac{GL}{2B}\right) - \frac{8}{\pi^2}\sum_{m=0}^{\infty}\frac{e^{-(\pi\kappa(2m+1)/2L)^2 t}}{(2m+1)^2}
$$

$$
\times\left[\frac{\sigma_{crit}}{B} - 16\frac{GL}{B}\sum_{n=0}^{\infty}\frac{e^{-(\pi\kappa(2n+1)/L)^2 t_{nuc}}}{\pi^2\left[(2m+1)^2 - 4(2n+1)^2\right]}\right]. \qquad (4.53)
$$

Here $V_{line} = LWH$ is the line volume, and W and H the linewidth and thickness. Figure 4.15a and b show the void evolution kinetics driven by the same DC currents for two cases: (a) an initially voidless line when a void was nucleated with the stress reaching σ_{crit} and (b) a line with an initially preexisting void following the volume growth kinetics deduced by He and Suo [33, 57]:

$$
V_{void}(t) = V_{sat}\left[1 - \frac{32}{\pi^3}\sum_{n=0}^{\infty}\frac{(-1)^n}{(2n+1)^3}\exp\left\{-\left(\kappa\frac{(2n+1)\pi}{2L}\right)^2 t\right\}\right]. \qquad (4.54)
$$

Here $V_{sat} = WH\frac{GL^2}{2B}$ is the void saturation volume developed when the atomic flux driven by EM is balanced by that from the stress gradient.

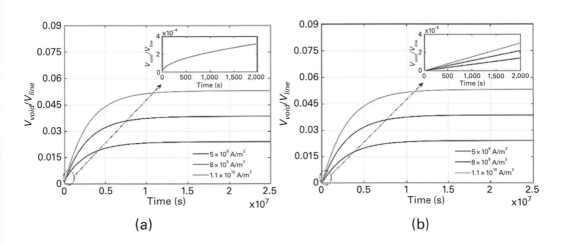

(a)   (b)

Figure 4.15 Evolution of void volume in the metal line at $T = 400$ K and $\sigma_T = 0$ under the same current density in (a) a voidless metal line and (b) a line with a preexisting saturated void at the cathode end. © 2016 IEEE. Reprinted, with permission, from [59].

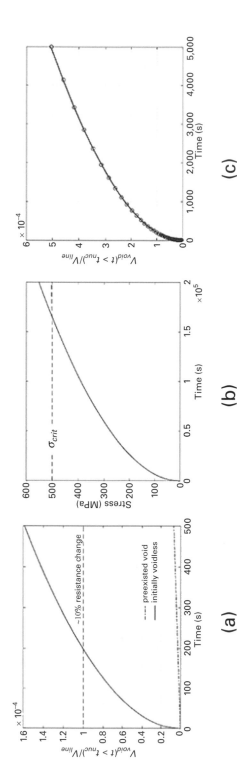

Figure 4.16 Void growth kinetics in (a) an initially voidless line and a line with a preexisting void, (b) the prevoiding kinetics of the stress buildup at the cathode end of line, and (c) an approximation of the initial growth kinetics by the square root time dependence (small dots). For all cases: $j = 1 \times 10^{10}$ A/m^2, $T = 400$ K and $\sigma_T = 0$. © 2016 IEEE. Reprinted, with permission, from [59].

Two important observations can be extracted from the preceding analysis: first, the void volume increases generally much faster in an initially voidless line, as shown in Figure 4.16a; and second, the initial void growth (up to 10^4 s for the chosen parameters) in a voidless line does not depend on the current density. This differs from the previous analysis for a line with a preexisting void that initially grows in proportional to the current density [58] (see insets to Figure 4.15a and b). Analysis of the initial void growth kinetics (4.53) in an initial voidless line shows a square root time dependence:

$$\frac{V_{void}(t)}{V_{line}} = \frac{\sigma_{crit}\Omega}{k_B T} \frac{\sqrt{D_a t}}{L},$$ (4.55)

which fits very well to the model as shown in Figure 4.16c.

4.9   Voiding-Induced Degradation of Resistance of Metal Lines

If we assume that the void grows through the line cross section and the Ta/TaN liner conducts the electric current in the line portion occupied by the void [33, 35], the change of the line resistance induced by the growing void can be presented as

$$\Delta R(t) = \frac{V_{void}(t)}{WH} \left(\frac{\rho_{TaN}}{h_{lin}(W + 2H)} - \frac{\rho_{Cu}}{WH} \right).$$ (4.56)

Here ρ_{TaN} and ρ_{Cu} are the resistivity of the Ta/TaN barrier and copper respectively, and h_{lin} is the thickness of the barrier layer. Figure 4.17 shows schematically a line with void formation at one end.

Shown in Figure 4.16, the time required for 10% increase in the line resistance calculated with $\rho_{lin} = 2.5 \cdot 10^{-6}$ Ωm, $\rho_{Cu} = 3 \cdot 10^{-8}$ Ωm, $H = 120$ nm, and $h_{lin} = 10$ nm is about three orders of magnitude faster (see Figure 4.15a) than the time required for void nucleation (see Figure 4.16b). An estimate based on (4.56) shows that a 10% resistance increase of a 100 µm long line is achieved when a void reached 10^{-4} fraction of the volume in the line cross section. Such a critical void length is only 10 nm, which can be formed in several minutes in an initially voidless line. For a line with a preexisting void, the time consumed by the void growth process to reach 10% resistance increase is about 10^4 s, which is about an order of magnitude less than the void nucleation time $\sim10^5$ s for a voidless line (see Figure 4.16b). As discussed in Section 4.6, such an estimate can be used to check the measured [46, 47] versus the predicted [59] current density exponents at the test temperature based on void nucleation times.

Void volume evolution kinetics (4.54) derived for a line with initially preexisting saturated void demonstrates that in the short time limit the void volume grows linearly with time. Indeed, expanding the exponential function in series and keeping just two first terms in (4.54), we get

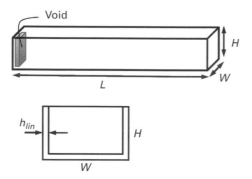

Figure 4.17 Schematics of the line geometry with void formation at one end. © 2016 IEEE. Reprinted, with permission, from [59].

$$V(t) \approx \frac{eZ\rho j D_a WHt}{k_B T}.$$ (4.57)

In this case, the stress plays a negligible role at initial void growth where the atomic flux reduces to $\Gamma_A = D_a eZ\rho j / \Omega k_B T$ while the void volume grows linearly with time, $V = \Gamma_A WH\Omega t$. In the long-time limit $(t \to \infty)$, (4.54) yields the same saturated volume as in an initially voidless line:

$$V_{SV} = \frac{eZ\rho j L^2 WH}{2\Omega B}.$$ (4.58)

This solution provides the saturated void volume achieved in a finite length line when the stress gradient–induced atomic back-flux compensates the EM-induced flux. This result can also be derived from (4.52), where in the steady-state limit the stress distribution varies linearly with the position along the line (4.35) and thus deduces the void-saturated volume given by (4.58). Evolution of the void volume is shown in Figure 4.18 as a function of time where the void saturation time coincides with that required for stress relaxation, reaching a steady-state $\tau_{SS} \approx L^2/2\kappa^2$, where κ is the previously defined effective "diffusivity" $\kappa^2 = D_a B\Omega/k_B T$.

Knowledge of the void-saturated volume V_{SV} developed in a metal line characterized by given geometries (L, W, and H), material properties (Ω, Z, ρ, and D_a), and an effective modulus, B, allows us to answer the question about line immortality. Traditional criterion of the line failure caused by EM is an increase of line resistance above the threshold ΔR_{th}, for example, 10% of its initial resistance. A relation between the resistance increases and the volume of the void responsible for this increase is described by (4.56). Figure 4.17 shows the schematics of the line with a void and the corresponding geometrical parameters. The void volume responsible for the threshold growth of the line resistance ΔR_{th} is called the critical volume, V_{crit}, and a relation between V_{SV} and V_{crit} allows us to define the state of line mortality. When $V_{SV} < V_{crit}$, a line will not fail since the void growth will stop before reaching a critical resistance increase. Equations (4.52) and (4.56) provide the immortality criterion as follows [23]:

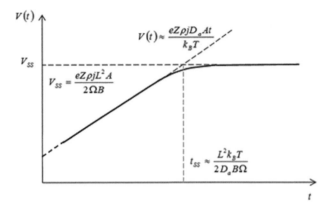

Figure 4.18 Void volume as a function of time. $A = WH$ is the line cross section.

$$jL^2 < \frac{2\Omega B}{eZ\rho_{Cu}} \frac{\Delta R_{th}}{\left[\frac{\rho_{lin}}{h_{lin}(W+2H)} - \frac{\rho_{Cu}}{HW}\right]} \qquad (4.59)$$

This can be considered as the Blech-limit immortality criterion if we introduce $R_{Cu} = \rho_{Cu}L/HW$, so

$$jL < \frac{2\Omega B}{eZ} \frac{\Delta R_{th}}{R_{Cu}} \frac{1}{\left[\rho_{lin} \frac{HW}{h_{lin}(W+2H)} - \rho_{Cu}\right]}. \qquad (4.60)$$

The opposite condition $V_{SV} > V_{crit}$ for a line failure states that a critical increase in the line resistance can occur before a void is saturated. It should be noted that the critical jL product in (4.60) depends on the failure criterion $\Delta R_{th}/R_{Cu}$, which can be arbitrary chosen. The original jL critical product, as discussed in Section 4.7, provides the condition for void nucleation, which can differ from the critical product defined for a specific resistance change.

So far, the results presented in this section on EM-induced line resistance change is based on void evolution kinetics in initially voidless lines. It should be mentioned that the predicted independence of the initial line resistance changes on the current density (see Figure 4.15a) was observed in experiment [54]. However, the observed absence of deviation in resistance changes of Cu lines tested with high ($3.6 \cdot 10^{10}$ A/m^2), and low ($3 \cdot 10^9$ A/m^2) current densities as shown in Figure 4.19 was explained by assuming that line resistance increase is mainly due to the penetration of Co from CoWP cap layer into the Cu line while the contribution from EM-induced line damage was small.

4.10   EM-Induced Void Migration

For both cases of Al and Cu metallizations, the experimental observations of the damage development, performed on lines with a variety of grain structures, have

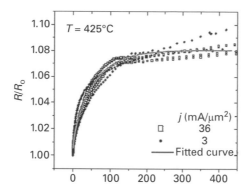

Figure 4.19 Plot of normalized line resistance change vs. log(time) for lines tested at a high current density of 36 mA/mm^2 (open squares) and lines tested at a low current density of 3 mA/mm^2 (solid circles) at 425°C. The solid lines are the least-square fitted lines. Reprinted from [55], with the permission of AIP Publishing.

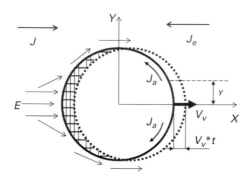

Figure 4.20 EM drives a void to move in an interconnect line by relocating material from one side of the void to the other.

demonstrated a nontrivial void evolution showing a complex sequence of void motion, growth, and shape changes [63, 64]. First, voids were not static but rather moved in the direction opposing the electron wind. This has been confirmed by in situ scanning electron microscopy (SEM) studies on Al and Cu lines [65, 66]

A phenomenological model of void drift in an isotropic conductor under an electric field was first developed by Ho more than 45 years ago [67]. The physical nature of void drifting is a directional atomic diffusion along the void surface under the action of the electric field. The tangential void surface component of the electric field E_t is responsible for the electron wind force F_E acting on the atoms located on void surface $F_E = -eZE_t$, as shown in Figure 4.20. Ohm's law relates E_t with the tangential component of the current density, which, is in the case of nonconductive void interior, represents the total current density due to vanishing of the normal current density component.

In this case, a circular void in the conductor migrates without shape changing under the action of a constant electric current. At the surface of the spherical void, the tangential component of the electric field is

$$E_t = -\frac{3E \cdot y}{2R_V} \tag{4.61}$$

and the mass conservation condition gives

$$J_a = \frac{v_V \cdot y}{\Omega}. \tag{4.62}$$

This in combination with the atomic mass transport under E, $J_a = \mu_S F_E = \frac{D_S}{k_B T} eZE_t$ gives

$$v_V = \frac{3D_S v e Z \rho j \Omega}{k_B T R_V} \tag{4.63}$$

Here v_V is the void drift velocity, R_V is the void radius, D_S is the atomic diffusivity along the void surface, and v is the effective number of atoms per unit surface area participation in diffusion. Despite the simple assumptions, this model provides a qualitatively correct description of the void drifting rate as a function of the void size and the current density. This was shown experimentally by direct observation of void motion that the small voids indeed drift faster than the larger one as predicted [68]. The presence of grain boundaries was found to affect the void drifting. Voids can be trapped by grain boundaries and triple points formed by intersections of GB and interfaces. Void pinning by GB and breaking away was observed in many experimental studies [68–70]. Suo, in his review paper [71], has discussed several phenomenological models describing the void attachment and breaking away from GBs, which established criteria for estimating the current densities needed for voids of varying size to break away from GB. He also described the diffusion mechanism for void shape change from an initially rounded void trapped by GB to a transgranular slit. A more general form for movement of the void surface, assuming an isotropic surface tension, is describing by the equation

$$\vec{n}\, \frac{\partial u(\vec{r}, t)}{\partial t} = \frac{D_S}{k_B T} \frac{\partial^2}{\partial s^2} (\Omega \gamma K - eZ\Phi). \tag{4.64}$$

Here $u(\vec{r}, t)$ is the displacement of the surface segment; $\vec{n}$ the unit vector normal to the void surface and directed into the metal interior; Φ the potential to provide the component of the electric field tangential to the void surface, $E_t = -\partial \Phi/\partial s$; s is the curve length along the void surface; K the void surface curvature; and γ the surface energy per unit area. Futher discussion on the evolution of the void shape using this approach is beyond the scope of 1D approximation for EM-induced degradation considered in this chapter. It will be discussed later together with the comprehensive analysis of the void shape evolution in Chapter 7.

Here, we discuss the void motion with a constant void volume taking place in a stress-free metal. If stress was introduced into the metal interior by, for example, thermal mismatch or under an electric current in a line due to atomic migration induced by the electron wind, the void will drift and change its shape simultaneously. This is driven by the difference in chemical potentials of vacancies located near the void surface and near the metal surface, at the interface under confinement, or near the GB. As discussed in Section 4.4, the creation of the pair of a vacancy and an atom at the stressed surface changes the potential energy by $E_V - \sigma\Omega$, where E_V is the energy of vacancy formation. Creation of the vacancy and atom pair at the void surface changes the potential energy by $E_V - 2\gamma\Omega/R$, where R is the void radius. The equilibrium vacancy concentrations near the stressed surface and near the void surface are respectively $\exp\{-(E_V - \sigma\Omega)/k_BT\}$ and $\exp\{-(E_V - 2\gamma\Omega/R)/k_BT\}$.

Comparing these vacancy concentrations, we can conclude that the void will grow if $\sigma > 2\gamma/R$, which is the same condition as deduced earlier in Section 4.6 (4.38). The void will grow to reach an equilibrium size of $R = 2\gamma/\sigma$ and establish a steady-state distribution of the vacancy concentration in the metal interior. The opposite condition $\sigma < 2\gamma/R$ will lead to a decrease in the void size and, in some cases, to void disappearance. Since the vacancy diffusion through the metal is a limiting step for void growth, we conclude that when a void moves with a velocity determined by the surface vacancy diffusivity, it would accompany a small reduction in the void size. In contrast, a noticeable change in the void size is to be expected when the void is pinned at the GB or triple points due to high vacancy diffusivity there. Voids may also grow as a result of EM, since the electric current may cause material to diffuse away from the voids along GB. A comprehensive review of the analyses of void drift, enlargement, and shape change can be found in [72].

To account for the size dependence of the void drift velocity, the analysis of the EM-induced degradation of the line resistance becomes even more complicated. In addition to void growth due to nucleation at different locations along the line where flaws of different sizes can be generated by processing, the drift of the voids toward the cathode, and their possible coagulation into larger voids, can affect the line resistance in a very complex way. This problem cannot be accurately analyzed in the 1D approximation.

4.11   Effect of Metal Microstructure on EM-Induced Stress Evolution: 1D Modeling

Numerous experimental and modeling studies have indicated that the EM-induced degradation and eventual interconnect failure depend on both the interface bonding and the microstructure of the copper interconnect [73–75]. These necessitate in situ SEM interconnect degradation studies, electron backscatter diffraction (EBSD)-based copper microstructure studies, and numerical simulations to show that EM-induced degradation mechanisms depend strongly on the bonding strength of interfaces [49, 76]. Different degradation mechanisms have been described for interfaces with weak and strong bonding strengths and for atomic transport along interfaces with different activation energies [75–77]. When the diffusivity for interface migration along strengthened interfaces becomes comparable to the diffusivity along copper grain boundary [64, 77–80], mass transport along grain boundaries will have to be considered for interconnects with polycrystalline microstructures. When interface diffusion is suppressed [64], microstructural parameters such as grain size and texture and stress behavior become even more important for assessing the effect of void migration on the reliability of copper interconnects [77, 81, 82]. Together, such studies have provided a better understanding of the factors controlling the copper microstructure and its role in stress and void evolution in Cu dual-damascene interconnects [26, 83–85].

In general, the microstructure can influence the mechanisms and kinetics of the stress evolution under EM. The most obvious and widely considered influences are the effect of elastic anisotropy on the grain structure and the state of stress [86–88] and the effect of fast atom migration along the grain boundaries on the overall stress evolution kinetics [4, 5, 8, 32, 89–91]. These effects are interrelated and when considered in the framework of the vacancy generation/annihilation kinetics, they can determine the final stress state for void formation or the steady state in a voidless regime. This has led to the development of a comprehensive 2D simulation model to provide a realistic picture of the strain–stress distribution under EM in a polycrystalline interconnect segment by considering specific stress generation and relaxation phenomena [26, 83, 84]. Such a 2D model describes both the prevoiding and postvoiding stress distributions and vacancy concentration in a polycrystalline interconnect segment embedded into an interlayer/intermetal dielectric (ILD/IMD) structure will be discussed in Chapter 7. Here we analyze the effect of microstructure on stress evolution and void formation in the 1D approximation.

In the 1D case, the analysis is simplified by introducing the stress through a variation of the matter density based on (4.8) in Section 4.3, where the effect of microstructure on the elastic properties is not included. The effect of microstructure on EM-induced degradation is considered only through the microstructure-dependent variation of the atomic diffusivity. Several attempts have been taken to solve this problem by considering the statistical distribution of the grain structure; see, for example, [4, 77, 85]. There, the general idea was based on a cluster of segments with continuous grain boundary paths along the line together with some segments of the bamboo-like structure as shown in Figure 4.21. Such segments, which are called as polygrain clusters, provide venue for much faster atomic diffusion in comparison with the bamboo segments. TEM observations have been performed to show that the length of polygrain clusters follows a lognormal distribution [91, 92]. It was assumed that

Bamboo segment   Polygrain cluster   Bamboo segment

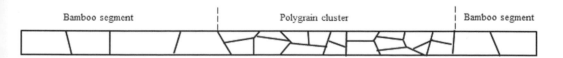

Figure 4.21 Grain morphology schematics in the line with a polygrain cluster surrounded by the bamboo-like segments.

bamboo grains can block the diffusion between polygrain clusters, and hence the stress evolution in each polygrain cluster can be considered in the same way as for a line with the flux-blocking boundary conditions (see Section 4.5). The central assumption is that the line will fail at the longest of the polygrain clusters because the stress gradient for back-diffusion will be smallest [4]. This cluster segment becomes the weakest link and thus determines the time to failure. In this way, the statistical EM lifetime for a metal line with such polygrain clusters connected in series can be determined based on the weakest link approximation. A detailed discussion of the statistical approach in the electromigration studies will be done in Chapter 8. In the following, we discuss some of the limitations of such statistical analyses of the EM lifetime.

The first reservation concerns the independent treatment of polygrain clusters in the same line. Indeed, when connected in series, these clusters as well as bamboo segments are loaded with the same current density. After a long time, a steady state is established, which is characterized by the vanished atomic flux everywhere along the line. At this point, the stress will be linearly distributed along the line with the slope determined by the current density as shown in Figure 4.22. This reflects the well-known fact that the diffusivity affects the kinetics of the stress evolution. The kinetics in the case of 10 times faster atomic diffusion along the polygrain cluster than in the bamboo segment is shown in Figure 4.22a, while the stress evolution along the same line when the diffusivity for bamboo segments was extended to the whole line is shown in Figure 4.22b. Clearly the final steady state is the same for both cases. The global stress evolution, however, shows a noticeable difference in local evolutions of stress distributions in the identical polygrain clusters at different locations; see, for example, the stress distributions near the cathode and anode ends of the line in Figure 4.23. The cathode cluster displays an essential tensile shift in comparison with the compressive shift in the anode cluster, as compared with the stress distribution at $t = 5 \cdot 10^5$ s shown in Figure 4.24. Two other curves demonstrate the stress distributions at initial time $t = 0$, and the steady state reached around $t = 8 \cdot 10^6$ s. It should

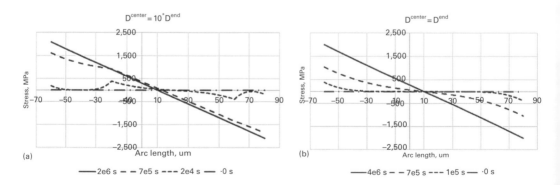

Figure 4.22 EM-induced evolution of the stress distribution in (a) the line with the grain structure depicted in Figure 4.21, and (b) with the bamboo-like grain structure only.

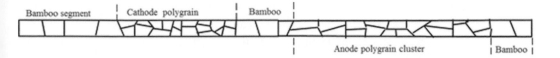

Figure 4.23 Schematics of the grain morphology in the line with two polygrain clusters located in the cathode and anode regions.

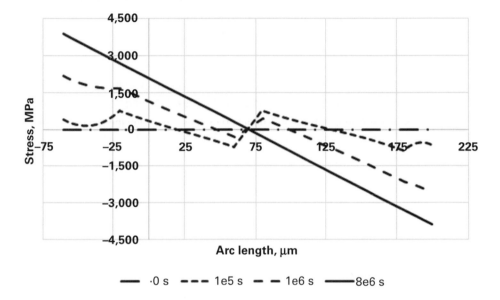

Figure 4.24 EM-induced evolution of the the stress distribution in the line with the grain structure depicted in Figure 4.23 taken at different instances in time.

be mentioned that different grain distributions existing in identical polygrain clusters can also result different stress evolution kinetics.

The kinetics of stress evolution along the line as discussed above was obtained from the solution of Korhonen's equations written for each and every bamboo and polygrain segment varying in lengths and diffusivities:

$$\frac{\partial \sigma_n}{\partial t} = \frac{\partial}{\partial x}\left[\kappa_n^2\left(\frac{\partial \sigma_n}{\partial x} + G\right)\right]. \tag{4.65}$$

As before, we use the following notations: $\kappa_n^2 = D_{eff}^n B\Omega/k_B T$, $G = eZE/\Omega$, and E the electric field. The boundary conditions reflecting the continuity of stress and atomic flux at each junction between bamboo and polygrain segments are as follows:

$$\sigma_n(x,t) = \sigma_{n+1}(x,t), \quad \text{at } x = x_n, t > 0 \tag{4.66}$$

$$\kappa_n^2\left(\frac{\partial \sigma_n}{\partial x} + G\right) = \kappa_{n+1}^2\left(\frac{\partial \sigma_{n+1}}{\partial x} + G\right), \quad \text{at } x = x_n, t > 0. \tag{4.67}$$

At both ends of the line, the standard zero-flux boundary conditions are employed:

$$\frac{\partial \sigma_1}{\partial x} + G = 0, \quad \text{at } x = 0, t > 0 \tag{4.68}$$

$$\frac{\partial \sigma_N}{\partial x} + G = 0, \quad \text{at } x = L, t > 0. \tag{4.69}$$

Here, N is the total number of the bamboo and polygrain segments in the line, and L is the line length. An analytical solution to the system of these PDEs can be in principle obtained in a way similar to that developed by Chen et al. [60]. Here we use the commercial finite element analysis (FEA) tool Comsol [93] to derive the stress evolution kinetics for two interesting cases.

The kinetics of the stress evolutions in the lines with the structures schematically shown in Figure 4.23 are shown in Figures 4.24 and 4.25. The results of the stress kinetics demonstrate that different kinetics can exist for void nucleation and evolution for polygrain clusters at different locations. Indeed, as shown in Figures 4.24a and 4.25a, the critical stress $\sigma_{crit} \approx 400-500$ MPa is developed at the cathode ends of the polygrain clusters. Differences in lengths and effective diffusivities, or microstructures, can result in slightly different nucleation times for these clusters. Following void growth, the line resistance will increase. Thus, in the 1D approximation, when a fast expansion of the nucleated void across the line cross-section is assumed, the initial line resistance change can be the result of a multivoid growth. Initially, all the voids are growing with almost the same pace and then varies in accordance with the local diffusivity variations. During void growth, the accumulation of compression at the anode half of line will start to suppress the growth of the local void clusters, and eventually will result in void disappearance. In Figure 4.24, we show a difference in the stress evolution taking place at the cathode edges of two polygrain clusters located near the line cathode and anode ends, respectively. Here, the examples are derived from the simulation of evolutions of stress distributions for the voidless cases. Since very large tensile and compression stresses cannot be developed, different stress relaxation processes will be initiated long before the steady state is achieved. However, void nucleation as a primary stress relaxation mechanism does change both the kinetics of the stress evolution and the final steady-state stress distribution, as shown in Sections 4.8 and 4.9. Here, the performed analysis aims to demonstrate that the effect of global stress evolution on the stress distribution kinetics in individual polygrain clusters has to be considered. It results in a correlated stress evolution in all clusters. It should also be mentioned that void-to-void interaction, which can delay the nucleation of an additional void by an existing void through a reduction of the tensile stress, has to be considered also [94].

To provide a complete picture of the microstructure effect on EM-induced line resistance evolution, we need to account for the drift of nucleated voids toward the cathode, and void coalescence there and at other locations, which can pin the drifted voids. Modeling of such degradation processes will be discussed based on the phase field method in 2D FEA analysis in Chapter 7.

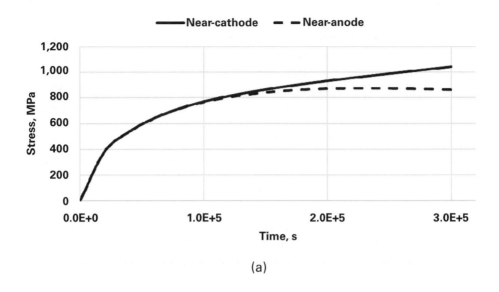

(a)

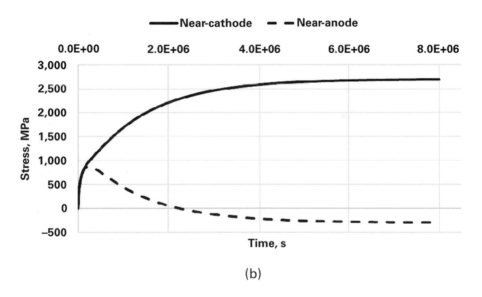

(b)

Figure 4.25 Kinetics of the EM-induced stress evolution occurring at the cathode edges of the polygrain clusters located at the near-cathode and near-anode regions of the line at (a) the initial time interval and (b) at the end of the steady state.

4.12   Summary

In this chapter, we summarize the major results that have been achieved with the 1D phenomenological model of EM. This model is capable of predicting the across-line transient and steady-state distributions of the vacancy concentration and the hydrostatic stress induced by EM. A void nucleation and its growth have also been described. A drift of small voids along a metal wire and the modification of their sizes were addressed for some specific cases. Results from the 1D model describe the concurrent growth of multiple voids generated in the multibranch interconnect segment and corresponding stress evolution. Despite the simplified 1D model, it is capable of addressing certain aspects of the confinement effect, including the effects of ILD/IMD dielectric and diffusion barriers on EM-induced degradation [12, 13], and to some extent, the effect of the grain structure [4, 8, 12, 112]. Nevertheless, not all questions can be adequately addressed. As we discussed, an initial void/hillock nucleation can be represented by different physical mechanisms, either a GB sliding or formation of wedge cracks, or an activation of the dislocation slip and climb at the GBs or interfaces, or formation of the jogs at the cut dislocations. To analyze the most probable source for damage evolution, all stress components should be included and properly analyzed. For example, the time evolution of the shear stress in a particular grain structure should be analyzed for possible grain slip. Most of the papers published so far have employed a statistical argument to account for the orientation of grain boundaries to model the stress as a scalar variable instead of a tensor. We should keep in mind, however, that since all grain boundaries have a 3D geometry, so some singularities in the stress field are ignored by the simplified 1D assumption. Gleixner and Nix [32] have attempted to fix these flaws by doing a semianalytic modeling of the atomic diffusion and stress evolution for complex grain structures. But the semi-empirical approach chosen for calculation of the stress field components has not been rigorously calibrated. Another problem, which is common for all 1D models, is to ignore the stress dependency of the atom/vacancy diffusivity. A proper picture of the void evolution, and thus an accurate analysis of the EM-induced change in the resistance of the interconnect segment taking into account the coalescence of the drifting voids, cannot be addressed by the 1D model.

FEA-based modeling has the capability to address these and many other questions regarding the material degradation caused by EM. A comprehensive review of such analyses will be done in Chapter 7.

References

1. I. A. Blech, Electromigration in thin aluminum films on titanium nitride, *Journal of Applied Physics* **47** (1976), 1203–1208.
2. I. A. Blech and C. Herring, Stress generation by electromigration, *Journal of Applied Physics Letters* **29** (1976), 131–133.
3. W. D. Nix and E. Arzt, On void nucleation and growth in metal interconnect lines and electromigration conditions, *Metallurgical Transactions* **23A** (1992), 2007–2013.
4. E. Arzt and W. D. Nix, A model for the effect of line width and mechanical strength on electromigration failure of interconnects with "near-bamboo" grain structure, *Journal of Materials Research* **6** (1991), 731–736.

5. R. Kirchheim, Stress and electromigration in Al-lines of integrated circuits, *Acta Metallurgica et Materialia* **40** (1992), 309–323.

6. R. Kirchheim and V. Kaeber, Atomistic and computer modeling of metallization failure of integrated circuits by electromigration, *Journal of Applied Physics* **70** (1991), 172–181.

7. C.-Y. Li, P. Borgesen and M. A. Korhonen, Electromigration-induced failure in passivated aluminum-based metallizations – the dependence on temperature and current density, *Applied Physics Letters* **61** (1992) 411–413.

8. M. A. Korhonen, P. Borgesen, K. N. Tu, and C.-Y. Li, Stress evolution due to electromigration in confined metal lines, *Journal of Applied Physics Letters* **73** (1993), 3790–3799.

9. C. Herring, Diffusional viscosity of a polycrystalline solid, *Journal of Applied Physics* **21** (1950), 437–445.

10. F. C. Larche and J. W. Cahn, Overview No. 41: the interactions of composition and stress in crystalline solids, *Acta Metallurgica et Materialia* **33** (1985), 331–357.

11. Y. Mishin and Chr. Herzig, Grain Boundary Diffusion in Metals, *Diffusion in Condenced Matter*, ed. J. Kärger, P. Heitjans, and R. Haberlandt (Braunschweig/Wiesbaden: Vieweg & Sohn Verlagsgesellschaft mbH, 1998), 33–366.

12. C. Herzig and S. V. Divinski, Grain boundary diffusion in metals: recent developments, *Materials Transactions* **44** (2003), 14–27.

13. J. C. Fisher, Calculation of diffusion penetration curves for surface and grain boundary diffusion, *Journal of Applied Physics* **22** (1951), 74–77.

14. M. Nomura and J. Adams, Self-diffusion along twist grain boundaries in Cu, *Journal of Materials Research* **7** (1992), 3202–3212.

15. M. Sorensen, Y. Mishin, and A. Voter, Diffusion mechanism in Cu grain boundaries, *Physical Review B* **62** (2000), 3658–3673.

16. H. Van Swygenhoven, D. Farkas, and A. Caro, Grain boundary structures in polycrystalline metals at the nanoscale, *Physical Review B* **62** (2000), 831–838.

17. K. M. Crosby, Grain boundary diffusion in copper under tensile stress. arXiv: cond-mat/0307065 (2003).

18. A. Suzuki and Y. Mishin, Atomistic modeling of point defects and diffusion in copper grain boundaries, *Interface Science* **11** (2003), 131–148.

19. P. S. Ho and T. Kwok, Electromigration in metals, *Reports on Progress in Physics* **52** (1989), 301–348.

20. M. J. Aziz, Thermodynamics of diffusion under pressure and stress: relation to point defect mechanisms, *Applied Physics Letters* **70** (1997) 2810–2812.

21. P. Shewmon, *Diffusion in Solids*, 2nd ed. (Switzerland: Pergamon International Publishers, 2016).

22. R. Rosenberg, and N. Ohring, Void formation and growth during electromigration in thin films, *Journal of Applied Physics* **42** (1971) 5671–5679.

23. Z. Suo, Reliability of interconnect structures. *Volume 8: Interfacial and Nanoscale Failure. Comprehensive Structural Integrity*, ed. W. Gerberich and W. Yang (Amsterdam: Elsevier, 2003), 265–324.

24. S. P. Hau-Riegea and C. V. Thompson, The effects of the mechanical properties of the confinement material on electromigration in metallic interconnects, *Journal of Materials Research* **15** (2000), 1797–1802.

25. M. E. Sarychev, Y. V. Zhitnikov, L. Borucki, C. L. Liu, and T. M. Makhviladze, General model for mechanical stress evolution during electromigration, *Journal of Applied Physics* **86** (1999), 3068–3075.

26. V. Sukharev, E. Zschech, and W. D. Nix, A model for electromigration- induced degradation mechanisms in dual-inlaid copper interconnects: effect of microstructure, *Journal of Applied Physics* 102 (2007), 053505 1–14.

27. J. J. Clement, Reliability analysis for encapsulated interconnect lines under dc and pulsed dc current using a continuum electromigration transport model, *Journal of Applied Physics* **82** (1997), 5991–6000.

28. M. Shatzkes and J. R. Lloyd, A model for conductor failure considering diffusion concurrently with electromigration resulting in a current exponent of 2, *Journal of Applied Physics* **59** (1986), 3890–3893.

29. J. J. Clement, Electromigration modeling for integrated circuit interconnect reliability analysis, *IEEE Transactions on Device and Materials Reliability* **1** (2001), 33–42.

30. B. M. Clemens, W. D. Nix, and R. J. Gleixner, Void nucleation on a contaminated patch, *Journal of Materials Research* **12** (1997), 2038–2042.

31. R. J. Gleixner, B. M. Clemens, and W. D. Nix, Void nucleation in passivated interconnect lines: effects of site geometries, interfaces, and interface flaws, *Journal of Materials Research* **12** (1997), 2081–2090.

32. R. J. Gleixner and W. D. Nix, A physically based model for electromigration and stress-induced void formation in microelectronic interconnects, *Journal of Applied Physics* **86** (1999), 1932–1944.

33. J. He, Z. Suo, T. N. Marieb and J. A. Maiz, Electromigration lifetime and critical void volume, *Applied Physics Letters* **85** (2004), 4639–4641.

34. S. J. Farlow, *Partial Differential Equations for Scientists and Engineers* (New York: Dover Publications, Inc., 1993).

35. X. Huang, T. Yu, V. Sukharev, and S. X.-D. Tan, Physics-based electromigration assessment for power grid networks, *Proceedings of the 51st Annual Design Automation Conference (DAC)* (San Francisco, CA: ACM, 2014), 1–6.

36. C. S. Hau-Riege, S. P. Hau-Riege, and A. P. Marathe, The effect of interlevel dielectric on the critical tensile stress to void nucleation for the reliability of Cu interconnects, *Journal of Applied Physics* **96** (2004), 5792–5796.

37. M. D. Thouless, Effects of the surface diffusion on the creep of thin films and sintered arrays of particles, *Acta Metallurgica et Materialia* **41** (1993), 1057–1064.

38. R. Huang, D. Gan, and P. S. Ho, Isothermal stress relaxation in electroplated Cu films. II. Kinetic modelling, *Journal of Applied Physics* **97** (2005), 103532 1–9.

39. P. A. Flinn, D. S. Gardner, and W. D. Nix, Measurements and interpretation of stress in aluminum-based metallization as a function of thermal history, *IEEE Transactions on Electron Devices*, **34** (1987), 689–699.

40. F. F. Abraham, *Homogeneous Nucleation Theory* (New York: Academic Press, 1974).

41. F. G. Yost, Voiding due to thermal stress in narrow conductor lines, *Scripta Metallurgica* **23** (1989), 1323–1328.

42. V. Sukharev, Beyond Black's equation: full-chip EM/SM assessment in 3D IC stack, *Microelectronic Engineering* **120** (2014), 99–105.

43. J. R. Black, Electromigration-a brief survey and some recent results, *IEEE Transactions on Electron Devices*, **16** (1969), 338–347.

44. M. Ohring, *Reliability and Failure of Electronic Materials and Devices* (San Diego: Academic Press, 1998).

45. J. R. Lloyd, New Models for interconnect failure in advanced IC technology, *Proceedings of the 14th International Symposium on the Physical and Failure Analysis of Integrated Circuits (IPFA)* (Piscataway, IEEE, 2008), 297–302.

46. M. Hauschildt, M. Gall, C. Hennesthal, et al. Electromogration void nucleation and growth analysis using large-scale early failure statistics, *Proceedings of the 13th International Workshop on Stress-Induced Phenomena and Reliability in 3D Microelectronics*, ed. P. S. Ho, C. K. Hu, M. Nakamoto, et al. (Kyoto: AIP Conference Proceedings 1601, 2014), 89–98.

47. M. Hauschildt, C. Hennesthal, G. Talut, et al. Electromigration early failure void nucleation and growth phenomena in Cu and Cu(Mn) interconnects, *Proceedings of the 2013 IEEE International Reliability Physics Symposium (IRPS)* (Anaheim: IEEE, 2013), 2C.1.1–6.

48. I. A. Blech and E. Kinsbron, Electromigration in thin gold films on molybdenum surfaces, *Thin Solid Films* **25** (1975), 327–334.

49. A. S. Oates, Strategies to ensure electromigration reliability of Cu/Low-*k* interconnects at 10 nm, *ECS Journal of Solid State Science and Technololgy* **4** (2015), N3168–N3176.

50. V. Sukharev and E. Zschech, A model for electromigration-induced degradation mechanism in dual-inlaid copper interconnects: effect of interface bonding strength, *Journal of Applied Physics* **96** (2004), 6337–6343.

51. Z. S. Choi, J. Lee, M. K. Lim, C. L. Gan, and C. V. Thompson, Void dynamics in copper-based interconnects, *Journal of Applied Physics* **110** (2011) 033505 1–9.

52. B. D. Knowlton, J. J. Clement, and C. V. Thompson, Simulation of the effects of grain structure and grain growth on electromigration and the reliability of interconnects, *Journal of Applied Physics*, **81** (1997), 6073–6080.

53. J. R. Lloyd, Nucleation and growth in electromigration failure, *Microelectronics Reliability* **47** (2007), 1468–1472.

54. C.-K. Hu and R. Rosenberg, Capping layer effects on electromigration in narrow Cu lines, *Proceedings of the 7th International Workshop on Stress-Induced Phenomena in Metallization*, ed. P. S. Ho, S. P. Baker, and C. Volkert (Austin: AIP Conference Proceedings 741, 2004), 97–111.

55. S. P. Hau-Riege and C. V. Thompson, Experimental characterization and modeling of the reliability of interconnect trees, *Journal of Applied Physics* **89** (2001), 601–609.

56. S. Chatterjee, V. Sukharev, and F. N. Najm, Fast physics-based electromigration checking for on-die power grids, *Proceedings of the 35th International Conference on Computer-Aided Design (ICCAD)* (Austin: IEEE/ACM, 2016), 110:1–8.

57. J. R. Black, Mass transport of aluminum by momentum exchange with conducting electrons, *Proceedings of the 6th IEEE International Reliability Physics Symposium (IRPS)* (Los Angeles: IEEE, 1967), 148–159.

58. J. He and Z. Suo, Statistics of electromigration lifetime analyzed using a deterministic transient model, *Proceedings of the 7th International Workshop on Stress-Induced Phenomena in Metallization*, ed. P. S. Ho, S. P. Baker, and C. Volkert (Austin: AIP Conference Proceedings 741, 2004), 15–26.

59. V. Sukharev, A. Kteyan, and X. Huang, Postvoiding stress evolution in confined metal lines, *IEEE Transactions on Device and Material Reliability* **16** (2016), 50–60.

60. H.-B. Chen, S. X.-D. Tan, X. Huang, T. Kim, and V. Sukharev, Analytical modeling and characterization of electromigration effects for multibranch interconnect trees, *IEEE*

Transactions on Computer-Aided Design of Integrated Circuits and Systems **35** (2016), 1811–1824.

61. X. Huang, A. Kteyan, S. X.-D. Tan, and V. Sukharev, Physics-based electromigration models and full-chip assessment for power grid networks, *IEEE Transactions on Computer-Aided Design of Integrated Circuits and Systems* **35** (2016), 1848–1861.

62. S. Chatterjee, V. Sukharev, and F. N. Najm, Power grid electromigration checking using physics-based models, *IEEE Transactions on Computer-Aided Design of Integrated Circuits and Systems* **37** (2018), 1317–1330.

63. 0. Kraft, S. Bader, J. E. Sanchez, Jr., and E. Arzt, Observation and modeling of electromigration-induced void growth in Al-based interconnects, *Materials Reliability in Microelectronics III, vol. 309, Materials Research Society Symposium Proceedings 309*, ed. W. F. Filter, H. J. Frost, P. S. Ho and K. P. Rodbell (San Francisco: Materials Research Society, 1993), 199–204.

64. E. Zschech, H.-J. Engelmann, M. A. Meyer, et al., Effect of interface strength on electromigration-induced inlaid copper interconnect degradation: Experiment and simulation, *Z. Metallkunde* **96** (2005), 966–971.

65. I. Vavra and P. Lobotka, TEM in-situ observation of electromigration in A1 stripes with quasi-bamboo structure, *Physica Status Solidi A* **65** (1981), K107–K108.

66. E. Zschech, M. A. Meyer, and E. Langer, Effect of mass transport along interfaces and grain boundaries on copper interconnect degradation, *Materials, Technology and Reliability Advanced Interconnects and Low-k Dielectrics, vol. 812, Materials Research Society Symposium. Proceedings*, ed. R. J. Carter, Proc. C. S. Hau-Riege, G. M., Kloster, T.-M. Lu, and S. E. Schulz (Warrendale: Materials Research Society, 2004), 361–372.

67. P. S. Ho, Motion of inclusion induced by a direct current and a temperature gradient, *Journal of Applied Physics* **41** (1970), 64–68.

68. P. R. Besser, M. C. Madden, and P. A. Flinn, In situ scanning electron microscopy observation of the dynamic behavior of electromigration voids in passivated aluminum lines, *Journal of Applied Physics*, **72** (1992), 3792–3797.

69. E. Arzt, O. Kraft, W. D. Nix, and J. E. Sanchez, Jr., Electromigration failure by shape change of voids in bamboo lines, *Journal of Applied Physics* **76** (1994), 1563–1571.

70. T. Marieb, P. Flinn, J. C. Brawman, D. Gardner, and M. Madden, Observations of electromigration induced void nucleation and growth in polycrystalline and near-bamboo passivated Al lines, *Journal of Applied Physics* **78** (1995) 1026–1032.

71. Z. Suo, Motion of microscopic surfaces in materials, *Advances in Applied Mechanics* **33** (1997), 193–294.

72. Z. Suo and W. Wang, Diffusive void bifurcation in stressed solid, *Journal of Applied Physics* **76** (1994), 3410–3421.

73. M. W. Lane, E. G. Liniger, and J. R. Lloyd, Relationship between interfacial adhesion and electromigration in Cu metallization, *Journal of Applied Physics* **93** (2003), 1417–1423.

74. E. Zschech and V. Sukharev, Microstructure effect on EM-induced copper interconnect degradation: experiment and simulation, *Microelectron. Engineering* **82** (2005), 629–638.

75. E. Zschech, M. A. Meyer, S. G. Mhaisalkar, et al. Effect of interface modification on EM-induced degradation mechanisms in copper interconnects, *Thin Solid Films* **504** (2006), 279–283.

76. E. T. Ogawa, K. D. Lee, V. A. Blaschke, and P. S. Ho, Electromigration reliability issues in dual-damascene Cu interconnects, *IEEE Transactions on Device and Materials Reliability* **51** (2002), 403–419.

77. L. Cao, K. J. Ganesh, L. Zhang, et al., Grain structure analysis and effect on electromigration reliability in nanoscale Cu interconnects, *Applied Physics Letters* 102 (2013), 131907, 1–4.

78. N. I. Peterson, Self diffusion in pure metals, *Journal of Nuclear Materials* **69, 70** (1978), 3–37.

79. *Handbook of Grain and Interphase Boundary Diffusion Data*, ed. I. Kaur and W. Gust (Stuttgart: Ziegler Press, 1989).

80. D. Gupta, C. K. Hu, and K. L. Lee, Grain boundary diffusion and electromigration in Cu-Sn alloy thin films and their VLSI interconnects, *Defect and Diffusion Forum* **143** (1997), 1397–1406.

81. E. Zschech, H. Geisler, I. Zienert, et al., Reliability of copper inlaid structures – geometry and microstructure effects, *Proceedings of the Advanced Metallization Conference (AMC)*, ed. B. M. Melnick, T. S. Cale. S. Zaima and T. Ohta (San Diego: Materials Research Society, 2002), 305–312.

82. M. A. Meyer, M. Grafe, H.-J. Engelmann, E. Langer, and E. Zschech, Investigation of void formation and evolution during electromigration testing, *Proceedings of the 8th International Workshop on Stress-Induced Phenomena in Metallization*, ed. E. Zschech, K. Maex, P. S. Ho, H. Kawasaki, and T. Nakamura (Dresden: AIP Conference Proceedings 817, 2006), 175–184.

83. A. Kteyan, V. Sukharev, M. A. Meyer, E. Zschech, and W. D. Nix, Microstructure effect on EM-induced Degradations in dual-inlaid copper interconnects, *Proceedings of the 9th International Workshop on Stress-Induce Phenomena in Metallization*, ed. S. Ogawa, P. S. Ho, and E. Zschech (Kyoto: AIP Conference Proceedings 945, 2007), 42–55.

84. V. Sukharev, A. Kteyan, E. Zschech, and W. D. Nix, Microstructure effect on EM-induced degradation in dual inlaid copper interconnects, *IEEE Transactions on Device and Materials Reliability* **9** (2009) 87–97.

85. M. A. Korhonen, P. Borgesen, D. D. Brown, and C.-Y. Li, Microstructure based statistical model of electromigration damage in confined line metallizations in the presence of thermally induced stresses, *Journal of Applied Physics* **74** (1993), 4995–5004.

86. J. R. Lloyd, C. E. Murray, T. M. Shaw, M. W. Lane, X.- H. Liu and E. G. Liniger, Theory for electromigration failure in Cu conductors, *Proceedings of the 8th International Workshop on Stress-Induced Phenomena in Metallization*, ed. E. Zschech, K. Maex, P. S. Ho, H. Kawasaki, and T. Nakamura (Dresden: AIP Conference Proceedings 817, 2006), 23–33.

87. T. Berger, L. Arnaud, R. Gonella, I. Touet, and G. Lormand, Electromigration character-ization of damascene copper interconnects using normally and highly accelerated tests, *Microelectronic Reliability* **40** (2000), 1311–1316.

88. L. Vanasupa, Y.-C. Joo, P.R. Besser, and S. Pramanick, Texture analysis of damascene-fabricated Cu lines by X-ray diffraction and electron backscatter diffraction and its impact on electromigration performance, *Journal of Applied Physics* **85** (1999), 2583–2590.

89. W. W. Mullins, Mass transport at interfaces in single component systems, *Metallurgical and Materials Transactions A* **26** (1995), 1917–1929.

90. S. Rzepka, E. Meusel, M. A. Korhonen, and C.-Y. Li, 3-D finite element simulator for migration effects due to various driving forces in interconnect lines, *Proceedings of the 5th International Workshop on Stress-Induced Phenomena in Metallization*, ed. O. Kraft, E. Arzt, C. Volkert, and P. S. Ho (Stuttgart: AIP Conference Proceedings 491, 1999), 150–162.

91. L. Zhang, J. P. Zhou, J. Im, et al. Effects of cap layer and grain structure on electromigration reliability of Cu/low-*k* interconnects for 45 nm technology node, *Proceedings of the 2010 IEEE International Reliability Physics Symposium (IRPS)* (Anaheim: IEEE, 2010), 581–585.

92. L. Zhang, M. Kraatz, O. Aubel, et al., Cap layer and grain size effects on electromigration reliability in Cu/low-*k* interconnects, *2010 IEEE International Interconnect Technology Conference Proceedings (IITC)* (Burlingame: IEEE, 2010), 1–3.
93. COMSOL, Inc., Burlington, MA. 8 New England Executive Park.
94. J.-H. Choy, V. Sukharev, S. Chatterjee, F. N. Najm, A. Kteyan, and S. Moreau, Finite-difference methodology for full-chip electromigration analysis applied to 3D IC test structure: simulation vs. experiment, *Proceedings of 2017 International Conference on Simulation of Semiconductor Processes and Devices (SISPAD)* (Kamakura: IEEE, 2017), 41–44.

5 Electromigration in Cu Interconnect Structures

5.1 Introduction

On-chip Cu wiring used in the back end of the line (BEOL) has been developed for several decades [1–3]. The number of integrated circuit (IC) chips manufactured with Cu BEOL has increased every year since first commercialized by IBM in 1997 [4]. The Cu interconnects improve electrical conductivity and reduce resistance and capacitance (RC) delay for interconnections as compared to the Al metallization. In addition, Cu interconnections have a longer electromigration (EM) lifetime with a higher EM activation energy as compared to the Al(Cu) metallization [5]. As described in Chapter 1, EM is defined as the movement of atoms under the influence of an electric potential gradient [6]. The EM-induced mass flow or the drift velocity is described by the Nernst–Einstein equation $(D_{eff}/k_BT)F_e$, where D_{eff} is the effective Cu diffusivity, k_B is the Boltzmann constant, T is the absolute temperature, and F_e is the EM driving force. D_{eff} varies from location to location in the metal line and depends on the local microstructure. The fast diffusion paths in the Cu interconnects were found to be at the grain boundaries and interfaces, which varied depending on the fabrication process and materials used. The imbalance of EM-induced atomic flux at some critical locations, particularly near the ends of the line/via interface with diffusion blocking boundaries, can induce the formation of voids or extrusions leading to interconnect failure.

Commercial on-chip Cu interconnections are fabricated by single- and/or dual-damascene processes, and reliable Cu interconnections require metal and insulator adhesion/diffusion barrier layers to form the multilevel lines and vias [4, 7]. A typical single-damascene level is fabricated by first depositing a planar dielectric stack, and then patterned and etched using lithographic and reactive ion etching (RIE) techniques and followed by a dry and/or wet clean to produce the desired wiring trench or via pattern. In dual-damascene processing, both vias and trenches are patterned in the dielectric layer before depositing the metals. To form the metal structures, physical vapor deposition (PVD), chemical vapor deposition (CVD) or atomic layer deposition (ALD) is used to form the metal liner, followed by a PVD or PVD Cu alloy seed layer, and then Cu overfills the remaining structures using an electrochemical plating deposition (ECD) technique. The excess metal in the field region, usually referred as the overburden layer, is removed using a chemical- or electro/chemical-mechanical polishing (CMP or e-CMP) process, leaving planarized wiring and vias embedded in the interlevel dielectric layer. Subsequent levels are fabricated by repeated applications of these processes. All wiring levels are planar at every level, which typically results in enhanced wafer yield over a nonplanar structure. For the 90 nm technology node and beyond, an interlevel dielectric SiO_2 ($\varepsilon = 4$) and a high dielectric constant SiN_x ($\varepsilon = 7$) Cu cap have been replaced by materials with a lower dielectric constant such as SiCOH with $\varepsilon = 2.7$ to 2.4 and an amorphous SiNO or a-$SiC_xN_yH_z$ cap layer with $\varepsilon = 4$–5, respectively. The material change is required to enable a lower effective dielectric constant for on-chip Cu interconnections. In these structures, the top surface of the Cu damascene line is covered with a thin dielectric diffusion barrier layer, e.g., SiN_x, or a-$SiC_xN_yH_z$, and the bottom surface and two sidewalls are covered with a metal liner/diffusion barrier layer, e.g., Ta-based [4, 8, 9]. The metal adhesion/diffusion barrier liner creates material dissimilarities at interlevel interfaces and is the cause for EM flux divergence if the Cu mass transport through the liner is negligible. For lines wider than 90 nm, the extension of the overburden grains into damascene trenches gives rise to a bamboo-like grain structure. The bamboo-like and polycrystalline structures are defined as single grain per linewidth or per via and more than two grains per linewidth, respectively. For Cu lines beyond the 65 nm node with linewidths reduced below 90 nm and a metal height/width aspect ratio (A.R.) > 1.2, significant changes in Cu microstructure were observed, where small random grains emerged to mix with the bamboo-like grain structure. In this case, multiple small grains were found to agglomerate in the lower Cu line from grain growth inside the line and invasion by grains originated from the overburden in the electroplating process. Only partial extension of the overburden grains will occur, where the grain structure in the line depends on the line height, width, and the thermal budget. Consequently, the Cu microstructure is controlled by the grain growth mechanism inside the line. In this case, sidewall and height pinning forces arising from preferential Cu grain orientations can neutralize grain boundary mobility and retard grain growth in the lines.

With continuous scaling, the electrical resistivity was found to increase (Chen and Gardner 1998) [10], as the Cu grain size approaches the mean free path length of the electrons, ℓ(39 nm at $T = 21°C$). This was attributed to the increase in electron scattering by grain boundaries, surface, and interfaces. In addition, the Cu damascene line required 2–3 nm thick metal liner to sustain EM reliability. Combine these two factors: increasing resistivity size effect and barrier/liner to Cu volume ratio, and increasing the interconnect R rapidly to significantly degrade its performance. Therefore, interconnect scaling presents great challenges not only for fabricating nanometer interconnects with the increasing aspect ratios without defects, it also significantly degrades the resistivity, reliability, and performance of Cu interconnects. Here the Cu microstructure and interface characteristics play an important role in controlling the resistance and EM reliability and must be carefully controlled. Together, these topics are reviewed in two chapters: first with Chapter 5 focusing on the EM reliability of Cu interconnects, followed by Chapter 6 to review the scaling effect on microstructure and the implication on interconnect resistivity. In Chapter 7, we will analyze the kinetics and microstructure effects on stress evolution and void formation driven by EM aiming to understand how scaling can continuously affect EM reliability.

This chapter reviews first some key findings on Cu microstructure and resistivity, and then focused on the scaling effect on EM reliability. It is organized into eight parts. First, we review in Sections 5.2 and 5.3 the Cu microstructure and resistivity for various CMOS technological nodes. In Section 5.4, we present the basic physics of the EM phenomenon, addressing EM mass transport, lifetime scaling rule, and damage formation in Cu damascene structures. This is followed with discussions on Blech short length in Section 5.5 and the EM scaling rule in Section 5.6. In Section 5.7, several approaches developed for improving EM lifetime using upper-level dummy vias, Cu surface treatments, alloys, and alternated liners/surface metal coating are discussed together with the effects of microstructure, ALD MnO$_x$ liner, and Cu/carbon nanotube composite line on EM. Finally, the EM lifetimes and activation energies through various technology nodes are presented in Section 5.8, and the chapter concludes with a summary in Section 5.9.

5.2 Microstructure

For Cu lines exceeding 100 nm linewidth but with a low metal height/width A.R. < 1, near bamboo-like Cu grain line structure was observed. Here smaller grains in the Cu damascene trenches were often found to dissolve in the large Cu grains in the electroplated (ECD) overburden to yield a bamboo-like grain structure. For the 65 nm node, as the linewidth reduced below 90 nm and A.R. > 1.2, significant changes in Cu microstructure were observed, where small random grains emerged to mix with bamboo-like grains [11–13]. Overall, the overburden effect was reduced, resulting in the grain structure dependent mainly on the linewidth and height A.R. and the thermal budget. Here the Cu microstructure was controlled more by the grain growth mechanisms in the line, where the sidewall and thickness pinning forces arising from preferential Cu grain orientations may neutralize grain boundary mobility to prevent grain growth in the line. Figure 5.1a–c are typical transmission electron microscope (TEM) images along 80 nm, 50 nm, and 28 nm wide lines, respectively. Figure 5.2 is the Cu grain orientation map along a 24 nm wide line with A.R. = 2.5 obtained using TEM precession-assisted diffraction to obtain grain structure information over a scanned area showing a high fraction of twin boundaries [15] and similar Cu grain maps [13]. Since the twin boundary is not a fast diffusion path, these images revealed that the grain structure in the Cu lines with width reducing to 24 nm is a mixture of bamboo and polycrystalline. In some sections, multiple small grains were observed through the thickness, but in some other sections, single Cu grains were observed extending through the entire line thickness, and occasionally more than two grains across the metal thickness were observed. The bamboo grains play a critical role in retarding Cu mass flow since the diffusivities at Cu–liner and Cu surface–metal cap interfaces are often many orders of magnitudes slower than the gain boundary diffusivity.

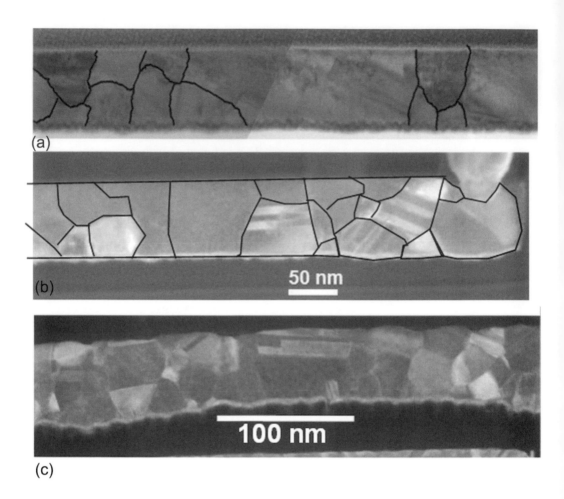

Figure 5.1 Cross-section TEM images along (a) Cu 80 nm, (b) 28 nm, and (c) 24 nm wide lines. Black lines on the images represent grain boundary locations. © 2013 MRS. Reprinted, with permission, from [14].

Figure 5.2 Cu orientation map along a 24 nm wide line using a technique that uses a focused TEM probe to obtain diffraction pattern information over a scanned area. © 2017 IEEE. Reprinted, with permission, from [15].

5.3   Resistivity

As scaling continues, the electrical resistivity, ρ, increases significantly when the Cu linewidth reduced below 0.1 μm. In addition, metal liners of 2–3 nm thick are required to sustain the EM reliability, which greatly reduced the volume of Cu occupied in the metal line. The increase in ρ with decreasing linewidth has become a major limiting factor for the speed of IC chips. The problem can be attributed to several factors, including increasing surface scattering from increasing surface area to line area ratio, increasing grain boundary scattering due to more small Cu grains being formed, and increasing impurity scattering from electroplating and/or Cu alloying. For EM tests, it is also important to establish the relationship between line area A and R/L, where R and L are the sample line resistance R and line length L, respectively, so a proper electric current density can be used.

The scaling effect on resistivity was extensively studied on Cu damascene lines processed in a 300 mm semiconductor processing line through several technological nodes. The resistivity ρ was measured using Cu line resistance structures of 50–1,000 μm long and 15 nm to 5 μm wide. The Cu lines were fabricated in low dielectric constant SiCOH or SiO_2 dielectric and typically capped with an amorphous $SiC_xN_yH_z$ or SiN_x barrier film. After processing, the samples were annealed at 400°C for 1 to 2 hours to simulate standard BEOL thermal processing and stabilize the Cu microstructure and impurity content. All the samples had either a PVD pure Cu or electroplated Cu with a Cu (0.5–2% Mn) seed.

The values of Cu resistivity were determined from TEM and temperature coefficient of resistance (TCR) [16] techniques. For the TEM technique, the Cu ρ was determined from the damascene line by using the equation, $R = \rho\,(L/A)$, where R was the measured Cu line resistance, A was the Cu cross-sectional area, and L was the line length, where A was determined using digital, TEM cross-section line images from samples prepared by an ex situ focused ion beam (FIB) lift-out. The Cu grain structure of some lines was determined from TEM samples prepared using FIB section along the line. Typically, the ρ test line was nested in a series of equal dimension lines and A was measured from TEM images taken at two separate magnifications of the test line, or from multiple TEM samples of the same test line. Due to the low Cu ρ, the resistance of the high-resistance TaN, Ta, or Co barrier/liner did not affect the line ρ measurements. Overall, the uncertainty of ρ measurement was found to be about 10%, coming mainly from the variation of line areas due to line edge roughness and RIE depth variation.

To apply the TCR method, the Cu ρ was measured as a function of temperature. Here only the electron–phonon scattering was temperature dependent, while the size effect from electrons-interfaces and grain boundaries scattering and electrons-impurity scattering was all independent of temperature. Hence the following equations were used to calculate the Cu interconnect resistivity:

$$\frac{dR}{dT} = \left(\frac{d\rho_{Bulk}}{dT}\right)\left(\frac{L}{A}\right),$$

(5.1a)

that is,

$$\rho = R\left(\frac{d\rho_{Bulk}}{dT}\right) \Big/ \left(\frac{dR}{dT}\right). \tag{5.1b}$$

This technique was often preferred by researchers since it required no TEM measurements. The bulk TCR coefficient was taken from literature [17].

The Cu resistivity on wide lines for w *and* $h > \ell$ was derived from the Cu bulk resistivity and the size effect on resistivity from electron scattering with interfaces was derived using the Fuchs–Sondheimer model [18, 19] and that with grain boundaries derived using the Mayadas–Shatzkes model [20], as follows:

$$\rho = \left(\frac{\rho_{Bulk}}{1 - (1 - P)\left[\frac{3\ell}{8}\left(\frac{1}{w} + \frac{1}{h}\right)\right]} - \rho_{Bulk}\right) + \left(\frac{\rho_{Bulk}}{1 - \frac{3}{2}\alpha + 3\alpha^2 - 3\alpha^3 \ln\left(1 + \frac{1}{\alpha}\right)}\right) + \rho_i,$$

$$\tag{5.2a}$$

where P is an electron specular reflection parameter at interface; ρ_{Bulk} is the resistivity of pure bulk Cu; ℓ is the mean free path in the bulk; $\rho_{Bulk} = 1.7\ \mu\Omega$ cm and $\ell = 39$ nm at 21°C, $\alpha = (\ell/d)\ R/(1 - R)$, where R is the electron reflection coefficient at grain boundaries, d is the diameter of Cu grain size, w is the Cu linewidth, and h is the line height; and ρ_i is the resistivity caused by electron-impurity scattering. The approximate coefficient of 3/8 in (5.2a) had to change, taking at w/ℓ and $h/\ell > 4$ based on the MacDonald and Sarginson model [21]. For data in ranges from $0.5 < w/\ell$ and $h/\ell < 3$, the factor of 3/8 in (5.2a) was replaced by a new coefficient of 0.474 [22]. Without knowing the physical constraint of the fitting parameters in the preceding equation, one can obtain a similar fitting curve with various values of fitting parameters with very different physical meanings [23–25]. One of methods to resolve this difficulty was to design an experiment to isolate each contribution in resistivity. For example, the interface scattering contribution was isolated by an experiment designed to vary the grain size distribution while maintaining a fixed area and aspect ratio of the Cu lines [22]. In this way, (5.2a) can be simplified into the following simple empirical equation [26]:

$$\rho = \rho_o + g\left(\frac{\rho_{Bulk}\ell}{A^{0.5}}\right) \tag{5.2b}$$

or replace $\rho_{Bulk} = (m_e v_F / N_{eff} e^2)/\ell$ in (5.2b), then

$$\rho = \rho_o + \frac{g m_e v_F / N_{eff} e^2}{A^{0.5}} = \rho_o + \frac{B}{A^{0.5}}, \tag{5.2c}$$

where m_e is the electron mass; v_F is the electron Fermi velocity; N_{eff} is the effective electron charge density; ρ_o is the sum of resistivity from bulk, impurity, and defect; and B and g are the size enhanced factor and also a function of w, h, P, R, d, etc.

Figure 5.3 shows a plot of Cu and Cu(Mn) resistivity with TaN/Co or TaN/Ta liner as a function of Cu line cross-sectional areas A at $T = 21$°C. For A below 4×10^4 nm^2, the ρ increased significantly as A decreased. Even though the data from TEM and TCR

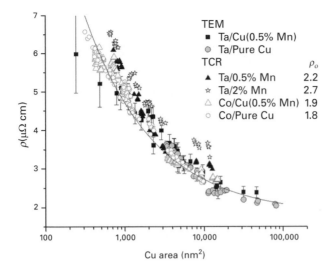

Figure 5.3 Cu line ρ vs. A_{Cu} for various damascene line dimensions. The solid line is the best fit to pure Cu data with Co liner using (5.2c).

techniques were obtained from various wafers and technological nodes, a smooth function of increasing ρ with decreasing A was found as shown. As shown, the effect of liners or Cu surface cap, e.g., Ta or Co liner and $SiC_xN_yH_z$ or Cu cap, on the Cu/ interface, resistivity was found not significant for samples fabricated using various liners and caps. Comparing the Cu ρ data derived from TEM to the TCR method, the data from both techniques are in reasonably good agreement. The ρ data are fitted to the empirical function in (5.2c) with $B/A^{0.5}$ to replace B/w [27]. A value of $\rho_o = 1.8$ $\mu\Omega$ cm was used for electroplated Cu as the sum of the bulk resistivity at 21°C (1.7 $\mu\Omega$ cm) and electroplated impurity contributions 0.1 $\mu\Omega$ cm. This value ρ_o of 1.8 $\mu\Omega$ cm is in good agreement with the measured value of 1.827 ± 0.004 $\mu\Omega$ cm obtained for 5 μm wide and 2.7 μm thick electroplated pure Cu damascene lines. The value of B was found to be 91.8 (10 Ω nm^2) for pure Cu with a Co or Ta liner. The Cu(Mn) alloy resistivity was obtained by adding the resistivity from electron-substitutional impurity Mn scattering ρ_i to the pure Cu resistivity using Matthiessen's rule. The Cu alloy data were fitted to the empirical function (5.2c) with a fixed vale of B 91.8 (10 Ω nm^2). The extracted values of ρ_o for Cu(0.5% Mn) seed with Co and Ta liner, and Cu(2% Mn) with Ta liner were 1.9 and 2.2, and 2.7 $\mu\Omega$ cm, respectively. Overall, the Co liner had a strong effect to reduce Cu(Mn) alloy resistivity, so it seems that the Co liner is a sink for Mn since the solid solubility limit of Mn in Co is high. The data along with the empirical best fit curves indicated that the Cu and Cu(Mn) ρ can be well described by an $(A_{Cu})^{-0.5}$ relationship. Such curves seem to fit well down to 5 nm technology node Cu lines. Thus, a high line height and width aspect ratio would be necessary for maintaining a low RC delay.

Figure 5.4 shows a plot of Cu, Co, and Ru R/L as a function of line cross-sectional areas, A, at $T = 21$°C. All data points were extracted from the TCR technique. In the

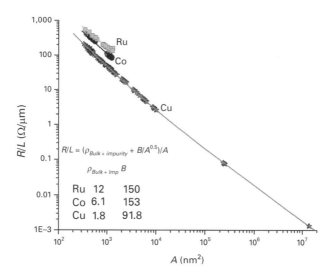

Figure 5.4 *R/L* vs. *A* for various damascene line dimensions and metals. The solid lines are the best fits to the experimental data. © 2017 IEEE. Reprinted, with permission, from [15].

extraction, the values of $d\rho_{Bulk}/dT$ used for Cu, Co, and Ru were 0.00685 [17], 0.031 [28], and 0.0314 μΩ cm/K [29, 30] based on their respective metal bulk resistivity, and the results were attributed to the electron–phonon scattering. The results on $R/L = \rho/A$ can be expressed as $(\rho_o + B/A^{0.5})/A$, where the curve fitting parameters ρ_o and B were deduced by Hu et al. [31]. Overall, the Cu resistivity data can be fitted to the empirical equation remarkably well for Cu area, reducing from 300 nm^2 to 1.4×10^7 nm^2. The coefficient of the size effect B on ρ was found to be 152 and 150 (10 Ω nm^2) for Co, and Ru, respectively. A similar size effect B was found for the three metals in this study, although the value of B for Cu, 92 (10 Ω nm^2) was the smallest. Since $\rho_{Bulk}\ell$ of Cu was well defined, the size effect enhancement factor g was found to be around 14. If $g = 14$ is also used to Co and Ru, then $\rho_{Bulk}\ell$ or $m_e v_f/(N_{eff}e^2)$ for Co and Ru were found to be 11 and 11 (10 Ω nm^2), respectively, which are in reasonably good agreement with the theoretical predictions of 9.0 and 7.1 (10 Ω nm^2) for Co and Ru, respectively [32, 33].

In Figure 5.5, the effective Cu ρ is compared for various liner to Cu volume ratios in Cu damascene lines and with Co line ρ without a liner. It is worth noting that the difference in the effective ρ between Co and Cu becomes small when no liner is used in Co lines while the Cu line has a liner occupying 40% of the trench. In addition to the size effect, the volume ratio of the liner plays a key factor to influence the line R. Only by using thinner or no liner, the resistivity of the Ru and Co wires can meet the Cu effective resistivity with a liner. For the case of 400 nm^2 of trench cross-section area, any metals with resistivity <13 μΩ cm and without liner will have R/L less than that of Cu with a 40% liner. As can be seen in Figure 5.5, the extendibility of Cu interconnection beyond 5 nm node will depend on how to scale down the liner to a Cu volume ratio below 40%.

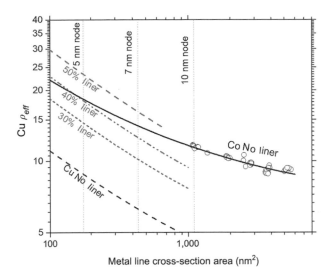

Figure 5.5 Comparison of effective Cu resistivity with various liner to Cu volume ratios in a damascene line with Co resistivity. © 2017 IEEE. Reprinted, with permission, from [26].

5.4 Mass Transport and Damage Formation in Cu Damascene Structures

5.4.1 EM Fundamentals: EM-Induced Mass Flow

EM describes atom diffusion under the influence of an electric potential gradient [6] as discussed in Chapter 1. In this section, we discuss EM-induced mass flow or the drift velocity in a Cu damascene line that can be directly measured with a diffusion blocking boundary at the cathode end of the line. The Cu damascene line has a top surface covered with a thin dielectric diffusion barrier, and the bottom Cu surface and two sidewalls are covered with TaN/Ta diffusion barriers, or Co or Ru liners. In this configuration, the multilevel on-chip interconnection is ideal for measuring the EM-induced mass flow from the void growth rate near the cathode end of a long line with a blocking boundary. The EM mass flow under a driving force F_e is $J_e = n\, v_d$, where n and v_d are the atomic density and drift velocity, respectively. The EM-induced drift velocity is expressed as $v_d = (D_{eff}/k_B\, T)F_e$, where $F_e = Z^*_{eff}\, e\rho\, j$. e is the absolute value of the electronic charge, Z^*_{eff} is the apparent effective charge number, ρ is the metallic resistivity, D_{eff} is the effective diffusivity of atoms diffusing through a metal line, T is the absolute temperature, and k_B is the Boltzmann constant. The sign of Z^*_{eff} for Cu is negative since Cu mass flow was found to follow the electron flow. Diffusivity is the dominant factor for the mass transport. The fast diffusion paths in Cu line would be Cu grain boundaries and the top Cu–dielectric cap interface, since the Cu diffusion along a good metal liner–Cu interface can be ignored for a nonmetal capped line. For a Cu polycrystalline or bamboo-like long line with a blocking boundary at the cathode end, the Cu EM drift velocity can be written as follows:

$$v_d = \left[Z_I^*(D_I/k_BT)\delta_I(1/h)\right]e\rho j \tag{5.3a}$$

or

$$v_d = \left[Z_{GB}^*(D_{GB}/k_BT)(\delta_{GB}/d)\right]e\rho j, \tag{5.3b}$$

where h and d are the line height and grain size, respectively; the subscripts I and GB refer to the Cu–dielectric cap interface and grain boundary (GB), respectively; δ and D denote the effective width and diffusivity, respectively; and f is a geometric factor.

For test structures consisting of bamboo–polycrystalline sections in the Cu line with a metal cap, the drift velocity becomes

$$v_d = \left[Z_{eff}^*(D_{eff}/k_BT)\right]e\rho j. \tag{5.3c}$$

In this case, v_d becomes rather complicated since the mass flows along interfaces and grain boundaries are significantly different but strongly coupled, where the mass moves in parallel or in series depending on the local microstructure. For example, the mass flow in a polycrystalline line between two bamboo sections will change under EM since the two bamboo grain sections act as mass-flow blocking boundaries due to the Blech short-length effect [34]. The fast mass flow in the polycrystalline section will decrease with time due to the buildup of the compressive stress at the cathode end, while the mass flow in the bamboo-like section continues to increase due to the stress gradient buildup in the surrounding polycrystalline sections [35]. The net drift velocity is then dependent on the lengths of the polycrystalline and bamboo sections, and the mass flow at the bamboo–polycrystalline boundaries. If the lengths of bamboo and polycrystalline sections are L_B and L_P respectively, the steady-state EM-induced drift velocity for Cu encased with a metal cap, Co or CoWP, becomes

$$v_d \cong n_I e\rho j Z^* \frac{D_I}{k_BT}\left(1+\frac{L_P}{L_B}\right), \tag{5.4}$$

where n_I is the number of atoms at the Cu–liner interfaces. In nanoscale lines, the atoms at the interface n_I with the interface diffusivity D_I are only a few percent.

5.4.2   EM Lifetime Void Growth Related to R Change

The EM-induced void growth in a long Cu line with a blocking boundary at the cathode end can be measured from the resistance R change as illustrated in Figure 5.6 for a test line with the cross-sectional TEM image. Here the void lengths ΔL_o and ΔL_d refer to the void grown over the W via and beyond the Cu–W overlapped region, respectively. The resistance change is shown in Figure 5.7, indicating that the resistance changed slowly first, followed by an abrupt step of resistance change and a period of rapid resistance increase. The void growth at/near the cathode end of the long line as shown in Figure 5.6 is directly correlated to EM-induced mass-flow rate v_d and the stress time. The total void length ΔL (void volume/Cu line cross-section area) at the cathode end of the Cu line is $\Delta L_o + \Delta L_d$ and can be expressed as follows:

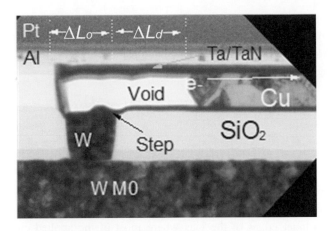

Figure 5.6 Cross-sectional TEM image of EM damaged sample for a Ta/TaN coated line shown in Figure 5.3. Pt and Al layers were coated on samples before FIB sectioning. © 2003 AIP, Reprinted, with permission, from [40].

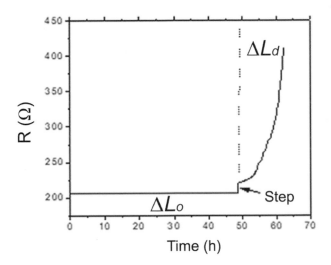

Figure 5.7 A typical line resistance curve vs. time for a sample with a Ta/TaN cap at 377°C with 35 mA/μm^2 for 62 h. ΔL_o and ΔL_d are the void sizes labeled in Figure 5.6. © 2017 IEEE. Reprinted, with permission, from [41].

$$\Delta L_{cr} = \int_0^{\tau} v_d dt. \qquad (5.5a)$$

For a constant v_d, the Cu lifetime τ for a long Cu line is

$$\tau = \Delta L_{cr}/v_d, \qquad (5.5b)$$

where ΔL_{cr} is the critical void length causing line failure and v_d is the drift velocity. Here the EM-induced mechanical backflow is ignored since for a long line, the products of the line length and current densities used in these studies were far greater than the threshold value of the Blech short-length product as reported in the range of 3,000–8,000 A/cm [36–39]. The rates of material depletion are related to the line resistance change ($\Delta R = R - R_o$) by the following relation:

$$\Delta R(t) = (\rho_{Liner}\Delta L)/A_{Liner} + \rho_{Cu}(L - \Delta L)/A_{Cu}, \qquad (5.6a)$$

and

$$\Delta R(t)/\Delta t = (\rho_{Liner}/A_{Liner} - \rho_{Cu}/A_{Cu})(\Delta L/\Delta t) \sim (\rho_{Liner}/A_{Liner})v_d. \qquad (5.6b)$$

Here A is the cross-sectional area of the test line, and L is the initial line length. The voltage change in the test structure was only sensitive to the line resistance change between the two inner edges of the vias where most of the applied current passed through the liner. Therefore, $\Delta R(t) \sim 0$, if $\Delta L < (\Delta L_o)$, where ΔL_o, as shown in Figure 5.6, is the distance that the via and the line overlap, including the line extension. Once the void grew beyond the Cu–W via overlapped region ($>\Delta L_o$), the exposed liner layer must sustain all the applied current to induce a high rate of resistance change. Here the period of a small R change is referred as the R incubation time, τ_I, where a nonlinear rapid R change versus the time indicated that the joule heating in the liner caused a voltage increase of v_d once the void grew beyond ΔL_o, and then a final liner open or melted was often observed.

For a physical stably exposed liner, we referred to it as a liner redundancy. Equation (5.6) indicates that the rate of change in line resistance is simply a linear function of the EM-induced mass flow, and that the liner resistivity inversely depends on the liner cross-sectional area. Equation (5.6) also shows that for the same mass-flow rate, a larger liner area and/or lower liner resistivity would result in a slower rate of resistance change and a better lifetime for a higher percent failure criterion. In Section 5.4.4, we will further discuss EM in Cu interconnects with various Ta liner thicknesses, and the relationship between void growth rate and EM lifetime to verify (5.6).

In Figure 5.8, we show line resistance change curves $\Delta R = R - R_o$ as a function of time obtained from testing 140 nm wide Cu M1 lines with TaN/Ta liner on W vias at 300°C with current density $j = 24$ mA/μm² in wafers C1 and G0. The top surfaces of the Cu M1 lines were coated with a 35 nm thick a-SiC$_x$H$_y$N$_z$ film deposited by CVD. The Cu M1 test line was 400 μm long and 0.14 μm wide, and the W via 0.12 μm in diameter. The liner consisted of a TaN/Ta dual-layer structure, where the Ta layer was deposited by conventional physical vapor deposition and the TaN layer was deposited by PVD technique for wafer C1 and by ALD for wafer G0 as described in Table 5.1. The liner deposition process consisted of a liner deposition, a sputter etch, and a second liner deposition. A thin PVD Cu seed layer was used for Cu electroplating.

The microstructure of the Cu line was examined by both FIB microscopy and TEM. The Cu M1 lines were found to have a bamboo-like grain structure with single-crystal Cu grains connecting along the length of the line. The dimension of the M1 Cu and Ta-based liner layers were measured using TEM images, and their elemental

Table 5.1. List of wafers corresponding to various liner processing steps, bottom Ta thickness, total liner, and Cu areas, Ta phases, and volume fractions of Cu ⟨111⟩ orientation. A sputter etch step was applied between two liner steps.

Wafer	First Liner	Second Liner	Bottom Ta (nm)	Total liner area (nm²)	Total Cu area (nm[2])	Ta-phase (XRD)	Volume fraction ⟨111⟩
C1	PVD TaN	PVD Ta	19	9056	18,222	α	0.066
G0	ALD TaN	PVD Ta	19	6768	18,527	β	0.12

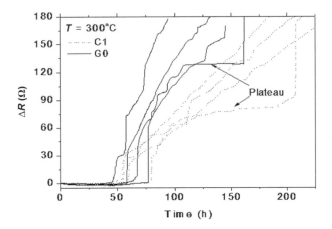

Figure 5.8 Plot of the 140 nm wide bamboo-like line resistance change as a function of time for wafers C1 and G0 with two different liners. © 2005 AIP, Reprinted, with permission, from [42].

compositions were analyzed by energy dispersive X-ray spectroscopy (EDS). The Ta phases and the Cu grain orientations were determined separately using a wafer-level in-line X-ray diffractometer for a test structure containing a large metal line maze with similar linewidth/space. The bottom liner thickness, total liner and Cu areas, Ta phases, and volume fraction of Cu ⟨111⟩ orientation are listed in Table 5.1.

The line resistance was found to vary slowly first, followed by an abrupt resistance jump, while in some samples there was a period of rapid linear resistance rise. The initial period of slow resistance change corresponds to a resistance incubation time, τ_i, which is associated with the lateral void growth (edge displacement) in the Cu/W overlapped region $\Delta L < \Delta L_o$, as discussed before, or with a vertical void growth (surface grain thinning) in a single Cu grain or in multiple grains from the top surface down. Cu void growth by surface grain thinning can be attributed to Cu atoms drifting along the Cu–dielectric interface fed by thinning of Cu grain directly upstream, where the surface atoms on the upstream grain diffuse to the void boundary and move up the step to feed the atoms left at the Cu–dielectric interface. This scenario was first reported in an in situ SEM study of EM in Cu damascene lines [43] and an EM study of Cu thin film lines [44] using SEM and TEM analyses. Similar surface grain thinning behavior has also been reported for Al [45, 46]. An abrupt step in the line resistance curve can occur either when a void growing by the edge displacement mechanism extends beyond the Cu/W overlapped area or when a void growing by surface thinning completely depleted Cu in a single- or multiple-grain region. The final period of resistance change can be attributed to continued void growth increasing the length of the high-resistance liner connecting the remaining Cu line and the W via. In case of void growth by edge displacement, the line resistance generally increased as the stress time increased. However, the resistance change curves showed first a plateau for some samples, followed by a large resistance jump. A similar step-like behavior in the resistance curve has also been reported [43–47]. Since the resistance plateau could be due to void growth by thinning of surface grains, so the tested line resistance should remain constant until the Cu grain or grains are completely thinned down to expose the liner and cause a resistance jump or step increase. The resistance change rate $\Delta R(t)/\Delta t$ in the final period depended on the exposed liner resistance and followed (5.6).

The samples from the C1 and G0 wafers had similar incubation time, τ_i, but the C1 samples had a slower resistance increase than G0, as shown in Figure 5.8. These results suggest that Cu lines on wafers C1 and G0 had the same EM-induced mass flow, but C1 had lines with a lower liner resistance than G0. X-ray diffraction (XRD) intensity profiles of the line cross section from wafer C1 and G0 show that the low-resistivity bcc α-Ta and high-resistivity tetragonal β-Ta diffraction peaks at various diffraction angles (2θ) were observed in the samples from wafers C1 and G0, respectively [42]. Also, the TEM images in Figure 5.9 indicate that the wafer C1 liner area is 13% larger than wafer G0. Combined, these two factors should give wafer C1 a lower liner resistance and a slower resistance change rate than wafer G0 as shown in Figure 5.8, even with the same EM-induced mass flow.

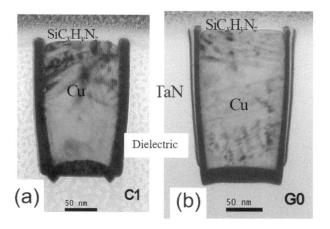

Figure 5.9 Cross-section TEM images of Cu damascene lines capped with $SiC_xH_yN_z$ for (a) C1, (b) G0, and (c) C0 wafers, respectively. © 2005 AIP, Reprinted, with permission, from [42].

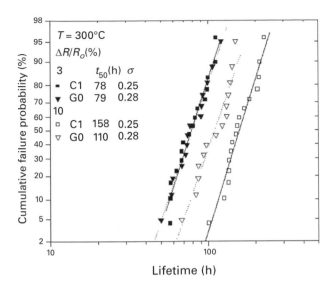

Figure 5.10 Cumulative lifetime probability of the Cu lines for various wafers plotted in a log-normal scale. The lifetime was determined at 3% tested line resistance increase failure criterion.

Figure 5.10 is a plot of the cumulative lifetime distribution based on failure criteria of $\Delta R/R_o = 3\%$ and 10% using a log-normal scale at a test temperature of 300°C and current density of 24 mA/μm^2. The observed lifetimes followed log-normal distributions with standard deviation σ around 0.3 for C1 and G0. The lifetime at 3% failure criteria was close to the resistance incubation times, τ_i, thus it was a good metric for comparing EM-induced void growth rates with less influence from the liner resistance. Depending on the void location, τ_i is the time to grow a void either larger than ΔL_o or to empty Cu out of a grain or grains. The similar values of t_{50} at 3% failure criteria for both wafers suggest that EM mass flow in the Cu lines were all about the same, despite the different Ta phases and thicknesses. However, for a 10% failure criterion, a longer lifetime would result from a line with a better liner resistance of α-Ta and larger liner area. The same behavior was also observed at test temperatures of 255°C, 320°C, and 348°C.

The average values of ΔL for samples terminated at a 20% resistance increase are 1.2 ± 0.3 and 0.8 ± 0.3 μm for C1 and G0, respectively. The void length is correlated to the liner area as predicted by (5.6), where a larger liner area results in a longer void length for the same ΔR. The lifetimes plotted in Figure 5.10 reflect the measurements for various void lengths with void growth rates of 0.4–0.6×10^{-2} μm/h determined from the slopes of void length versus time. In the present case of a line connected to a blocking boundary, the void growth rate was closely approximating to the mass-flow rate or drift velocity. The similar values of drift velocity in the various samples suggest that the effective diffusivity of the Cu lines was about the same and not influenced by the Cu/Ta phase interfaces. In the case of a Cu line with a SiN$_x$ cap, one can assume the dominant diffusion path to be along the Cu/a-SiC$_x$N$_y$H$_z$ interface for the line with

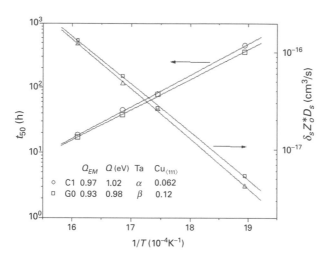

Figure 5.11 Plot of t_{50} and $\delta_s Z_o^* D_s$ vs. (1/T) with extracted values for Cu activation energy of EM and diffusion for various Ta liner phases and volume fractions of Cu with texture $\langle 111 \rangle$. © 2005 AIP, Reprinted, with permission, from [42].

bamboo-like grains, so the variations along the slower Cu/Ta diffusion path cannot affect the overall Cu effective diffusivity.

The median lifetimes t_{50} with 3% failure criteria for Cu damascene lines with various Ta phase liners and Cu textures are plotted as a function of $1/T$ in Figure 5.11. The solid lines are the least-squares fits. The median lifetime t_{50} is expressed as $t_{50} = \tau_o e^{(Q_{EM}/kT)}$, where τ_o is a constant and Q_{EM} is the activation energy for EM. The value of Q_{EM} for these Cu damascene lines can be extracted from the slopes of the lines in Figure 5.11 and found to be 0.93–0.97 eV with an uncertainty of about 0.04 eV. Similar Q_{EM} between different wafers further supports the result that the mass-flow rate is independent on the Ta phase in the Cu lines. The value of t_{50} from a 3% failure criterion closely represents the resistance incubation time τ_t and corresponds to the time for a critical void length ΔL_{c}, of 0.5 μm. The drift velocity can then be estimated using (5.6) and $v_d = \Delta L_{cr}/\tau_t$ and the values obtained were found to be in good agreement with the data deduced from the slops of void size versus time. Equation (5.6) can be rewritten as $(\delta_s Z^* D_s) = \Delta L_{cr} h k_B T/(\tau_t e \rho j)$ for the bamboo-like grain structure with the dominant fast diffusion path along the Cu/dielectric cap interface, where D_S and δ_S are the diffusivity and effective width along Cu/dielectric interface, respectively, and k_B is the Boltzmann constant. It is customary to express $Z^* = Z_{el}^* - [a/\rho(T)]$, where Z_{el}^* is the effective charge number from the electrostatic force and $a/\rho(T)$ is the effective charge number from the electron wind force [6]. In general, the electron wind force is much larger than the electrostatic force in metals, so it is reasonable to assume that the temperature dependence of $Z^*(T) \sim -a/\rho(T) = -(a/\rho_o)/(1+bT) = -Z_o^*/(1+bT) x \rho(T) = \rho_o(1+bT)$, where b is the temperature coefficient of the resistivity, and Z_o^* and ρ_o are Z^* and ρ at 0°C, respectively. Thus, the extracted activation energy from $\delta_s Z_o^* D_s$ versus $1/T$ would be the activation energy of diffusion. The plot of $(\delta_s\, Z_o^* D_s) = \Delta L_{cr}\, h\, kT/(\tau_f e \rho_a j)$ as a function of $1/T$ is plotted in Figure 5.11, where the activation energies for Cu interface diffusivity Q of 0.98–1.02 eV were obtained from the line slopes. The difference between the activation energy for EM and diffusion is about 0.05 eV. The Q_{EM} values in the Cu lines of 0.70 and 0.75 eV for the β-Ta liner, and 0.84 and 0.98 eV for the α-Ta liner, respectively, have been reported [48, 49], here the measured values for samples with the β- or α-Ta liners are about the same and similar to the values obtained for Cu lines capped with SiN_x [50]. The present results with similar activation energies and void growth rates were extracted from different wafers, indicating that the dominant diffusion path in these bamboo-like Cu lines were along the $Cu/a\text{-}SiC_xN{+}_y H_z$ interface, and not the Ta/Cu interface. Therefore, the EM behavior of these samples was controlled by the diffusivity along the Cu/dielectric interface in bamboo-like lines, and not affected by the Ta phases in the liner, nor by the use of an ALD TaN/Ta barrier, unless the liner was heavily contaminated. The Cu growth rate at the cathode end of the Cu line connected with a blocking boundary seemed not influenced by the Cu texture for volume fractions of Cu $\langle 111 \rangle$ varying from 0.06 to 0.12.

As on-chip Cu interconnection dimension scaled down below 64 nm pitch, a thin PVD TaN/Ta, <2–3 nm thick barrier/liner on trench sidewalls was used to reduce the line resistance R, which was often found to be not a reliable EM interconnect, so either TaN/CVD Ru or TaN/CVD Co had to replace the PVD TaN/Ta liner. The CVD metal deposition could provide a continuous layer and better step coverage than the PVD liner on the trench sidewalls. Figure 5.12 shows the 24 nm wide line without a metal cap resistance change curve as a function of time at $T = 300°C$. The long resistance incubation time τ_i observed in Cu samples with PVD TaN/Ta liner was not observed in the Cu line with Co liner, and the Cu line with Co liner showed little

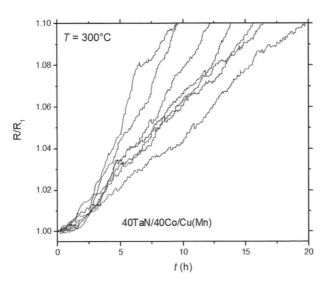

Figure 5.12 Plot of the 24 nm wide line resistance change as a function of time for a wafer with TaN/Co liner.

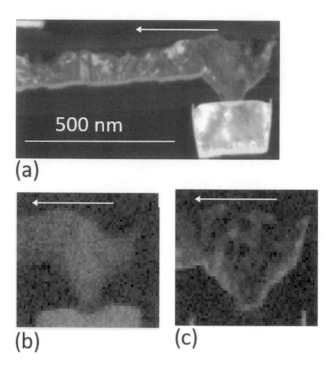

Figure 5.13 TEM and EDX dot maps along Cu 24 nm wide EM-tested line with Co liner without Co cap at 300°C for 1.5 hours: (a) is a TEM image, and (b) and (c) are EDX elements for Cu and Co, respectively. Arrows show the direction of the electron flow.

resistance incubation time. Figure 5.13a–c are TEM images and EDX spectroscopy dot maps along the cathode end of a Cu 24 nm wide EM-tested line at 300°C for 1.5 hours, respectively. Although a large surface void >1 µm in length was formed, the line R only increased by 0.3% and the EDX Cu dot map indicated that M2/V1 to M1 was still well connected, and the surface void did not have a large impact on R.

Figure 5.14a and b shows a TEM image and EDX element dot maps respectively, along the cathode end of a Cu 24 nm wide line after being EM-tested at 300°C for 3 hours. The R change in this sample increased by 3% and the bottom Cu V1-to-M1 contact area was reduced, as shown in Figure 5.14b. Under EM, the Cu atoms migrated along the Cu–dielectric interface and/or at GBs and left behind Cu vacancies that were filled by Co atoms from the Co liner and resulted in a detectable R change. For the longer resistance incubation time observed in the Cu lines with the Ru or Ta liner, the result suggests that the Cu atoms above the Cu via filled in these voids, which resulted in little or no detectable R change.

5.4.3   EM Lifetime and Q_{EM} in a Defected Line

In an early wafer fabrication development, one could often measure the wafer with structure defects. In this section, we present a case study of metal lines with voids at a

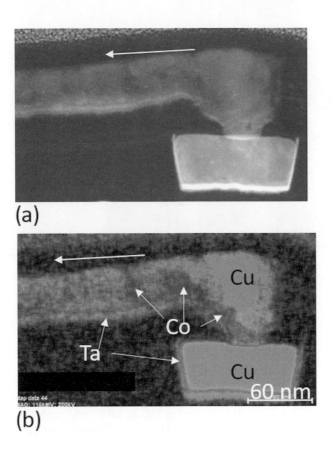

Figure 5.14 (a) TEM image; and (b) EDX dot maps along Cu 24 nm wide EM-tested line at 300°C for 3 hours, combined elements Cu, Co, and Ta, respectively. Arrows on the tops showed the direction of the electron flow. © 2017 IEEE. Reprinted, with permission, from [26].

defected sidewall or a discontinued liner and show how the defected lines influenced EM lifetime and activation energy Q_{EM}. Figure 5.15a and b shows cross-section TEM images of 70 nm wide single-damascene M1 lines with two PVD TaN/Ta liner splits [11]. Figure 5.16a and b shows scanning transmission electron microscope (STEM) image and electron energy loss spectroscopy (EELS) element map of Ta, Cu, and Co on a 36 nm pitch M1 line with PVD TaN/CVD Co liner, respectively [51]. The sidewall voids shown in Figures 5.15b and 5.16a and the discontinued Co liner shown in Figure 5.16b are probably the cause for the fast EM mass flow. For the 70 nm wide single-damascene Cu line, the metal liner deposition process for Cu interconnects was multistepped, consisting of a PVD TaN/Ta liner deposition, a sputter etch, and an additional PVD Ta layer deposition [52]. Here, the effect of the sputter etch step on liner step is clearly observed in the metal line cross sections shown in Figure 5.15a and b, where the TaN$_x$ layer on the bottom of the metal line was removed and the metal line recessed about 10 nm into the dielectric. These images show a thin TaN$_x$ liner at the sidewall interface with the thicknesses of TaN of about 3 nm, while the thicknesses

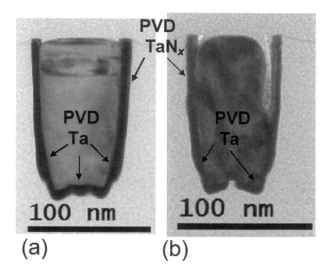

Figure 5.15 Cross-section TEM images from EM Cu M1 test lines: (a) and (b) PVD TaN/ sputter etch/Ta liner in 70 nm wide lines, respectively. © 2007 ECS, Reprinted, with permission, from [51].

of the Ta liners at the bottom and middle sidewall of the trenches were 9 nm and 4 nm, respectively. The volume fraction of the liner in the metal lines was found to be about 25% in both cases. The sidewall void seen in Figure 5.15b was probably created due to a discontinuous PVD Cu seed deposition on heavily contaminated Ta creating local voids in electroplating Cu fill. The number of sidewall voids in the fine lines can be estimated from the SEM line cross-section images in a large, fine 70 nm wide line maze containing a huge number of parallel fine lines or EM test line structures consisting of three fine 70 nm wide lines and two wider outer lines. The two adjacent fine lines near the EM-tested line were used as extrusion monitors. Only a few percent ~ 6% of sidewall voids on the lines in the line maze were observed. However, sidewall voids on the three fine lines in the EM test structure were often observed in sample sections, giving a probability of sidewall void on the three fine lines in the EM structure greater than 80%. These observations suggested that the generation of the sidewall voids was strongly dependent on the structure designed for deposition of the PVD Ta line seed. The sidewall voids can also be seen in the trenches with a barrel shape profile, where the reentrance of sidewalls made it difficult for metal liner and Cu seed depositions. For Co capped 18 nm wide Cu lines with a thin PVD TaN/CVD Co liner used to reduce the line resistance, a discontinuous TaN or Co layer found on trench sidewalls resulted in a poor Cu fill. Consequently, sidewall voids were formed on the Cu line and the EM lifetime was dropped by more than $100\times$ at $T = 380°C$ as compared with samples without sidewall voids.

The cumulative lifetime distributions of 70 nm wide M1 lines are shown in Figure 5.17 using a log-normal scale at a sample temperature of 300°C, where the data points with open and solid star symbols are the samples with sidewall voids. The lifetime was determined by the failure criterion of $\Delta R/R_o = 20\%$, and the data were

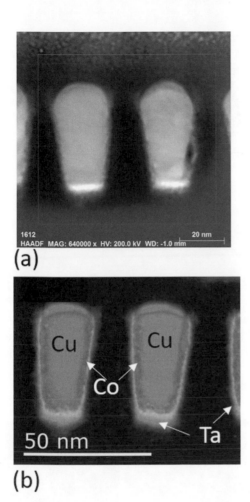

(a)

(b)

Figure 5.16 Line cross-section images from 36 nm pitch EM Cu M1 test lines: (a) a TEM image showing a sidewall void; and (b) a STEM/EELS element map for Ta, Cu, and Co, respectively. © 2017 IEEE. Reprinted, with permission, from [31].

fitted to the log-normal distribution well. In general, the log-normal or multi-log-normal plot was often used to analyze the EM lifetime distribution with the distributions of the critical void size and void growth rate being related to the diffusivity following an $\exp(-Q/kT)$ relationship. The solid curves and the dotted curves represent the samples measured with the electron flow from W CA to Cu M1 and Cu V1 to Cu M1, respectively, where there was no significant difference in their failure lifetime. The observed lifetimes follow log-normal distributions and with values of deviation σ, around 0.3–0.4. These results are consistent in showing that EM-induced mass flow does not depend on the direction of electron flow. The similar lifetimes between the Cu mass flow away from both W and Cu vias suggested that the Ta at the bottom of Cu via was a good diffusion barrier layer. However, a lower lifetime was observed for the PVD samples with sidewall voids as shown with open and solid stars in

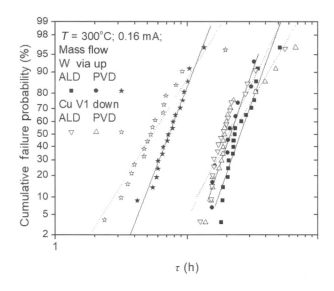

Figure 5.17 The cumulative failure probability vs. a 70 nm wide single-damascene M1 lifetime for the PVD TaN/Ta samples. The data points with open and solid star symbols are the samples with sidewall voids. © 2007 ECS, Reprinted, with permission, from [51].

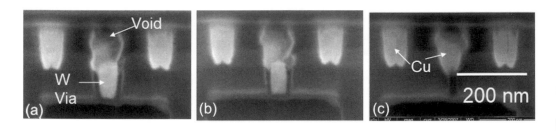

Figure 5.18 Cross-section SEM images of an EM-tested line at 300°C for 5 hours, taken at (a) near the end of cathode end of the M1 line, (b) located above the middle of W via, and (c) slightly beyond the W via. The center line is the EM-tested line, and two adjacent lines are extrusion monitors for detecting the electrical short. © 2007 ECS, Reprinted, with permission, from [51].

Figure 5.18 and cross-section TEM images in Figure 5.15b. Here the M1 lines had voids on the upper sidewalls between the Cu and the Ta liner; consequently, there was poor Cu filling, and Cu sidewall voids were formed. Attempting to observe the void morphology on the EM-damaged samples with a failure criterion of $\Delta R/R_o = 20\%$ was not successful since many failure lines were observed to be open, which obscured the EM-induced void. For this reason, a new set of experiments was performed to preserve the EM-induced void. The experiment was tested at a sample temperature of 300°C and was terminated when the sample lifetime reached $\Delta R/R_o = 1\%$.

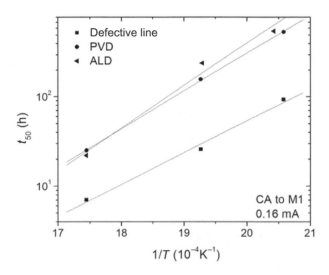

Figure 5.19 Plot of median lifetime vs. $1/T$ for the samples with various liner processing steps. © 2007 ECS, Reprinted, with permission, from [51].

Figure 5.18a–c is a sequence of SEM line cross-section images on samples from the wafer with sidewall voids, starting from the cathode end of the line. The sample had a lifetime of 5 hours when it reached $\Delta R/R_o = 1\%$. The lifetime of 5 hours is about a factor of 4 shorter than samples without sidewall voids at the same test temperature. Figure 5.18a showed the void above the W via with Cu background of Cu. Figure 5.18b was an image taken at a Cu line above the middle of a W via, showing a larger void near the left-hand sidewall. Figure 5.18c was taken at the end of W via and showed the top surface void with a shallow slope. These images seem to indicate that void growth was preferred along one sidewall and the sidewall void enhanced the void growth. The existing sidewall void provided a faster diffusion path, which was probably the reason for the poor EM results.

The median lifetime t_{50} at 20% failure criteria for the Cu M1 mass flow away from W via is plotted as a function of 1/T in Figure 5.19, where the solid lines are the least-squares fits. The Q_{EM} in these Cu damascene test lines can be extracted from the slopes of the lines and were found to be around 0.90 ± 0.05 eV for the samples with the microstructure of Figure 5.15a and 0.80 ± 0.05 eV for the samples with the Figure 5.15b microstructure, respectively. Similar Q_{EM} was found for ALD and PVD TaN/PVD Ta wafers, indicating that the mass-flow rate was independent of the ALD and PVD TaN_x in the Cu lines since the Cu–liner interfaces for these lines are the same Ta/Cu. The fact that similar lifetimes and activation energies were extracted from the ALD and PVD TaN_x/Ta wafers is consistent in showing that the dominant diffusion paths in these Cu lines are not sensitive to the PVD Ta; instead the EM mass transport in these samples was controlled by the diffusivity along the Cu dielectric interface and/or the grain boundaries, and not greatly affected by ALD or PVD TaN_x with the PVD Ta. However, the samples with the upper sidewall voids in Figure 5.15b had a lower activation energy and a lower lifetime. An unacceptable on-chip Cu

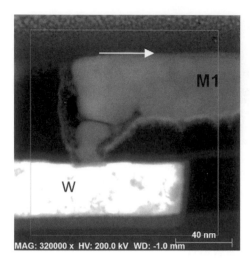

Figure 5.20 TEM image along an EM-damaged 18 nm wide EM-tested Cu line. Voids along the via and trench sidewalls and a slit void in the Cu via were observed. © 2018 IEEE. Reprinted, with permission, from [15].

interconnect lifetime was estimated from these defective samples. Nitrogen was detected in the sidewall voids by EDX, suggesting that $a\text{-SiC}_x\text{H}_y\text{N}_z$ was deposited in the void, the void was created prior to $a\text{-SiC}_x\text{H}_y\text{N}_z$ deposition, and not every void at Cu GBs or interfaces was filled by Co atoms. The large defects and imperfect $a\text{-SiC}_x\text{H}_y\text{N}_z$ coverage due to these defects would provide an additional fast diffusion path to reduce the EM lifetime.

For the defective 18 nm wide Cu line with a thin PVD TaN/CVD Co liner and Co cap shown in Figure 5.16b, a discontinuous TaN or Co layer was found on the trench sidewalls, resulting in a poor Cu fill. The thin liner was used to reduce the line resistance; consequently, the Cu lines were formed with sidewall voids and a discontinued liner. For these samples, the EM lifetime was dropped by more than $100\times$ at $T = 380°C$ as compared with samples without a sidewall void and a discontinued Co liner. Figure 5.20 is a TEM image along the cathode end of the failed Cu line, where fast diffusion path was believed to be along the poor Cu/liner interface. A slit void between two Cu grains and interface voids were observed to cause a large increase in R and a very short lifetime.

However, some defects did not have a large impact on EM-induced mass flow, e.g., isolated voids. Figure 5.21 is a TEM image of an unstressed sample taken along a 24 nm wide line with a Co cap, where many defects of circular voids and surface voids were found in this untested sample. The surface voids were probably from CMP, and the isolated voids close to the bottom of trench were from Cu fill voids. The EM lifetimes from these defected samples were found to be close to the samples without visible voids. Figure 5.22a and b shows TEM images of the defected sample along the cathode end of V0/M1 line with Co cap after being EM stressed at $T = 380°C$ with $100\ \text{mA/μm}^2$ for 128 hours. The EM-induced voids at the cathode end of the bottom via V0 and M1 line are shown in Figure 5.22a and b. In addition to the via void in (a),

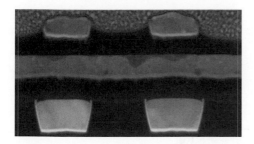

Figure 5.21 TEM image along an unstressed 24 nm wide Cu line with defects. Line surface voids and isolated circular voids are shown. © 2018 IEEE. Reprinted, with permission, from [15].

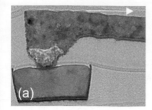

Figure 5.22 TEM image along a 24 nm wide Cu line after being EM stressed at $T = 380°C$ with 100 mA/μm^2 for 128 hours. (a) The cathode end of V1/M2. (b) the line void on the left is about 1 μm from the cathode end V1. The arrows show the direction of electron flow. The unstressed sample from the same wafer was shown in Figure 5.21. © 2018 IEEE. Reprinted, with permission, from [15].

a large line void was found located about 1 μm from the cathode end of the line. Some circular voids seemed to grow larger and did not migrate.

5.4.4   A Liner That Is Not a Completely Blocking Boundary

5.4.4.1   Thin TaN/Ta Liner

In this section, we discuss EM in multilevel Cu interconnects where the liner between Cu via and Cu line was not a completely blocking boundary for Cu migration. In this case, the EM drift velocity would be difficult to measure, making it difficult to use the Blech short length [34] to determine a high current density for the short-length lines. EM lifetime is prolonged by an artifact. Figure 5.23 is a schematic diagram of a two-level Cu interconnect structure, Cu M1/Cu V1/Cu M2, where Figure 5.23a shows the top view, and Figure 5.23b the cross-section view of one side of test structure, with M1 and M2 being the first- and second-level metal lines, respectively, and V1 the interlevel via or bar for connecting M1 to M2. Here M1 and M2 are imbedded in low interlevel dielectric (ILD) constant material, and M1 and V1/M2 are single- and dual-damascene interconnects, respectively. The thicknesses of M1 and M2 are 0.31 μm and 0.35 μm, respectively. M1 is 3 μm wide and 12 μm long, and M2 consists of one

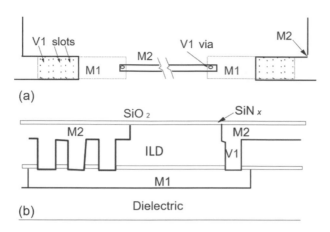

(a)

(b)

Figure 5.23 Schematic diagrams of the test structure: (a) the top view, and (b) the cross-section view of one side of test structure.

0.21 μm wide fine test line and two 3 μm wide lines on either side of test line. The fine M2 line is 375 μm long, and each end of the 0.21 μm wide M2 connects to one end of M1 through a V1 via with top and bottom diameters of 0.35 μm and 0.2 μm, respectively. The other end of M1 is connected to the 3 μm wide M2 using three 3 μm long V1 slots. The widths of the top and bottom V1 slots are 0.46 μm and 0.2 μm, respectively. The microstructures of the Cu line/via were examined by both TEM and FIB microscopy. The thicknesses of the liner at the M1/V1 via and M1/V1 slot interfaces were measured by TEM micrographs. Void growth and extrusion of the tested lines were also examined by FIB. A direct current (dc) of 1.6 mA was applied in the test lines, which resulted in current densities, j, in the 0.21 μm wide fine M2, 3 μm wide M2 and M1 of 22 mA/μm^2, 2.3 mA/μm^2, and 2.6 mA/μm^2, respectively.

Figure 5.24a and b shows FIB top view images of 3 μm wide and 0.21 μm wide M2 taken at an ion beam angle of 10°, respectively, and Figure 5.24c is a TEM cross-section view of the M2/V1/M1. A near-bamboo grain structure was observed in the 3 μm wide M2 with its overall microstructure very similar to that observed in Figure 5.24a. A bamboo-like microstructure was also found in the 0.21 μm wide Cu damascene M2 and Cu V1 vias/slots. (The bamboo-like and polycrystalline structures are defined as a single grain across the linewidth or the via and more than two grains across the linewidth, respectively.) The large Cu grain sizes in the damascene lines, slots, and vias were due to the dual-damascene process and abnormal grain growth in the electroplated Cu [53]. The barrier/liner thicknesses at the bottom of V1 via and V1 slot interfaces were around 8 nm and 19 nm as measured from the TEM micrographs, respectively. The barrier/liner thickness on the via sidewall was thinner than the liner on the via bottom, and its variation was due to the variation of PVD step coverage depending on the test structures and locations in the wafer.

When the barrier/liner at the via–line interface provided complete blocking boundaries for Cu migration, the void growth rate became the same as the drift velocity, since the Cu atoms would drift away from the liner at the contact interface (the

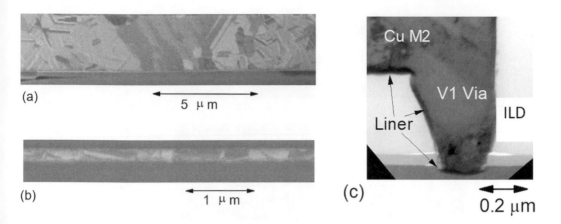

(a)

5 μm

(b)

1 μm

(c)

0.2 μm

Figure 5.24 (a) and (b) are FIB images of 3 μm and 0.21 μm wide M2 lines taken at an ion beam angle of 10°, respectively, and (c) is a TEM image of a cross-sectional view of the M1/V1 via/M2.

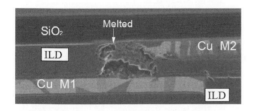

Figure 5.25 FIB image of the tested sample for failure time of 64 hours from the short lifetime group. The open failure at M2/V1 via is shown.

cathode end of the line) to create a void as described in Section 5.4.1.3. On the other hand, Cu could also drift to the anode end of the line to generate an extrusion. The lifetime τ was calculated by (5.5b) with τ defined by the amount of time required to grow a critical void of ΔL_{cr} size and with enough $\Delta R/R$ to cause line failure. In the case of partial blocking boundaries, voids and extrusions in metal lines appeared due to an imbalance of Cu fluxes at a location, x, following the equation,

$$-\frac{\partial n}{\partial t} = \frac{\partial J}{\partial x} = (J_{in} - J_{out})/\Delta x,$$ (5.7)

where J_{in} and J_{out} represent the Cu flux entering and leaving at x. The EM lifetime distributions of these samples were found to be in two distinguishable groups of different failure times. The median lifetime of the short lifetime group was about an order of magnitude less than the long lifetime group. Figure 5.25 is an FIB image of the 0.21 μm wide M2 line from the short lifetime group with a lifetime of 64 hours, where a void at V1/M2 induced an open failure. The lifetime of 64 hours at 298°C in the short lifetime group was consistent with the time for growing a 0.3 μm length void

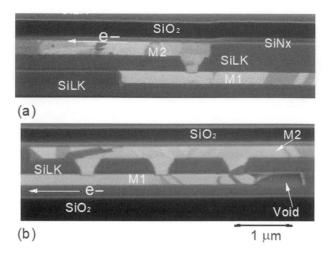

(a)

(b)   1 μm

Figure 5.26 FIB images of the tested line at $T = 298°C$ for 585 hours: (a) M1/V1/M2 interface, and (b) the wide M1 short line, respectively. Note that the void at end of wide M1 line with no electric field was shown.

at the cathode end of the 0.21 μm wide line at a blocking boundary as observed in bamboo-like Cu/SiO$_2$ structures. In contrast, the voids were not observed in the long lifetime group at either end of the 0.21 μm wide M2 in most cases. Figure 5.26a and b shows FIB images of the wide and short M1 to V1 via to 0.21 μm wide M2, and wide M2 feeder line/multiple V1 bars to the wide and short M1, respectively. No visible void at V1 via/M1 was found in Figure 5.26a, even after stressing the line for a longer time. A void in the 3 μm wide M1 is shown in Figure 5.26b. For an EM drift velocity at 298°C of 0.006 μm/h for the 0.21 μm wide line, a 3.5 μm long void would be expected in the 0.21 μm wide M2 line after testing for 600 hours if the liner at V1/M1 interface were a blocking boundary. However, since no void was observed near the M1/V1 via/M2 interfaces in most samples, this suggests that Cu migration in the 0.21 μm wide M2 was replenished by the wide M1 through the V1 via. Instead, some voids were found in the 3 μm wide M1, V1 and M2 from other EM-tested samples (not shown), and some at the end of the wide M1 section where there was little or no electrical field, as shown in Figure 5.26b. Similar behaviors have also been reported in Al/SiO$_2$ [54, 55] and Cu/SiO$_2$ [56, 57]. During EM, the Cu atoms drifted along the Cu/SiN$_x$ interface (the top Cu surface), and the depletion of atoms was responsible for creating voids as expected to occur at the cathode end of the fine M2 lines. However, a large portion of the tested samples showed no voids at the fine M2/V1 via interface in the long lifetime group. The mass transport of Cu toward the anode along the top M2 surface generates excess vacancies and tensile stress near the cathode, the vacancy concentration, and stress gradients that generate a vacancy flux. The flux moves to vacancy sinks such as interfaces, grain boundaries, or the ends of the M1 and M2 lines to form voids, where vacancies piling up at the cathode end of the M2 line due to the blocking boundary liner and migrating through to M1, resulting in short and long lifetimes, respectively. If the liners at the V1/M1 interfaces were good blocking

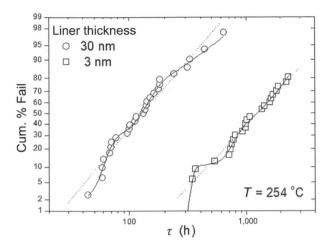

Figure 5.27 Plot of cumulative percent fail vs. τ for samples with 30 nm and 3 nm thick liners at the bottom of V1 with electron flow from M1 to V1/M2. The dotted lines and solid curves are the least-squares fitted lines using a single log-normal and triple-log-normal functions. © 2002 ECS, Reprinted, with permission, from [56].

boundaries, mass accumulation should occur at the anode end of the M1 lines to generate a stress gradient in a line to cancel out the EM driving force fulfilling the Blech critical length criterion [34] as $Z^{*}epj = \Omega(\Delta\sigma/L_c)$, where Ω, $\Delta\sigma$ and L_c are the Cu atomic volume, EM-induced stress, and the critical length, respectively. No interconnect resistance R increase should occur in a metal line length below L_c estimated to be about 77 μm with j at 2.6 mA/μm² (jL_c of 200 mA/μm). However, voids were found in the 12 μm long M1 lines under 2.6 mA/μm², indicating that most of the thin liners at M1/V1 interfaces in the test structures cannot withstand the incoming Cu EM flux, so Cu migrated continuously through M1 to V1. Thus, the failure mechanism in this structure was similar to a fine test line connecting to two large reservoirs at both line ends, where continuous or partial flow from the reservoirs will enable a very long lifetime without the Blech short-length effect.

As discussed previously, a leaky barrier layer can artificially increase the test line lifetime, even if the EM-induced mass flow remained the same. Figure 5.27 shows the EM lifetime τ plot as percent of cumulative failure for a 5 μm wide Cu M1 to Cu 0.27 μm diameter V1 via/0.27 μm wide M2 test structure obtained from two wafers with different V1 barrier/liner thicknesses. The barrier/liner thicknesses at the bottom of V1 were 30 nm for one wafer and 3 nm for the other as measured from TEM micrographs. The distributions of lifetime were analyzed with a single log-normal, or triple-log-normal function. Clearly, the Cu lifetime was enhanced by about an order of magnitude when the thinner liner was used. Apparently, the incoming Cu flux from the 5 μm wide M1 even at $j = 1$ mA/μm² can generate sufficient EM-induced compressive stress under the V1 liner to force the Cu atoms to punch through the 3 nm thick V1 liner to M2 but not the 30 nm thick liner, with the latter serving as a completely blocking boundary at the cathode end of the Cu line. As mentioned in the previous section, the Cu drifted away along the top M2 Cu–silicon nitride interface, creating excess Cu vacancies that then piled up at the end of the M2 line or V1–M1 liner interface. A short lifetime would be expected in the latter case because only a small void at the V1–M1 interface was needed to cause the line to fail. In the thin liner case, however, the vacancies piled up at the V1–M1 interface or the top M2 line surface at the cathode end of V1/M2 can be filled by Cu atoms from M1, where a compressive stress would occur in Cu M1 beneath V1, rendering (5.5b) no longer valid ($J_{in} \neq 0$), and (5.7) should be used instead. The failure mechanism was then similar to a single line with large reservoirs at both ends of the line to enable a very long lifetime. However, in spite of the variation in liner thickness, if vacancies pile up at the end of the M2 line to reduce stress and vacancy concentration gradients for both cases, the lifetime would be about the same. Therefore, lifetimes of about 300 hours were observed in the late failure group in the 30 nm thick liner and the first failure group of 3 nm thick liner. The results are consistent with the time of 300 hours required to grow a critical void size of 0.4 μm at 254°C from the known Cu drift velocity [56].

Following the discussion on EM mass flow from Cu V1 up to Cu M2 in the previous section, here we discuss the case of a single-damascene M1 with either a completely blocking boundary, W CA via, or a leaky thin liner at V1. Figure 5.28 shows the plot of normalized line resistance as a function of time with current density of 22 mA/μm^2 at a sample temperature of 350°C. The solid curves and the dotted curves represent the samples measured with the electron flow from W M0/W CA to Cu M1 and Cu M2/V1 to Cu M1 to W CA/M0, respectively. A marked difference in line resistance change was observed between electron flow from CA to M1 and V1 to M1 in this study. Initially, the line resistance slowly decreased then increased, followed by a period of rapid rise. The initial resistance decrease can be attributed to impurity migration in Cu and the reduction of contact resistance, although the initial

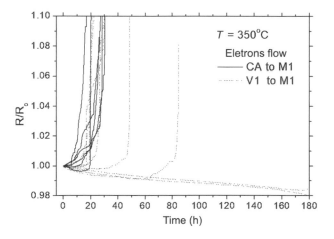

Figure 5.28 Line resistance vs. time. The solid lines represent electron flow from W to Cu lines, and the dotted lines represent electron flow from the Cu V1 via to the Cu M1 line. © 2002 ECS, Reprinted, with permission, from [56].

Cu void growth did not contribute to the R increase. The following slow line resistance increase was due to surface thinning of Cu grains and/or the void growth within the line or via, then a sharp resistance increase occurred when a void grew completely across the line or a via as we described before. Here, a drastic increase in interconnect resistance appeared as the current had to pass through the high-resistance liner, eventually causing metal line melting and an open circuit because of the high current density in the liner. A rather tight failure time distribution (t_{50} of 26 hours and σ of 0.3) was observed first in the liner case, then the open circuit showed a wide range of failure times. Only 50% of the samples with the open circuit had a similar lifetime of around 26 hours as the liner case with a Cu diffusion blocking boundary. The other half of the samples showed a much longer lifetime where some even had no resistance increase after 180 hours. Since the EM drift velocity in the Cu line should be the same, independent of the direction of electron flow, if both ends of M1 were contacted to a completely blocking boundary, mass depletion at the cathode end should occur together with mass accumulation at the anode end of the line. Thus, for the case of electron flow from a blocking boundary W CA to M1, the void growth rate in M1 near the W CA should follow (5.5b). On the other hand, if the liner at the V1/M1 interface was only a partial blocking boundary for electron flow from V1 to M1, the migration of Cu M1 beneath the V1 liner can be replenished from M2/V1 Cu and resulted in a very long lifetime. Figure 5.29a and b shows the SEM images of the two dual-damascene Cu samples tested at 294°C after 271 and 2,200 hours, respectively, where the electron flow was from V1 to M1 for both samples. A sharp line resistance increase was observed after testing for 271 hours in the former, but no line resistance increase was observed in the latter, even though the testing time was about eight times longer. Figure 5.29a shows that a void did grow (edge displacement) beneath the bottom V1 and beyond the V1 via/M1 line overlapping section after testing for 271 hours. This void location resulted in a sharp resistance increase and caused the line

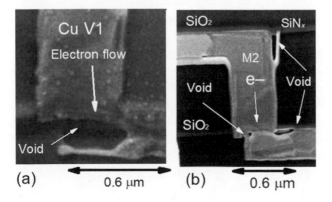

(a)   0.6 μm   (b)   0.6 μm

Figure 5.29 SEM micrographs of the M2/V1 to M1. The lines were tested for 271 hours in (a) and over 2,000 hours without fail in (b). The arrows show the direction of electron flow. © 2002 ECS, Reprinted, with permission, from [56].

failure, since the current had to pass through the thin liner to connect the remaining Cu line and via. Here the void growth rate follows (5.5b) and is consistent with the Cu EM drift velocity estimated for a completely blocking W boundary. The result indicated that a good diffusion barrier layer was formed in the sample in Figure 5.29a. However, two tiny voids were observed on the surface of a single-crystal M1 near V1 and at the anode end of M2 as shown in Figure 5.29b, indicating that Cu accumulation and the compressive stress at the anode ends of M2 would occur if the bottom V1 liner were blocking boundary. Surprisingly, voids, not hillocks, grew at the anode end of M2 /V1 in Figure 5.29b. Moreover, a void of about 10 μm size should have been observed in M1 under V1 based on (5.5b) and the measured EM drift velocity, but only two small voids on the surface of a single-crystal M1 under V1 and the end of the M2 line were seen. These results suggest that Cu atoms at M2 moved through bamboo-like grains in V1 via, resulting in a long lifetime >1,000 hours, and the small voids, shown in the micrograph Figure 5.29b, were not the cause of the large resistance change seen in EM testing, since they did not grow completely across the line or via. According to (5.5b), the drift velocity was mainly determined by the fast diffusion and not by current crowding. Only atoms diffusing along the fast path Cu/SiN$_x$ interface will control the atom movement under EM, where the Cu atoms drifted along the Cu/SiN$_x$ interface (the top Cu surface) with an activation energy of around 1 eV [7]. The depletion of atoms created the void, and they are expected to occur at the top surface of M1. Surprisingly, voids under or between the bamboo grains and in the single crystalline region where there is little electric field (the ends of the M1 and M2 lines) were observed. Similar observations of void formation in a section outside of the region carrying current were also shown in Figures 5.26b and 5.29b as previously reported [55], The mass transport of Cu along the top M1 surface driven by the applied current generated excess vacancies and tensile stress. The vacancy flux moved to vacancy sinks, such as interfaces, grain boundaries, or the ends of the lines, and formed voids to relieve stress, where the excess vacancies easily migrate through because of the low migration energy. One can estimate the time required for vacancies to diffuse through the Cu line using the vacancy diffusivity: $D = D_o \exp(-Q_m/ k_B T)$, where the prefactor $D_o = 0.16$ cm^2/sec, and values of the vacancy migration energy Q_m are reported to be 0.71 eV [58] or 0.78 eV [59]. The time required for vacancy diffusion at 295°C over a 2 μm diffusion length is estimated to be within a minute [60].

Figure 5.30 is an FIB micrograph of EM-tested V1 to M1 to W CA at the anode end of the line. A huge Cu extrusion occurred since W CA is a blocking boundary, and the large compressive stress due to the mass accumulation induced by the EM driving force cracked the SiO$_2$ dielectric to allow Cu to extrude in the crack.

5.4.4.2   Imbedded Via

Let us consider another case of EM mass flow from a wide Cu M1 through an imbedded via V1 to a fine Cu M2. The V1 imbedded into M1 was fabricated by multiple PVD liner sputter etch and deposition steps, where the thin liner on the via sidewall connected to Cu M1 often resulted in a nonblocking boundary for Cu. In this

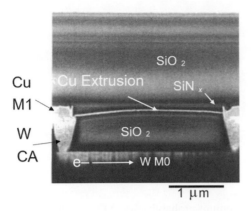

Figure 5.30 FIB micrograph of the tested samples at the anode ends of M1. Extrusion caused by a blocking boundary was observed. © 2002 ECS, Reprinted, with permission, from [56].

case, a wide range of EM failure time was observed. The EM test structure consisted of M1/V1/M2/V2/M3, where the wide M3 was at the third level and the V1 and V2 vias connected M1 to M2 and M2 to M3, respectively. The M1 and M2 test line had the same width 0.07 μm while the V1 and V2 vias had the same dimension of 0.08 μm in diameter. The Cu M1 were fabricated by a single-damascene process, and the upper levels of V1/M2 and V2/M3 were fabricated by a dual-damascene process. For the metal liners, the TaN_x liner was deposited either by a PVD or ALD technique. The metal liners were fabricated by multistepped processes consisting of an ALD or PVD TaN liner deposition, a PVD Ta layer, a sputter etch, and an additional Ta layer deposition. The sputter etching step was used to remove the high resistivity ALD TaN layer, existing Cu oxides, or contamination at the bottom of open vias. This step created trench recess into the dielectric material and via recess into the line below. The via recess improved the interconnect via resistance and reliability. The top surfaces of the Cu lines were coated with a 30 nm thick chemical vapor deposition $a\text{-}SiC_xH_yN_z$ film, where the Cu surface was subjected to a plasma treatment prior to $a\text{-}SiC_xH_yN_z$ deposition. The final Cu lines were passivated with SiO_2/SiN_x and Al(Cu) metallization was used to coat the bonding pads only. The microstructure of the Cu line was examined by both FIB microscopy and TEM with the dimensions of the M1 Cu and PVD and ALD TaN_x liner layers measured from TEM images. EDX was used to analyze the elemental composition of the samples. Figure 5.31a shows the cumulative failure probability in a log-normal plot, and Figure 5.31b is a TEM image of EM-tested sample without failure after testing for 742 hours. M1 was a 3.8 μm long and wide line and M2 was a 0.07 μm wide and long line. A nonlinear behavior was observed in a log-normal probability plot, where the failure time distributions can be analyzed by a double log-normal (bimodal) or triple-log-normal (trimodal) [61] function. The results show that 50–60% of the samples had a similar lifetime of around 20 hours, and a similar lifetime was also observed for the W via to Cu M1 line.

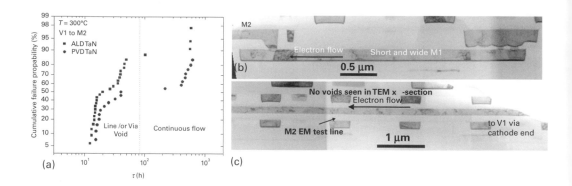

Figure 5.31 (a) The cumulative failure probability vs. M2 lifetime for electron flow from Cu V1 up to Cu M2. © 2007 ECS, Reprinted, with permission, from [51]. (b) and (c) are TEM images of M2/M1 tested for 742 hours without failure in which no voids were observed at the cathode end of M2 and (c) in a 10 μm long cathode end of M2, respectively.

The remaining samples showed a much longer lifetime, and some had no resistance increase even after testing for 1,000 hours. As shown in Figure 5.31b, no void was observed along the cathode end of M2. In general, Cu mass flow in a Cu line with similar Cu microstructure and interfaces should be identical under the same current density regardless of the test structures. If the end of the M2 line was contacted to a completely blocking boundary, the same mass depletion at the cathode end of the line should occur with the same and relatively short lifetime, as observed in one group of the test lines. However, if the PVD Ta coverage at the via bottom sidewall was too thin, a partial blocking boundary could be created in some samples, where Cu migration in the M2 line can be replenished from the M1 Cu contacted to the thin bottom V1 sidewalls. Such continuous or partial mass flow at the boundary would result in a very long lifetime, which was observed in another group of the test lines as shown in Figure 5.31a.

5.4.4.3   MnO$_x$ as a Liner

The Cu effective resistance has increased rapidly as the interconnect size is reduced and the barrier/liner to interconnect cross-sectional area ratio is increased. The barrier/liner with thickness less than 2–3 nm was required for wafer fabrication and reliability. A thin ALD liner would be an attractive choice for decreasing the volume fraction of liner in the metal line and reducing interconnect resistance. For this purpose, an ALD MnO$_x$ layer is a good choice if it could preserve good EM Cu reliability. In this section, we describe the effect of MnO$_x$ liner on Cu EM. The test samples contained a two-level EM test structure, M1-V1-M2 and with the Cu line surfaces capped with a-SiC$_x$N$_y$H$_z$. All M1, V1, and M2 were fabricated by a single-damascene process. M1 was a short length and wide line and connected to M2 of 2 μm width through via V1. For this study, the M2 liner was either TaN/Ta, MnO$_x$/Ta, or MnO$_x$ while the liners of M1 and V1 were TaN/Ta.

Figure 5.32 shows the cumulative failure probability of the Cu M2 lines obtained as a function of lifetime for 2 μm wide lines either with TaN/Ta, MnO$_x$/Ta, or MnO$_x$

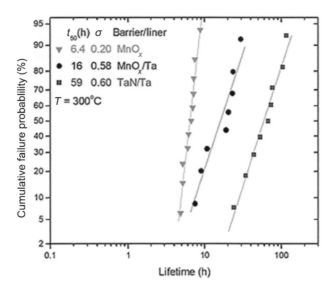

Figure 5.32 Cumulative lifetime probability for Cu M2 lines fabricated with TaN/Ta, MnO_x/Ta, or MnO_x liner.

(a)   (b)

Figure 5.33 TEM images along the cathode end of the EM M2 tested lines: (a) MnO_x/Ta liner and (b) MnO_x liner wafers. © 2018 IEEE. Reprinted, with permission, from [15].

liner at 300°C with current density of 50 mA/μm^2. The medium lifetimes were found to be 56, 16, and 6.4 hours for TaN/Ta, MnO_x/Ta, or MnO_x liner wafers, respectively, so the EM lifetimes were degraded by 4 and 9× for the MnO_x/Ta and MnO_x wafers as compared with the TaN/Ta samples. This indicated that the interfaces of MnO_x/Cu and MnO_x/Ta/Cu with Ta deposited on MnO_x became a faster diffusion path than the Cu/dielectric interface, resulting in a shorter lifetime. Figure 5.33a and b shows TEM images of EM-tested lines with either MnO_x/Ta or MnO_x liner, respectively. The usual M2 cathode end void as shown in Figure 5.33a indicated that the MnO_x/Ta liner was a good Cu diffusion blocking boundary. However, the observation of EM-induced V1 void in Figure 5.33b showed a non-Cu diffusion blocking boundary for M1 with ALD MnO_x liner. The Cu atoms in V1 can migrate through MnO_x to Cu M2, so the ALD

MnO$_x$ liner was a leaky barrier and the Blech short-length rule cannot be used for on-chip short-length lines to boost the current density for the short-length lines.

The results discussed here illustrate that the exact condition of the barrier/liner at the Cu via and Cu line interface and test structures can influence the EM lifetime. The void growth rate and the use of the Blech short length are strongly influenced by a partially or completely blocking boundary; therefore, applying the short-length effect for short-length Cu interconnects should be proceeded with caution. As the interconnection dimensions continue to be scaled down together with the metal barrier/liner to minimize the effective Cu resistivity, a thin ALD TaN, MnN or MnO$_x$, graphene, a-TiWSiC, and self-forming barrier layer, etc., have been proposed [54, 60, 62–64]. The experiments performed so far have shown that the EM lifetime can be enhanced by the thin liner even with an increase in the EM-induced mass flow at the poor thin liner–Cu interface. It is important to caution that if the large increase in lifetime was due to a continuous flow hinged on the availability of a Cu reservoir and not truly mitigation of the EM mass flow, then it would be impractical to design a reservoir for every contact in a typical BEOL structure and lose the use of the high current density based on the Blech effect for the short-length lines.

5.4.5   In Situ SEM Cu Line Surface Void Movement and Void Growth

The mass transport in Cu interconnects occurs mainly at interfaces and/or grain boundaries that depend on the Cu microstructure. The migration of Cu atoms along these fast diffusion paths is accompanied by either edge displacement or surface grain thinning mechanisms, leading to void formation. However, in some cases, the initial stage of void formation was found at a distance away from the cathode end of the line, and then moved to the cathode end of the line/via or got trapped at certain favorable locations [65–67]. Here we illustrate one of the cases reported in [65]. Figure 5.34a–g is a series of SEM images taken at the cathode end of a 0.8 µm wide M1 line during an

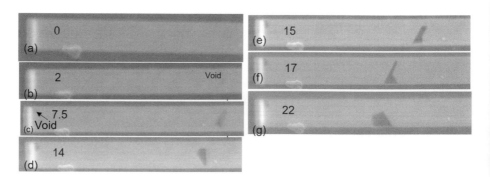

Figure 5.34 In situ images of cathode end of a 0.8 µm wide strip for EM test times of (a) 0 h, (b) 2 h, (c) 7.5 h, (d) 14 h, (e), 15 h, (f) 17 h and (g) 22 h, respectively. Electron flow was left to right. The bright structures at the cathode are W-bar vertical connectors located under the Cu line to provide electrical continuity. © 2020 IEEE. Reprinted, with permission, from [65].

EM test in an in situ SEM experiment showing the dynamics of void evolution. The sample structure used in this study consisted of Cu single-damascene (M1) drift type structure, capped with 10 nm thick $Si_xC_yN_z$. Figure 5.34b shows that the first void became visible at 6 μm away from the W bar at $t = 2$ hours, then the surface void moved toward the cathode end of the line, accompanied by void growth and shape change during the EM test. Figure 5.34c shows a detectable void at the cathode end of the Cu line above the W bar at $t = 7.5$ hours, while the first void shown in Figure 5.34b continued to grow. In some cases, the moving void would reach the cathode end of the line, while in some other cases the moving void stuck at certain locations, as shown in Figure 5.34f and g. After the void stopped move, subsequent void growth occurred either by edge displacement or by surface grain thinning mechanisms, leading to a larger void size and finally ending in a line failure. The observation of surface void movement ceased after the void moving to the cathode end only in some samples was probably because the initial stage of void movement was due to the agglomeration of small surface voids in the defected samples. After the visible surface voids being nucleated, they can diffuse along the Cu line surface opposite to the electron flow direction, where the initial voids can change their shapes and sizes during evolution. In other cases, the voids got trapped at some locations and then grow larger by either edge displacement or surface grain thinning until failure occurs.

5.5 The Blech Short-Length Effect

Under EM, the atom flow consists of two opposing transport mechanisms operated simultaneously: atom migration due to the EM force, and atom backflow due to an EM-induced stress gradient [34]. The stress gradient occurs because atoms, which are driven out of the cathode end of the conductor, causing tensile stresses, accumulate at the anode end, where the atomic density becomes higher, causing compressive stresses. The one-dimension mass-flow equation with EM driving force can be written as follows:

$$\frac{\partial n}{\partial t} = -\left(\frac{\partial}{\partial x}\left\{nD_{eff}\left[\frac{1}{n}\left(\frac{\partial n}{\partial x}\right)\right] - n\frac{D_{eff}}{kT}Z^*e\rho j\right\}\right). \tag{5.8}$$

Combining the EM force and backflow effects, the net drift velocity is as follows:

$$v_d = v_e - v_d = \frac{D_{eff}}{k_BT}[Z^*e\rho j - (\Delta\sigma\Omega)/\Delta x], \tag{5.9}$$

where v_e is the EM drift velocity and v_d, is the mechanical backflow velocity. An important implication of these effects is that for sufficiently short lines, L, or low current densities, the driving force from the stress gradient can completely suppress EM mass transport to vanish v_d at a time of $\sim L^2/D_{eff}$. We can define a threshold value of $jLc \propto \Delta\sigma$ at a given j with a critical line length L_c (Δx in (5.9)), below which the net mass transport vanishes ($v_d = 0$). The magnitude of the EM-induced stress $\Delta\sigma$ depends on the EM force and should be less than the fracture strength $\Delta\sigma_c$ of the passivation

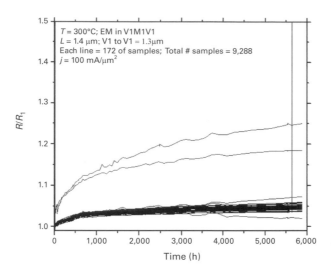

Figure 5.35 Change in normalized line resistance as a function of time with $j = 100$ mA/μm^2. Each line contains 172 of 1.4 μm long line links. Total number of the tested lines is 9,288.

layer and barrier/liner or the adhesion strength between the line surface and the cap layer. In addition to the mechanical strength of the dielectric material and barrier/liner, the anode end of the line must connect to a complete blocking boundary in order to generate the short-length effect. A threshold value of jL_c from 300 to 800 mA/μm has been reported for Cu damascene lines with electron flow to a blocking boundary [36–39, 68]. Here we present one of EM in short-length experiments. The test structure for each sample consists of 172 M1 lines of 1.4 μm long and 32 nm wide linked in a series. The EM mass flow in this case is from V1 to M1 to V1 to M1 and the diameter of V1 is 32 nm. Figure 5.35 shows the normalized line resistance change as a function of time for 9,288 of 1.4 μm long M1 lines after testing at 300°C with a current density $j = 100$ mA/μm^2 for 6,000 hours. The EM damage in this group of 1.4 μm long lines was found to be negligible at $j = 100$ mA/μm^2 after 6,000 hours, as compared with another test for 100 μm long line with $j = 25$ mA/μm^2, where the median lifetime was only 10 hours. The results correspond to $jL = 140$ mA/μm, and thus reflect the short-length effect predicted by (5.9), although one failure did occur at $t = 5,630$ hours that might have been caused by a defective sample.

5.6   EM Lifetime Scaling Rule

For the bamboo line structure, the EM flux is constrained to the top interface in an area of $\delta_I w$, where δ_I is the effective thickness of the interface region and w is the linewidth. The relative amount of flux, at constant line current density, flowing through the interface region is proportional to the ratio of the interface area to the line area, $\delta_I w$ / (wh), or δ_I /h, where h is the line thickness. The lifetime τ for a near-bamboo-like

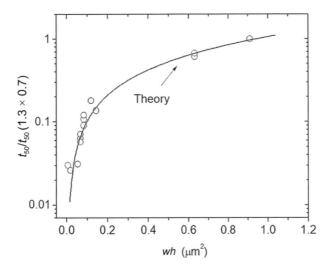

Figure 5.36 Normalized median lifetime t_{50} vs. wh for various Cu interconnect generations. © 2017 IEEE. Reprinted, with permission, from [41].

structure with a blocking boundary at the cathode end of the line, (5.5b), can be written as follows:

$$\tau = \Delta L_{cr}/v_d = \Delta L_{cr}hk_BT/(\delta_s D_{eff}F_e), \tag{5.10}$$

where ΔL_{cr} is the critical void length to cause the line to fail, δ_s and D_{eff} are the effective surface width and diffusivity, respectively. In the present case, ΔL_{cr} is very close to the via diameter, w, which is about the same as the line–via overlap length. Then the lifetime would be proportional to $w\,h$, i.e., the line area in (5.10). The ratio of the median lifetime t_{50} relative to the 1 μm technology with 1.3 μm wide × 0.7 μm thick lines is plotted in Figure 5.36 as a function of $\Delta L_{cr}h$ for each CMOS generation from 45 nm to 1 μm Cu technology. The t_{50} clearly decreases in each CMOS generation correlating to the decrease in the line area wh. The solid line in Figure 5.36 is the curve according to (5.10), with no adjustable parameter. The experimental data points follow the theoretical curve closely, which supports the validity of the model. So far, the EM test results showed the EM lifetime for on-chip interconnect structures generally decreased by half for every new interconnect generation when tested at the same current density [41, 50]. This EM lifetime scaling rule can be explained by two factors: first, the fraction of atoms on fast diffusion paths increases for every new generation, and second, scaling of the via size will reduce the critical void length to cause faster line failure. The same scaling rule can also apply to Cu nanowires with or without a metal cap.

Here the results clearly illustrate the nature of the EM reliability problem for Cu interconnects, so the pure Cu interconnect structures must be changed in order to develop a reliable Cu interconnect technology for 65 nm technology node and below, either by implementing Cu alloy interconnects or a large via size, or by improving the

Cu/dielectric interface. In general, the Cu alloy and thick liner approach offers reliable Cu interconnects at the expense of Cu line conductivity. Alternately, large vias provide a better connection between line levels to give a longer EM lifetime, since a large void is needed to cause line failure, and a large via has a better chance of connecting exposed liners between levels when a void has formed at the via–line interface. However, the interconnect density as well as the transistor density would be adversely affected by the large via size design. So far, the best method for maintaining high conductivity, high performance, and interconnect reliability is to alter the Cu/dielectric interface to reduce the Cu transport along interfaces. This can be achieved by a selective metal cap on top of the Cu line. Modification of the Cu conductor surface is performed after Cu CMP by a selective metal deposition process, whereby a thin layer selectively coats the Cu conductor surface but not the dielectric surface. The metal coating should be sufficiently thin but still able to reduce atomic interface transport of the Cu conductor in order to provide maximum reliability without overly complicating the fabrication process. Thus, the pure Cu on-chip interconnect used up to the 32 nm node was changed to the Cu alloy for the 32 to 14 nm nodes, and a metal cap was used in the 10 nm node and beyond in integrated circuit (IC) production. Such methods for improving Cu interconnect EM reliability are discussed in the following section.

5.7   Methods to Improve EM Reliability of the Cu Interconnect

5.7.1   Test Structure Effect with Upper-Level Dummy Vias

In this section, we discuss a technique to modify the interconnect structure for improved EM lifetime by adding upper-level dummy vias on top of a Cu line, where the dummy vias interrupt the Cu mass flow along the interface to improve the EM lifetime. This structure also provides a powerful tool for understanding the role of Cu microstructure and fast diffusion path on EM reliability. Figure 5.37 shows a custom-designed structure that has either a single upper dummy via bar 2 μm away from the contact via or repeating dummy via bars every 5 μm or 10 μm. The dummy vias, 70 nm to 1.7 μm wide, covering the full M1 linewidth and with M2 on the top Cu line, were used to understand the effect of the upper dummy via on Cu EM. The Cu M1 test line was connected to the underlying

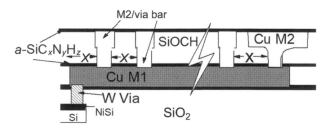

Figure 5.37 Schematic diagrams of test structure. x is the distance between the dummy vias. © 2009 AIP, Reprinted, with permission, from [69].

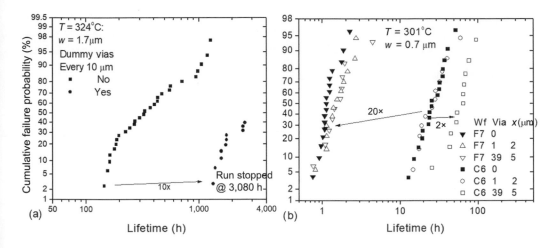

Figure 5.38 Cumulative failure probability vs. lifetime for (a) 1.7 μm and (b) 70 nm wide lines, respectively. The total number of dummy vias and the via spacing, x, on the 200 μm M1 test line is also described. © 2009 AIP, Reprinted, with permission, from [69].

NiSi line and M2 through W and Cu contact vias, respectively, at each end of the Cu M1 line. The dual-damascene dummy via/M2 bar was 40% wider than the Cu M1 line and placed on top of the M1 line with no other connections.

The cumulative failure (lifetime) probability of the Cu lines is plotted in Figure 5.38a and b as a function of lifetime tested at 324°C for 1.7 μm wide lines and 301°C for 70 nm wide lines with and without upper-level dummy vias. The applied current density was less than the Blech critical current density, j_c, assuming dummy via distance x is equivalent to the critical short length ($jx < 120$ mA/μm). When compared with the no dummy via structure, enhancements of 10× for 1.7 μm and 2–3× for a 70 nm wide line were observed. In this case, if the upper dummy vias on the top surface of the Cu lines can block the Cu mass flow along the top surface interface, the lifetime should be significantly improved. However, if the dummy via were on top of the polycrystalline Cu sections or the distance between dummy vias on bamboo-like Cu grains was longer than the Blech length, the lifetime improvement would diminish. Here the larger improvement in 1.7 μm wide line samples can be attributed to this line having more bamboo sections and larger grains than the 70 nm wide lines. The data for the one dummy via structure (open circles in Figure 5.38b), showed little or no EM improvement as compared with the lifetimes using standard no dummy via structure samples. This is because the present structure with the embedded via enabled the EM failure voids to form by emptying the Cu line, so any improvement in EM lifetime by transferring the EM-induced voids beyond the dummy via was small. Figure 5.38b also includes data from samples with abnormally short lifetimes being reduced by 20×. These abnormally short lifetime wafer F7s came from a different wafer lot that was fabricated with a metal liner/seed deposition deviated

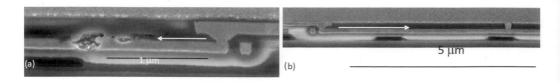

Figure 5.39 SEM line cross-section images of samples tested for 210 hours. The distance between the dummy via, x, is every 5 μm for both (a) and (b). The arrows show the electron flow directions. © 2009 AIP, Reprinted, with permission, from [69].

from the standard liner/seed processing steps. These samples were tested with and without dummy vias, and the results plotted as solid and open triangles, respectively, and the result showed a negligible lifetime improvement. This suggests that the abnormal samples had a faster Cu/liner interfacial diffusivity than that at the interface of the Cu/dielectric cap. Thus, the upper-level dummy via structures have indeed provided additional understanding of the role of Cu microstructure and helped to distinguish the fast diffusion path in these samples.

Figure 5.39a and b shows SEM images of EM-tested lines with short-length effects, where the usual void under or near the contact via for mass flow from the via down to the line was not seen. Instead, Figure 5.39a shows the void located away from via/line contact area beyond the effective EM short length. Some lines after testing for a long time showed no line resistance increase as shown in Figure 5.39b, reflecting the fact that most of bamboo–polycrystalline section lengths in these lines were less than the Blech length, again showing the Blech short-length effect.

5.7.2 Cu Lines with Surface Treatment

Improved EM lifetimes with a H_2-based plasma preclean on top of the Cu surface before a high-density plasma (HDP) [70] silicon nitride have been reported. The effect of plasma clean for a plasma-enhanced chemical vapor deposition (PECVD) a-SiC$_x$N$_y$H$_z$ cap film [71] on Cu EM has also been reported. In these studies, a trade-off between improved EM and increased copper resistivity was shown. In this section, we describe a study of hydrogen-based plasma cleans and SiH_4 soaks prior to a-SiC$_x$N$_y$H$_z$ cap film deposition. In Figure 5.40, we show the EM lifetime distributions from samples precleaned with NH_3/N_2 plasma and H_2 plasma/SiH$_4$ soak/NH$_3$/N$_2$ plasma at a sample temperature of 227°C. The samples with H_2 plasma preclean on the Cu line surface followed by SiH_4 soak and NH_3/N_2 plasma exposure at low temperature showed a moderate 2× EM lifetime improvement with ~8% increase in line resistance, as compared with samples only with NH_3/N_2 plasma clean. In a previous study [70], an improvement more than 10× in the median lifetime was reported together with a 2× increase in initial line resistance after annealing at 350°C for 14 hours. The improved Cu lifetime for this set of samples was believed to be due to the formation of a $CuSi_xN_y$ interfacial layer; however, the copper resistivity was also increased due to silicon diffusion into the copper grains from the

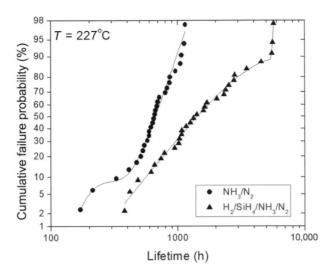

Figure 5.40 Plot of cumulative lifetime probability for Cu lines from samples with various Cu surface treatments. The solid curves are the least-squares bimodal fits. © 2009 AIP, Reprinted, with permission, from [69].

cap–copper interface. The transport of silicon was reduced by exposure to NH_3/N_2 plasma after Cu silicide formation.

5.7.3 Effect of Microstructure (ECD vs. CVD Cu) and Nonmetallic Impurities

In another study, the effect of microstructure for electroplated ECD and CVD Cu and nonmetallic impurities was investigated. In Figure 5.41, we show the resistance change of 2 µm wide lines with ECD and CVD Cu damascene lines as a function of EM stress time. The CVD Cu damascene lines were fabricated by filling the trench/via with CVD Cu after PVD liner/Cu seed depositions. The EM damage on the CVD Cu line was faster than that of the electroplated (ECD) Cu line. In addition, the observed EM activation energy for CVD Cu was 0.8 eV, lower than the value of 1.1 eV for the ECD-Cu damascene lines. The superior EM reliability of an ECD Cu versus CVD or PVD Cu was attributed to either improved microstructure (near-bamboo for ECD Cu vs. polycrystalline for CVD Cu) and/or to more impurities in the ECD Cu [5]. The possibility of electroplating impurity effect on Cu EM was further investigated using two test structures with electroplating solutions A and B for Cu trench and via fill. The concentrations of Cl, O, C, and S in the electroplated Cu films measured by second ion mass spectrometer (SIMS) are shown in Figure 5.42. The Cu films using recipe A had an order of magnitude higher Cl, O, S, and C impurity concentrations than films from recipe B.

Figure 5.43 is the cumulative failure probability plotted on a log-normal scale as a function of time for 65 nm and 2 µm wide Cu M2 test structures. The tests were performed at a sample temperature of 275°C for M1-V1-M2 line structures with M2

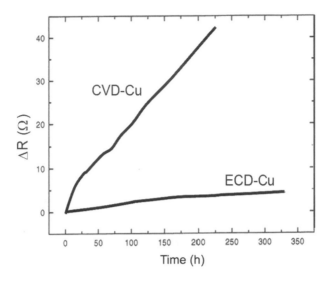

Figure 5.41 Cu line resistance as a function of EM stresses time for CVD Cu and ECD Cu. © 2010 AIP, reprinted, with permission, from [72].

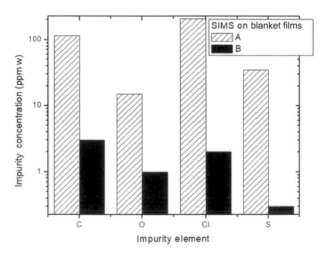

Figure 5.42 Impurity concentrations obtained by SIMS on ECD-Cu films using recipes A and B, respectively. © 2010 AIP, reprinted, with permission, from [72].

deposited using ECD recipes A and B and a V1-M2 dielectric ultralow dielectric constant (ULK) of $k = 2.4$. The lifetime data followed a log-normal distribution function, and the straight lines are the least-squares fits to a log-normal distribution function. There was a significant difference in lifetime between 2 μm and 65 nm wide lines with an improvement of 10× in lifetime and 0.5 reduction in σ for the 2 μm wide lines as compared with the 65 nm wide lines. Since the lifetime distributions for 2 μm

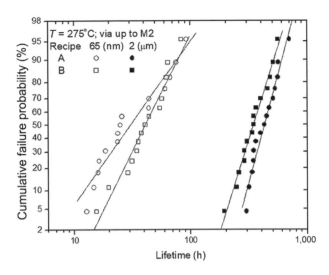

Figure 5.43 Cumulative lifetime probability for 65 nm and 2 μm wide Cu lines fabricated with ECD Cu recipes A and B. © 2010 AIP, reprinted, with permission, from [72].

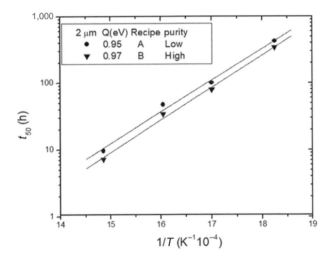

Figure 5.44 Plot of t_{50} vs $1/T$ for high and low levels of impurities in 2 μm wide lines. © 2010 AIP, reprinted, with permission, from [72].

wide damascene lines in ULK were similar to those observed for Cu/LK structures, the 2 μm wide lines were used to evaluate the plating impurity effect. Figure 5.44 shows the EM median lifetime and EM activation energy Q_{EM} for the 2 μm wide lines with Q_{EM} of 0.95 ± 0.05 eV and 0.97 ± 0.05 eV measured for the 2 μm lines of recipe A and B, respectively. The small lifetime difference between the two cases could be due to a small variation in test structure fabrication, such as line area and line–via overlapping area. Thus, we concluded that little or no effect of ECD impurities, Cl, C, O, and S on Cu EM was found. These results are contradictory to a previous study reporting a large nonmetallic impurity effect on Cu EM [73]. We believe that the amount of impurities ($<0.1\%$) in the plated Cu is too low to make a significant contribution to the interface and GB mass transport.

The EM improvement of ECD-Cu over CVD- or PVD-Cu can be attributed to the better microstructure and not due to more nonmetallic impurities in the ECD-Cu line. ECD-Cu lines with a larger grain structure than CVD- or PVD-Cu lines were the result of abnormal grain growth in ECD-Cu during Cu damascene line fabrication. Polycrystalline only and near-bamboo-like grain structures were observed in 2 μm wide Cu damascene lines for PVD/CVD-Cu and ECD-Cu, respectively. Therefore, grain boundary dominated in PVD/CVD-Cu, and instead interface diffusion dominated in ECD-Cu. The grain size of CVD- and PVD- Cu may have been limited by the Cu film thickness. The measured EM activation energy of 0.8 eV from CVD-Cu was also consistent with the results obtained by isotope-tracer Cu tracer GB diffusion [74] and EM in polycrystalline ECD-Cu, which will be discussed in Section 5.7.5.

5.7.4   Cu-CNT Composite Interconnects

In this section, we discuss EM in the Cu/carbon nanotubes (CNT) composite damascene lines. The EM test structure consisted of two-level M1-V1-M2. M1, V1, and M2 that were processed by a single-damascene recipe. M1 was a wide and short line, and M2 was 2 μm wide line. For the M2 Cu/CNT wafer, the Cu and CNT were filled into the M2 trenches using a coelectroplating technique. Both Cu and Cu/CNT single-damascene M2 lines were passivated with a 30 nm thick a-SiC$_x$N$_y$H$_z$ layer and annealed at 400°C before wafer dicing and EM test. EM tests were carried out in a forming gas (5% H$_2$) ambient.

The median lifetimes t_{50} for the Cu M2 mass flow away from V1 via are plotted as a function of $1/T$ in Figure 5.45 with the solid lines being the least-squares fits. The EM activation Q_{EM} in Cu and Cu/CNT composite damascene lines can be extracted from the slopes of the lines and are found to be around 1.0 eV for the pure Cu and 0.70 eV for the Cu/CNT composite lines, respectively. These results were opposite to a previous study reporting that Cu/CNT void growth was reduced and the lifetime was improved by 100× and $Q_{EM} = 2$ eV [75]. The discrepancies could be due to the variations in the Cu/CNT composite and EM sample test structures. EM samples used in the study by [75] were Cu lines without a passivation layer, so their results could be influenced by the free Cu surface diffusion. In addition, only samples of a single stripe were tested and under an extremely high current density [75] $j = 720 \times 10^6$ A/cm^2 as compared to the typical EM test structure with Cu lines connected to a blocking boundary at the cathode end and $j = 0.5$ to 15×10^6 A/cm^2. Unlike the multilevel EM test structure, the single stripe connected two large reservoirs on both ends of the line so void formation might not cause a large line R increase since the CNT incorporated into the Cu line can serve as a redundancy line to carry electrical current.

Figure 5.46a and b shows top-down SEM images of Cu and Cu/CNT composite 2 μm wide lines respectively used in our study after the a-SiC$_x$N$_y$H$_z$ passivation layer

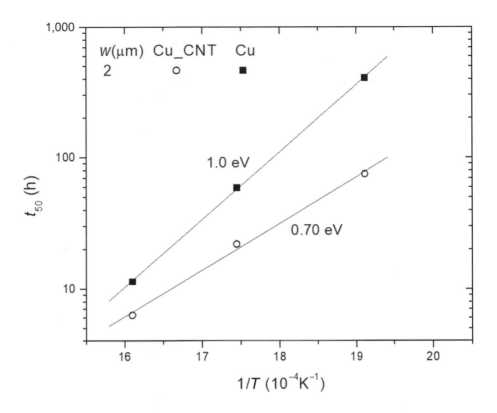

Figure 5.45 Plot of t_{50} vs. $1/T$ for Cu and Cu–CNT composite 2 μm wide lines.

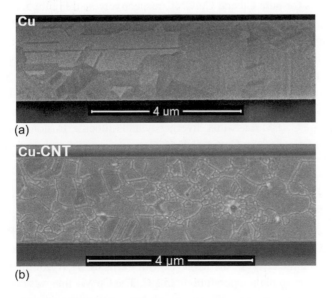

Figure 5.46 Top-down SEM images with a-SiC$_x$N$_y$H$_z$ cap removal by FIB: (a) Cu and (b) Cu/CNT composite line, respectively. © 2018 IEEE. Reprinted, with permission, from [15].

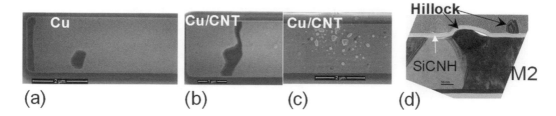

Figure 5.47 (a)–(c) Top-down SEM images along the cathode end of the EM M2 tested lines: (a) pure Cu composite lines; (b) and (c) Cu/CNT composite lines. (d) TEM image sectioning along the Cu/CNT composite line.

was removed by FIB. A large near-bamboo-like structure was observed in the pure Cu line while small polycrystalline Cu grain structures were observed in the Cu/CNT composite line. The results of Q_{EM} of 1.eV and 0.7 eV for pure Cu and Cu/CNT samples, respectively, can be explained by the two distinct microstructures. In the pure Cu bamboo-like line case, the migration of Cu atoms along the Cu–a-SiC$_x$N$_y$H$_z$ interface controlled the EM-induced mass flow with an activation energy of 1 eV. On the other hand, for the Cu/CNT line, the CNT inhibited the grain growth in the electroplated Cu, resulting in a polycrystalline structure. As a result, the mass transport was dominated by diffusion along Cu grain boundaries, yielding an activation energy of 0.7 eV.

The void characteristics in pure Cu and Cu/CNT lines were examined as shown in Figure 5.47, where Figure 5.47a and Figure 5.47b and c are the top-down SEM images of EM-tested lines for pure Cu and Cu/CNT, respectively, and Figure 5.47d is a TEM sectioning image along the M2 Cu/CNT composite line. Here typical EM-induced voids were found near the cathode end of test lines, as shown in Figure 5.47a and b, where many voids were formed together with hillocks in the Cu/CNT line as shown in Figure 5.47c, and then the hillocks extruded and buckled out the passivation a-SiC$_x$N$_y$H$_z$ layer, as shown in Figure 5.47d. Such formation of voids and hillocks in proximity suggested that the Cu/CNT lines had a nonuniform Cu grain distribution resulting in a large Cu EM mass-flux divergency and hillock extrusion to damage and reduce the EM lifetime of the Cu/CNT lines.

5.7.5   Metal Cap Layer Effects: Ru, Ta, CoWP

5.7.5.1   Selective CVD Ru Cap

In this section, we report the study on the effect of Ru, Ta, and CoWP metal cap layers. Figure 5.48 shows the plots of cumulative failure probability on a log-normal scale by comparing the effect of the Ru layer with the a-SiC$_x$N$_y$H$_z$ for the 65 nm wide Cu M1 lifetime at a sample temperature of 252°C. The Cu M1 line was connected to a Cu diffusion blocking boundary W via at both ends of the line. In this case, the Cu void growth rate at the cathode end of the line was directly correlated to the EM mass-flow rate v_d, and the lifetime data exhibit a log-normal distribution function with the straight lines indicating the least-squares fits of the data. The values of sigma in the

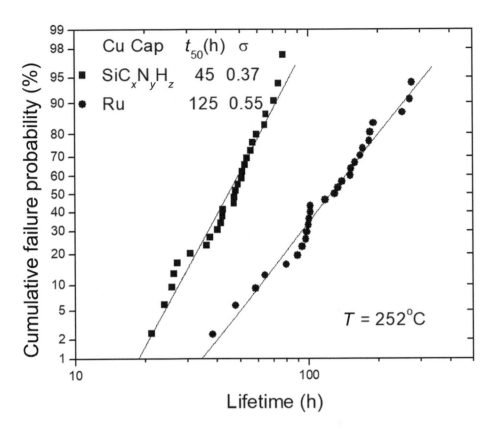

Figure 5.48 Plot of lifetime vs. cumulative lifetime probability for samples without and with the Ru cap used (log-normal scale). © 2009 AIP, reprinted, with permission, from [69].

log-normal distribution function were found to be 0.37 and 0.55 for Cu capped without and with a thin selective CVD Ru layer, respectively. The low sigma value for the samples without a Ru cap was due to the electron flow across a blocking boundary (W via) to the Cu line subjected to a physically stable redundant liner. The higher sigma value for the Ru capped samples reflects the variability of Cu–Ru interface quality and uniformity. However, a 3× improvement in t_{50} for the Ru cap samples suggests a slower void growth rate with corresponding slower mass flow at the Cu–Ru interface as compared with the Cu–a-SiC$_x$N$_y$H$_z$ interface. The fact that no Ru was observed on the top of the Cu surface with EDX indicated that the thickness of the Ru layer was below the detectability of EDX. This could be one of the reasons that the EM lifetime enhancement from the Ru capped samples was far less than for thin CoWP, Co, or Ta/TaN-capped samples [35, 76, 77]. The EM median lifetimes for 65 nm wide lines are plotted as a function of $1/T$ in Figure 5.49, with the solid lines being the least-squares fits. The activation energies for Cu EM lifetime, Q_{EM}, can be extracted from the slopes of the lines and are found to be 0.87 eV and 0.88 eV for Ru cap and no Ru cap samples, respectively. The similar EM activation energies obtained from both types of samples suggest that the Ru cap layer did not have a significant

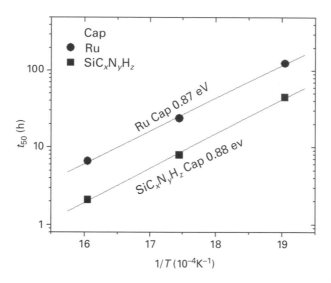

Figure 5.49 Plot of t_{50} vs $1/T$ for with and without the Ru cap for 65 nm wide lines. © 2009 AIP, reprinted, with permission, from [69].

impact on Cu mass flow along the interface. These results also suggest that the use of a selective CVD Ru cap will require further development to improve EM reliability.

5.7.5.2   The Effect of Ta and CoWP Caps

For this study, the first set of samples had a test structure consisting of a two-level damascene M2/V1/M1/V1/M2 interconnect. The fine M1 test line was 400 μm long, 0.10 μm wide, and 0.14 μm thick, and the diameter of the Cu V1 via was about 0.12 μm. The second set of samples has a test structure consisting of M0/CA/M1/CA/M0, where M0 and CA were single-damascene W lines and vias, respectively. The M0 W lines were connected to the M1 Cu lines through CA W vias. The fine M1 Cu test line was 100 μm long, 0.16 μm wide, and 0.16 μm thick, and the diameter of the CA W via was 0.25 μm. The M1 and M2 had a near-bamboo grain structure, thus the interface diffusivity should determine the EM-induced mass flow. The top surfaces of the Cu M1 and M2 lines were coated with a Co (3 at.%W, 6 at.%P) cap using a selective electroless deposition process. The concentrations of W and P in the CoWP film were determined by Rutherford backscattering analyses. The top surfaces of the M1 Cu lines were coated with Ta(24 nm)/TaN(20 nm) in the second set of wafers. For the CowP cap and no metal cap wafers, a thin diffusion barrier layer of $SiC_xN_yH_z$ was first placed on top of the M1 and M2 Cu lines before the next-level dielectric layer was deposited. Next, two more levels of Cu interconnections were fabricated on top of the M2 lines. The final passivation layer consisting of a SiN_x/SiO_2/polyimide cap on the interconnect structures and Al(Cu) was deposited to cover the Cu bonding pads only. Figure 5.50 shows a plot of normalized line resistance versus time for Cu near-bamboo grain M1 line/W via structures Cu M1 capped with various overlayers. The

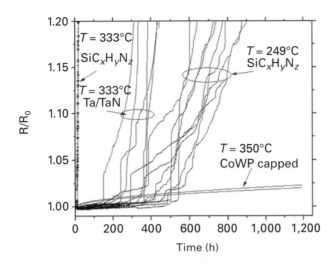

Figure 5.50 Test line resistance change vs. time for the $SiC_xH_yN_z$, Ta/TaN, and CoWP capped samples. © 2003 Cambridge University Press. Reprinted, with permission, from [78].

Cu lines with a $SiC_xH_yN_z$ cap were tested at 249°C and 333°C, with a Ta/TaN cap at 333°C, and with a CoWP cap tested at 348°C. The results show that for the Ta/TaN and $SiC_xH_yN_z$ capped samples, the initial line resistance changed slowly, followed by a period of rapid resistance rise. As discussed in the previous sections, the slow resistance change period was probably associated with void growth within the Cu–W overlap region. The rapid increase of resistance was attributed to void growth when $\Delta L > \Delta L_o$, where the current passed through the thin, high-resistance liner connecting the remaining Cu line to the W via. The curvatures in the resistance change curves for the Ta/TaN and $SiC_xH_yN_z$ capped samples in Figure 5.50 indicate that the void growth rate for $\Delta L > \Delta L_o$ increases slightly as the void growth progresses. When stressed at 333°C, abrupt line resistance increases occurred in the $SiC_xH_yN_z$- and Ta/TaN-coated samples at times <20 hours extending over several hundred hours. The $SiC_xH_yN_z$ capped samples had to be tested at a lower temperature, 249°C, in order to obtain lifetimes similar to the Ta/TaN capped samples tested at 333°C. Interestingly, no significant resistance increase was observed for the selective electroless CoWP capped samples even after testing at 348°C for over a thousand hours. The gradual line resistance increase over time in CoWP-capped samples was due to Co bulk diffusion into the Cu grains causing reduced Cu conductivity from impurity scattering, which will be discussed in detail in the next section. Figure 5.50 clearly shows that the capped metal samples provided significant improvement against EM damage when compared to samples with a $SiC_xH_yN_z$ cap, and CoWP-capped samples had longer lifetimes than Ta/TaN capped samples. Figure 5.51a–c shows focused ion beam images taken at an ion beam angle of 45° for tested lines with $SiC_xH_yN_z$, Ta/TaN, and CoWP caps stressed at 250°C, 377°C, and 377°C with 35 mA/µm² and lifetimes

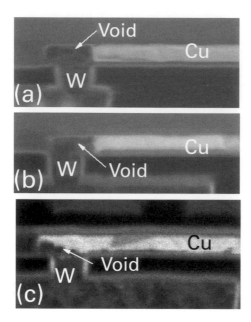

Figure 5.51 FIB images of the EM M1 failed samples for (a) a $SiC_xH_yN_z$ capped line at 250°C with a lifetime of 600 hours; (b) a Ta/TaN coated line at 377°C for 64 hours; and (c) a CoWP capped line at 377°C for 1,220 hours. © 2017 IEEE. Reprinted, with permission, from [41].

of 600, 62, and 1,220 hours, respectively. From the sample temperatures and lifetimes, these images show that the void growth rates in the Cu lines are drastically reduced for the line capped with Ta/TaN and further reduced for the line capped with CoWP. Unlike usual void growth by either edge displacement and/or Cu surface grain thinning from the top surface, the void growth was found to be initiated from the bottom Cu line–W via interface in the CoWP capped samples. The image shows that the Cu atoms drifted along the electric current field lines in the Cu grain and suggests that bulk diffusion in the lines could be occurring.

To study the Cu void growth rate, the W M0/W CA/Cu M1/W CA/W M0 test structures were used. Here, the Cu M1 test line is connected to W vias completely blocking Cu diffusion at both ends of the line, making the Cu void growth rate closely related to the EM drift velocity. The lifetime of the Cu line can be expressed as $\tau = \Delta L_{Cr}/v$, where the drift velocity $v = (D_{eff}/kT)Z^*eE$, and ΔL_{Cr} is the critical void length at the cathode end as described before. Unlike the single-damascene Cu lines capped with the dielectric or Ta/TaN [77], voids did not start to grow from the cathode end of the line or from the top of the Cu grains, but grew mostly from the Cu line–W via contact interface, as shown in Figure 5.51c or shown in Figure 4 of [76]. This result suggests that Cu bulk diffusion could be occurring in the bamboo-like grain lines coated with CoWP. Since Co diffusion into Cu can cause appreciable line resistance increases and the Co solubility in Cu is dependent on sample temperature, the Cu lifetime is determined using a conventional resistance increase percent (e.g., 1–5%),

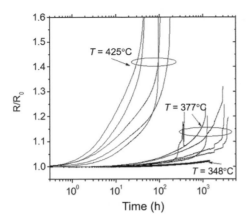

Figure 5.52 A plot of the normalized line with CoWP cap resistance as a function of time at various sample temperatures.

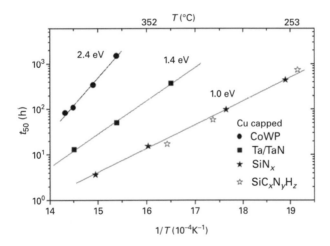

Figure 5.53 Plot of lifetime vs. $(1/T)$ with extracted value for EM activation energy for CoWP-coated samples in this present study (solid circles), Ta/TaN-capped Cu lines (solid squares), and dielectric-coated Cu lines (stars). © 2004 AIP, reprinted, with permission, from [76].

which would result in an erroneous calculation of the activation energy for EM. Thus, the failure criterion was set at a lifetime τ when an abrupt line resistance change rate $(1/R_o)(\Delta R/\Delta t) \geq 7\%$/h occurred as shown in Figure 5.52, corresponding to a time when a void grew large enough to connect the Cu line and W damascene via through the highly resistive Ta liner. The median lifetime for Cu near-bamboo-like grain lines with a CoWP cap as a function of $1/T$ is plotted in Figure 5.53. For comparison, the results from EM stressing of Cu near-bamboo-like grain lines with Ta/TaN and dielectric caps [77] are included in Figure 5.53. The symbols represent the data points and the solid lines are the least-squares fits of the data. The Q_{EM} for Cu near-bamboo

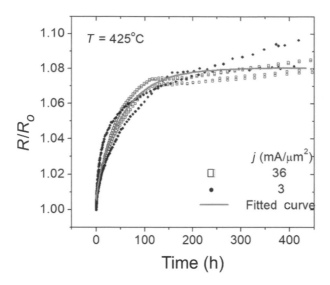

Figure 5.54 Plot of line resistance change vs. time for lines tested at a high current density of 36 mA/μm^2 (open squares) and lines tested at a low current density of 3 mA/μm^2 (solid circles) at 425°C. The solid lines are the least-square fitted lines. © 2004 AIP, reprinted, with permission, from [76].

lines with CoWP caps was found to be 2.4 ± 0.2 eV. This is higher than those obtained from dielectric (0.9–1 eV) and Ta/TaN (1.4 eV) capped Cu lines, and even higher than the values of 2.07 eV or 2.19 eV for Cu bulk diffusion activation energy [79] and the Q_{EM} of 1.9 eV [77] obtained from via bottom void growth in CoWP capped, dual-damascene Cu lines with bamboo-like grain structures tested with a current density of 112 mA/μm^2 at the via bottom. We attribute the high Q_{EM} of 1.9 to 2.4 eV mainly to Cu bulk or Cu/Ta liner interface diffusion with some additional effects from Co diffusion in Cu where vacancies are repelled by the substitutional Co atoms [80] to cause an increase in the activation energy for Cu volume diffusion [81]. However, the Co solute concentration in the Cu lines was less than 0.06 at.%, so such an impurity effect should be small. Instead, the large uncertainty in the measured Q_{EM} determined from Figure 5.53 was partly due to the narrow test temperature range 377°C to 425°C, the uncertainty in measuring the void size, the small sample size, and the nonuniformity of the CoWP coating on the samples. The increase in the value of Q_{EM} from Ta/TaN- to CoWP-capped samples also suggested that diffusion along the Cu/Ta liner sidewall–bottom interface and the top Cu surface–Ta cap interface could be different; even the test samples with either CoWP or Ta/TaN caps contained a similar Cu/Ta liner sidewall–bottom interface. Overall, the result suggests that slight variations in Ta/Cu preparation could cause drastically different interface diffusion.

Figure 5.54 shows the normalized line resistance as a function time for samples stressed at 425°C with a wide M2 connecting to a V1/fine M1/V1 test structure. No large M2 line resistance increase (symbols with open squares) was observed even after

400 hours at current densities of 3 and 36 mA/μm^2. Instead, a large line R change or failure was observed in the fine Cu M1 connected to a blocking boundary W via under the same EM test conditions. Since the EM-induced mass flow under the same j should be the same for M1 and M2, but no large R increase was observed in Cu M1 to Cu M2 and large R change in W to M1, the result suggested that TaN was no longer a diffusion barrier layer for the CoWP-capped Cu bamboo grain at $T = 425°C$. In this case, with a leaky barrier at the M1/V1 interface, the M1 would act as a reservoir and provide the mass transport to replenish M2 as discussed in Section 5.4.4. The sample temperature and long test time are correlated to the estimated Cu drift velocity, suggesting that a Cu line with a CoWP cap has a continuous flow of Cu through the leaky barrier TaN/Ta at the V1/M1 interface, so no EM damage for the CoWP-capped bamboo Cu line was observed in M1 to V1/M2 after testing at $T = 425°C$ for 400 hours. For comparison, Figure 5.54 also shows the line resistance (solid circles) as a function of time using a low current density of 3 mA/μm^2, where the EM damage in these samples was minimal. No significant deviation between the high and low current density line resistance data also suggests that the line resistance increase was mainly due to impurity penetration into the Cu line, although the contribution from EM V1 to M1 line damage was small due to the leaky barrier layer at the V1/M1 interface. The curves turn after about 100 hours, indicating that the impurities in Cu were approaching saturation. The source of the impurity diffusion was most likely the CoWP cap, where the W and P impurities probably did not cause resistance increases during test because the solubility of W in Cu was extremely limited [82] and volume diffusion of P in Cu had a low activation energy of 1.4 eV [83]. One can estimate a diffusion length $(2Dt)^{1/2}$ of 0.5 μm for P in Cu at 400°C for 1 hour, far thicker than 0.14 μm, the thickness of the Cu line, h. In the test samples, the Cu M1 lines were heated above 400°C for several hours during the wafer fabrication. Therefore, the resistance increase from P diffusion into the Cu line would have been stabilized before EM testing. The Co impurity was the most probable source for the line resistance increase while testing. For the short sample annealing time $t \ll h^2/D$, the dissolution of Co in the Cu line can be considered as constant surface composition diffusion in an infinite medium, and the solution of such a diffusion problem is a complementary error function. However, for the more difficult case of diffusion in a finite medium with thickness of h, the problem can be solved by the method of images as a plane sheet of symmetry at $x = 0$ of thickness of $2h$ and surfaces at $x = h$ and $-h$ maintained at constant concentration C. The atomic concentration $c(x,t)$ of Co in Cu as a function of time t at location x between $0 \leq x \leq h$ is given by a series of complementary error functions [84]:

$$c(x,t) = C \sum_{n=0}^{\infty} (-1)^n erfc \frac{(2n+1)h - x}{2\sqrt{D_{Co}t}} + C \sum_{n=0}^{\infty} (-1)^n erfc \frac{(2n+1)h + x}{2\sqrt{D_{Co}t}}, \quad (5.11)$$

where C_o and D_{Co} are the atomic concentration in percent of the solubility limit at the Cu surface and the bulk diffusivity of Co in Cu, respectively. The diffusivity can be expressed by $D_{Co} = D_o e^{(-\frac{Q_{Co}}{kT})}$, where D_o and Q_{Co} are the prefactor and activation energy of Co volume diffusion in Cu, respectively. The residual resistivity of the alloys is a linear function of the dilute solute concentration. The Co solute in Cu increases the resistivity by 6.9 $\mu\Omega$ cm/at.% Co [85], so the resistivity of the Cu(Co) alloy can be expressed by Mattiessen's rule $\rho_o(1 + \alpha T) + 6.9c(x,t)$, where ρ_o is the resistivity of pure Cu at 0°C, and α is the temperature coefficient of resistivity of the uncoated Cu line. The values of ρ_o and α for these samples were estimated to be 2 $\mu\Omega$ cm and 0.32%, respectively. The effective resistivity of the Cu line with the CoWP cap as a function of the annealing time, ρ_{eff}, can be obtained by assuming a series of parallel resistors $R_i = \rho_i L/(w\,\Delta x)$ in the line, where L and w are the length and width of the Cu line, respectively, and Δx the thickness of the Cu(Co) alloy section with a resistivity ρ_i at a concentration $c(x_i,t)$. With Δx assumed to be $h/2{,}000$, the equation for the effective resistivity of a set of 2,000 parallel resistors is given by $1/\rho_{eff} = \sum 1/\rho_i$. The data in Figure 5.54, obtained using low current testing, were fitted to the estimated curves using a $\chi^{[2]}$ minimization method with two adjustable parameters C and D_{Co}. The values of $c(x,t)$ were determined by the Co penetration profile using (5.11). The activation energy Q_{Co} and preexponential factor D_o of Co diffusivity in Cu are estimated to be 2.2 ± 0.2 eV and 2.0 cm²/s, respectively, which are in good agreement with the published values [86] of $Q_{Co} = 2.19$ eV and $D_o = 0.39$ cm²/s for Co bulk diffusivity in Cu measured between 700–800°C. The best values of C versus T obtained from nonlinear least-square fits are plotted in Figure 5.55 along with the values obtained earlier [82]. The present results are reasonably consistent with measurements taken at high temperatures by the earlier investigators. The atom fraction solubility of Co in Cu was found to be 18 exp[−0.57 eV/kT] from fitting all the data points in Figure 5.55.

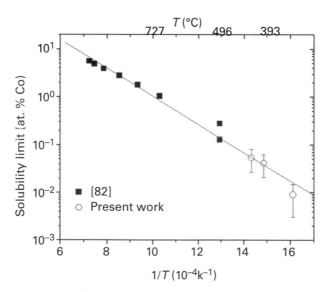

Figure 5.55 Solubility limit for Co diffusing into Cu vs. (1/T). Open symbols represent data generated from this present study, and solid symbols represent data from earlier workers. © 2004 AIP, reprinted, with permission, from [76].

5.7.5.3 Polycrystalline Line with CoWP Cap

In Section 5.7.5.2, we reported an improvement in Cu EM lifetime with a thin (1–20 nm) surface layer of electroless CoWP or a Co cap at the Cu surface, where the measured EM activation energy in the bamboo-like Cu damascene line with CoWP was found to be close to that of Cu bulk diffusion. In this section, we report a study on the effect on EM due to incorporating Co in the Cu GB and the metal cap in Cu polycrystalline lines, which is of increasing importance since there are more polycrystalline sections in Cu nanowiring. To investigate this effect, a 2 μm wide fine-grained polycrystalline Cu line capped with $CoWP/a\text{-}SiC_xN_yH_z$ was used for this study. The polycrystalline damascene lines were achieved by skipping the post-ECD Cu annealing and carrying out the CMP step directly after Cu ECD, where Co distributed in the Cu grain boundaries was achieved by the high temperature $a\text{-}SiC_xN_yH_z$ deposition step and a final 400°C anneal for 2 hours.

Figure 5.56a is a plan-view TEM image of the 2 μm wide Cu polycrystalline damascene lines, where the hand-traced lines on the graph illustrate the location of Cu grain boundaries. Fine grain sizes and polycrystalline line structure with a network of GB paths are clearly shown in Figure 5.56a. Figure 5.56b and c are the plots of cumulative failure probability versus Cu lifetime and median lifetime versus $1/T$ for the samples with and without a CoWP cap, respectively. Unlike Q_{EM} of 2 eV observed in Cu bamboo-like grains capped with CoWP, the EM activation energies were found to be 0.79 ± 0.05 eV and 0.72 ± 0.05 eV from samples with and without a CoWP cap, respectively. The measured Q_{EM} of 0.79 eV is in good agreement with the activation energy of Cu grain boundary diffusion [74]. Slightly reduced Cu lifetimes and Q_{EM} in Cu lines with a CoWP cap were observed, suggesting that the Co addition in Cu grain boundaries slightly enhanced Cu GB diffusion and hence reduced EM lifetime.

5.7.6 Effect of Alloying on Cu Electromigration

The EM lifetime scaling rule data shown in Figure 5.36 indicate that the pure Cu interconnect structures must be improved in order to obtain a reliable Cu interconnect beyond the 65 nm technology node, either by implementing Cu alloys or metal capping on the Cu line surface. Dilute metal impurities in Cu lines (Cu-alloys), such as Al [87], Mn [88], Pd [89], Sn [54], Ti [90, 91], etc., can enhance the EM lifetime and have a negligible effect on Cu ρ since pure Cu ρ below the 45 nm node increased by more than 2× as shown in Figure 5.3. To achieve EM reliability improvement as shown in Figure 5.53, a selective electroless CoWP or CVD Co cap was used to deposit on the Cu line surface, although this process is more complicated and expensive than the Cu alloy approach. In practice, the implementation of selective CoWP or CVD-Co in wafer fabrication is quite challenging, thus the Cu alloying technique became the favorite choice in IC manufacturing for 65 to 14 nm nodes to meet the requirement of maximum current densities used in the BEOL metallization for these nodes.

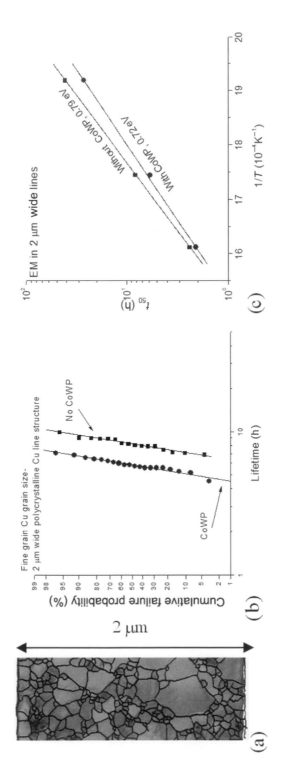

Figure 5.56 (a) Top-view TEM image and corresponding drawings of grain boundaries for a 2 μm wide line; (b) plot of lifetime vs. cumulative lifetime probability for the samples with and without a CoWP cap; and (c) plot of t_{50} vs. $1/T$ for with and without CoWP cap on the polycrystalline lines. © 2010 AIP, reprinted, with permission, from [72].

5.7.6.1   Cu(Al) Alloying on Near-Bamboo, Bamboo–Polycrystalline, and Polycrystalline Lines

The use of Cu(Al) to improve EM reliability in 65 nm and 1.5 μm wide lines is described in this section. The Cu(Al) damascene interconnections were fabricated using a Cu(1–2 at.% Al) seed layer to replace the pure Cu seed on the line trenches or vias before the electroplating Cu step. A polycrystalline–bamboo microstructure was obtained in the 65 nm wide lines following a 100°C post-ECD Cu annealing step, and near-bamboo and polycrystalline-only structures were obtained in the 1.5 μm wide lines with and without the 100°C post-ECD Cu annealing step and CMP Cu right away, respectively. The median lifetime t_{50} based on a 3% failure criterion for the 65 nm wide polycrystalline–bamboo Cu damascene lines using a pure Cu seed or Cu (1%Al) seed is plotted as a function of $1/T$ in Figure 5.57. (Al impurity enhanced Cu EM lifetime results were also observed in 1.5 μm wide with 100°C postplating Cu anneal, but the data are not shown). Comparing these samples to pure Cu samples, more than a $10\times$ enhancement factor for EM in Cu(Al) lines was observed at a sample temperature of 400°C, as shown in Figure 5.57, where the straight lines are the least-square fits. Q_{EM} in these Cu damascene lines can be extracted from the line slopes in Figure 5.57 and found to be 0.85 ± 0.05 eV and 1.15 ± 0.1 eV for pure Cu and Cu(Al), respectively. Similar observations have also been reported [87].

During this investigation, a set of large-grained wafers were exposed to air for a short time while the wafers were still hot to yield oxidized Cu(Al) lines, then followed by a plasma preclean before the a-SiC$_x$H$_y$N$_z$ dielectric cap deposition. The Cu lifetimes of these samples were greatly decreased, indicating that a slight Cu(Al) surface oxidation can have a large impact in degrading the Cu/a-SiC$_x$H$_y$N$_z$ interface.

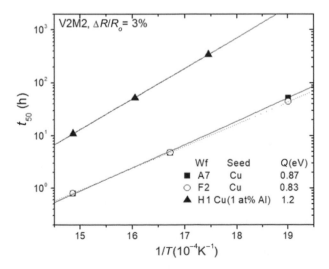

Figure 5.57 Median lifetime vs. $1/T$ for 65 nm wide lines from samples using pure Cu and Cu(1% Al) seeds. © 2010 AIP, reprinted, with permission, from [72].

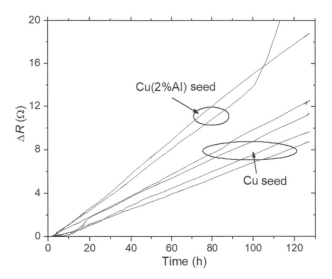

Figure 5.58 Plot of the line resistance change for 1.5 μm wide Cu(2% Al) and pure Cu seed lines. Both Cu(Al) and pure Cu are polycrystalline lines. © 2010 AIP, reprinted, with permission, from [72].

Figure 5.58 shows the line resistance change versus EM stress time for pure Cu and Cu(2% Al) seed drift velocity samples. These samples were 1.5 μm wide, 100 μm long, and with a polycrystalline fine-grain microstructure. These polycrystalline damascenes lines were fabricated by skipping the post-ECD Cu annealing and with the CMP step performed directly after Cu ECD and capped with a 35 nm thick a-SiC$_x$H$_y$N$_z$. Voids were typically observed near and/or at the cathode end of the Cu test line. In Figure 5.59a and b, the void length is compared for pure Cu and Cu(Al) 100 μm long EM-tested samples and found to be 15 μm and 23 μm, respectively. Similar EM behavior was also observed in Cu(1%Al) seed samples. Overall, the results show that Al-enhanced Cu grain boundary diffusion increases the void growth rate under EM. This behavior was consistent with that reported previously [92], where extrusions were seen near or at the anode end of the line, as shown in Figure 5.59c. The Cu mass accumulated at the anode end of the line driven by the EM driving force to induce a compressive stress, which was relieved by cracking the a-SiC$_x$N$_y$H$_z$ capping film to form a Cu extrusion. Interestingly, no EM line damage was observed in lines less than or equal to 10 μm in length, manifesting the Blech critical length effect [34] at the current density applied in these tests. The faster EM-induced void growth and larger line resistance degradation rates in the Cu(Al) samples suggest that Al enhanced the Cu grain boundary diffusion and increased the EM-induced void growth rate.

These results suggested that Al mitigated the Cu migration along the Cu interfaces but not at the grain boundaries. The fact that enhanced Cu lifetimes in Cu(Al) were obtained in 65 nm and 1.5 μm wide near-bamboo and polycrystalline–bamboo Cu(Al) lines but not the 1.5 μm wide polycrystalline fine-grained lines indicated that the bamboo grains play a key role in slowing down void growth rate. We believe, as previously discussed, that the bamboo grains act as Cu diffusion blocking boundaries

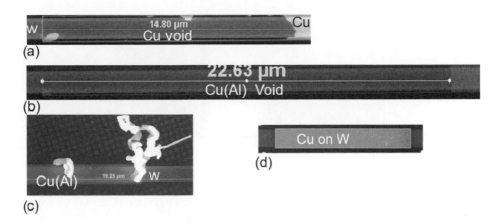

Figure 5.59 Top-down SEM images of EM-tested 1.5 μm wide 100 μm long (a) pure Cu [72], (b) and (c) Cu(2%Al) ([72] [88]) respectively), and (d) 10 μm long Cu(Al) [72] line segments on W line at 300°C with total test times of 127 hours, respectively. The electron flow direction is from left to right. Large void lengths at the cathode end (a) and (b), and extrusion at the anode end of the line (c) were observed in 100 μm long lines, and no visible damage was observed in 10 μm long lines (d). © 2010 AIP, reprinted, with permission, from [72]. © 2012 AIP, reprinted, with permission, from [88].

for GB mass flow and generate a mechanical backflow to further reduce the EM GB mass flow, as evidenced by the presence of bamboo grains within 1 μm lengths in TEM cross-section images of the 65 nm wide lines. A sub-1 μm polycrystalline length section is shorter than the estimated Blech critical length of 10 μm in Cu, and the result is consistent with our observations of no EM line damage in 5 or 10 μm long lines.

5.7.6.2   Cu(Ti) Alloy Seed

In this section, we discuss the effect of a Cu(Ti) seed layer on EM reliability. The study was carried out using a Cu(0.5 wt.% Ti) or Cu(2.5 wt.% Ti) seed layer in 0.22 μm wide Cu dual-damascene lines. The initial volume fraction occupied by the Cu alloy seed in the metal line was about 30%. Comparing to pure Cu lines, the line resistance was found to increase by 5% with the Cu(0.5 wt.% Ti) and 17% with the Cu (2.5 wt.% Ti) seeds after annealing at 400°C for 2 hours. In Figure 5.60a and b, we show the TEM cross-sectional images of the Cu via/Cu line and Cu line, respectively, from wafers with a PVD Cu(2.5 wt.% Ti) seed layer, which was used to replace the pure Cu seed at the line trenches/vias before the ECD Cu processing step. The wafers were annealed at 100°C after Cu electroplating and before CMP, during which the initial fine-grain structure of a pure PVD Cu seed layer would have dissolved into large electroplated Cu grains due to abnormal grain growth. However, Figure 5.60a shows that the same process did not dissolve the Cu(Ti) seed layer covering the via bottom and sidewalls into large Cu grains and retained as a fine-grain Cu(Ti) layer as shown. This indicates that the Ti impurity pinned the PVD Cu grain boundary movement with no grain growth. The fine Cu(Ti) grains, however, disappeared into

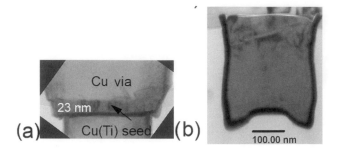

Figure 5.60 Cross-section TEM images of (a) Cu via/Cu line with a fine-grain Cu(2.5%Ti) alloy seed layer; and (b) Cu(2.5%Ti) line after 400°C anneal. © 2007 AIP, reprinted, with permission, from [11].

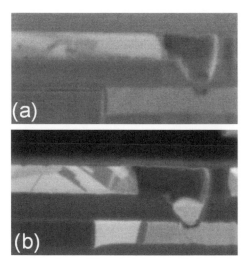

Figure 5.61 FIB images of EM-tested pure Cu lines with SiCHN cap at 267°C with total test times of (a) 839 hours and (b) 1,011 hours, respectively. © 2007 AIP, reprinted, with permission, from [11].

the large-plated Cu grain after 400°C annealing, as shown in Figure 5.60b. The Ti concentrations in the Cu grains and at the surface were below the detection limit of EDX; instead, Ti was detected in the Ta liner, indicating that most of the Ti impurity was dissolved into the Ta liner and not in Cu, which may explain why no EM enhancement was observed in the sample with the Cu(0.5 wt.% Ti) seed.

EM enhancement was observed in the Cu line using a Cu(2.5 wt.% Ti) seed layer. Figure 5.61a and b shows FIB images of dual-damascene pure Cu failed lines, where EM-induced void formation was observed in the line and/or at the via with electron flow upstream from the Cu via to the Cu line. For these lines, a median EM lifetime of 870 hours and a deviation (σ) of 0.4 were measured. In comparison, no failure was observed in the Cu(2.5% Ti) alloy lines even after testing for 2,600 hours. During the

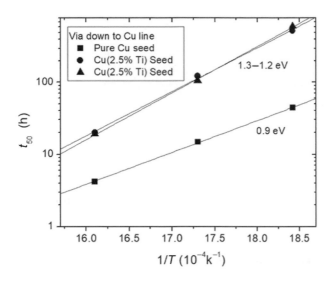

Figure 5.62 Plot of t_{50} vs. $(1/T)$ for pure Cu and Cu(Ti) samples. © 2007 AIP, reprinted, with permission, from [11].

EM test, the Cu(Ti) line resistance decreased, suggesting that an impurity purification process has taken place. In Figure 5.62, we compare the median lifetime t_{50} obtained with a 3% resistance increase failure criterion for the Cu and Cu(Ti) damascene lines as a function of $1/T$. The solid lines are the least-squares fits. The activation energy for EM, Q_{EM}, was found to increase from 0.91 eV for the Cu lines to 1.3 eV for the Cu(Ti) lines, with the uncertainties of about 0.05 eV. The results illustrate an improvement in the chip lifetime for the Cu(2.5 wt.% Ti) alloy, where the activation energies are in good agreement with previously reported values [90].

5.7.6.3   Effect of Mn on Cu Grain Boundary Diffusion

The Mg [54] and Al impurities have been found to increase the Cu grain boundary diffusivity in polycrystalline lines. However, in samples with low concentration of Ti, Al, Mn, or Co, no significant effect on Cu grain boundary diffusion was observed. Instead, the addition of Ti, Mn, or Al solutes was found to cause a reduction in Cu interface diffusion and an increase in the EM activation energy for Cu-alloy bamboo–polycrystalline lines as compared to pure Cu. In this section, we present a study on the effect of Mn impurity on Cu grain boundary diffusion using 1.5 μm wide Cu(Mn) polycrystalline line samples with grain boundary structures as shown in Figures 5.63 and 5.64. The polycrystalline damascene lines were fabricated using a pure Cu or Cu(1–3% Mn) seed for ECD Cu, skipping the post-ECD Cu anneal, and performing the CMP step directly after Cu ECD. The Cu alloy line segments were 5, 10, 30, 60, and 100 μm long and connected to lower W lines. The Mn was dispersed in the Cu grain boundaries by a final 400°C anneal for 2 hours, then the samples were passivated with a 25–30 nm thick a-SiC$_x$H$_y$N$_z$ layer to eliminate Cu surface diffusion. The samples were tested at current

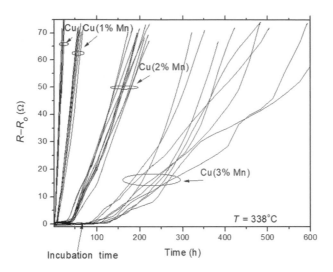

Figure 5.63 The line resistance change vs. time for four different Cu metallizations using pure Cu, Cu(1% Mn), Cu(2% Mn), and Cu(3% Mn) seeds. A resistance incubation time, a period with no resistance increase, is also plotted. © 2014 AIP, reprinted, with permission, from [35].

density of 50 mA/μm^2 as a function of temperature and time to observe the void and hillock growth kinetics. Void formation was found in Cu lines longer than 5 μm, suggesting that the critical length for these samples was between 5 and 10 μm.

In Figure 5.63, we show the line resistance change observed at 338°C as a function of time for 1.5 μm wide polycrystalline lines of pure Cu and with Cu (1–3% Mn) seeds. The results show typically a period of no resistance increase (incubation period), then followed by an increase in the line resistance. Figure 5.64a–d compares the edge displacement (void length) of pure Cu and Cu(2% Mn) seed lines as a function of time at 338°C, while Figure 5.64e is an image of the TEM cross section along the anode end of the pure Cu line, showing a hillock formation as a results of epitaxial Cu growth on the original Cu line. Figure 5.65 shows the growth of the void length as a function of time at a rate for the Cu(Mn) alloy line considerably lower than the pure Cu. In comparison, both the resistance increase rate and the drift velocity were reduced with increasing Mn concentrations. The incubation period was close to zero for pure Cu samples and increased with increasingly Mn concentrations, which can be attributed to the fact that Mn was depleted in the cathode end of the line before Cu could be significantly migrated. The hillocks started to form beyond the critical length (~6–8 μm) from the cathode end of the line as shown in Figure 5.64b and continued to grow at the interface due to the fast Cu mass flow in the Mn depleted zone but slow in the Cu(Mn) zone. The location of the hillock growth was found to follow the Mn depleted edge in the direction of electron flow as shown in Figure 5.64b–d. These observations are similar to the case of EM in Al(2 at.% Cu) [93].

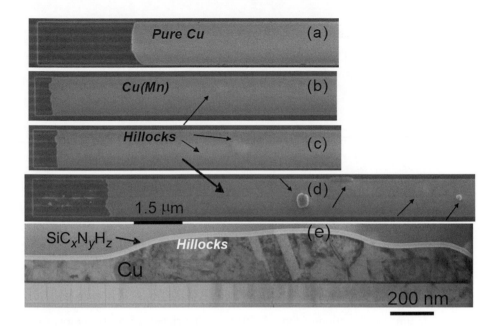

Figure 5.64 Top-view SEM images of EM-tested drift velocity samples near the cathode end of the lines at 338°C: (a) pure Cu seed; (b–d) Cu(2%Mn) seed with various EM stress times of (a) 11, (b) 24, (c) 48, and (d) 84 hours, respectively. (e) TEM cross section along the anode end of the pure Cu line. Arrows indicate the hillock locations and electrons flow from left to right. © 2014 AIP, reprinted, with permission, from [35].

5.7.6.4 Trapping Model

The reduction of Cu mass flow or diffusivity in Cu(Mn) alloys can be qualitatively accounted for by a trapping model [94]. Based on this model, the mobility of Cu at the GB is reduced by the presence of Mn such that the free Cu travels rapidly through GB until it meets the Mn where it is trapped. Then one can apply the mass action law to set the relationship between the number of trapped Cu atoms n_t and the number of free Cu atoms n_f as follows:

$$n_t = n_f x_{gb} z e^{\left(-\frac{F}{k_b T}\right)},$$ (5.12)

where z is the number of degenerate positions at the trapping the Mn atom, x_{gb} is the Mn concentration in gb, and F is the free energy $F = -B - TS$. Here B is the binding energy between solute and Cu vacancy or interstitialcy, which depends on diffusion mechanism in gb, and S is the associated entropy. Assuming the trapped Cu makes no significant contribution to diffusion, one obtains the following:

$$\left[\left(\frac{D}{D_x}\right) - 1\right] \Big/ x_{gb} = [z \exp (S/k)] \exp (B/(k_B T)).$$ (5.13)

The drift velocities of Cu and Cu(Mn) related to diffusivities were obtained by using data for the void growth rate within the 1.5 μm void length. The ratios of effective

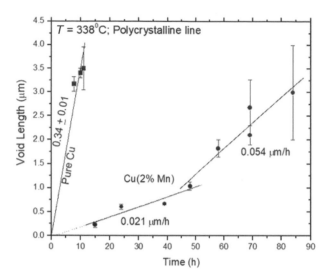

Figure 5.65 Edge displacements of Cu and Cu(Mn) measured at 338°C and 50 mA/μm^2 as a function of time. © 2014 AIP, reprinted, with permission, from [35].

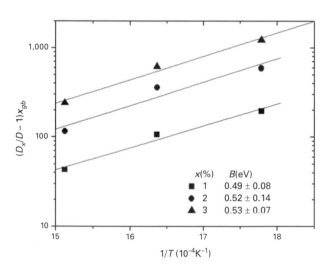

Figure 5.66 Plot of $D/D_x - 1$ vs. $1/T$ using polycrystalline only samples. x is the percentage of the Mn concentration Cu alloy seed layer. © 2014 AIP, reprinted, with permission, from [35].

diffusivity D/D_x in Cu gb between pure Cu D and Cu(Mn) alloy D_x were obtained by using drift velocity from Figure 5.65 and (5.3b) with the assumption of similar values of grain size d and $Z^*\rho$ in pure Cu and Cu(Mn) alloys. The left-hand side of (5.13) is plotted versus $1/T$ in Figure 5.66. Systematically low values of $[(D/D_x) - 1]/x_{gb}$ for the lower Mn concentration samples are shown and are presumably due to the estimated errors from the values of x_{gb}, which were assumed to be linearly proportional to the seed layer concentration $x\%$ Mn in these calculations. As a result, a larger fraction of Mn would piled up at interfaces with the lower Mn concentration. Consequently, the low Mn concentration samples had overestimated x_{gb} values as compared to the high Mn concentration samples. However, a constant and a strong binding energy of 0.5 ± 0.05 eV was obtained from the slopes of these samples. In contrast rather, weak Cu vacancy–Mn binding energy in the bulk was observed by positron annihilation technique [95]. These results suggest that the mechanistic effects of solute on diffusion in the grain boundaries are fundamentally different than found in the bulk and can be unpredictable, e.g., adding Al solute in Cu, which appears to increase the GB diffusivity.

5.8   EM through Various Technology Nodes

In this section, we summarize the discussions presented so far on the EM performance for Cu damascene lines in various CMOS technology generations. The measured EM lifetimes are summarized as a function of $1/T$ in Figure 5.67 for Cu BEOL with continued scaling in line widths. Pure Cu interconnects were generally used down to the 45 nm node, and then Cu alloy interconnects were employed from the 32 to the 14 nm nodes. Beyond the 14 nm node (64 nm linewidth and space pitch), a metal cap was employed on the top surface of the Cu line to further improve the EM

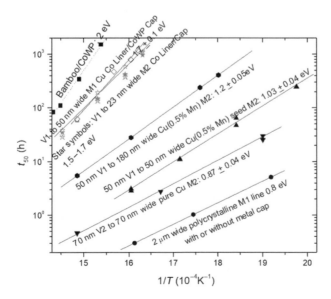

Figure 5.67 Plot of t_{50} vs. $1/T$ for various Cu microstructures: bamboo, bamboo–polycrystalline, and polycrystalline-only line structures with and without a metal cap. All data points are bamboo–polycrystalline lines, unless specified. The lines are the least-square fits to the data. © 2018 IEEE. Reprinted, with permission, from [15].

performance in meeting the current density requirement for high-performance ICs. As the line structure changed with continued scaling, the improvement in EM performance became evident from the increase in the EM activation energy Q_{EM} with corresponding changes in the grain structure. In Figure 5.67, we show that the value of Q_{EM} improved from 1.0–0.9 eV for pure Cu lines without a metal cap above and below the 90 nm width to about 2 eV for 0.2 μm wide Cu lines with TaN/Ta liners and CoWP caps with near-bamboo grains (solid squares) and 1.7 eV for 50 nm wide bamboo–polycrystalline lines (open diamonds). For the 23 nm wide lines with bamboo–polycrystalline grains and TaN/Co liners and a selective CVD Co cap (open and solid stars), the activation energy was found to be 1.7–1.5 eV. For Cu(Mn) alloy lines without a metal cap, the alloying effect improved EM to yield a Q_{EM} of 1.2 and 1.0 eV for 50 nm wide lines with near-bamboo and bamboo–polycrystalline grains, respectively. The alloying effect in mitigating the EM-induced mass transport can be explained by the trapping model as discussed in Section 5.7.6.4. This model showed that the number of free diffusing Cu atoms in the fast diffusion paths were reduced by the impurity traps, depending on the binding between Cu and impurity atoms in the fast diffusion paths. For pure Cu polycrystalline lines, Q_{EM} values as low as 0.75 eV were observed with or without a CoWP cap and with no measurable effect for EM in the polycrystalline lines. Such results were in sharp contrast to the 2eV for 0.2 μm Cu lines and the 1.7 eV for 50 nm Cu lines with a TaN/Ta liner and CoWP cap and near-bamboo and bamboo–polycrystalline grains, which showed significantly improved EM performance.

Overall, the EM data in Figure 5.67 illustrate that the Cu grain structure and the diffusivities at the Cu/liner or Cu/capping layer interfaces do play a significant role in improving the overall Cu EM lifetime with Q_{EM} increased from 0.8 eV to 2 eV. It is worth noting that the enhanced Cu lifetimes in Cu lines with a CoWP or Co cap were obtained only in bamboo grain lines or bamboo–polycrystalline lines, and not in polycrystalline-only lines. These results suggest that the Cu bamboo grain sections together with Cu/liner and Cu/cap interface are important in reducing the Cu diffusivities in the lines to slow down the EM-induced void growth rate. The improvement can be attributed to the Blech short-length effect, where the bamboo grain sections in the Cu line served as diffusion blocking boundaries to retard the GB mass flow and generated a mechanically induced backflow to reduce the EM GB mass flow. However, short-length models simply based on GB mass flow to estimate the lifetime $\tau = \Delta L_{cr}/v_d$ either as $\tau = \Delta L_{cr}/[p_{GB}v_{GB} + (1 - p_{GB})v_B] \sim \Delta L_{cr}/(p_{GB}v_{GB})$, or $\tau = p_{GB}\Delta L_{cr}/v_{GB} + (1 - p_{gb})$ $\Delta L_{cr}/v_B \sim (1 - p_{gb}) \Delta L_{cr}/v_B$, where p_{gb}, v_{GB} and v_B are the fraction of polycrystalline grain sections in the line, Cu grain boundaries, and bulk drift velocities respectively, could not account for the results of the bamboo–polycrystalline lines. This is because neither of these equations include the Cu mass transport along the Cu–liner interface or along the metal cap–Cu interface, which is important in contributing to the overall Cu mass flow since the volume fraction of atoms at interfaces increases rapidly in the nanowires, with the mass transport being increasingly controlled by the liner–Cu interface. Taking this effect into account, the net EM drift velocity in a Cu damascene line with a metal cap and bamboo-like or bamboo–polycrystalline grains was derived as (5.4). In this equation, the mass flow v_d in the polycrystalline section between two Cu bamboo grains with a metal cap is retarded by the EM-induced mechanical backflow due to the Blech short-length effect. Thus, v_d in bamboo–polycrystalline lines is dominated by the mass flow with Q_I equal to 1.5–1.7 eV in the bamboo sections with a Cu–Co interface instead of the Q_{EM} of 0.8 eV along the grain boundaries. With the short-length effect, a Cu line 20–24 nm wide with a TaN/Co liner and a Co cap, t_{50} is predicted to exceed several thousand years at 140°C with $j = 1.5 \times 10^7 A/cm^2$ [15]. Thus, highly EM reliable interconnects can be achieved by capping the Cu line surface with a Co cap together with a good Cu–liner interface.

5.9 Summary

In this chapter, we have discussed the effects of scaling on microstructure, resistivity, and EM performance for Cu damascene interconnects through various technology nodes. Transmission electron microscopy (TEM) with the precession diffraction technique was used to characterize the microstructure in the Cu lines. With decreasing line widths, the grain structure in Cu interconnects was found to vary from near-bamboo-like to polycrystalline mixing with single-crystal bamboo grains. The mass transport in Cu interconnects was found to occur mainly by interface and grain boundary diffusion, strongly dependent on the Cu microstructure. When Cu atoms migrate along these diffusion paths, the mass transport can generate line edge displacement or activate surface grain thinning mechanisms to induce void formation. The shape evolution and location of the EM-induced voids were correlated to the line resistance change as a function of time. Initially, the line resistance changed slowly, followed by an abrupt stepwise resistance change and a period of rapid resistance rise. In some cases, the void was found to form at a distance away from the cathode end of the line, and then moved toward the end via or got trapped at certain locations without a detectable line resistance change. After being trapped, the void movement mostly ceased, then continued to grow, and finally resulted in the line failure.

In general, the kinetics of void growth and resistance change that determine the EM lifetime depend on the net rate of Cu diffusion driven by the current and the distribution of structural defects, where the grain structure in the Cu line plays a crucial role. In the previous section, we show that with continued scaling, the mass transport at the top surface of the Cu line became increasingly important. This has led to the use of CoWP or Co cap layers to improve the EM lifetime by reducing the interface mass transport, particularly below the 20–24 nm linewidth. The improvements, however, were obtained only in bamboo grain lines or bamboo–polycrystalline lines, and not in polycrystalline-only lines, which was attributed to the Blech short-length effect, with the bamboo grains serving as diffusion blocking boundaries. Similarly, the fast diffusion in Cu lines can be greatly reduced with the addition of metallic impurities, such as Ti, Al, and Mn. Using Cu(1 at.% Al) or Cu(2.5 wt.% Ti) seed layers for the 65 nm and 250 nm wide lines, the EM lifetimes were enhanced, with Q_{EM} increasing from 0.9 to 1.15 eV and 0.9 to 1.3 eV, respectively. However, EM studies of 1.5–2 μm wide polycrystalline-only lines with Al and Co impurities yielded a reduced lifetime with a higher void growth rate due to the lack of bamboo grains serving as blocking boundaries. Thus, a line can become immortal if the EM test condition meets the criteria of the Blech short-length effect, with the current density-length product jL below a threshold value and in a line with completely blocking boundaries for Cu diffusion.

The alloying process can improve EM lifetimes but at the expense of metal conductivity. Hydrogen-based plasma precleans prior to Cu passivation with a silicon carbon nitride has also found to reduce EM failure in Cu interconnects, but as in the case of alloys, there is a trade-off between reliability and performance. The addition of nonmetallic impurities of Cl, C, S, and O has also been used, but no difference in Cu lifetime was observed, suggesting that the microstructure plays a key role in the superiority of ECD Cu compared to CVD or PVD Cu. There are other structure features that can improve the EM lifetime. For examples, an artificially long lifetime can be obtained if the via is connected through a leaky barrier layer to the Cu line with a reservoir. An upper dummy via structure can also enhance the EM lifetime by interrupting the Cu mass flow along the top interface without increasing the Cu resistivity. With dummy via structures, wide lines with near-bamboo microstructure have shown improved EM lifetimes as compared with fine lines with a larger number of polycrystalline grain sections. Such structures can be used to understand the role of the Cu microstructure and to distinguish different fast diffusion paths in the line.

Recently, the use of Cu/CNT composite lines [75] and Cu lines with MnO_x liners [96] has generated considerable interest. The EM lifetimes were found to be highly degraded due to the CNT pinning on Cu grain growth and the fast diffusivity along MnO_x/Cu interface. Further techniques for improving Cu nanowire microstructure and reducing defects, e.g., the ALD barrier layer/CVD liner and Cu seed reflow process [97], have been suggested to develop 5 nm node and beyond technologies. The initial results showed that interface sidewall voids and/or discontinuous thin barrier/liner on sidewalls were probably responsible the poor EM lifetime observed. At this time, the barrier/liner thickness on sidewalls ≥ 3 nm is probably too thick for the future nodes. The extendibility of Cu BEOL for future nodes will rely on the improvement of Cu microstructure for the high metal height and width aspect ratio line, the thin barrier/liner ≤ 3 nm thick, Cu seed, and ECD Cu processes.

References

1. C.-K. Hu, M. B. Small, F. Kaufman, and D. J. Pearson, Copper-polyimide wiring technology for VLSI circuit, *Proceedings of the VLSI V* (Warrendale: Materials Research Society, 1990), 369.
2. J. Paraszczak, D. Edelstein, S. Cohen, E. Babich, and J. Hummel, High performance dielectrics and processes for ULSI interconnection technologies, *Tech. Digest, IEEE International Electron Devices Meeting* (Piscataway: IEEE, 1993), 261.

3. D. Edelstein, C. Uzoh, C. Cabral, Jr., et al., A high performance liner for copper damascene interconnects, *IEEE International Interconnect Technology Conference* (2001), 9–11.

4. D. Edelstein, J. Heidenreich, R.D. Goldblatt, et al., Full copper wiring in a sub-0.25 μm CMOS ULSI technology, *Tech. Digest, IEEE International Electron Devices Meeting* (Piscataway: IEEE, 1997), 773.

5. C.-K. Hu and B. Luther, Electromigration in two-level interconnects of Cu and Al alloys, *Materials Chemistry and Physics* **41** (1995), 1.

6. H. B. Huntington, *Electromigration in Metal in Diffusion in Solids: Recent Developments*, ed. A. S. Nowick and J. J. Burton (New York: Academic, 1974), 303.

7. C.-K. Hu, and J. M. E. Harper, Copper interconnection fabrication and reliability, *Material Chemistry and Physics* **51** (1998), 5.

8. C-K. Hu, S. Chang, M. B. Small, and J. E. Lewis, Diffusion barrier studies for Cu, *Proceedings of the 3rd IEEE VMIC Conference* (1986), 181.

9. J. M. E. Harper, E. G. Colgan, C.-K. Hu, J. P. Hummel, L. P. Buchwalter, and C. E. Uzoh, Materials issues in copper interconnections, *Materials Research Society Bulletin* **19** (1994), 23.

10. F. Chen, and D. Gardner, Influence of line dimensions on the resistance of Cu interconnections, *IEEE Electron Device Letters* **19** (1998), 508–510.

11. C.-K. Hu, L.M. Gignac, B. Baker, et al., Electromigration reliability of advanced interconnects, *AIP Conference Proceedings* **945**, (2007), 27–41.

12. L. Zhang, J. Im, and P. S. Ho, Line scaling effect on grain structure for Cu interconnects, *AIP Conference Proceedings* **1143**, (2009), 151–155.

13. S.-T. Hu, L. Cao, L. Spinella, and P. S. Ho, Microstructure evolution and effect on resistivity for Cu nanointerconnects and beyond, *2018 IEEE International Electron Devices Meeting* (Piscataway: IEEE, 2018), 5.4.1.

14. C.-K. Hu, E. G. Liniger, L. M. Gignac, G. Bonilla, and D. Edelstein, Materials and scaling effects on on-chip interconnect reliability, MRS Online Proceedings Library (2013), 1559. mrss13–1559-aa07–01 doi:10.1557/opl.2013.872.

15. C.-K. Hu, L. Gignac, G. Lian, et al., Mechanisms of electromigration damage in Cu interconnects, *2018 IEEE International Electron Devices Meeting* (Piscataway: IEEE, 2018), 5.2.1–5.2.4.

16. H. A. Schafft, S. Mayo, S. N. Jones, and J. S. Suehle, An electrical method for determining the thickness of metal films and the cross-sectional area of metal lines, *Proceedings of IEEE International Integrated Reliability Workshop (IRWS)* (1994). doi:10.1109/IRWS.1994.515820.

17. D. E. Gray, *American Institute of Physics Handbook*, 2nd ed. (New York: McGraw-Hill Book Company Inc., 1957).

18. K. Fuchs, The conductivity of thin metallic films according to the electron theory of metals, *Proceedings of the Cambridge Philosophical Society* **34** (1938), 100.

19. E. H. Sondheimer, The mean free path of electrons in metals, *Advances in Physics* **1** (1952), 1.

20. A. F. Mayadas, and M. Shatzkes, Electrical-resistivity model for polycrystalline films: the case of arbitrary reflection at external surfaces, *Physical Review* **B 1** (1970), 1382.

21. D. K. C. MacDonald and K. Sarginson, Size effect variation of the electrical conductivity of metals, *Proceedings of the Royal Society* **203** (1950), 223.

22. R. S. Smith, E. T. Ryan, C.-K. Hu, et al., An evaluation of Fuchs–Sondheimer and Mayadas–Shatzkes models below 14nm node wide lines, *AIP Advances* **9** (2019), 025015.

23. S. Maitrejean, R. Gers, T. Mourier, A. Toffoli, and G. Passemard, Experimental measurements of electron scattering parameters in Cu narrow lines, *Microelectronics Engineering* **83** (2006), 2396–2401.
24. Q. Huang, C. M. Lilley, M. Bode, and R. Divan, Surface and size effects on the electrical properties of Cu nanowires, *Journal of Applied Physics* **104** (2008), 023709-6.
25. T. Sun, B. Yao, A. P. Warren, K. Barmak, M. F. Toney, R. E. Peale, and K. R. Coffey, Dominant role of grain boundary scattering in the resistivity of nanometric Cu films, *Physical Review B* **79** (2009), 041402R.
26. C.-K. Hu, J. Kelly, J. H-C. Chen, et al., Electromigration and resistivity in on-chip Cu, Co and Ru damascene nanowires, *2017 IEEE International Interconnect Technology Conference (IITC)* (Piscataway: IEEE, 2017), 1–3. doi: 10.1109/IITC-AMC.2017.7968977.
27. W. Zhang, S. H. Brongersma, Z. Li, D. Li, O. Richard, and K. Maex, Analysis of the size effect in electroplated fine copper wires and a realistic assessment to model copper resistivity, *Journal of Applied Physics* **101** (2007), 063703.
28. R. Newnham, *Properties of Materials: Anisotropy, Symmetry, Structure* (Oxford: Oxford University Press, 2004).
29. E. Justi, Electrical properties of ruthenium, *Z. Naturforsch* **49** (1949), 472.
30. R. J. Tainsh and G. K. White, Resistivity of Ru, *Canadian Journal of Physics* **42** (1964), 208–209.
31. C.-K. Hu, J. Kelly, H. Huang, et al., Future on-chip interconnect metallization and electromigration, *IEEE International Reliability Physics Symposium (IRPS)* (Piscataway: IEEE, 2018), section 4F.1.
32. M. F. Mott and H. Jones, *The Theory of the Properties of Metals and Alloys* (New York: Dover Publications, 1958), chapter 7.
33. E. Lee, N. Truong, R. Prater, and D. Morales, Novel ruthenium-based materials and ruthenium alloys: their use in vapor deposition or atomic layer deposition and films produced therefrom. US20080274369 (2006).
34. A. Blech, Electromigration in thin aluminum films on titanium nitride, *Journal of Applied Physics* **47** (1976), 1203–1208.
35. C.-K. Hu, L. G. Gignac, J. Ohm, et al., Microstructure, impurity and metal cap effects on Cu electromigration, *AIP Proceedings* **1601** (2014), 67.
36. M. H. Tsai, R. Augur, V. Blaschke, et al., Electromigration reliability of dual damascene Cu/CVD SiOC interconnects, *Proceedings of the IEEE 2001 International Interconnect Technology Conference* (Piscataway: IEEE, 2001), 266.
37. S. Tokogawa, and H. Takizawa, Electromigration induced incubation, drift and threshold in single-damascene copper interconnects, *Proceedings of the IEEE 2002 International Interconnect Technology Conference* (Piscataway: IEEE, 2002), 127.
38. C. S. Hau-Riege, A. P. Marathe, and V. Pham, The effect of line length on the electromigration reliability of Cu interconnects, *Conference Proceedings of ULSI XVIII* (Warrendale: Materials Research Society, 2002), 169.
39. S. Thrasher, M. Gall, C. Capasso, et al., Examination of critical length effect in copper interconnects with oxide and low-*k* dielectrics, *AIP Conference Proceedings 741* (2004), 165.
40. C.-K. Hu, L. Gignac, E. Liniger, et al., Comparison of Cu electromigration lifetime in Cu interconnects coated various caps, *Applied Physics Letters* **83** (2003), 869.

41. C.-K. Hu, D. Canaperi, S. T. Chen, et al., Effects of overlayers on electromigration reliability improvement for Cu/low-*k* interconnects, *Proceedings of the IEEE International Reliability Physics Symposium, IRPS* (2004), 222–228.

42. C.-K. Hu, L. M. Gignac, E. Liniger, C. Detavernier, S. G. Malhotra, and A. Simon, Effect of metal liner on electromigration in Cu Damascene lines, *Journal of Applied Physics* **98** (2005), 124501–124508.

43. E. Liniger, C.-K, Hu, L. Gignac, and S. Kaldor, Effect of liner thickness on electromigration lifetime, *Journal of Applied Physics* **92** (2002), 1803.

44. B. H. Jo and R. W. Vook, Dependence of electromigration rate on applied electric potential, *Applied Surface Science* **89** (1995), 237.

45. R. Augur, F. Van den Elshout, and R. A. M. Wolters, Interface diffusion and electromigration failure in narrow aluminum lines with barrier layers, *AIP Conference Proceedings* **373** (1996), 279.

46. J. S. Huang, T. L. Shofner, and J. Zhao, Direct observation of void morphology in step-like electromigration resistance behavior and its correlation with critical current density, *Journal of Applied Physics* **89** (2001), 2130.

47. Q. Guo, A. Krishnamoorthy, N. Y. Huang, and P. D. Foo, Resistance degradation profile in electromigration of dual-damascene Cu interconnects, *Conference Proceedings VLSI XVIII* (Warrendale: Material Research Society, 2002), 191.

48. S. Demuynck, Z. S. Tokei, C. Bruynseraede, J. Michelon, and K. Max, Alpha-Ta formation and its impact on electromigration, *Proceedings of the Advanced Metallization Conference 2003, AMC XIX* (Warrendale: Materials Research Society, 2004), 355.

49. J. C. Lin, S. K. Park, K. Pfeifer, et al., Electromigration reliability study of self-ionized plasma barriers for dual damascene Cu metallization, *Proceedings of the Advanced Metallization Conference 2002, AMC XVIII* (Warrendale: Materials Research Society, 2003), 233.

50. C.-K. Hu, R. Rosenberg, H. S. Rathore, D. B. Nguyen, and B. Agarwala, Scaling effect on electromigration in on-chip Cu wiring, *Proceedings of the IEEE International Interconnect Technology Conference, IITC* (Piscataway: IEEE, 1999), 267–269.

51. C.-K. Hu, L. Gignac, E. Liniger, et al. Comparison of EM in Cu interconnects with ALD or PVD TaN liners, *Journal of the Electrochemical Society* **154** (2007), H755.

52. G. B. Alers, R. T. Rozbicki, G. J. Harm, S. K. Kailasam, G. W. Ray, and M. Danek, Barrier-first integration for improved reliability in copper dual damascene interconnects, *Proceedings of the IEEE 2003 International Interconnect Technical Conference* (Piscataway: IEEE, 2003), 27.

53. J. M. E. Harper, C. Cabral Jr., P. C. Andricacos, et al., Mechanisms for microstructure evolution in electroplated copper thin films near room temperature, *Journal of Applied Physics* **86** (1999), 2516.

54. C.-K. Hu, K. Y. Lee, C. Cabral, Jr., E. G. Colgan, and C. Stains, Electromigration drift velocity in Al-alloys and Cu-alloys lines, *Journal of the Electrochemical Society* **143** (1996), 1001.

55. S. Shingubara, T. Osaka, S. Abdeslam, H. Sakue, and T. Takagi, Void formation mechanism at no current stressed area, *AIP Conference Proceedings* **418** (1998), 159.

56. C.-K. Hu, L. Gignac, E. Liniger, and R. Rosenberg, Electromigration in on-chip single/dual damascene Cu interconnections, *Journal of the. Electrochemical Society* **149** (2002), G408.

57. C-K. Hu, L. Gignac, S. G. Malhotra, R. Rosenberg, and S. Boettcher, Mechanisms for very long electromigration lifetime in dual-damascene Cu interconnections, *Applied Physics Letters* **78** (2001), 904.

58. P. G. Shewmon, *Diffusion in Solids* (New York: McGraw-Hill, 1963).

59. G. Gupta, Some formal aspects of diffusion: bulk solids and thin films, *Diffusion Phenomena in Thin Films and Microelectronics Materials,* ed. D. Gupta and P. S. Ho (Park Ridge: Noyes, 1998), 23.

60. C.-K. Hu, L. Gignac, and R. Rosenberg, Electromigration in Cu thin film, *Diffusion Processes in Advanced Technical Materials,* ed. D. Gupta (New York: William Andrew, Inc., 2005), 449.

61. R. F. Liu, C.-K. Hu, L. Gignac, et al. Effects of failure criteria on the lifetime distribution of dual-damascene Cu line/via on W, *Journal of Applied Physics* **95** (2004)**,** 3737.

62. L. Li and H.-S. Wong, Integrating graphene into future generations of interconnect wires, *2018 IEDM* (Piscataway: IEEE, 2018), 5.05.

63. T Nogami, B. Briggs, S. Korkmaz, et al., Through-Co self forming barrier for Cu/ULK BEOL, *IEEE IEDM* (Piscataway: IEEE, 2015), 8.1.1.

64. R. Wongpiya, J. Ouyang, T. R. Kim, et al., Amorphous thin film TaWSiC as a diffusion barrier for copper interconnects, *Applied Physics Letters* **103** (2013), 022104.

65. C. Witt, V. Calero, C.-K. Hu, and G. Bonilla, Electromigration: void dynamics, *Transactions on Device and Materials Reliability* **16** (2016), 446.

66. A. V. Vairagar, S. G. Mhaisalkar, A. Krishnamoorthy, et al., *In situ* observation of electromigration-induced void migration in dual-damascene CuCu interconnect structures, *Applied Physics Letters* **85** (2004)**,** 2502–2504.

67. E. Zschech, M. A. Meyer, and E. Langer, Effect of mass transport along interfaces and grain boundaries on copper interconnect degradation, *Materials Research Society Symposium Proceedings* **812** (2004), 361–372.

68. M. H. Lin and A. S. Oates, An electromigration failure distribution model for short-length conductors incorporating passive sinks/reservoirs, *IEEE Transactions on Device Materials Reliability* **13** (2013), 322–326.

69. C.-K. Hu, L. Gignac, E. Liniger, et al., Electromigration challenges for nanoscale Cu wiring, *AIP Conference Proceedings* **1143** (2009), 3–11.

70. A. K. Stamper, H. Baks, E. Cooney, et al. Damascene copper integration impact on EM and stress migration, *Proceedings of the Advanced Metallization Conference* (Warrendale: Materials Research Society, 2005), 727.

71. E. T. Ryan, J. Martin, G. Bonilla, et al. H-base plasma modifications of SiCN/Cu interface to mitigate EM failure, *Journal of the Electrochemical Society* **154** (2007), H604–H610.

72. C-K. Hu, M. Angyal, B. C. Baker, et al, Effect of impurity on Cu EM, *AIP Conference Proceedings* **1300** (2010), 57–67.

73. M. Stangl, M. Lipták, J. Acker, V. Hoffmann, S. Baunack and K. Wetzig, Influence of incorporated non-metallic impurities on electromigration in copper damascene interconnect lines, *Thin Solid Films* **517** (2009), 2687–2690.

74. T. Surholt and C. Herzig, Grain boundary self-diffusion in Cu polycrystals of different purity, *Acta Materialia* **45** (1997), 3817.

75. C. Subramaniam, T. Yamada, K. Kazufumi, et al., One-hundred-fold increase in current carrying capacity in a carbon nanotube-copper composite, *Nature Communication* **4** (2013), Article no 2202.

76. C.-K. Hu, L.M. Gignac, R. Rosenberg, et al., Atom motion of Cu and Co in Cu damascene lines with a CoWP cap, *Applied Physics Letters* **84** (2004), 4986–4988.

77. C.-K. Hu, L. Gignac, R. Rosenberg, et al., Reduced Cu interface diffusion by CoWP surface coating, *Microelectronic Engineering* **70** (2003), 406.

78. C.-K. Hu, D. Canaperi, S. T. Chen, et al. A study of electromigration lifetime for Cu interconnects coated with CoWP, Ta/TaN, or $SiC_xN_yH_z$, *2003 Proceedings of the Advanced Metallization Conference* **19** (2004), 253.

79. N. L. Peterson, Self-diffusion in pure metals, *Journal of Nuclear Materials* **69–70** (1978), 3.

80. A. D. Le Claire, On the theory of impurity diffusion in metals, *Philosophical Magazine* **7** (1962), 141.

81. F. J. Bruni and J. W. Christian, The chemical diffusion coefficient in dilute copper-cobalt alloys, *Acta Metallurgica* **21** (1973), 385–390.

82. P. M. Hansen and K. Anderko, *Constitution of Binary Alloys*, 2nd ed. (New York: McGraw-Hill, 1958).

83. P. Spindler and K. N. Nachtrieb, Lattice and grain-boundary diffusion of phosphorus in commercially-pure copper, *Metallurgical and Materials. Transactions* **A9** (1978), 763.

84. J. Crank, *The Mathematics of Diffusion*, 2nd ed. (New York: Oxford University Press, 1975), 21.

85. F. J. Blatt, *Physics of Electronic Conduction in Solids* (New York: McGraw-Hill, 1968), chapter 7.

86. C. A. Macklift, Diffusion of iron, cobalt, and nickel in single crystals of pure copper, *Physical Review* **109** (1958), 1964.

87. S. Yokogawa and H. Tsuchiya, Effects of Al doping on the electromigration performance of damascene Cu interconnects, *Journal of Applied Physics* **101** (2007), 013513.

88. C.-K. Hu, J. Ohm, L. M. Gignac, et al., Electromigration in Cu(Al) and Cu(Mn) damascene lines, *Journal of Applied Physics* **111** (2012), 093722.

89. C. W. Park and R. W. Vook, Electromigration-resistant Cu-Pd alloy films, *Thin Solid Films* **226** (1993), 238.

90. T. Tonegawa, M. Hiroi, K. Motoyama, H. Fujii, and H. Miyamoto, Suppression of bimodal stress-induced voiding using high-diffusive dopant from Cu-alloy seed layer, *Proceedings of IEEE International Interconnect Technology Conference* (Piscataway: IEEE, 2003), 216.

91. C.-K. Hu, L. Gignac, and R. Rosenberg, Electromigration of Cu/low dielectric constant interconnects, *Microelectronics Reliability* **46** (2006), 213.

92. N. L. Michael and C. U. Kim, Electromigration in Cu thin films with Sn and Al cross strips, *Journal of Applied Physics* **90** (2001), 4370–4377.

93. C-K. Hu, P. S. Ho, and M. B. Small, Electromigration in Al/W and Al(Cu)/W two-level structure, *Journal of Applied Physics* **72** (1992), 291.

94. C.-K. Hu and H. B. Huntington, Electromigration and diffusion of impurities in lead solders, *Diffusion Phenomena in Thin Films and Microelectronic Materials*, ed. D. Gupta and P. S. Ho (Park Ridge: Noyes Data Corporation, 1988), chapter 10.

95. H. Fukushima, and M. Doyama, The formation energies of a vacancy in pure Cu, Cu-Si, Cu-Ga and Cu- gamma Mn solid solutions by positron annihilation, *Journal of Physics F: Metal Physics* **6** (1976), 677.

96. T. Watanabe, H. Nasu, T. Usui, et al., Self-formed barrier technology using CuMn alloy seed for copper dual-damascene interconnect with porous-SiOC/porous-PAr hybrid dielectric, *2007 IEEE International Interconnect Technology Conference* (Piscataway: IEEE, 2007), 7–9.
97. M. Naik, Interconnect trend for single digit nodes, *2018 IEEE IEDM* (2018), 5.6.1.

6   Scaling Effects on Microstructure of Cu and Co Nanointerconnects

6.1   Introduction

In Chapter 5, we showed that continued scaling can significantly degrade electromigration (EM) reliability of Cu interconnects, and the problem can be traced to the increasing demand in current density and the structural evolution of the Cu/low-k damascene interconnects at the nanoscale. To prevent Cu diffusing into low-k dielectrics, barriers and liners are needed at the sidewalls that are less conducting than Cu and effectively reduce the Cu line cross section to increase the line resistivity. Under EM, the mass transport that controls the void growth and the lifetime occurs primarily at the interfaces and grain boundaries in the Cu line. Prior to the 45 nm node, Cu damascene lines have a bamboo-like grain structure where the mass transport is dominated by diffusion along the Cu/dielectric cap interface. Beyond the 45 nm node, the scaling effect on EM reliability became more complex, depending on the characteristics and control of the Cu grain structure. Hu et al. [1] showed that agglomerates of small grains emerged in 70 nm Cu lines intermixing with bamboo grains to increase the grain boundary mass transport and degrade the EM lifetime (see Sections 5.4 and 5.5 in Chapter 5). A subsequent study [2] of 70 nm Cu lines found that with CoWP capping to suppress the interface diffusion, the EM lifetime became affected by the grain structure, and increased with larger grain size but with wider statistics. The result was attributed to the Blech short-length effect, also discussed in Chapter 5.

With continued scaling, the interconnect cross section continues to decrease while the grain structure becomes more irregular with decreasing grain size. Both these structural trends degrade the EM lifetime and have generated great interest in the study of the scaling effect on grain structures and the effect on EM reliability. The problem was reviewed in Chapter 5, focusing on the role of the microstructure and interfaces in controlling the EM reliability of Cu damascene interconnects. We showed that by improving the interfaces with metal capping or the grain boundaries with alloying, the EM lifetime can be significantly improved, extending well into the nanoscale. In this chapter, we discuss the scaling effect on microstructure evolution and the implication on EM reliability for Cu damascene lines. The scaling effect on microstructure was investigated using transmission electron microscopy (TEM) with a high-resolution precession microdiffraction technique to characterize the grain structure in Cu damascene lines down to a 22 nm linewidth for the 14 nm node. The TEM results showed a systematic trend of microstructure evolution in Cu damascene lines with continued scaling. The TEM study was supplemented by a Monte Carlo simulation to investigate grain growth in interconnects extending to the nanoscale based on total energy minimization. The results from the simulation enabled us to understand the scaling effect and how the interface energy counteracts the strain and grain boundary energies to control the microstructure evolution in Cu lines as the linewidth continues to decrease. The simulated microstructures are used to project the scaling effect on EM reliability for Cu interconnects.

The continued scaling of Cu low-k technology is facing serious challenges imposed by basic limits from materials, process, and reliability. This has generated great interest recently to develop alternatives to Cu, particularly Co and Ru nanointercon-nects beyond the 10 nm node. In this chapter, we extend the study of the scaling effect on microstructure evolution for Co damascene lines following the approach used to study the Cu interconnects.

This chapter is organized into three parts. First, we introduce the TEM precession electron diffraction technique for microstructure analysis and present the results from studies of the scaling effect on the microstructure of Cu and Co nanolines. Then we present the results from the Monte Carlo simulation to project the scaling effect on microstructure evolution for Cu and Co nanolines. Finally, we discuss the implication of scaling on microstructure evolution for EM reliability of Cu and Co nanolines, which is based on the Blech short-length effect.

6.2   Precession Electron Diffraction Microscopy

In this chapter, we describe the high-resolution diffraction scanning TEM (D-STEM) together with precession electron diffraction (PED) for orientation mapping of grain structures in Cu and Co nanolines. This technique was developed by Rauch et al. [3] based on precession electron diffraction introduced by Vincent and Midgley in 1994 [4]. This high-resolution microdiffraction technique was capable of providing quantitative and statistical measurements on the grain orientation (crystallographic texture), grain size, and grain boundary characteristics. The precession electron diffraction is schematically shown in Figure 6.1 using an electron path diagram where the incident beam is first tilted off the TEM optic axis by a precession angle (α in Figure 6.1, usually ranging from 0.1°–3°), and precessed around the optic axis ($\theta = 2\pi$) at a constant angle, forming an effective hollow cone upon the specimen and then scanned on the sample to acquire diffraction patterns. As the beam is rotated conically on the specimen surface, the diffracted intensities generate a rotating diffraction pattern. To compensate for the movement of the diffraction pattern and keep the diffraction pattern stationary, the diffracted intensities are descanned in a complementary manner with respect to the tilt scan signal to restore the spots to their default locations. It can be seen in Figure 6.1 that Circle I is created by beam tilt scan and then by descanning to collapse Circle I to Spot II, creating a stationary spot [5].

As illustrated, the observed diffraction pattern is an integration of all the diffraction patterns within the precession cone and can be directly interpreted as a conventional

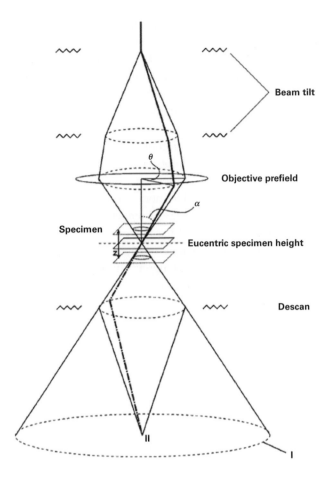

Figure 6.1 Schematic electron ray diagram of precession electron diffraction. Reprinted from [5], with the permission of AIP Publishing.

diffraction pattern. Since the diffraction pattern contains integrated intensities from all the incident beams from off-axis directions, fewer beams are excited simultaneously, so much of the dynamical scattering that is strong at the exact Bragg condition is avoided, and much of the dynamical effects is reduced to achieve quasikinematical conditions. In addition, using the precession technique allows more higher-order Laue reflections to be excited under a kinematical condition and enable a collection of increased number of diffraction intensities, which is useful for eliminating the index-ing ambiguity of spot diffraction patterns [3]. Precession is also able to obtain symmetrical diffraction patterns even when the crystal orientation is off zone axis by as much as 1°, hence there is no need for perfect zone orientation when acquiring the diffraction pattern [6]. These features render the precession electron diffraction as a promising technique for acquiring highly reliable orientation information from the collected spot diffraction patterns and is used to analyze the scaling effect on micro-structures of Cu and Co nanolines reported in this chapter.

The cross-sectional Cu line samples used for PED analysis were prepared using a FEI Strata™ DB235 focused ion beam (FIB) system. For this work, a D-STEM lens configuration was employed on a 200 kV JEOL ARM200F TEM to create a 1–2 nm near-parallel beam with convergence angle <1 mrad. The probe was scanned across the sample with a step size of 1.5 nm while precessing at an angle of 0.4° at each step using the TopSpin™ system from NanoMEGAS. The details of this technique are described elsewhere [7]. The PED technique allows us to determine a complete three-dimensional orientation of each individual grain, which can be projected onto three orthogonal directions as orientation pole plots at each direction. The orientation data acquired after indexing were then parameterized in terms of the Euler angles and exported into the TexSEM Laboratories Orientation Imaging (TSL OIM™) software for further analysis [8]. After this procedure, the reduced parametric data are analyzed based on a series of generalized spherical harmonics to obtain the orientation distribution function. In this fashion, quantitative orientation maps are deduced and presented in units of multiples of random distribution (MRD), where MRD = 1 represents a random distribution of crystal orientations [9].

6.2.1   Microstructure Analysis by Precession Electron Diffraction

Prior to reviewing the scaling effect on microstructure of Cu damascene lines, we first present the results obtained from the 22 nm line to illustrate the capability of the PED technique for microstructure analysis [10]. Here we show the plane-view TEM bright field image of Cu damascene interconnects with 22 nm linewidth in Figure 6.2. While the grain structure can be observed in the TEM micrograph, the image does not provide the details needed for quantitative analysis of the grain structure characteristics.

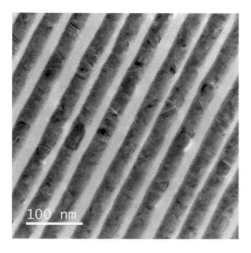

Figure 6.2 Bright field TEM image of periodic Cu interconnects with linewidth of 22 nm. Reprinted from [10] with the permission of S. T. Hu.

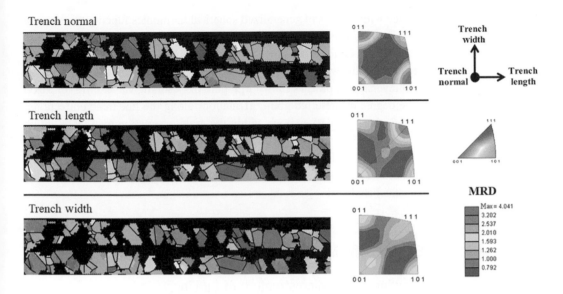

Figure 6.3 On the left, representative color-coded grain orientation maps for Cu interconnects of 22 nm linewidth along the trench normal, trench width, and trench length directions overlaid with reconstructed grain boundaries. On the right, inverse pole figure plots of texture strength along corresponding directions are indicated in units of multiples of random distribution (MRD). Reprinted from [10], with the permission of S. T. Hu. A black and white version of this figure will appear in some formats. For the color version, please refer to the plate section.

In comparison, we show in Figure 6.3 results from the grain orientation and texture analysis obtained by the high-resolution PED technique. Here on the right are color-coded inverse pole figure (IPF) orientation maps along the trench normal, trench width, and trench length directions where the three principal directions, [100], [110], and [111], are coded red, green, and blue, respectively. On the left of the IPF orientation maps, we show the reconstructed grain distribution maps marked with grain boundaries in the corresponding trench normal, trench width, and trench length directions for two 22 nm wide Cu lines.

In Figure 6.3, the three inverse pole plots on the right quantify the grain textures along the trench normal, the trench width, and the trench length, respectively. The colored numbers adjacent to the inverse pole plots indicate the frequency of occurrence of a specific orientation. The grain distribution maps on the left show that the distribution of the grain orientation is statistical in nature, as evidenced by comparing the variations in the grain size and orientation along the line trench orientations between the two Cu lines. The orientation statistics are represented by an orientation distribution function (ODF) f along the line trench orientations in units of MRD, and together show the overall grain orientation distributions in the interconnect line. The ODF was obtained by first parameterizing the orientation data collected from the scanned area in terms of the Bunge–Euler angles, then the parameterized data were binned and fitted with a series of generalized spherical harmonics functions to become the ODF. The detailed procedure can be found in the literature [11]. The ODF was normalized such that a uniform intensity indicates a completely random distribution of grain orientations. In this way, the ODFs project the 3D grain orientation distributions onto three 2D inverse pole plots as shown in Figure 6.3 with the value of $f(g)$ or MRD to indicate the degree of preferred grain orientations along the line trench directions.

In Figure 6.3, there are dark regions within the grain orientation map representing the grains with an indexing reliability lower than 5. The problem can be attributed to the overlapping grains with superimposed diffraction patterns, where the index software was unable to determine a unique crystal orientation. Such grains were excluded from the statistical analysis.

6.3   Scaling Effect on the Microstructure of Cu Interconnects

6.3.1   Previous Studies of the Microstructure of Cu Lines

Grain growth in polycrystalline thin films has been extensively studied and reported [12–15]. In polycrystalline thin films, grain growth is driven by minimization of the Gibbs free energy coming from the grain boundary energy, the elastic strain energy, and the surface and interface energies. Grain growth in Cu thin films has generated considerable early interest in studying its growth mechanisms as reviewed by Thompson [12] and later due to the use of Cu for on-chip interconnects [15]. Since then, the studies on grain growth have extended to Cu interconnects fabricated by electroplating in a damascene or inlaid structure [15, 16]. Most of the early studies have used diffraction techniques based on X-ray, TEM, and scanning electron backscattering diffraction (EBSD) to characterize the grain texture and orientation in the Cu lines. An early study by Lee et al. [17] examined the grain texture change in electroplated Cu lines with width ranging from 6 μm to 0.2 μm using EBSD. They reported that in the 6 to 2 μm wide lines, the {111} grains grew normal to the trench bottom with $\langle 110 \rangle$ or $\langle 112 \rangle$ orientation along the trench length. As the linewidth decreased to 0.24 to 0.2 μm, the {111} fiber character became less dominant, and instead the {111}$\langle 110 \rangle$ texture with minor twin components of {115}$\langle 110 \rangle$ and {115}$\langle 141 \rangle$ orientations emerged. The authors attributed the observed texture change to the elastic anisotropy of Cu, where the overall strain energy is minimized by aligning the minimum Young modulus direction to the direction of the maximum stress.

Similar results have been reported by Besser et al. [18] in a study of the microstructure evolution in electroplated Cu inlaid lines. They used the (111) pole plots from X-ray diffraction (XRD) of Cu lines with linewidths ranging from 1.06 μm to 0.35 μm to show that the grains nucleated and grew from the trench bottom normal with a preferred (111) texture and a [110] type orientation parallel to the sidewalls. When the linewidth decreased from 1.06 μm to 0.35 μm, the {111} texture grains started to emerge from the sidewall as they were being nucleated from both trench bottom and trench sidewalls. The observation of the sidewall (111) grains was attributed to grain growth driven by minimization of the elastic and surface energies in the confined line structure.

Another comprehensive study was reported by Cho et al. [19] using XRD and EBSD to investigate the annealing effect on the microstructure evolution for Cu lines with 2 μm to 0.14 μm in width. They observed that all the Cu lines contained grains with dominant $\{111\}\langle110\rangle$ orientation components along the trench normal direction, but the amount of the fiber-like texture decreased as the linewidth decreased from 0.5 μm to 0.14 μm. In addition, a weak $\{111\}$ texture sidewall component was observed to emerge in the 0.5 μm wide line and increased in intensity as the linewidth decreased to 0.14 μm. Upon annealing, the intensity of the $\{111\}$ sidewall component was found to decrease and followed a trend of increasing intensity with decreasing linewidth. The grain boundary characteristics were analyzed in this study, and it was found that the fraction of the $\Sigma3$ coherent twin boundaries decreased as the linewidth decreased from 0.5 μm to 0.14 μm (Figure 6.4b), while the grain structure in the 2 μm line changed from polycrystalline to a bamboo-like grains in the 0.14 μm line. The observed microstructure characteristics are consistent with the kinetics of grain growth in line structures driven by total energy minimization. In wider lines, the surface energy is dominant to enhance the $\{111\}\langle110\rangle$ texture components along the trench normal direction. As the linewidth decreases, the sidewall surface energy becomes more important, enabling the $\{111\}$ sidewall texture to emerge and increase with decreasing linewidth. Upon annealing, the elastic strain energy increases to compete with the surface energy as reflected by the increase in the $\langle100\rangle$ direction with a reduction of the $\{111\}$ sidewall texture as shown in Figure 6.4a.

The capability of the diffraction techniques used in the early studies was limited to Cu lines wider than 140 nm due to the limited spatial resolution, the difficulty in interpreting the dynamic diffraction patterns, and the lack of automation for statistical analysis of grain structures in a large grain ensemble. These problems were overcome with the development of the precession electron diffraction microscopy technique as described in Section 6.2. The PED technique together with automated data processing was first applied by Ganesh et al. [20, 21] to analyze the evolution of the grain structure in Cu damascene lines of 1.8 μm to 70 nm in width. For the 1.8 μm Cu lines, a strong $\{111\}$ fiber texture along the trench normal was observed. As the linewidth decreased to 180 nm, the $\{111\}$ fiber texture along the trench normal became weaker and barely observable at the 120 nm linewidth. At this point, a $\{111\}$ texture along the trench width direction emerged and intensified as the linewidth decreased to 70 nm. A corresponding decrease in the amount of coherent twin boundaries was observed with decreasing linewidth, although the mechanism was not explained. In a separate study, the electrical resistance was measured in Cu nanolines, where the resistance increase was found to be far greater for random grain boundaries (GBs) than coherent GBs due to the increase in GB scattering [22]. These results raised interesting questions regarding the scaling effect on grain structure and the impact on EM reliability, particularly beyond the 70 nm linewidth. Such questions are addressed in this chapter using precession electron diffraction microscopy and Monte Carlo simulation to investigate the scaling effect on microstructure evolution in Cu and Co interconnects.

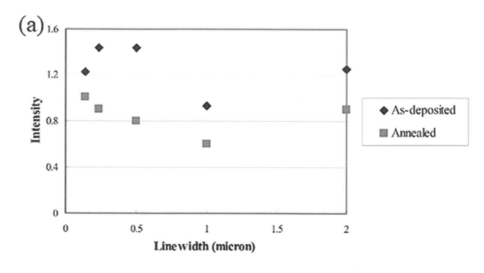

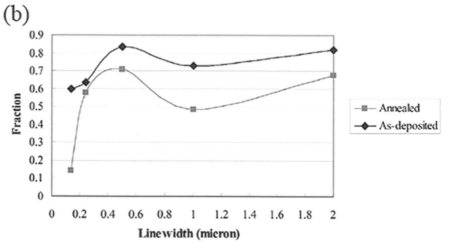

Figure 6.4 (a) Intensity of sidewall {111} texture component of the as-deposited and annealed sample as a function of linewidth. (b) Fractions of $\Sigma 3$ CSL boundaries in Cu interconnect lines as a function of linewidth as reported by Cho et al. Reprinted from [19] with permission from Springer Nature.

6.3.2   Effect of the Overburden Layer

In the damascene process, the Cu interconnects are formed by electroplating Cu into prepatterned trenches covered with diffusion barriers on the sidewalls and at the trench bottom. This approach required an excellent hole-filling capability of the deposition process to ensure defect-free interconnects. The electroplating process was developed to meet this requirement to provide a high deposition rate for low cost and good process control [23]. In electroplating, special electrochemical additives are used to facilitate Cu filling into the underlying line trench by forming a thick coating layer

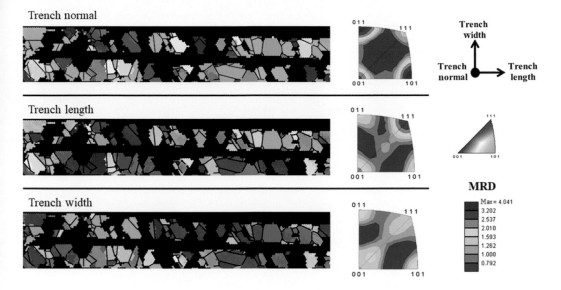

Trench normal

Trench length

Trench width

Trench
width

Trench
normal

Trench
length

MRD

Max = 4.041
3.202
2.537
2.010
1.593
1.262
1.000
0.792

Figure 6.3 On the left, representative color-coded grain orientation maps for Cu interconnects of 22 nm linewidth along the trench normal, trench width, and trench length directions overlaid with reconstructed grain boundaries. On the right, inverse pole figure plots of texture strength along corresponding directions are indicated in units of multiples of random distribution (MRD). Reprinted from [10], with the permission of S. T. Hu.

Technology	Line Width (nm)	Representative Grain Orientation Mapping
90nm	120	
45nm	70	
28nm	45	
22nm	40	

Figure 6.6 Representative color-coded inverse pole figure grain orientation maps along the trench width direction for Cu lines of four technology nodes with 120, 70, 45, and 40 nm linewidths. Color keys for grain orientations are represented in the standard stereographic triangle on the right and overlaid with reconstructed grain boundaries. © 2018 IEEE. Reprinted with permission from [27].

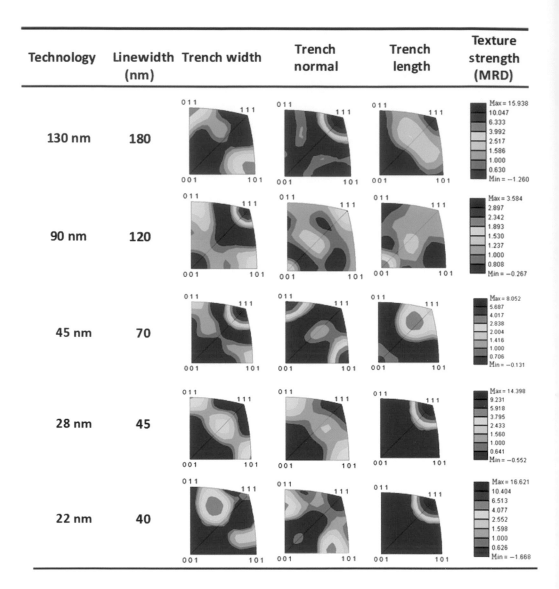

Figure 6.7 Color-coded inverse pole figure maps for grain orientation distribution along trench normal, trench width, and trench length directions for Cu lines of 180, 120, 70, 45, and 40 nm linewidths. Respective texture strengths along trench normal, trench length, and trench width direction are indicated in units of MRDs on the right. © 2018 IEEE. Reprinted with permission from [27].

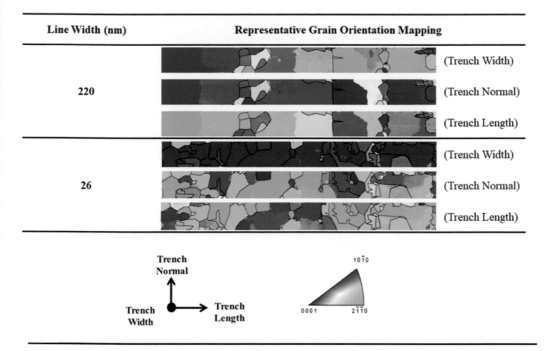

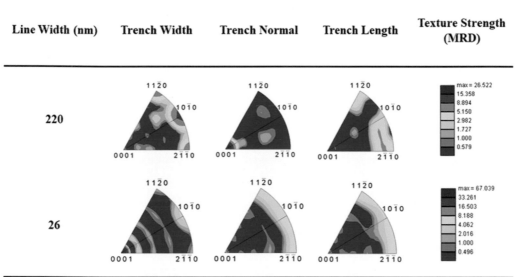

Figure 6.12 Top, the representative color-coded grain orientation maps, and bottom, the color-coded IPFs for grain orientation distribution for Co lines with linewidths of 220 nm and 26 nm plotted along the trench width, trench normal, and trench length directions.

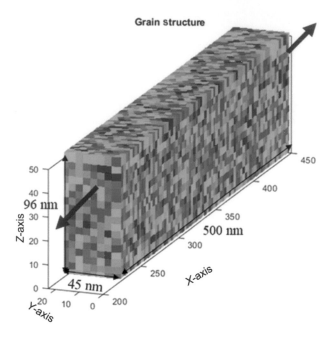

Figure 6.18 A schematic image of microstructure in the interconnect at the initial state. Reprinted from [10] with permission from S. T. Hu.

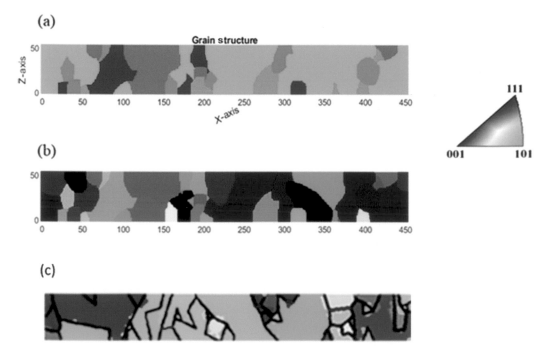

Figure 6.20 The cross-sectional view of a 45 nm Cu interconnect with (a) a grain orientation map along the trench normal direction, (b) the individual grain map after 500 MCS, and (c) the grain orientation map along the trench width orientation acquired using the PED technique. Reprinted from [10] with permission from S. T. Hu.

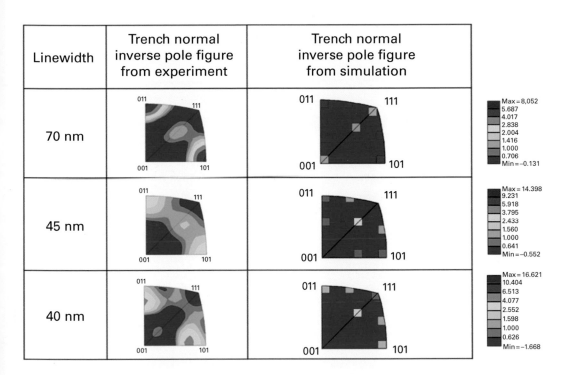

Linewidth	Trench normal inverse pole figure from experiment	Trench normal inverse pole figure from simulation	
70 nm			Max = 8,052 5.687 4.017 2.838 2.004 1.416 1.000 0.706 Min = −0.131
45 nm			Max = 14.398 9.231 5.918 3.795 2.433 1.560 1.000 0.641 Min = −0.552
40 nm			Max = 16.621 10.404 6.513 4.077 2.552 1.598 1.000 0.626 Min = −1.668

Figure 6.21 The inverse pole plots from PED at trench normal direction are compared with results from simulation for Cu lines with 70 nm, 45 nm and 40 nm widths. The shading codes on the right represent the volume fraction for each crystal orientations after 500 MCS. Reprinted from [10] with permission from S. T. Hu.

22 nm Cu line

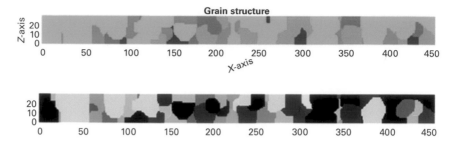

15 nm Cu line

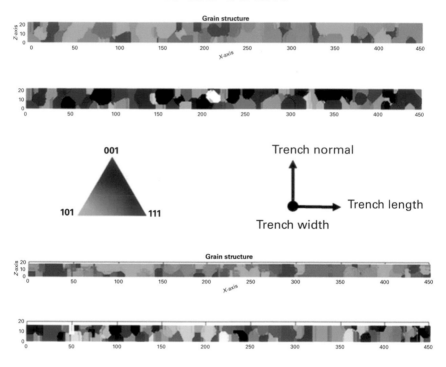

8 nm Cu line

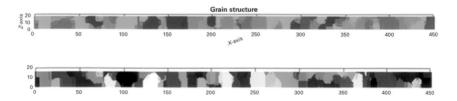

Figure 6.22 The cross-sectional view of representative Cu interconnects with linewidths of 22 nm, 15 nm, 11 nm, and 8 nm. The top image shows the grain orientation map along the trench normal direction, and the bottom image the grain distribution map. Reprinted from [10] with permission from S. T. Hu.

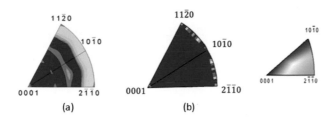

Figure 6.25 Comparison of the inverse pole plots acquired from (a) PED and (b) Monte Carlo simulation along the trench normal direction for Co interconnect with 26 nm linewidth. Reprinted from [10] with permission from S. T. Hu.

22 nm Co line

15 nm Co line

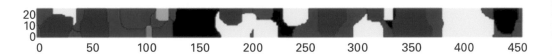

11 nm Co line

8 nm Co line

Figure 6.26 The grain distribution maps obtained by Monte Carlo simulation in Co interconnects with linewidths of 22 nm, 15 nm, 11 nm, and 8 nm. Reprinted from [10] with permission from S. T. Hu.

Figure 6.5 A schematic cross-sectional view of Cu microstructure with linewidth of 700 nm (left) and 30 nm (right) after annealing and before CMP. Reprinted from [19] with permission from Springer Nature.

above the trench, which is called an overburden layer, and the process is called a "superfill" process (Figure 6.5). With continued scaling, the superfill process becomes increasingly important for defect-free electroplating of Cu into line trenches as the linewidth decreases, while the aspect ratio increases to ~2. In the as-deposited state, the grain size of Cu is usually of the order of few tens of nanometers as it is limited by pinning of the electrochemical additives [15]. An annealing treatment about 150°C to 250°C is required after electroplating to promote grain growth by desorbing the additives to stabilize the Cu microstructure. The overburden layer is then polished away by a chemical mechanical polishing (CMP) process (see steps 5 and 6 in Figure 3.11).

Upon annealing, the grain growth process in Cu interconnects generally starts from the overburden layer to propagate into the line trenches [24]. In lines above 300 nm in width, an almost complete invasion from the overburden layer to the trench is commonly observed. This results in the formation of columnar grains in the trenches and the overburden layer very similar to that observed in thin films [25, 26]. However, as the line narrows, grain growth from the overburden layer cannot fully propagate into the trenches even at an elevated annealing temperature. The problem becomes more severe when the linewidth continues to decrease as the invasion depth from overburden layer decreases [25]. In this way, the overburden layer is important in control grain growth in a Cu damascene line, although its effect diminishes with continued scaling. This effect is shown schematically in Figure 6.5 by comparing grain growth in a 700 nm wide line to a 30 nm narrow line [26].

In addition to additive desorption, grain growth in the overburden layer is constrained by the sidewalls and the trench bottom. With decreasing linewidth, the rate of additive desorption is reduced due to increasing pinning at the sidewalls, while abnormal grain growth to minimize the interface energy of the sidewall can occur before the grain growth in the overburden reaches the trenches. This leads to a grain structure with growth of small {111} texture grains from the sidewalls and trench bottom intermixing with some larger grains propagating from the overburden layer. Since no two $\langle 111 \rangle$ directions can be perpendicular to each other and as the effect of trench sidewall become more dominant with decreasing linewidth, clusters of the small grains growing from the trench bottom remain trapped while the {111} grains at the sidewall continue to grow. As a result, the {111} grain texture from the sidewall becomes dominant, changing the overall grain orientation from the trench normal to the trench width direction, while the overall grain size becomes smaller as the linewidth continues to

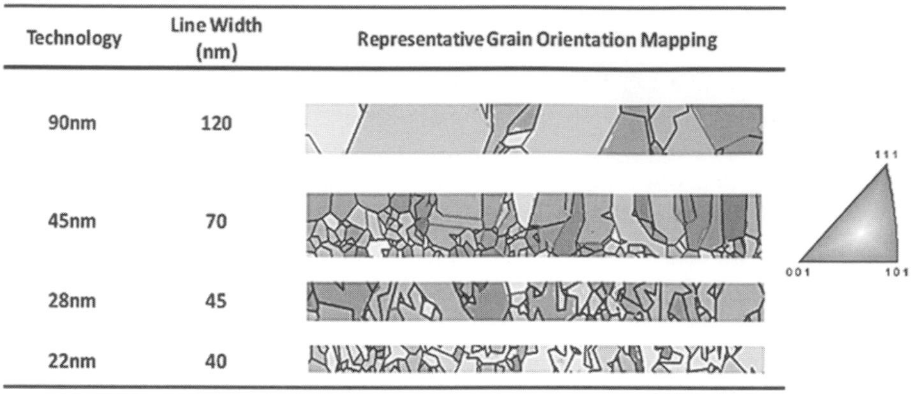

Technology	Line Width (nm)	Representative Grain Orientation Mapping
90nm	120	
45nm	70	
28nm	45	
22nm	40	

Figure 6.6 Representative color-coded inverse pole figure grain orientation maps along the trench width direction for Cu lines of four technology nodes with 120, 70, 45, and 40 nm linewidths. Color keys for grain orientations are represented in the standard stereographic triangle on the right and overlaid with reconstructed grain boundaries. © 2018 IEEE. Reprinted with permission from [27]. A black and white version of this figure will appear in some formats. For the color version, please refer to the plate section.

decrease. In the next section, we show that such a trend was observed in Cu damascene lines using the PED technique when the linewidth decreased to 70 nm.

6.3.3   Microstructure Evolution in Cu Nanolines

In this section, we review the results from PED studies [10, 28] on the scaling effect on microstructure evolution in Cu damascene lines, extending from 180 nm linewidth of the 130 nm node to 22 nm linewidth of the 14 nm node. The Cu lines used in these studies were fabricated using a well-controlled dual-damascene process to ensure the uniformity of the sidewall Ta/TaN barrier layers and the control of the electroplating process. The microstructure study is followed by a Monte Carlo simulation in the next section to model grain growth based on total energy minimization and to project microstructure evolution beyond the PED studies.

To understand the scaling effect on microstructure evolution, it is useful to briefly review the grain growth characteristics in thin films to contrast with the line structures, a topic that has been extensively studied [12–14]. In general, grain growth is driven by minimization of the Gibbs free energy in the film deriving from grain boundaries, surfaces, interfaces, stress, and impurities, which in turn depend on the film thickness, substrate materials, and annealing temperatures. Upon annealing, the grains in a film start to grow to minimize the grain boundary energy, increasing the average grain size to about the film thickness and with a log-normal or a Weibull size distribution, which are typical characteristics of normal grain growth. Upon further annealing, the surface, interface, and strain energies come into play, competing with the grain boundary energy to drive grain growth, leading to the growth of a subgroup of grains with a bimodal distribution in the grain size, which is characterized as an abnormal grain growth. For this growth mode, the effect of the surface and interface energies depends on the film and the substrate materials while the strain energy depends on the thermal mismatch with the substrate and together, the growth characteristics depend on the film thickness, substrate materials, and processing temperatures. With temperatures at or below 0.5 T_m (the absolute melting point), the surface energy generally dominates, yielding {111} texture grains in Cu films. With increasing temperatures, the strain energy comes into play due to increasing thermal mismatch between Cu and the Si substrate, driving the growth of the {100} texture grains to minimize the elastic strain energy and compete with the {111} texture grains in the film.

Comparing to films, the Cu damascene line is a complex structure encapsulated with metal or dielectric barriers at the sidewall and the line bottom to prevent Cu diffusion into the silicon. For Cu lines, the microstructure evolution is increasingly constrained by the materials and the geometry of the sidewalls and the trench bottom in the dual-damascene structure, particularly as scaling advances into the nanoscale. The Cu face-center cubic (fcc) lattice structure has the lowest surface energy (2,534 mJ/m^2) for the closest-packed (111) planes [29] to favor grain growth with the {111} texture. The elastic property of Cu is highly anisotropic with the elastic modulus about three times higher along the $\langle 111 \rangle$ direction than the $\langle 100 \rangle$ direction [30], so minimization of the strain energy would favor growth of the {100} texture grains to compete with the {111} texture grains and affect the overall grain structure in abnormal grain growth. For Cu damascene lines, there are two annealing steps in fabrication to affect grain growth: one around 250°C after electroplating to desorb the plating additives, and the other at about 400°C as a final anneal to stabilize the interconnect structure [30]. Starting with the 250°C anneal, the {100} texture grains emerge to compete with the {111} texture grains in the line, then increase in population during the 400°C anneal. The competition of these energies alters the kinetics of the abnormal grain growth and is reflected in the overall grain size distribution. Here the scaling effect is complicated since both interface and strain energies depend on the line aspect ratio and the materials on the sidewalls and the trench bottom. The effect becomes more pronounced as the interconnect downscaling continues to increase the surface-to-volume ratio and thus the contribution of the surface/interface energy for grain growth in ultrafine Cu interconnects. With increasing aspect ratios, the interface energy from the sidewalls exerts more influence on Cu grain growth than at the trench bottom, leading to the preferential growth of a subpopulation of grains with {111} texture at the sidewall. This was observed in the Cu lines with linewidth of 70 nm or less. In addition, more twins were formed being triggered by minimization of the strain energy to convert some of the {111} texture grains to the {100} orientation in a confined Cu line [27–28, 32–33].

In Figure 6.6, we show representative grain orientation maps overlaid with grain boundaries for Cu lines for four technology nodes with linewidths from 120 to 40 nm, where for simplicity only the results along the trench width direction are shown [27]. Here the orientations and boundaries of grains less than 3 nm in size were clearly resolved, demonstrating the resolution of the PED technique. In Figure 6.7, we show the inverse pole plots for grain orientations along trench width, trench normal, and

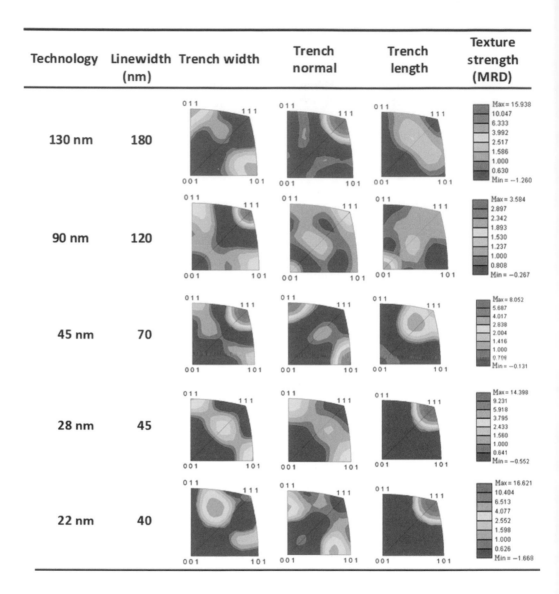

Technology	Linewidth Trench width (nm)	Trench normal	Trench length	Texture strength (MRD)
130 nm	180			Max = 15.938 10.047 6.333 3.992 2.517 1.586 1.000 0.630 Min = −1.260
90 nm	120			Max = 3.584 2.897 2.342 1.893 1.530 1.237 1.000 0.808 Min = −0.267
45 nm	70			Max = 8.052 5.687 4.017 2.838 2.004 1.416 1.000 0.708 Min = −0.131
28 nm	45			Max = 14.398 9.231 5.918 3.795 2.433 1.560 1.000 0.641 Min = −0.552
22 nm	40			Max = 16.621 10.404 6.513 4.077 2.552 1.598 1.000 0.626 Min = −1.668

Figure 6.7 Color-coded inverse pole figure maps for grain orientation distribution along trench normal, trench width, and trench length directions for Cu lines of 180, 120, 70, 45, and 40 nm linewidths. Respective texture strengths along trench normal, trench length, and trench width direction are indicated in units of MRDs on the right. © 2018 IEEE. Reprinted with permission from [27]. A black and white version of this figure will appear in some formats. For the color version, please refer to the plate section.

trench length directions from 120 nm linewidth to 40 nm linewidth for the corresponding technology nodes [27].

Here the results in Figures 6.6 and 6.7 together with Figure 6.3 show a systematic trend of scaling effect on microstructure evolution in Cu damascene lines, extending from 180 nm to 22 nm in linewidth. Starting from the 180 nm linewidth in Figure 6.7, the grain orientation in the lines exhibited a strong {111} fiber texture along the trench normal, reflecting dominant grain growth normal to the trench bottom for the wide Cu lines. In this width range, the microstructure evolution was found to start from the overburden layer, then propagated into the Cu line trench upon annealing at 250°C. (Similar grain structure was observed in the 1.8 μm line but is not shown.) This is similar to the Cu thin films [13, 14], where the grain growth is controlled by minimizing the grain boundary and surface energies normal to the trench bottom. When the linewidth is reduced to 120 nm, the {111} fiber texture shifted from the trench normal to the trench width direction. This is due to the increase in the surface-to-volume ratio, making the surface energy of the TaN/Ta trench sidewalls more dominant over the trench bottom to control the grain growth. The shift in the {111} fiber texture to the trench width direction has also been observed in previous studies [18, 19]. Here the shift of the {111} texture orientation is not strong with MRD only 4 normal to the trench width, and near random distribution for other grain orientations. In Figure 6.6, we show that the grain structure at the 120 nm linewidth consisted predominantly of large bamboo-like grains across the line thickness.

With scaling to 70 nm linewidth, the IPF plots show that the {111} grain texture normal to the sidewalls becomes stronger with MRD of 8 and orthogonal ⟨110⟩ grain orientations along the trench normal (Figure 6.7). This indicates that the interfacial energy of the sidewall became more dominant as its area increased relative to the trench bottom due to scaling. Here the geometrical restriction due to the decrease in linewidth limited the effect of grain growth from the overburden layer. This led to an overall reduction in the grain size, particularly for the bamboo-like grains and the emerging growth of a subpopulation of smaller grains near the trench bottom (Figure 6.6). Such changes in the grain structure were first reported by Hu et al. [1] at the 70 nm linewidth and found to degrade the EM reliability of Cu damascene line (see Section 5.5 in Chapter 5). By analyzing the grain size distribution, we found the changes can be attributed to abnormal grain growth as shown in Figure 6.8, extending to about 20% of the small grains at the low end. As a result, the overall grain structure can be delineated as a series of bamboo–polygrain segments with log-normal length statistics. Such grain structures form a series of Blech short-length segments, which can improve the EM lifetime; the implication for interconnect reliability is discussed in Section 6.8.

At the 45 nm linewidth, the orientation of the {111} texture changed again, switching to along the line length direction. Here the change in the texture orientation was relatively strong with MRD of 14 although the other grain orientations remained random. At the 40 nm linewidth, the strength of the ⟨111⟩ texture further increased to MRD of 17 although the other grain orientations remained relatively random (Figure 6.7). For both 45 and 40 nm linewidths, the {111} grains grew predominantly from the sidewall along the trench width and oriented under the geometrical confinement along the trench length. With scaling increasing the interface-to-volume ratio, interfacial pinning at the sidewalls and the trench bottom became more effective in anchoring small grains. This favored the formation of small grains near the line bottom as the linewidth decreases.

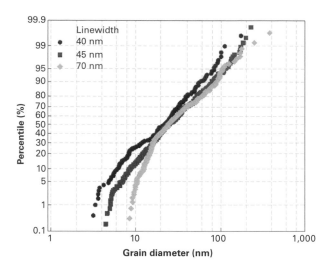

Figure 6.8 Log-normal statistical distributions of grain size in Cu lines with linewidths of 70, 45, and 40 nm. Abnormal grain growth was observed as deviation at the low end starting below 5% at the 40 nm linewidth and increased to about 30% at the 70 nm linewidth. © 2018 IEEE. Reprinted with permission from [27].

The size distributions of grains as measured in terms of equivalent circle diameter are summarized in Figure 6.8 for 40 nm, 45 nm, and 70 nm Cu lines. Overall, the grain size distribution followed the log-normal statistics for normal grain growth but started to deviate from the lower percentile to reflect abnormal grain growth. The abnormal grain growth led to the growth of a subgroup of grains with bimodal log-normal statistics. This occurred at the lowest 30% for the 70 nm lines and diminished to about 5% for the 40 nm wide lines (Figure 6.8). The driving forces for abnormal grain growth in interconnect lines are similar to that of the thin films, arising from the surface/interface and strain energies to compete with the grain boundary energy. For the Cu lines, the abnormal grain growth is also affected by the constraint of the line geometry, which becomes more important with continued scaling, as shown in Figure 6.8. Near the trench bottom, the contribution of surface/interface energy first became more evident by changing the {111} texture orientation normal to the trench sidewall, but as the trench aspect ratio increased, the sidewall and the bottom surfaces exerted more influence on Cu grain growth. This led to the preferential growth of a subgroup of grains with predominant {111} orientations, starting near the bottom of the 70 nm line and extended to the 45 and 40 nm lines. However, the geometrical constraint due to the reduction in the linewidth caused the grains to impinge on each other, limiting the abnormal grain growth in the 45 and 40 nm lines (Figure 6.6). During abnormal grain growth, some of the {111} grains were flipped to the ⟨100⟩ orientation to minimize the strain energy during the 400°C anneal in the damascene process. The kinetics for flipping of the grain orientation was facilitated through a series of intermediate twin configurations that require less strain energy in a confined Cu line [27, 28, 33].

The trend continues with scaling to the 22 nm linewidth for the 14 nm technology node, where the IPFs reveal a strong trend of $\{111\}\langle 110\rangle$ oriented grains along the line length direction in the 22 nm Cu lines (Figure 6.3). An analysis of the grain distribution of the 22 nm Cu lines [8] shows that almost 70% of the grains had sizes smaller than 40 nm, which is the electron mean free path of Cu (~39 nm) at room temperature. The presence of more small grains will contribute to mass transport and electron scattering at grain boundaries, thus affect the EM reliability and electrical resistivity beyond the 22 nm linewidth, as discussed later in Section 6.8.

So far, we report that the dominant $\{111\}$ texture orientation switched from trench normal to trench width at 120 nm linewidth [18–20] and switched again to along the trench length for the 45 and 40 nm Cu lines [28, 32], then the trend continued with the 22 nm line (Figure 6.3). Such changes of the dominated grain orientation have also been observed in Co nanolines but different in details, as discussed later in Section 6.4. This is a distinctive feature of the scaling effect on microstructure evolution of interconnects at the nanoscale, which together with the change in the overall grain size distribution induced by abnormal grain growth can significantly increase the electrical resistivity. In addition, the formation of the bamboo–polygrain segments and the scaling effect on their statistical distribution can affect the grain boundary mass transport and the EM lifetime based on the Blech short-length effect. These topics are discussed in subsequent sections, first on microstructure evolution derived from Monte Carlo simulations in Section 6.5, followed by the scaling effect on EM reliability in Section 6.8.

6.3.4 Scaling Effect on Twin Formation in Cu Nanolines

In this section, we discuss the formation of twin boundaries in Cu interconnects, which is of interest because twin boundaries can affect the mechanical, electrical, and diffusional characteristics and impact the reliability and performance of Cu interconnects [20, 35, 36]. Twin boundaries are special high-angle boundaries with a specific orientation relationship to allow joining of the two adjoining lattices with relatively little distortion and hence low grain boundary energy. The twin boundary structures can be characterized using the coincidence site lattice (CSL) model [37]. In the CSL model, twin boundaries are described using the Σ value, which is defined as the ratio of the area enclosed by a unit cell of the coincidence sites and the standard unit cell. In this way, Σ describes the reciprocal density of the coincident sites shared by the two grains [38]. The $\Sigma 3$ twin boundaries are of particular interest due to their unique structure with low lattice disorder, which can be formed in fcc lattices by rotating the orientation of a parent grain 60° about the $\langle 111\rangle$ axis. The coherent $\Sigma 3$ twin boundary has the orientation of the boundary plane parallel to the $\{111\}$ twin plane, while the incoherent twin boundaries with the boundary plane are not coincided with the $\{111\}$ twin plane. For Cu, the energy of the coherent twin boundary is the lowest (24 mJ/m^2) among all types of grain boundaries, while the average energy of the incoherent twin boundary is 498 mJ/m^2, close to that of the high-angle boundary (625 mJ/m^2) [39, 40]. Therefore, the formation of a coherent twin boundary can effectively reduce the total grain boundary energy to facilitate grain growth because its energy is significantly lower than that of normal boundaries. In addition to the low boundary energy, the electrical resistivity of coherent twin boundary is about an order of magnitude smaller than other boundaries [22]. Such grain boundaries can reduce the contribution of the grain boundary scattering to Cu resistivity as the line dimension approaches the mean free path of electron scattering. It is worth noting that the special properties of $\Sigma3$ boundaries such as low energy and low electron scattering are valid only for the coherent $\Sigma3$ boundaries; as for the incoherent $\Sigma3$ boundaries, they are treated as regular high-angle boundaries.

In general, twin boundary is induced by single-grain deformation during grain growth, so annealing twins are most prevalent in large, plated Cu grains where many individual grains show multiple twin sites. The large number of coherent twin boundaries observed in Cu lines with width of 1.8 µm to 120 nm (Figure 6.9) indicated that the strain energy can outweigh the surface/interface energy in contributing to twin formation in this width range. Upon annealing, some {111} texture grains were flipped to the {100} texture to minimize the elastic energy through the formation of twin boundaries as an efficient step to reduce the overall strain energy. In confined Cu lines, multiple twinning of {111} grains was found to be an efficient twinning process, where several twin orientations were formed to facilitate switching from the ⟨111⟩ orientation to the ⟨100⟩ orientation to reduce the overall strain energy. This process weakens the overall texture, and only some of the twins are coherent twins. With continued scaling, twin formation becomes more limited with decreasing grain size as the plastic deformation becomes more difficult due to the Hall–Petch relationship [41]. This is consistent with our observations of a decrease in the twin population, particularly the coherent twins and a corresponding increase in the ⟨111⟩ texture strength.

The PED orientation data from this study were used in the grain boundary trace analysis to characterize the $\Sigma3$ coherent twin boundaries, and the results are shown in Figure 6.9. Here all the Cu lines have been subjected to the 400°C anneal to increase the elastic strain energy to promote twin formation. In the 1.8 µm lines, up to 42% of the grain boundaries were found to be coherent twins, indicating that the annealing twins are most prevalent in large, plated Cu grains where many individual grains show multiple twins. With scaling, the reduction in the line dimensions promotes the formation of small grains near the trench bottom and the sidewalls and continues to reduce the density of the coherent twins. For 180 nm and 120 nm lines, the fraction of coherent twin boundaries was reduced to about 25% and 18%, respectively, and continued to reduce to about 13% for the 70 nm Cu lines. For 45 nm wide Cu lines, the length fraction of these boundaries was further reduced to only about 2% (Figure 6.9) This trend of decreasing coherent twins with scaling continued to the 40 nm linewidth with less than 1% and reduced further to only ~0.2% for the 22 nm lines. It is worth noting that for the 40 nm and 22 nm lines, the amount of the coherent $\Sigma3$ boundaries was undercounted due to the difficulty in the orientation analysis as shown by the dark region in the orientation map for the 22 nm lines (Figure 6.3), which had an indexed reliability lower than 5. The difficulty was due to the overlapping grains with superimposed diffraction patterns, where the index software was

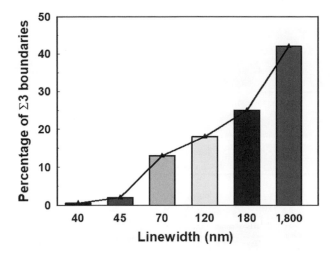

Figure 6.9 Scaling effect on twin formation in Cu interconnect lines. © 2018 IEEE. Reprinted with permission from [27].

unable to determine a unique crystal orientation and such grains were excluded from the statistical analysis. The characteristics of twin formation shown here are typical for Cu damascene lines and can vary depending on the materials and processing used in fabrication.

Coherent twin boundaries are low-energy boundaries and not fast diffusion paths for EM mass transport. With fewer twin boundaries in Cu nanolines to slow down the grain boundary mass transport, the EM reliability will degrade, a problem that was discussed in Chapter 5. Also, the resistivity of coherent twin boundary is one order of magnitude smaller than other grain boundaries [22]. Therefore, with fewer twin boundaries, the contribution of grain boundary scattering on Cu resistivity will increase, particularly as the line dimensions approach the mean free path for electron scattering (39 nm at room temperature). The scaling effects on microstructure evolution are investigated by simulation in Section 6.6 and then on EM reliability in Section 6.8.

6.4   Scaling Effect on the Microstructure of Co Interconnects

Cobalt was recently proposed as an alternative material to replace Cu for the advanced technology nodes [42–44]. Even though Co has a higher bulk resistivity than Cu, but because of its ability to incorporate with very thin or even no barrier layer, its electrical resistivity is less affected by scaling and can be a promising alternative material for future interconnects. With the interest in using Co as interconnects in the nanoscale, the microstructure evolution of the Co interconnects was investigated using the precession electron diffraction technique [10, 45].

The Co interconnects used in the study were fabricated using a damascene process, and the grain structure was analyzed by the PED technique, similar to the Cu

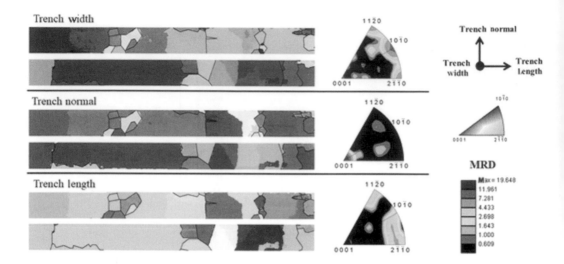

Figure 6.10 Grain structures obtained from precession electron diffraction for two 220 nm Co lines. On the right are orientation texture plots represented by inverse pole figures along the trench width, trench normal, and trench length directions with corresponding texture strength distributions. The corresponding grain structures are shown on the left overlaid with reconstructed grain boundaries. Reprinted from [10] with permission from S. T. Hu. Please refer to the color plate of Figure 6.12 for details.

interconnects. The Co lines available for this study were limited to 220 nm and 26 nm linewidths. The results were used to guide Monte Carlo simulation to investigate the scaling effect on the grain structure of Co lines at the nanoscale.

The grain orientation maps obtained from PED analysis for the 220 nm Co lines are shown on the left side of Figure 6.10 and superimposed on the right with the reconstructed grain distribution maps along the trench width, trench normal, and trench length directions. Since Co has a hexagonal closed packed (hcp) structure, the orientation maps are represented by the hcp color-coded stereographic triangle along the three principal orientations of [0001], $[10\bar{1}0]$, and $[2\bar{1}\bar{1}0]$. For quantitative texture analysis, the grain orientation distributions along the trench width, trench length, and trench normal are plotted on the right side of Figure 6.10, where three corresponding IPF plots are constructed based on the hcp structure. Here the results for two Co lines show the typical variations of the grain structure observed. The IPF plots reveal that the 220 nm Co lines have a strong {0001} texture along the trench normal direction (MRD = 19) combined with a mixed $\{10\bar{1}0\}$ and $\{2\bar{1}\bar{1}0\}$ textures at the trench width direction. The observed texture distribution for the Co line can be delineated into two different sets of biaxial textures, one with the $\{0001\}\langle10\bar{1}0\rangle$ texture orientation and the other with the $\{0001\}\langle2\bar{1}\bar{1}0\rangle$ texture orientation, because in an hcp structure, the (0001) plane is perpendicular to both $\langle10\bar{1}0\rangle$ and $\langle2\bar{1}\bar{1}0\rangle$ directions. The grain structures shown on the left of Figure 6.10 are overlaid with reconstructed grain boundaries along the trench width, trench normal, and trench length directions. The grain structures exhibit predominant large bamboo-like grains

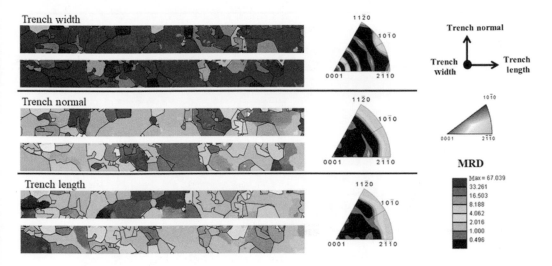

Figure 6.11 Grain structure analysis for two 26 nm Co interconnects: (left) representative inverse pole figure orientation maps in trench width, trench norma,l and trench length directions overlaid with reconstructed grain boundaries; (right) texture plots and their respective texture strengths along the trench width, trench length, and trench normal directions. Reprinted from [10] with permission from S. T. Hu. Please refer to the color plate of Figure 6.12 for details.

in most regions and interspersing with some clusters of smaller grains. Here the microstructure characteristics with a strong {0001} texture along the trench normal are similar to the Cu 120 nm interconnects with a dominant {111} texture in the fcc Cu lattice (Figure 6.6).

The results of the grain structure obtained by the PED technique for Co interconnects of 26 nm linewidth are shown in Figure 6.11. On the left of Figure 6.11, we show the grain structure maps along the trench width, trench length, and trench normal directions. Along with some short bamboo grains, the boundary and orientation for grains less than 5 nm were observed using the PED technique. The statistical distributions of grain orientations along the trench width, trench length, and trench normal directions are plotted on the right of Figure 6.11. The IPF plots reveal a strong {0001} texture with an MRD value of 67 along the trench width direction, indicating a dominant growth of the {0001} grains from the sidewall of the Co line. Along both the trench normal and the trench length directions, the textures appear to be between the $\{10\bar{1}0\}$ and the $\{2\bar{1}\bar{1}0\}$ orientations, both of which are represented by the $\{hki0\}$ type Miller–Bravais indices and perpendicular to the {0001} orientation along the trench width direction. In the 26 nm line, the large bamboo-like grain structures in the 220 nm Co line are no longer observed; instead, the grain structures become polycrystalline intermixing with some short bamboo-like grains.

For convenience of discussion, the grain orientation maps for Co lines of 220 nm and 26 nm widths (one each) are replotted in Figure 6.12 with the grain orientation distributions on top and grain textures shown in the bottom. Both maps are plotted along the trench width, trench normal, and trench length directions.

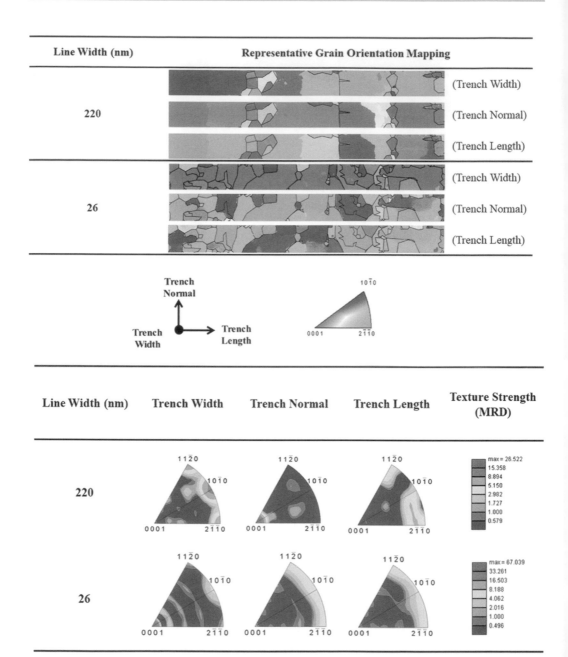

Figure 6.12 Top, the representative color-coded grain orientation maps, and bottom, the color-coded IPFs for grain orientation distribution for Co lines with linewidths of 220 nm and 26 nm plotted along the trench width, trench normal, and trench length directions. A black and white version of this figure will appear in some formats. For the color version, please refer to the plate section.

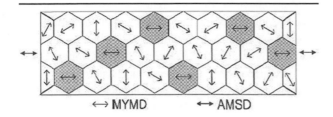

$\longleftrightarrow$ MYMD $\longleftrightarrow$ AMSD

Figure 6.13 Dark grains represent the grains aligned with minimum Young's modulus direction (MYMD) parallel to the absolute maximum stress direction (ASMD). All such grains will grow preferably during annealing to minimize the elastic strain energy. Reprinted from [17] with permission from Springer Nature.

Overall, the scaling effect on microstructure evolution of Co interconnects is similar to the Cu interconnects and can be understood by considering the grain growth driven by overall energy minimization. For the 220 nm Co lines, the inverse pole figures and grain textures show a strong {0001} fiber texture along the trench normal (MRD 19), indicating a strong {0001} grain texture from the trench bottom. The characteristics are similar to the 180 nm Cu line shown in Figure 6.7, with a strong {111} texture emerging from the trench bottom since the {0001} plane in the hcp structure is equivalent to the {111} plane in the fcc structure and both have the lowest interfacial energy. Along the trench width direction, we observed the $\{10\bar{1}0\}$ and $\{2\bar{1}\,\bar{1}0\}$ texture orientations, indicating grain growth with $\{0001\}\langle10\bar{1}0\rangle$ and $\{0001\}\langle2\bar{1}\,\bar{1}0\rangle$ type orientations. The characteristics can be explained based on the mechanism proposed by Lee et al. [17] to account for the texture evolution of Cu damascene interconnect during annealing as shown in Figure 6.13. As illustrated, each grain has a different crystal orientation, thus different elastic modulus and elastic strain energy; the ones with the minimum elastic modulus orientation when aligned with the maximum stress direction will grow preferably during annealing to minimize the elastic strain energy. For hcp Co, the orientation dependence of interfacial energy follows the order of $\gamma_{0001} < \gamma_{10\bar{1}0} \cong \gamma_{\bar{2}110} < \gamma_{hkl0}$ and elastic modulus $E_{10\bar{1}0} \cong E_{\bar{2}110} \cong E_{hkl0} < E_{0001}$. As the strain energy competes with the interface energy to drive abnormal grain growth at 400°C, the grains with the $\{0001\}\langle10\bar{1}0\rangle$ or $\{0001\}\langle2\bar{1}\,\bar{1}0\rangle$ type orientation will grow preferably to minimize the overall interface and strain energies. As a result, we observe the abnormal grain growth with large bamboo-like grain structures interspersing with some clusters of smaller grains in the 220 nm Co interconnect (Figure 6.12). This is similar as the Cu lines in forming the Blech short-length structure with bamboo–polygrain segments with similar implications on EM lifetime as discussed in Section 6.8.

When the linewidth scaled down to 26 nm, a switch of the {0001} texture orientation from the trench bottom to the trench width direction was observed, which is similar to that observed in the 70 nm Cu lines. This indicates that for the 26 nm lines, the increase in the sidewall interfacial energy outweighed that from the trench bottom to induce a switch of the {0001} texture orientation. Interestingly, the Co interconnect with 26 nm linewidth did not exhibit the second switch of the {111} texture from the trench width direction to the trench length direction as observed in the 45 nm Cu lines but rather continued with the {0001} texture along the trench width direction. This suggests that the interfacial energy is not as dominant over the elastic strain energy in Co interconnects as in Cu interconnects at 26 nm linewidth at 400°C. The higher strain energy observed in Co lines can be attributed to the elastic modulus of Co, which is between 186.62 GPa and 280.9 GPa, which is higher than 67 GPa and 192 GPa for Cu although not as anisotropic. The effect of strain energy in Co is also reflected in the IPF figures in Figure 6.12, where a very strong {0001} texture (MRD = 67) was observed at the trench width direction with the {$hki0$} type texture along the trench normal. This is because the strain energy can be minimized by aligning the {$hki0$} type texture, which is perpendicular to the {0001} orientation along the trench normal direction. However, the growth of such grains is limited since this increases the surface energy along the trench normal direction. Nevertheless, the abnormal grain growth yields some small clusters of {$hki0$} texture grains at the low end of the grain size distribution for both 220 nm and 26 nm Co lines. In Figure 6.14, the overall grain size distributions are presented for the 220 nm and 26 nm Co lines, where bimodal lognormal distributions are observed at the high end of the grain size distribution (black arrows), indicating a preferential growth of the {0001}$\langle 10\bar{1}0 \rangle$ and the {0001}$\langle 2\bar{1}\bar{1}0 \rangle$ type grains. At the low end (the gray oval), there are small grains

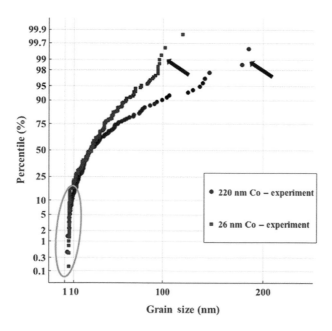

Figure 6.14 Grain size distributions obtained from precession electron diffraction for Co interconnects with linewidths of 220 nm and 26 nm. The results show the abnormal grain growth at high end (black arrows) and at the low end (the gray oval). Reprinted from [10] with permission from S. T. Hu.

indicating a limited growth of the {hki0} type grains. Both grain size distributions manifest an abnormal grain growth where the strain energy is important in competing with the interface energy to control the microstructure evolution of Co interconnects. The abnormal growth of the small grains will contribute more to the mass transport and resistivity increase than the large grains in Co lines.

6.5 Monte Carlo Simulation of the Microstructure Evolution in Nanointerconnects

So far, we show that the scaling effect can significantly influence microstructure evolution in Cu and Co interconnects up to the 22 nm linewidth. Since the microstructure is important in controlling EM performance, we are interested to investigate the scaling effect on microstructure evolution and EM performance beyond the 22 nm linewidth for future technology nodes. In this section, a Monte Carlo simulation is used to extend the study on the scaling effect on microstructure evolution for Cu and Co interconnects. Monte Carlo simulations are commonly used to study grain growth in two and three dimensions. These included the Monte Carlo Potts model [46–48], cellular automata method [49, 50], the phase field model [51, 52], the vertex model [53, 54], the front tracking model [55], and finite element models [56]. A series of studies by Srolovitz et al. [47, 57, 58] based on a 2D Monte Carlo model have simulated normal grain growth by minimization of the grain boundary energy. The simulation results showed good agreement in predicting normal grain growth kinetics of $r = K \cdot t^n$, where r is the grain radius, t the annealing time, and K a constant. The simulation verified that the parameter n can be less than the classical value of 0.5, depending on the impurity concentration but not due to the second phase particles or preferred orientation, as once thought. In addition, the anisotropy of the grain boundary energy can yield different values of n ranging from 0.25 to 0.42, and abnormal grain growth was observed due to the surface energy anisotropy [57, 58]. In a 3D Monte Carlo simulation, Jung et al. [46] found that microstructure evolution in Cu damascene lines depends on the trench aspect ratio, and the bamboo structure cannot evolve in trenches with high-aspect ratios due to the residual grain growth effect. The sidewall texture also affected the grain structure where a strong seeded sidewall texture led to a polygranular microstructure while a random seeded texture led to a bamboo structure instead.

The Monte Carlo method developed in this section is a 3D simulation based on total energy minimization aiming to study the scaling effect on microstructure evolution in Cu and Co nanolines. In this study, the interfacial and elastic strain energies are specified in detail as a function of grain orientation since they are important in controlling microstructure evolution, particularly for projecting abnormal grain growth. The material parameters used in the simulation are verified and adjusted if needed, so the simulation results would be consistent with that obtained from PEDs. The model developed is used to project the scaling effect on microstructure evolution beyond the 22 nm linewidth.

6.5.1   Energetics for Grain Growth in Line Structures

To develop the simulation algorithm, we need to specify the energetics for grain growth in line structures that are derived from polycrystalline thin films. In thin films, grain growth is driven by minimization of the Gibbs free energy associated with grain boundaries, surfaces, interfaces, and elastic strains [12–15]. The grain boundary energy is minimized by reducing the number of the small grains to increase the grain size. In this process, the migration of the grain boundary is driven by reducing the free energy ΔG when atoms jump from a high-energy grain to a lower-energy grain. The driving force per unit area of boundary can be expressed as follows:

$$F = \frac{\Delta G}{V_m} \mathrm{Nm}^{-2},$$

where V_m is the molar volume. The velocity of grain boundary migration is related to the driving force $\Delta G/V_m$ as

$$v = m \cdot \frac{\Delta G}{V_m}, \tag{6.1}$$

where m is the grain boundary mobility. When ΔG comes only from the grain boundary, the process is described as a normal grain growth with a monomodal grain size distribution. In the interconnect line, grain growth is more complex since it also involves the surface, interface, and elastic strain energies due to the dimensional constraint and the damascene structure and fabrication process. All these energies are included in the simulation, as discussed in the following subsections.

6.5.1.1   Grain Boundary Energy

A grain boundary in a single-phase polycrystalline material refers to an interface separating two crystalline grains with different orientations. Its characteristics depend on the misorientation of the two adjoining grains (three degrees of freedom) and the orientation of the boundary plane (two degrees of freedom), altogether with five degrees of freedom (DOF) for one grain boundary. In general, grain boundaries can be characterized into three types: tilt boundaries, twist boundaries, and mixed boundaries. As illustrated in Figure 6.15, a tilt boundary is formed when the axis of rotation is parallel to the boundary plane, whereas a twist boundary is formed when the rotation of axis is perpendicular to the boundary, and a mixed boundary combines the twist and tilt characteristics.

The grain boundary energy is the excess free energy associated with the boundary interface. To estimate the grain boundary energy, a simple assumption can be made by relating the total grain boundary energy γ_{GB} to the dislocation density ($1/D$) per unit area of the boundary, where D is the spacing between the dislocations [59]. The dislocation spacing in a low-angle grain boundary can be described as

$$D = \frac{b}{\sin \theta} \simeq \frac{b}{\theta},$$

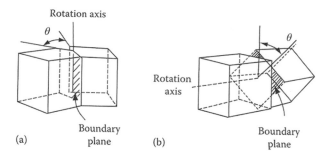

Rotation axis

θ

Rotation
axis

θ

Boundary
plane

(a)

(b)

Boundary
plane

Figure 6.15 The schematics showing the relative orientations of the crystals and the grain boundary plane in forming (a) a tilt boundary and (b) a twist boundary. Reprinted from [10] with the permission of S. T. Hu.

where b is the Burgers vector of the dislocation and θ is the angular misorientation across the boundary. Therefore, the grain boundary energy γ_{GB} is proportional to the dislocation density as $\gamma_{GB} \propto \theta$.

As the misorientation angle θ increase above 10–15° for high-angle boundaries, the grain boundary structure changes from atoms in the lattice sites with only slightly distorted atomic bonds to a poorly fitted area with a highly distorted or even broken bond to increase the boundary energy. For high-angle boundaries, the strain fields induced by the dislocations cancel each other, resulting in a grain boundary energy nearly independent of the boundary misorientation. However, not all high-angle boundaries are highly disordered with high boundary energy; in particular, twin boundaries are special high-angle boundaries with special orientation relationship to allow the two adjoining grains to fit together with relatively little distortion and hence a low boundary energy. The Σ3 coherent twin boundaries discussed in Section 6.3.4 belong to this special type boundary with highly coincided boundary lattice sites and low grain boundary energy. Such types of twin boundaries have minimal effects on electrical resistivity and EM reliability.

The fcc Cu has a high elastic anisotropy, making the grain boundary energy not only dependent on the grain misorientation but also on the inclination of the boundary plane. For fcc metals, Bulatov et al. have determined the grain boundary energies as a function of grain misorientation and the type of the grain boundary planes [60]. The results for GB energy of Cu are shown in Figure 6.16 for twist and symmetric tilt boundaries relating to the $\langle 100 \rangle$, $\langle 110 \rangle$, and $\langle 111 \rangle$ axes. In this study, we adopted the approach of Bulatov et al. to specify the grain boundary energy based on the orientation between the grain boundary plane and the three principal symmetry axes: $\langle 100 \rangle$, $\langle 110 \rangle$, and $\langle 111 \rangle$. To calculate grain boundary energy, we first determined the orientations of specified grain boundaries from the principal axes, then input into an open-source MATLAB code, *GB5DOF.m,* based on the grain boundary energy functions in Figure 6.16. (More details can be found in the paper by Bulatov et al. [60].)

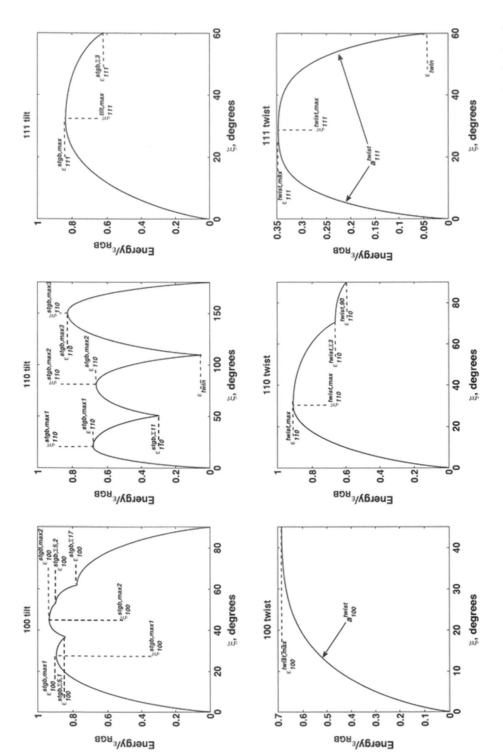

Figure 6.16 Parameterization of GB energy variations with different twist and symmetric tilt boundaries in the ⟨100⟩, ⟨110⟩, and ⟨111⟩ axis. Reprinted from [60], with permission from Elsevier.

Once the grain boundary energy γ_{GB} is determined, there is a driving force γ_{GB}/r acting on the curved-shape boundary toward its center of curvature, where r is the grain radius. The rate of grain growth v is related to the grain boundary mobility and the driving force as

$$v = \frac{dr}{dt} = m\gamma_{GB}\kappa,$$ (6.2)

where κ is the curvature of the grain, or $1/r$ after integration. This yields a grain growth rate as $r = K \cdot t^n$, where K is a temperature-dependent rate constant with a classical result of $n = 0.5$. When the driving force comes only from the grain boundary energy, the characteristic is the normal grain growth.

6.5.1.2 Surface/Interface Energy

In metal lines, as grains grow by minimizing the grain boundary energy, two other energy components from the surface/interface and the elastic strain energies come into play as the line dimensions are scaled down. Together this leads to an abnormal grain growth with a bimodal grain size distribution, as observed in the PED studies.

For a fcc material such as Cu, the variation of surface energy with respect to the orientation generally follow the trend of $\gamma_{111} < \gamma_{110} < \gamma_{100}$ for low Miller index planes. This order can be roughly explained by the broken bond model by counting the free energy associated with the bonds broken to form the surface. Since the {111} plane in an fcc crystal structure is the close-packed plane, the least bonds are broken when a {111} surface is formed with the lowest surface energy.

For damascene interconnects, the trench sidewalls provide additional interfaces contributing interface energy to affect the microstructure in the line. To include the interface energy from the trench sidewall in the driving force, (6.2) becomes

$$v = m\left(\gamma_{GB}\kappa + \frac{\Delta\gamma_{s/i}}{h} + \frac{\Delta\gamma_i^{SW}}{w}\right),$$ (6.3)

where $\Delta\gamma_{s/i}$ represents the change in the interfacial energy at the top and bottom trenches, $\Delta\gamma_i^{SW}$ is the interfacial energy difference at the sidewalls, h is the trench height, and w the trench width. At advanced technology nodes, the interfacial energy from the sidewalls and the trench bottom can become dominant to control grain growth. As a result, there is a preferential growth of small grains initiated from the sidewalls to minimize the sidewall interfacial energy.

6.5.1.3 Elastic Strain Energy

After Cu electroplating, the interconnect structure is subject to two anneals at 250°C and 400°C to expedite the grain growth and stabilize the grain structure. During annealing, the Cu line is subjected to thermal strains due to the CTE mismatch between Cu and the Si substrate. Under a plane stress condition, the strain energy density can be expressed as

$$E_\varepsilon = \varepsilon^2 M_{hkl}, \tag{6.4}$$

where ε is the biaxial strain between the film and the substrate, and M_{hkl} is the effective biaxial modulus, which depends on the (hkl) direction normal to the plane of strain. The effective biaxial modulus has been derived by Murikami and Chaudhari [61] as

$$M_{hkl} = C_{11} + C_{12} + K - \frac{2(C_{12} - K)^2}{C_{11} + 2K}$$

$$K = (2C_{44} - C_{11} + C_{12})\left(h^2 k^2 + k^2 l^2 + h^2 l^2\right) \tag{6.5}$$

$$h^2 + k^2 + l^2 = 1,$$

where C_{11}, C_{12}, and C_{44} are the elastic constants. For Cu, C_{11} = 168.3 GPa, C_{12} = 122.1 GPa and C_{44} = 75.7 GPa. As shown in Figure 6.17, Cu is highly elastic anisotropic with the lowest elastic modulus along the $\langle 100 \rangle$ direction equal to about 1/3 of the highest elastic modulus along the $\langle 111 \rangle$ direction. The orientation dependence of the strain energy was included in the Monte Carlo simulation.

Finally, by including the strain energy, the grain growth rate becomes

$$v = m\left(\gamma_{GB}\kappa + \frac{\Delta\gamma_{s/i}}{h} + \frac{\Delta\gamma_i^{SW}}{w} + \varepsilon^2 \Delta M_{hkl}\right), \tag{6.6}$$

where ΔM_{hkl} is the difference in the biaxial moduli between two adjacent grains. Overall, grain growth in the interconnect line is driven by minimization of the total energy, but each energy component changes differently with scaling, depending on the materials and the fabrication process. For Cu, the {111} grains have the lowest surface energy but the highest strain energy, while the {100} grains have the lowest strain energy but the highest surface energy. Therefore, the texture evolution in the Cu

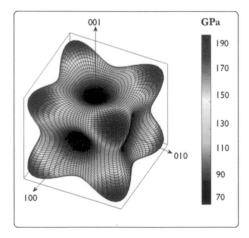

Figure 6.17 Elastic moduli of Cu as a function of crystal orientation, with the modulus along the $\langle 111 \rangle$ orientation almost three times the modulus along the $\langle 100 \rangle$ orientation. Reprinted from [30], with open access permission from Springer Nature.

line is controlled by the competition between the {111} and the {100} grains together with other energy components to minimize the overall energy. As a result, microstructure evolution in Cu lines is characterized as abnormal grain growth with twin formation and continues to change with scaling.

6.5.2   Simulation Algorithm

In this section, the algorithm of Monte Carlo simulation is developed for 3D grain growth in Cu and Co interconnects by considering all the energy components and their anisotropic and orientational characteristics. The grain orientations of the boundary planes are expressed in terms of the Miller indices for Cu and the Miller Bravais indices for Co as reference orientations of the crystal lattices. The simulation algorithm was programmed in MATLAB code, and the programming flow was based on a code published in a MathWorks file exchange [62]. For this study, the code was modified and extended to treat a rectangular interconnect geometry based on minimization of the grain boundary, surface, and elastic strain energy components. The code also evaluated the effect of postprocessing thermal treatments on grain size distribution and texture characteristics.

The Monte Carlo simulation started by setting up a 3D interconnect geometry with user inputs of the linewidth, the line height, and the line length. The initial state was defined in arbitrary volume units corresponding to an initial distribution of the grain size and lattice number in the interconnect, which was then discretized as small lattice sites and mapped onto a 3D geometry. An example is illustrated in Figure 6.18 defining an interconnect geometry of 45 nm × 96 nm × 500 nm (linewidth × line height × line length) for an initial state of microstructure consisting of eight grains × 17 grains × 90 grains with 24 sites × 51 sites × 270 lattice sites. Together this structure has an initial grain size of 5.625 nm × 5.625 nm × 5.625 nm, with each grain containing 3 × 3 × 3 lattice sites. For different linewidths, the initial grain size is varied to scale with the linewidth to yield a comparable number of initial lattice sites to maintain similar initial statistics in the study of the scaling effect.

In the initial state, the grain orientations are first assigned using 23 unique crystal orientation families starting with the low Miller indices planes of {100}, {110}, and {111}, then followed by generating all possible permutations of each orientation family in a 3D space. Altogether there are 602 possible orientations, which are stored in a reference list. Each grain is then randomly assigned to a crystal orientation family, where a unique permutation of that family orientation is randomly chosen to form the initial state with the same orientation for all the lattice sites inside the grain. This is set up in order not to favor a particular orientation family with more permutations. The gray arrow in Figure 6.18 represents a periodic boundary condition being applied to prevent geometry side effects. In this study, the overburden layer is not considered since it would not affect the microstructure evolution in interconnects with linewidth beyond 45 nm, as discussed in Section 6.3.3.

After the initial assignment of orientation to each grain, the Monte Carlo loops based on energy minimization begin by randomly choosing certain lattice sites within

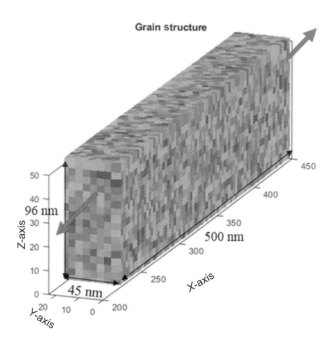

Figure 6.18 A schematic image of microstructure in the interconnect at the initial state. Reprinted from [10] with permission from S. T. Hu. A black and white version of this figure will appear in some formats. For the color version, please refer to the plate section.

the interconnects. The energy E_i refers to all the energy components associated with a certain site I, and according to the modified Potts model [46–48], it can be described by the Hamiltonian:

$$E_i = j_i \sum_{nn} \left(1 - \delta_{s_i s_j}\right) + H_i, \qquad (6.7)$$

where j_i is the grain boundary energy; δ is the Kronecker delta; s_i and s_j represent the orientation number in site i and its neighbors j, respectively; H_i is the lattice site energy, including both the surface energy and the elastic strain energy; and the subscript nn indicates a summation over all the nearest neighbors. For the simulation of line structures, all the energy components in (6.6) are normalized to the line geometry.

To calculate the total energy E_i we started with the grain boundary energy following the approach of Bulatov et al. [60]. This started by first determining the grain orientation from the principal axes, then input into Bulatov et al.'s open-source MATLAB code, *GB5DOF.m*, to calculate the grain boundary energy using the GB energy functions shown in Figure 6.16. The grain boundary energy obtained was then multiplied by the local grain curvature to determine the grain boundary driving force (6.2). The local grain curvature was determined using a model proposed by Mason [63] based on a 3D square lattice. To calculate the local curvature κ, the 3D square lattice containing two layers of neighboring lattices was set up with a total of 124 neighbor sites, including 26 near-neighbors and some third-nearest-neighbor sites, and according to Mason as

$$\kappa = \exp\left(\frac{\beta^2\lambda^2}{2\sigma^2}\right)\left[\frac{a_v D}{4\pi\sigma^4} - \frac{\sqrt{\pi}}{\sqrt{2}\sigma}\,\mathrm{erf}\left(\frac{\beta\lambda}{\sqrt{2}\sigma}\right)\right],$$ (6.8)

where β is a constant equal to 0.5, λ is the spacing between the lattice, and a_v is the lattice volume. In this way, the nonlinear relationship of the grain boundary energy as a function of the grain misorientation can be computed and was included in the simulation.

To derive the anisotropic surface energies of Cu, we surveyed through the literature and found most of the previous studies were based on analytical models [64–67]. Although the surface energy from different models can vary up to 27%, it generally followed a trend with the (111) orientation possessing the lowest surface energy and the (110) orientation the highest surface energy. In this study, the surface energies derived by Rodriguez et al. [64] were used since their estimates included the surface energy anisotropy for all the 23 surface orientations of fcc crystals. In the simulation, we first identified the orientation normal to the trench bottom and the trench sidewalls, then determined the surface and interface energies for the corresponding orientations. In the simulation algorithm, the surface energy was normalized to the (111) surface energy and then multiplied by a surface energy factor, making it easier for future reference for specific surfaces. An initial surface energy of 1.81 J/m^2 based on the broken bond assumption was used to start the simulation and then adjusted by fitting to the grain size distribution observed by PED.

Finally, the elastic strain energy that does not depend on the neighboring lattices can be computed straightforwardly. We assumed that the thermal strain arises only from the CTE mismatch between the metals and the substrate after the final 400°C annealing in damascene fabrication. The derivation followed the approach for thin films as in (6.4), except that the interconnect has a triaxial stress state with an effective modulus as follows [66]:

$$\frac{1}{E_{hkl}} = \frac{1}{E_{\langle 100\rangle}} - 3\left(\frac{1}{E_{\langle 100\rangle}} - \frac{1}{E_{\langle 111\rangle}}\right)(\alpha^2\beta^2 + \beta^2\gamma^2 + \alpha^2\gamma^2),$$ (6.9)

where α, β, and γ are the directional cosine of the $\langle hkl\rangle$ grain orientation, and for Cu $E_{\langle 111\rangle} = 192$ GPa and $E_{\langle 100\rangle} = 67$ GPa. The strain energy density was then calculated as a function of grain orientation using (6.4) with the result varying by a factor of 2.5 depending on the elastic anisotropy of Cu.

In the simulation, the total energy E_1 associated with the lattice site i was first evaluated with the orientation of site i randomly chosen to flip to one of its neighbor orientations. Then the energy E_2 with the new orientation is evaluated to determine the change in the system energy $\Delta E = E_1 - E_2$ for reorienting the site. The probability p for the reorientation is chosen as 1 if $\Delta E \leq 0$, or as 0 if $\Delta E > 0$, where a flip is allowed if the reorientation can lower the overall site energy. In this way, each Monte Carlo step represents a random probing of all the lattice sites in the interconnect, regardless whether the step is allowed energetically. After sufficient steps, the total energy of the system will approach a steady state, then the result is analyzed to quantify the grain growth characteristics. The grain size distribution is determined using the grain boundary intercept method to construct the grain size distribution function. The grain orientation and texture distributions in the interconnect are extracted and verified with the PED results. In the simulation, the input parameters can be modified in order to examine the effect of specific energy components on grain growth or the effect of new materials incorporated in the damascene structure, such as sidewall barriers or alloy additions.

6.6   Simulation Results for Copper Interconnects

6.6.1   Verification of the Monte Carlo Model

To simulate the microstructure evolution in Cu nanolines, we first verified the Monte Carlo model by comparing with the observed PED results. This was carried out in two steps: first to check the effect of the individual energy components on grain growth, and then to verify the statistics of the overall grain size distribution. The effect of the grain boundary energy on grain growth was examined in a 45 nm Cu line with an aspect ratio of 2 after 300 Monte Carlo steps (MCS). In this case, the growth of large grains was found to occur at the expense of small grains with an overall monomodal log-normal size distribution. This confirms that when only the grain boundary energy is considered, a normal grain growth would occur in the interconnect as expected. Next we evaluated the individual contributions of the surface energy and the strain energy components in the same 45 nm Cu line. By considering only the surface energy, we found a dominant {111} grain texture normal to the trench sidewall with a corresponding (110) orientation along the trench normal. And by considering only the elastic strain energy, we obtained a microstructure dominated by {100} grain orientations after 16 MCS in simulation. These results are also as expected.

To complete the model verification, we applied all the energy components to check the overall growth characteristics in a 40 nm wide Cu line. With all the energy components included, the results show the growth of large grains with a {111} texture along the trench width and a (110) orientation normal to the trench bottom and with large grains extending to the trench bottom. The results show bimodal abnormal grain growth characteristics as expected but the grain growth is too dominated by the sidewall interface energy and not consistent with the PED results shown in Figures 6.7 and 6.8, where the growth of the {111} texture grains is more limited with many smaller grains being pinned at the trench bottom. The problem is traced to the interface energy of 1.8 J/m^2 used in the simulation, which is too large to exaggerate the contribution of the sidewall interface energy, leading to the growth of large {111} texture grains along the trench width. In Cu damascene structures, the sidewalls are formed with a PVD Ta barrier, which has a good lattice match with Cu [68], so it is reasonable to have an interface energy less than 1.8 J/m^2 estimated from a broken

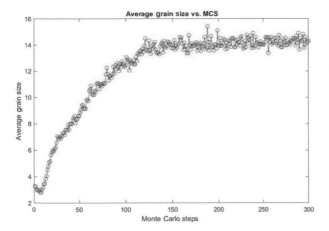

Figure 6.19 Average grain size evolution of the 22 nm Cu interconnect plotted with respect to each Monte Carlo step. Reprinted from [10] with permission from S. T. Hu.

bond model. We found that an interface energy of 0.25 J/m² would give a better match and is used in subsequent simulations of Cu nanolines.

Finally, we evaluated how the simulation evolves to approach a stable final state. In Figure 6.19, we show the evolution of the average grain size in a 22 nm Cu line with increasing Monte Carlo steps. Here the grain size steadily increases during grain growth with Monte Carlo steps until 100 MCS, then it approaches an asymptotic plateau where the grain size becomes stabilized after 500 MCS, reaching a steady state in the growth process. Subsequent simulations show that the time steps required to reach a steady grain size generally depend on the annealing temperature, as expected from the grain growth kinetics. Even though the Monte Carlo simulation is unable to accurately specify the time scale, the model can provide an estimate of the relative annealing time required to reach a stable microstructure. Based on this analysis, all the simulations were performed with 500 MCS for Cu interconnects as a function of the linewidth.

6.6.2 Simulation of Microstructure Evolution in Cu Interconnects

After verification, we simulated the microstructure evolution in Cu nanolines using an interface energy of 0.25 J/m² and 500 MCS, first for the 70 nm, 45 nm, and 40 nm Cu lines, and then extended the study to the 22 nm lines and beyond to 8 nm to investigate the scaling effect on microstructure evolution. For the first group of simulations, we illustrate the results for the 45 nm lines, which were derived using 24 sites × 450 sites × 51 sites and an initial grain size of 5.625 nm. The grain orientation map is presented in Figure 6.20a, where for simplicity we show the orientations only along the trench normal direction. The complementary grain distribution map is shown in Figure 6.20b, where each grain in the grain map has a unique shade regardless of its orientation. Together, the simulation results show that the grain structure in the 45 nm line consists of very short bamboo sections intermixing with polycrystalline grains

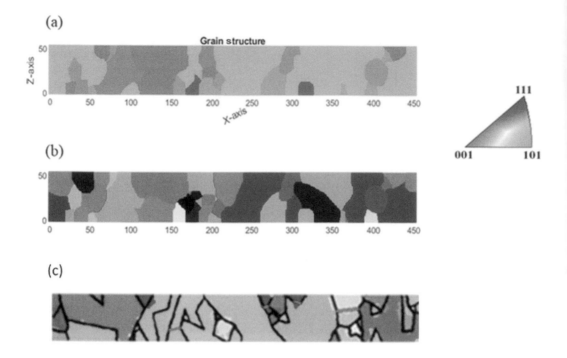

Figure 6.20 The cross-sectional view of a 45 nm Cu interconnect with (a) a grain orientation map along the trench normal direction, (b) the individual grain map after 500 MCS, and (c) the grain orientation map along the trench width orientation acquired using the PED technique. Reprinted from [10] with permission from S. T. Hu. A black and white version of this figure will appear in some formats. For the color version, please refer to the plate section.

with predominant {110} texture along the trench normal. The results are consistent with the PED grain orientation map in Figure 6.20c showing predominant {110} texture grains with comparable grain size and short sections of bamboo-like grains. For the 45 nm lines, the simulation also found predominant {111} texture along the trench length for the corresponding {110} grains along the trench normal, which is consistent with PED results shown in Figures 6.6 and 6.7.

In Figure 6.21, we compare the simulated grain structures for the 70 nm, 45 nm, and 40 nm lines along the trench normal with the PED results. The agreement is good as shown by the inverse pole plots from PED and simulation. For the 70 nm line, the (110) orientation is dominant along the trench normal and with corresponding {111} texture orientation along the trench width. As the linewidth decreases to 45 nm and 40 nm, the {110} texture strength along the trench normal is reduced while the {111} texture orientation shifts from along the trench width to along the trench length.

After validating the simulation model with the PED results up to the 40 nm linewidth, the simulation was extended to project the scaling effect on microstructure for Cu lines with widths from 22 nm to 8 nm. The results for the 22, 15, 11, and 8 nm Cu lines are

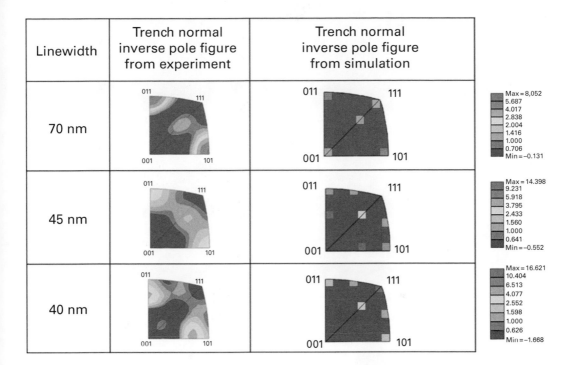

Linewidth	Trench normal inverse pole figure from experiment	Trench normal inverse pole figure from simulation	
70 nm			Max = 8,052 5.687 4.017 2.838 2.004 1.416 1.000 0.706 Min = −0.131
45 nm			Max = 14.398 9.231 5.918 3.795 2.433 1.560 1.000 0.641 Min = −0.552
40 nm			Max = 16.621 10.404 6.513 4.077 2.552 1.598 1.000 0.626 Min = −1.668

Figure 6.21 The inverse pole plots from PED at trench normal direction are compared with results from simulation for Cu lines with 70 nm, 45 nm and 40 nm widths. The shading codes on the right represent the volume fraction for each crystal orientations after 500 MCS. Reprinted from [10] with permission from S. T. Hu. A black and white version of this figure will appear in some formats. For the color version, please refer to the plate section.

shown in Figure 6.22, where for each linewidth, an orientation map and a grain size distribution map are plotted for the simulated microstructure after the 400°C anneal.

Here the results show that as scaling continues to the 22 nm linewidth, the {110} grain texture continues to dominate along the linewidth direction and with corresponding {111} grain texture along the line length direction. This indicates that the interface energy from the sidewall dominates the abnormal grain growth in the 22 nm lines. Beyond the 22 nm linewidth, the interface energy increases relative to the other energy components due to further scaling of the line geometry, leading to a microstructure with smaller grain size and shorter bamboo sections, as shown in the individual grain distribution maps in Figure 6.22. To further demonstrate the grain growth characteristics, the simulated grain size distributions are plotted in Figure 6.23 for Cu lines with linewidths from 45 to 8 nm. Here the grain size distributions clearly show abnormal grain growth statistics with bimodal log-normal distributions at the high end and at the low end of the grain size (indicated by the oval). Overall, these characteristics are consistent with the PED results, indicating that the simulation based on energy minimization is capable of projecting microstructure evolution in Cu lines at the nanoscale.

22 nm Cu line

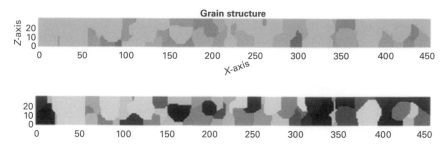

15 nm Cu line

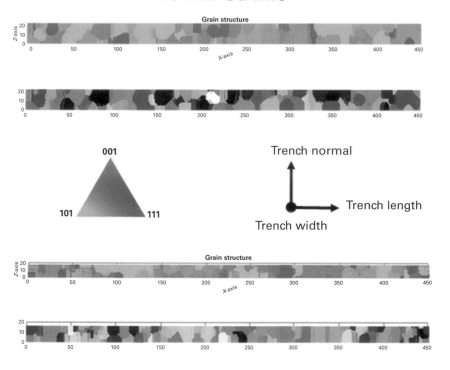

8 nm Cu line

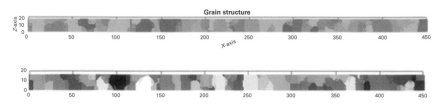

Figure 6.22 The cross-sectional view of representative Cu interconnects with linewidths of 22 nm, 15 nm, 11 nm, and 8 nm. The top image shows the grain orientation map along the trench normal direction, and the bottom image the grain distribution map. Reprinted from [10] with permission from S. T. Hu. A black and white version of this figure will appear in some formats. For the color version, please refer to the plate section.

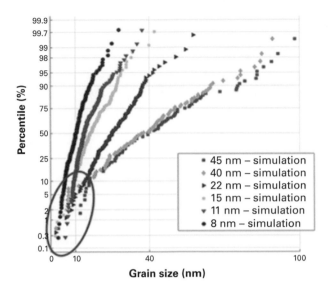

Figure 6.23 Results from Monte Carlo simulation on the grain size distributions for various linewidths of Cu Interconnects showing bimodal log-normal statistics with abnormal grain growth at the high end and the low end (the oval) of the grain size distribution, where more small grains are present due to the dimensional modulation of grain growth characteristics. Reprinted from [10] with permission from S. T. Hu.

The trend presented here show that the serial bamboo–polygrain segments persist with continued scaling in the Cu lines, which can be beneficial for EM reliability, as discussed later in Section 6.8. The line scaling and the decrease in grain size will increase electron scattering and mass transport at surfaces and grain boundaries and impact the electrical resistivity and EM reliability of the Cu lines.

6.7   Simulation Results for Cobalt Interconnects

Following the Monte Carlo simulation of Cu interconnects, the study was extended to investigate the scaling effects on Co interconnects. For Co interconnects, the simulation algorithm for grain growth was similarly based on total energy minimization. However, since Co has an hcp structure, the crystal orientation has to be modified from the Miller indices (*hkl*) to the Miller–Bravais indices (*hkil*) based on the hcp crystal symmetry, where 57 different crystal orientations were generated with all the possible permutations to yield a total of 632 crystal orientations in the 3D space.

Literature search was conducted to find reliable experimental data for the surface and grain boundary energies of Co. The results of grain boundary energy deduced from the disclination-structural unit model are summarized in Figure 6.24, where specific crystal misorientations of coincident sites display local minima of grain boundary energies [69]. The data were transformed into a piecewise function and

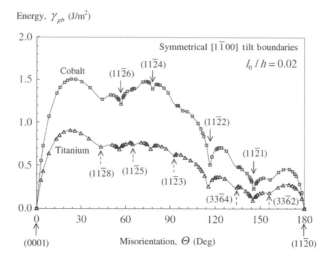

Figure 6.24 The hcp Co grain boundary energy as a function of misorientation between two adjacent crystals deduced from the disclination-structural unit model. Reprinted from [69] with permission from Taylor & Francis Ltd.

then multiplied by the grain boundary curvature to calculate the driving force from the grain boundary energy.

The anisotropic surface energy has been analyzed for hcp Co using different analytical methods [70–72]. In general, the (0001) orientation of the base plane of the Co hcp structure is found to have the lowest surface energy and, similar to Cu, Co does not show a specific relationship between surface energies and crystal orientations. For this study, the Co surface energies evaluated by Zhang et al. [70] and Luo and Qin [72] were incorporated into the simulation algorithm to estimate the surface/interface energy contribution. And as with Cu, the surface energies of Co used in the simulation were normalized with respect to the (0001) orientation, then multiplied by a surface energy factor. An initial surface energy of 2.5 J/m^2 derived from a broken bond model was used to start the simulation.

To deduce the strain energy component, we again assumed that the strain energy comes only from the CTE mismatch between the Co line and the Si substrate during thermal processing. The effective triaxial elastic modulus for different crystal orientation E_{hkl} in a hexagonal system can be described as follows:

$$\frac{1}{E_{hkl}} = \left(1 - \ell_3^2\right)s_{11} + \ell_3^4 s_{33} + \ell_3^2\left(1 - \ell_3^2\right)\left(2s_{13} + s_{44}\right), \tag{6.10}$$

where ℓ_3 represents the directional cosine of the (hkl) plane. The compliance constants for Co are $s_{11} = 4.99 \times 10^{-12}$ Pa^{-1}, $s_{13} = -0.87 \times 10^{-12}$ Pa^{-1}, $s_{33} = 3.56 \times 10^{-12}$ Pa^{-1} and $s_{44} = 14.08 \times 10^{-12}$ Pa^{-1} [65]. In this way, the elastic anisotropy of hcp Co was included in the simulation algorithm to calculate the strain energy contribution.

Following the simulation for Cu lines, we first verified the values of the energy components used in the simulation by checking the grain growth characteristics of the

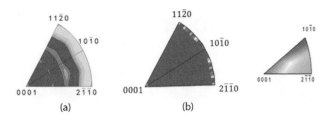

Figure 6.25 Comparison of the inverse pole plots acquired from (a) PED and (b) Monte Carlo simulation along the trench normal direction for Co interconnect with 26 nm linewidth. Reprinted from [10] with permission from S. T. Hu. A black and white version of this figure will appear in some formats. For the color version, please refer to the plate section.

26 nm Co line. The results of the grain size distribution deduced with different interface energies are compared with the PED results. The simulation results show a bimodal log-normal grain size distribution as expected for abnormal grain growth, but the interface energy of 2.5 J/m^2 seems to yield a grain size distribution that is too large compared to the PED results. This is because the 2.5 J/m^2 surface energy is deduced from the broken bond model, which is too large for the sidewall barrier interface. Judging from the PED results, an interface energy of 0.5 J/m^2 for the (0001) orientation would provide a better match between the simulated and the observed grain size distributions and was used in subsequent simulations for the Co lines. The grain orientation distribution of the 26 nm Co line along the trench normal direction is compared with the PED results in Figure 6.25. Overall, the agreement is good where the inverse pole figures show a dominant grain orientation between the $(10\bar{1}0)$ and the $(2\bar{1}\bar{1}0)$ planes along the trench normal direction and a corresponding strong {0001} grain texture along the trench width direction [10].

After verification, Monte Carlo simulations were carried out to project the microstructure evolution in Co interconnects with linewidths of 22 nm, 15 nm, 11 nm, and 8 nm, and the results of grain distribution maps are shown in Figure 6.26. With continued scaling, the microstructure in the Co lines evolves with decreasing grain size but retains some short bamboo grain sections. The simulated grain size distributions are plotted in Figure 6.27, showing bimodal log-normal statistics of abnormal grain growth at the high end as well as at the low end (indicated by the oval) of the grain size, which is similar to Cu lines. This indicates that the abnormal grain growth observed by PEM for the 26 nm Co line is extended beyond the 22 nm linewidth. It is worth noting that in contrast to the Cu interconnects, the Co interconnects did not show a switch from the trench width direction to the trench length direction beyond the 26 nm linewidth. Instead, it continued with the trend of the 26 nm Co line. This indicates that the interface energy in Co is not as dominant over the elastic strain energy to drive the abnormal grain growth as in Cu narrow lines. This is because the elastic modulus of Co is higher than Cu although not as anisotropic, so the elastic strain energy is larger and can compete with the interface energy to drive abnormal grain growth.

22 nm Co line

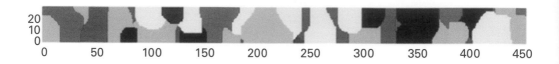

15 nm Co line

11 nm Co line

8 nm Co line

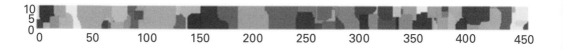

Figure 6.26 The grain distribution maps obtained by Monte Carlo simulation in Co interconnects with linewidths of 22 nm, 15 nm, 11 nm, and 8 nm. Reprinted from [10] with permission from S. T. Hu. A black and white version of this figure will appear in some formats. For the color version, please refer to the plate section.

6.8   Simulated Bamboo–Polygrain Structures and Implications for EM Reliability

In the Monte Carlo simulation, we found that the grain structures of the Cu and Co lines generally consist of a series of polycrystalline grain clusters intermixing with bamboo-like grain segments. An example of the simulated serial bamboo–polygrain structures for the 40 nm Cu line is shown in Figure 6.28. This is of interest for EM reliability, as they form the Blech short-length segments that can improve the EM lifetime. The Blech

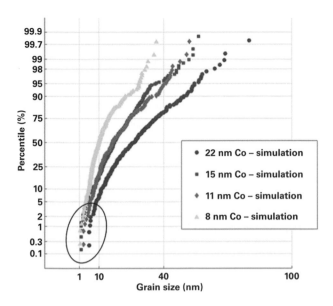

Figure 6.27 The simulated grain size distributions for Co interconnects with linewidths of 22 nm, 15 nm, 11 nm, and 8 nm. The results show abnormal grain growth at the high end as well as the low end of the grain size distribution. Reprinted from [10] with permission from S. T. Hu.

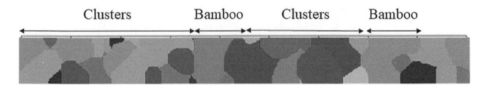

Figure 6.28 The shading-coded grain distribution map of a Cu 40 nm line presents the microstructure consisting of serial segments of polygrain clusters mixing with bamboo grains. Reprinted from [10] with permission from S. T. Hu.

short-length effect was discussed in Chapter 5, referring to the reduction of EM mass transport due to the buildup of a backflow stress gradient in a serial bamboo–polygrain structure. In such a structure, EM is through grain boundaries in the polygrain section while in the bamboo–grain segment, it is controlled by interface diffusion that is much slower. With the bamboo-like grains serving as blocking boundaries, a stress gradient $\Delta\sigma/\Delta x$ can be built up in the polygrain section, driving a mass backflow v_b against EM v_e to slow down the overall mass transport v_d as follows:

$$v_d = v_e - v_b = \frac{D_{eff}}{kT}\left[Z^*e\rho j - \frac{\Delta\sigma\Omega}{\Delta x}\right]. \tag{6.11}$$

This is (5.9) in Chapter 5, where D_{eff} is the effective diffusivity, k the Boltzmann constant, T the absolute temperature, and $Z^*e\rho j$ the EM driving force, with Z^*e the effective charge, ρ the electrical resistivity, j the current density, and Ω the atomic volume. At a critical length L_c, the backflow can be sufficient to make $v_d = 0$ when

$$jL_c = \frac{\Delta\sigma\Omega}{Z^* e\rho}.$$
(6.12)

Thus, the Blech short-length effect depends on the length of the bamboo–polygrain section; if it is less than the critical length L_c, the backflow can overcome the EM driving force, making the line immortal. In Section 5.5, we show that the EM lifetime was significantly improved for Cu damascene lines with serial bamboo–polygrain segments by incorporating a CoWP or Co cap at the interface, increasing the activation energy Q_{EM} from 0.8 eV, typical for Cu lines, to 1.5–1.7 eV of the capped interface.

To evaluate the Blech short-length effect, we examined the statistical distribution of the bamboo–polygrain clusters obtained from Monte Carlo simulation for Cu and Co interconnects beyond 22 nm linewidth. We found that the serial mixed grain segments continue to exist with log-normal statistics in length distribution for Cu interconnects from 22 to 8 nm in linewidth. The average cluster length (evaluated at 50%) continues to decrease from 145 nm for the 22 nm line, 120 nm for the 15 nm line, 60 nm for the 11 nm line, to 57 nm for the 8 nm line, as shown by the cluster lengths at 50% for different linewidths in Figure 6.29. Therefore, we expect that the Blech short-length effect would continue for Cu interconnects by incorporating a proper interface cap such as a CoWP or Co cap. Based on (6.12), the value of the critical jL_c product depends on the backflow stress $\Delta\sigma$ that can be built up in the bamboo–polygrain segment. While such a stress has

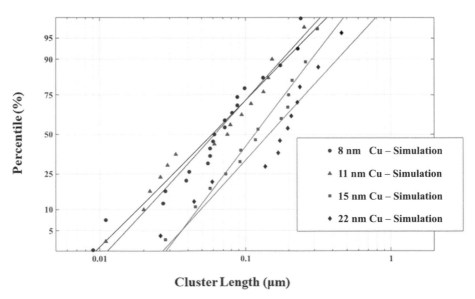

Figure 6.29 Cluster length distribution for simulated Cu interconnects with linewidth beyond 22 nm. The solid lines represent log-normal curve fitting. Reprinted from [10] with permission from S. T. Hu.

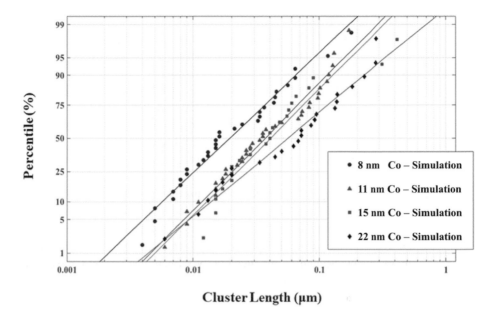

Figure 6.30 Cluster length distributions deduced from Monte Carlo simulation for Co interconnects with linewidths of 22 nm and beyond. The solid lines represent lognormal curve fitting. Reprinted from [10] with permission from S. T. Hu.

not been measured in Cu nanolines for the range of linewidth discussed here, we show in Section 3.4 in Chapter 3 that the hydrostatic stress in Cu nanolines is relatively independent of the linewidth down to the 200 nm range. If this trend continues with scaling, we expect that the Blech short-length effect would continue to exist to improve EM performance for future technology nodes of Cu nanolines.

From the simulated grain structures of Co interconnects, we found similar bamboo–polygrain clusters with log-normal statistics of cluster length decreasing with linewidth, as shown in Figure 6.30. The average cluster length of Co lines at 50% seems to be smaller than that of Cu: 56 nm for the 22 nm line, 40 nm for the 15 nm line, 36 nm for the 11 nm line, and 20 nm for the 8 nm line. Therefore, we expect that the Blech short-length effect would work also for Co interconnects with continued scaling to improve EM performance.

6.9   Summary

In this chapter, we investigated the scaling effect on microstructure evolution and EM reliability of Cu and Co interconnects. The scaling effect on microstructure evolution was first measured using a high-resolution TEM-based precession electron diffraction (PED) technique. With the PED technique, we were able to quantify the microstructure evolution to 22 nm linewidth for Cu interconnects and to 26 nm linewidth for Co interconnects. Based on the PED results, we developed the Monte Carlo simulation based on total energy minimization, taking into account the orientation-dependent grain boundary, strain, and interface energies to project the scaling effect on grain growth for future technology nodes.

Summarizing the results from the PED studies, a consistent picture of the scaling effect on microstructure evolution in Cu and Co interconnects has emerged. In general, microstructure evolution is controlled by minimization of the surface, inter-face, grain boundary, and elastic strain energy components. For Cu lines, when the linewidth exceeded 180 nm, the microstructure was dominated by {111} texture grains along the trench normal, indicating that grain growth was controlled by the surface and interface energies at the bottom trench. With the linewidth decreased to 120 nm, the growth of the {111} texture grains shifted from the trench normal to the trench width direction, indicating that grain growth became dominant by the surface energy from the sidewalls. The sidewall surface energy continued to dominate to the 45 nm linewidth, where the growth of the {111} grain texture shifted again to along the trench length direction (Figures 6.6 and 6.7), a trend that continued to the 22 nm linewidth (Figure 6.3).

In Cu lines, the surface, interface, and strain energies compete with the grain boundary energy to drive grain growth, resulting in abnormal grain growth with a bimodal log-normal distribution of the grain size. Since Cu is highly anisotropic, the minimization of the strain energy has led to the formation of twin boundaries, which was quantified using grain boundary trace analysis. A large number of coherent twin boundaries were observed up to 25% in 180 nm Cu lines (Figure 6.9), as a result of converting {111} grains to {100} grains to minimize the elastic energy during the 400°C anneal. With continued scaling, the geometric confinement in narrow Cu lines limits the grain growth and thus the twin formation, which continues to decrease to ~1% at 22 nm linewidth. At the same time, small grains emerge near the trench bottom to compete with the growth of the {111} texture grains driven by the surface energy, altering the overall microstructure in Cu narrow lines.

The PED results of the Co interconnects provide a similar but somewhat different picture on microstructure from the Cu interconnects. At the 220 nm linewidth, a {0001} grain texture along the trench normal becomes dominant, and when scaling down to 26 nm, the {0001} grain texture switched from along trench normal to along trench width direction. This indicates an abnormal grain growth controlled by the interface energy at the sidewall, which is similar to the {111} grain texture in Cu lines. However, the Co lines did not show a switch from the trench width direction to the trench length direction when scaling down to 26 nm linewidth as in Cu lines of 22 nm linewidth. The result suggests that microstructure evolution in Co lines is less dominant by the interfacial energy over the strain energy due to the higher elastic modulus of Co.

The shift in the orientation of the {111} grain texture is a distinctive feature of the scaling effect on microstructure evolution of the Cu and Co interconnects. This together with the change in the overall grain size distribution due to abnormal grain growth can significantly affect the line resistivity and the mass transport under EM and thus the EM reliability. The implication of the scaling effect on EM reliability was projected based on the Blech short-length effect. We found that the Blech short length of mixed bamboo–polygrain segments would continue to exist with scaling, but the average segment length continues to decrease. We expect that the Blech short-length effect would continue to improve EM performance for Cu and Co nanolines.

References

1. C. K. Hu, L. Gignac, and R. Rosenberg, Electromigration of Cu/low dielectric constant interconnects, *Microelectronics Reliability* **46** (2006), 213–231.
2. L. Cao, K. J. Ganesh, Lijuan Zhang et al., Grain structure analysis and effect on electromigration reliability in nanoscale Cu interconnects, *Applied Physics Letters* **102**, (2013), 131907. https://doi.org/10.1063/1.4799484.
3. E. F Rauch, M. Veron, J. Portillo, D. Bultreys, Y. Maniette, and S. Nicolopoulos, Automatic crystal orientation and phase mapping in TEM by precession diffraction, *Microscopy and Analysis–UK* 128 (2008), S5–S5.
4. R. Vincent and P. Midgley, Double conical beam-rocking system for measurement of integrated electron diffraction intensities, *Ultramicroscopy* **53** (1994), 271–282.
5. C. S. Own, L. D. Marks, and W. Sinkler, Electron precession: a guide for implementation, *Review of Scientific Instrument* **76** (2005), 033703.
6. C. S. Own, System design and verification of the precession electron diffraction technique, unpublished PhD dissertation, Northwestern University, 2005.
7. E. F. Rauch and M. Veron, Coupled microstructural observations and local texture measurements with an automated crystallographic orientation mapping tool attached to a TEM, *Material Science and Engineering Technology* **36** (2005), 552–556.
8. A. D. Darbal, K. J. Ganesh, X. Liu, et al., Grain boundary character distribution of nanocrystalline Cu thin films using stereological analysis of transmission electron microscope orientation maps, *Microscopy and Microanalysis* **19** (2013), 111–119.
9. D. Viladot, M. Veron, M. Gemmi, et al., Orientation and phase mapping in the transmission electron microscope using precession-assisted diffraction spot recognition: state-of-the-art results, *Journal of Microscopy* **252** (2013), 23–34.
10. S. T. Hu, Scaling effects on microstructure and resistivity for Cu and Co nanointerconnects, unpublished PhD dissertation, University of Texas at Austin, 2019.
11. V. Randle and O. Engler, *Introduction to Texture Analysis: Macrotexture, Microtexture and Orientation Mapping* (Boca Raton: CRC Press, 2000).
12. C. V. Thompson, Grain growth in thin films, *Annual Review of Materials Science* **20** (1990), 245–268.
13. C. V. Thompson and R. Carel, Texture development in polycrystalline thin films, *Material Science and Engineering B* **32** (1995), 211–219.
14. C. V. Thompson, Structure evolution during processing of polycrystalline films, *Annual Review of Materials Science* **30** (2000), 159–190.
15. J. M. E. Harper, C. Cabral Jr., P. C. Andricacos, et al., Mechanisms for microstructure evolution in electroplated copper thin films near room temperature, *Journal of Applied Physics* **86** (1999), 1.371086.

16. R. Rosenberg, D. C. Edelstein, C.-K. Hu, and K. P. Rodbell, Copper metallization for high performance silicon technology, *Annual Review of Materials Science* **30** (2000), 229–262.

17. H. J. Lee, H. N. Han, and D. N. Lee, Annealing textures of copper damascene interconnects for ultra large scale integration, *Journal of Electronic Materials* **34** (2005) 1493–1499. https://doi.org/10.1007/s11664–005-0156-8.

18. P. R. Besser, E. Zschech, W. Blum et al., Microstructural characterization of inlaid copper interconnect lines, *Journal of Electronic Materials.* **30** (2001), 320-330.

19. J. Y. Cho, H. J. Lee, H. Kim, and J. A. Szpunar, Textural and microstructural transformation of Cu damascene interconnects after annealing, *Journal of Electronic Materials* **34** (2005), 506–514.

20. K. J. Ganesh, A. D. Darbal, S. Rajasekhara, et al., Effect of downscaling nano-copper interconnects on the microstructure revealed by high resolution TEM-orientation-mapping, *Nanotechnology* **23** (2012), 135702.

21. K. J. Ganesh, Effect of downscaling copper interconnects on the microstructure revealed by high resolution TEM orientation mapping, unpublished PhD dissertation, University of Texas at Austin, 2011.

22. T. H. Kim, X. G. Zhang, D. M. Nicholson, et al., Large discrete resistance jump at grain boundary in copper nanowire, *Nano Letters* **10** (2010), 3096–3100.

23. P. C. Andricacos, C. Uzoh, J. Dukovic, J. Horkans, and H. Deligianni, Damascene copper electroplating for chip interconnections, *IBM Journal of Research and Development* **42** (1998), 567–574.

24. C. Lingk, and M. E. Gross, Recrystallization kinetics of electroplated Cu in damascene trenches at room temperature, *Journal of Applied Physics* **84** (1998), 5547–5553.

25. S. Brandstetter, E. F. Rauch, V. Carreau, et al., Pattern size dependence of grain growth in Cu interconnects, *Script Materialia* **63** (2010), 965–968.

26. V. Carreaua, S. Maitrejeana, M. Verdierb, et al., Evolution of Cu microstructure and resistivity during thermal treatment of damascene line: Influence of line width and temperature, *Microelectronic Engineering* **84** (2007), 2723–2728.

27. S. Hu, L. Cao, L. Spinella, and P. S. Ho, Microstructure evolution and effect on resistivity for Cu nanointerconnects and beyond, *2018 IEEE International Electron Devices Meeting (IEDM), San Francisco, CA* (2018), 5.4.1–5.4.3, doi: 10.1109/IEDM.2018.8614547.

28. L. Cao, Effects of scaling on microstructure evolution of Cu nanolines and impact on electromigration reliability, unpublished PhD dissertation, University of Texas at Austin, 2015.

29. M. McLean, Determination of the surface energy of copper as a function of crystallographic orientation and temperature, *Acta Metallurgica* **19** (1971), 387–393.

30. A. Basavalingappa, M. Y. Shen, and J. R. Lloyd, Modeling the copper microstructure and elastic anisotropy and studying its impact on reliability in nanoscale interconnects, *Mechanics of Advanced Materials and Modern Processes* **3** (2017) 10.1186/s40759–017-0021-5.

31. G. C. Schwartz and K. V. Srikrishnan, *Handbook of Semiconductor Interconnection Technology*, 2nd ed. (London: Taylor & Francis Group, 2006), 388–461.

32. L. Cao, L. Zhang, P. S. Ho, P. Justison, and M. Hauschildt, Scaling effects on microstructure and electromigration reliability for Cu and Cu(Mn) interconnects, *2014 IEEE International Reliability Physics Symposium, Waikoloa, HI* (2014), 5A.5.1–5A.5.5.

33. M. Hauschildt, Effects of barrier layer, annealing and seed layer thickness on microstructure and thermal stress in electroplated Cu films, unpublished M.S. thesis, University of Texas at Austin, 1999.

34. A. Volinsky, M. Hauschildt, J. B. Vella, et al., Residual stress and microstructure of electroplated Cu films on different barrier layers, *Materials Research Society Symposium Proceedings* 695 (2001), https://doi.org/10.1557/PROC-695-L1.11.1

35. K. Lu, L. Lu, and S. Suresh, Strengthening materials by engineering coherent internal boundaries at the nanoscale, *Science* **324**, (2009), 349–352.

36. K. C. Chen, W. W. Wu, C. N. Liao, et al., Observation of atomic diffusion at twin-modified grain boundaries in copper, *Science* **321** (2008), 1160777.

37. S. Ranganathan, On the geometry of coincidence site lattices. *Acta Crystallographica* **21** (1966), 197–199.

38. V. Randle, *The Role of Coincident Site Lattice in Grain Boundary Engineering* (Cambridge: Cambridge University Press, 1996).

39. L. E. Murr, *Interfacial Phenomena in Metals and Alloys* (Reading: Addison-Wesley Publishing Company, 1975).

40. V. Randle and V. Randle, *The Measurement of Grain Boundary Geometry* (Philadelphia: Institute of Physics Pub., 1993).

41. N. J. Petch, The cleavage strength of polycrystals, *Journal of the Iron and Steel Institute* **174** (1953), 25–28.

42. F. W. Mont, X. Zhang, W. Wang, et al., Cobalt interconnect on copper barrier process integration at the 7nm node, *2017 IEEE International Interconnect Technology Conference (IITC), Hsinchu* (2017), 1–3.

43. N. Bekiaris, Z. Wu, H. Ren, et al., Cobalt fill for advanced interconnects, *2017 IEEE International Interconnect Technology Conference (IITC), Hsinchu* (2017), 4–6.

44. C.-K Hu, J. Kelly, H. Huang, et al., Future on-chip interconnect metallization and electromigration, *2018 IEEE International Reliability Physics Symposium (IRPS), Burlingame* (2018), 4F.1-1–4F.1-6.

45. S. Hu and P. S. Ho, Scaling effects on microstructure and implication on resistivity of Co nanointerconnects, *2020 IEEE International Interconnect Technology Conference (IITC), San Jose* (2020).

46. J. K. Jung, N. M. Hwang, and Y. C. Joo, Microstructure evolution in damascene interconnects, *Journal of the Korean Physical Society* **40** (2002), 90–93.

47. D. J. Srolovitz, M. P. Anderson, P. S. Sahni, and G. S. Grest, Computer simulation of grain growth, *Acta Metallurgica* **32** (1984), 783–802.

48. M. P. Anderson and G. S. Grest, Computer simulation of normal grain growth in three dimensions, *Philosophical Magazine B* **59** (1988), 293–329.

49. Z. Li, J. Wang, and H. Huang, Grain boundary curvature based 2D cellular automata simulation of grain coarsening, *Journal of Alloys and Compounds* **791** (2019), 411–422.

50. D. Raabe, Introduction of a scalable 3D cellular automation with a probabilistic switching rule for the discrete mesoscale simulation of recrystallization phenomena, *Philosophical Magazine A* **79** (1999), 2339–2358.

51. J. Gao, M. Wei, L. Zhang, Y. Du, Z. Liu, and B. Huang, Effect of different initial structures on the simulation of microstructure evolution during normal grain growth via phase-field modeling, *Metallurgical and Materials Transactions A* **49** (2018), 6442–6456.

52. C. E. Krill and L. Q. Chen, Computer simulation of 3-D grain growth using a phase-field model, *Acta Materialia* **50** (2002), 3057–3073.

53. K. Kawasaki, T. Nagai, and K. Nakashima, Vertex models for two-dimensional grain growth, *Philosophical Magazine B* **60** (1989), 399–421.

54. D. Weygand, Y. Bréchet, J. Lépinoux, and W. Gust, Three-dimensional grain growth: a vertex dynamics simulation, *Philosophical Magazine B* **79** (1999), 703–716.

55. H. J. Frost, C. V. Thompson, C. L Howe, and J. Whang, A two-dimensional computer simulation of capillarity-driven grain growth: preliminary results, *Scripta Metallurgica* **22** (1988), 65–70.

56. B. Sun, Z. Suo, and W. Yang, A finite element method for simulating interface motion, *Acta Materialia* **45** (1977), 1907–1915.

57. G. S. Grest, D. J. Srolovitz, and M. P. Anderson, Computer simulation of grain growth – IV. Anisotropic grain boundary energies, *Acta Metallurgica* **33** (1985), 509–520.

58. D. J. Srolovitz, G. S. Grest, and M. P. Anderson, Computer simulation of grain growth – V. Abnormal grain growth, *Acta Metallurgica* **33** (1985), 2233–2247.

59. D. A. Porter, K. E. Easterling, and M. Y. Sherif, *Phase Transformations in Metals and Alloys*, 3rd ed. (Boca Raton: CRC Press, 2009).

60. V. V. Bulatov, B. W. Reed, and M. Kumar, Grain boundary energy function for fcc metals, *Acta Materialia* **65** (2014), 161–175.

61. M. Murakami and P. Chaudhari, Thermal strain in lead thin films I: dependence of the strain on crystal orientation, *Thin Solid Films* 46 (1977), 109–115.

62. S. Anandatheertha, Monte Carlo simulation of three dimensional grain growth code version No.1 (basic), MathWorks File Exchange, *MathWorks* (2012), http://mathworks.matlabcentral/fileexchange.

63. J. K. Mason, Grain boundary energy and curvature in Monte Carlo and cellular automata simulations of grain boundary motion, *Acta Materialia* **94** (2015), 162–171.

64. A. M. Rodriguez, G. Bozzolo, and J. Ferrante, Multilayer relaxation and surface energies of fcc and bcc metals using equivalent crystal theory, *Surface Science* **289** (1993), 100–126.

65. S. G. Wang, E. K. Tian, and C. W. Lung, Surface energy of arbitrary crystal plane of bcc and fcc metals, *Journal of Physics and Chemistry of Solids* **61** (2000), 1295–1300.

66. D. Tromans, Elastic anisotropy of HCP metal crystals and polycrystals, *International Journal of Research and Reviews in Applied Sciences* 6 (2011), 462–483.

67. Y. Lu and R. Qin, Influences of the third and fourth nearest neighbour interactions on the surface anisotropy of face-centered-cubic metals, *Surface Science* **624** (2014), 103–111.

68. S. Wong, C. Ryu, H. Lee, and K. Kwon, Barriers for copper interconnections, *Materials Research Society Proceedings* **514** (1998), 75–81.

69. M. S. Wu, A. A. Nazarov, and K. Zhou Misorientation dependence of the energy of symmetrical tilt boundaries in hcp metals: prediction by the disclination-structural unit model, *Philosophical Magazine* **84**, 8 (2004), 785–806.

70. J. M. Zhang, D. D Wang, and K. W Xu, Calculation of the surface energy of hcp metals by using the modified embedded atom method, *Applied Surface Science* **253** (2006), 2018–2024.

71. Q. Fu, W. Liu, and Z. L Li, Calculation of the surface energy of hcp-metals with the empirical electron theory, *Applied Surface Science* **255** (2009), 9348–9357.

72. Y. Luo and R. Qin, Surface energy and its anisotropy of hexagonal close-packed metals, *Surface Science* **630** (2014), 195–201.

7 Analysis of Electromigration-Induced Stress Evolution and Voiding in Cu Damascene Lines with Microstructure

7.1 Introduction

In Chapter 4, we analyzed EM-induced stress generation and the effect on mass transport and void formation in 1D confined line structures. The analysis laid the foundation to understand how stresses evolve under EM and interact with mass transport to affect void growth and damage formation in confined lines. There are serious limitations in the 1D approach, making it difficult to address a range of problems and results observed in Cu damascene interconnects as presented in Chapter 5. There we showed that Cu damascene interconnects are 3D multilevel wiring structures consisting of metal lines of finite lengths in each wiring level and connecting to vias and contacts. The Cu interconnect lines are fabricated by the damascene process to have specific structural elements consisting of bamboo-like grains linked to polycrystalline grain segments and enclosed by interface and sidewall barriers. The results presented in Chapter 5 show that the 3D configuration and structural elements of the wiring structure can significantly affect the EM characteristics and, if properly optimized, could yield robust and reliable interconnect structures. Several approaches to improve EM reliability have been reviewed, including microstructure optimization, the use of the Blech short-length effect, alloying at grain boundaries, and optimization of the interface barrier.

In this chapter, the analyses on EM and stress characteristics are generalized from 1D to 3D damascene interconnects, where the formulation is more complex without some of the simplified assumptions used in Chapter 4 for the 1D analysis. In particular, the local vacancy-stress equilibrium used in the 1D case can lead to a simultaneous equilibrium of vacancy concentrations in the grain interior with that at grain boundaries under EM stressing. In particular, it ignores the atomic migration in the grain interior as part of the mechanism for vacancy generation/annihilation, and thus would lead to inaccurate kinetics of stress relaxation. This chapter is organized into three parts, with the first part started by reviewing the underlying physics in 3D models already developed to analyze the processes involved in EM and stress evolution in interconnect lines. This is followed by a general 3D analysis of stress evolution, developed based on the thermal stress analysis formulated by Landau and Lifshitz, taking into account the vacancies and plated atoms generated under EM as local volumetric strains in the grain interior and at GBs. In the second part, the discussion is focused on the microstructure effects by combining a physics-based simulation together with in situ SEM experiments to observe and analyze stress evolution and void formation in polycrystalline copper interconnects segments. The simulation is carried out for two copper interconnect segments based on experimentally extracted grain structures, taking into account the variations of the GB diffusivities and elastic properties of individual grains to analyze the microstructure effects leading to copper interconnect degradation. In the third part, we discuss EM-induced void nucleation, migration, growth, and shape evolution studied by SEM/TEM experiments and physics-based modeling. In the analysis, the effect of current crowding on void shape change and the role of the grain size and texture distributions on void evolution are included.

7.2   EM-Induced Mass Transport

As discussed in Chapter 3, electromigration is the directional migration of lattice atoms and defects driven by a direct force from the electric field and an electron wind force originated from scattering with the charge carriers in the metal. The forced migration of the atoms can generate nonuniform chemical composition and elastic/plastic deformation to induce stresses in a confined metal line. The directional driving force can be expressed as the gradient of a potential energy $F(\vec{r}) = -\nabla U(\vec{r})$ with a corresponding velocity $\vec{v} = \mu F(\vec{r})$, where the mobility μ is related to the atomic diffusivity D_a through the Einstein relation [1]: $D_a = \mu k_B T$, where k_B and T are the Boltzmann constant and the absolute temperature, respectively. This leads to an atomic flux $\Gamma(\vec{r})$ denoting the number of atoms crossing unit area normal to the direction of the atomic flow per unit time:

$$\Gamma(\vec{r}) = \mu F(\vec{r})N_A(\vec{r}) = -N_A(\vec{r})\frac{D_a}{k_B T}\nabla U(\vec{r}). \qquad (7.1)$$

Here $N_A(\vec{r})$ is the coordinate dependent atomic concentration.

The potential energy depends on the physical state of the system defined by temperature, chemical composition, mechanical stress, electrical field, etc. Under EM, the atomic flux can be expressed as follows:

$$\Gamma_{EM}(\vec{r}) = -N_A(\vec{r})\frac{D_a}{k_B T}eZ^*\rho j(\vec{r}). \qquad (7.2)$$

Here Z^* is the effective valence presenting the EM driving force coming from two opposite forces, one from the electric field, and the other from the electron wind force due to scattering by the conduction electrons; $\rho(\vec{r})$ is the metal resistivity; and $j(\vec{r})$ the current density. For Al and Cu, the driving force is dominated by the electron wind force, where the effective valence Z^*e has been extensively studied and the results were reviewed in Chapter 2.

The EM-induced mass transport results in an inhomogeneous distribution of the atoms, which induces an additional atomic flux due to the concentration gradient:

$$\Gamma_N(\vec{r}) = -D_a \nabla N_A. \tag{7.3}$$

When the electric current passes through a metal line, it generates a volumetric strain due to the redistribution of the migrating atoms. When confined to a rigid surrounding, the volumetric strain generates a nonuniform distribution of stresses, which in turn induces additional atom migration due to the inhomogeneous elastic energy associated with a hydrostatic stress σ_{Hyd} as [2, 3]:

$$\Delta U(\vec{r}) = U_{Hyd}(\vec{r}) = -\Omega \sigma_{Hyd}(\vec{r}). \tag{7.4}$$

Here Ω is the atomic volume, and the hydrostatic stress σ_{Hyd} is the average of the normal stress components, $\sigma_{Hyd} = (\sigma_x + \sigma_y + \sigma_z)/3$. The elastic energy is the work done by the stress field when an extra atom with volume Ω is created in a confined metal line. The atomic flux caused by the inhomogeneous hydrostatic stress can be written as follows:

$$\Gamma_\sigma(\vec{r}) = N_A(\vec{r}) \frac{D_a \Omega}{k_B T} \nabla \sigma_{Hyd}(\vec{r}). \tag{7.5}$$

A nonuniform temperature distribution generated by inhomogeneous heat release during IC chip operation can also generate an atomic flux driven by the temperature gradient via the Soret effect [4]:

$$\Gamma_T(\vec{r}) = -N_A(\vec{r}) \frac{Q^* D_a}{k_B T^2(\vec{r})} \nabla T(\vec{r}), \tag{7.6}$$

where Q^* is the heat of transport.

Mass transport in Cu is dominated by the vacancy mechanism, consisting of two independent steps: vacancy formation and migration, as discussed previously in Section 4.2. Vacancy formation at a lattice site is accompanied by removing the atom occupying the original site to another site with minimum energy required, which can be at the surface, interfaces, or grain boundaries (GBs). The displaced atom can stay as a surface atom or build (plate) to continue the lattice at the interface or GB where the atom came from. In this way, the vacancies are generated or dissipated in pairs, with the atoms being plated at the interfaces and GBs. It should be mentioned that for GBs, only the incoherent type, characterized by incompatible lattices of neighboring grains, is energetically favored to accommodate the displaced atoms [3]. The coherent boundaries characterized by compatible lattices of neighboring grains – for example, the twin boundaries – cannot accommodate the plated atoms, and thus do not contribute to the dynamics of lattice defects.

It is known that an infinite crystalline solid in the state of zero stress is characterized by an equilibrium concentration of vacancies: $N_0^{ZS} \equiv N_0(T) = \Omega^{-1} \exp\{-E_V/k_B T\}$, where E_V is the energy of vacancy formation, and T is the temperature. Stressing the solid with various types of loads (electrical, mechanical, thermal) will disturb the state of thermochemical equilibrium of the lattice vacancies, and then the system would evolve to a new state of equilibrium accordingly. As we discussed in Section 4.2,

Herring [2] and Larche and Cahn [3] showed that the effect of stress is to yield a new equilibrium concentration of vacancies, as follows:

$$N = \Omega^{-1} \exp\left\{-\frac{E_V - f\Omega\sigma_{Hyd}}{k_B T}\right\} = N_0(T)\exp\left\{\frac{f\Omega\sigma_{Hyd}}{k_B T}\right\}. \quad (7.7)$$

Here, as it was discussed in Section 4.2, $f = \Omega_V/\Omega$ is the ratio of the vacancy volume (Ω_V) to the atomic volume and $f\Omega$ is the vacancy formation volume. $f\Omega\sigma_{Hyd}$ is the work done by or against the hydrostatic stress during vacancy formation, which is obtained by combining $\Omega\sigma_{Hyd}$ as the work done against the pressure in plating a lattice atom at a grain boundary or interface site with the work $-(1-f)\Omega\sigma_{Hyd}$ gained in relaxing the lattice surrounding the newly formed vacancy. Here we have simplified by replacing the normal stress components used by Herring [2] by the hydrostatic pressure σ_{Hyd} in calculating the work required for grain boundary or interface deformation.

For vacancy mechanism of atomic diffusion, there are equal fluxes of atoms and vacancies. As discussed in Section 4.2, we can consider the vacancy as a quasispecies migrating through the crystal lattice, so that a multiparticle problem can be approximated as the migration of a single particle, i.e., the vacancy [5]. Thus, we can write

$$N_a D_a = ND, \quad (7.8)$$

where $N_a = \Omega^{-1}$ is the concentration of the lattice cites, N is the vacancy concentration, and D is the vacancy diffusivity that is described as follows:

$$D = D_0 \exp\left\{-\frac{E_{VD} - \Omega^m \sigma_{Hyd}}{k_B T}\right\}. \quad (7.9)$$

Here E_{VD} is the activation energy of vacancy diffusion, and Ω^m is the migration volume, which represents the volume change at the saddle point in the vacancy migration path, as illustrated in Figure 4.7 [6, 7]. Combining (7.7)–(7.9) gives the effective atom diffusivity as follows:

$$D_a = D_0 \exp\left\{-\frac{E_{aD} - \Omega^* \sigma_{Hyd}}{k_B T}\right\}, \quad (7.10)$$

where the activation energy $E_{aD} = E_V + E_{VD}$ is the sum of the activation energies of vacancy formation and migration, and the activation volume $\Omega^* \approx 0.95\Omega$ is the combined volumes of vacancy formation and migration [6]. More detailed discussion of the stress effect on atomic diffusion can be found in [7].

7.3 Evolution of Vacancy and Plated Atom Concentrations in a 3D Confined Metal Line under Electric Current Stressing

In the following, we formulate the kinetics of stress evolution driven by an electric current in a metal line segment embedded in a rigid confinement. Following Chapter 4, we model a 3D Cu damascene line as a collection of Cu grains, characterized by different sizes, orientations, and crystallography and separated by GBs characterized by high atomic diffusivity and high vacancy concentration in comparison to the grain interior, as shown in Figure 4.1. All metal components in the multilayer interconnect, including the lines, vias, and pads, are covered by special liners formed by refractory metals (Ti, Ta, W) to prevent Cu atoms from diffusing into the surrounded dielectrics. The top interface dielectric barrier is formed after chemical mechanical polishing (CMP), which generates high defect density at the interface dominating the mass transport. The Cu grains, the barrier/interface and the vias and pads are important structural elements, as shown in Figure 4.1b.

Under EM, the atoms are driven by the electrons to migrate toward the anode end of line (see Figure 4.1). This occurs mainly through grain boundaries (GBs) and interfaces due to their disordered structures, characterized by a high concentration of vacancies and macroscopic defects, such as dislocation edges, allowing fast atom migration (see Figures 4.4 and 4.5). Migration of atoms through the grain interior is much slower due to a much lower concentration of vacancies available for diffusion of atoms. Overall, depletion and accumulation of atoms occur in a damascene line at the cathode and anode ends of polygrain clusters, respectively, where the atomic migration is blocked by bamboo-like grains, as shown in Figure 4.1. As a result, tension and compression stress components are generated at the cathode and anode ends-of-line segment containing the polygrain clusters. This leads to local stress gradients to induce atomic flux in a direction opposing the electron flow, which is known as the back-stress flux [8]. Therefore, the stress evolution has a dual character: the atomic flux divergence generates a volumetric stress, which in turn initiates a back-flux of atoms along the line segment where the flux divergence occurs. In a damascene line, the atomic flux driven by the electric current will deplete the atoms at the cathode end and accumulate at the anode end near the diffusion barriers at the line ends as well as in a line segment with blocking bamboo-like grains. There is a corresponding accumulation and depletion of vacancies at the GBs at the ends of these segments, as shown in Figure 4.5. Such volumetric transformations due to interaction with the rigid confinement generate the corresponding tensile or compressive stresses. Formulating in this way, we have a self-consistent problem where the kinetics of stress evolution in a line segment depends on the formation and migration of vacancies and plated atoms driven by the electric current, where the stress components in turn generate a backflow atomic flux against the EM-induced flux.

To solve this self-consistent problem, a simultaneous determination of the kinetics of stress evolution and the migration of vacancies and plated atoms in a metal line segment driven by an electric current is required. Here the evolution of the concentrations of vacancies and plated atoms is described by mass continuity equations correlating the flux divergence of the diffusing species with the rate of its local concentration change. Based on the Gauss–Ostrogradsky divergence theorem [9], a general continuity equation can be written as

$$\frac{\partial \rho}{\partial t} + \nabla \Gamma = R,$$

$$(7.11)$$

where ρ is the amount of the species per unit volume, which in our case is the concentration of the vacancies or the plated atoms, Γ is the corresponding flux, t is time, and R is the generation rate of the vacancy-plated atom pairs per unit volume per unit time. In terms that generate ($R > 0$) or remove ($R < 0$), these pairs are referred to as "sources" and "sinks," respectively. We should mention that the continuity equation for vacancy concentration written for the grain interior differs from that for the interfaces and GBs by the absence of the term R, because the vacancy-plated atom pair generation/annihilation takes place only at the grain boundaries and interfaces. Thus, combining all vacancy fluxes induced by electric stressing, (7.2), vacancy gradient (7.3), and stress gradient (7.5), and replacing the atomic concentration and diffusivity (N_a, D_a) with the vacancy characteristics (N, D), we obtain the continuity equations for vacancy concentration evolution in the grain interior (7.12), and at the GBs/interfaces (7.13) as follows:

$$\frac{\partial N}{\partial t} + \vec{\nabla}\left(-D\vec{\nabla}N - \frac{DN}{k_B T}\left((1-f)\Omega\vec{\nabla}\sigma_{Hyd} + eZ\vec{\nabla}V\right)\right) = 0 \qquad (7.12)$$

$$\frac{\partial N}{\partial t} + \vec{\nabla}\left(-D_{int/GB}\vec{\nabla}N - \frac{D_{int/GB}N}{k_B T}\left((1-f)\Omega\vec{\nabla}\sigma_{Hyd} + eZ_{int/GB}\vec{\nabla}V\right)\right) + R = 0. \qquad (7.13)$$

Here, the factor $(1-f)$ in the vacancy flux induced by the stress gradient is associated with the lattice free energy for vacancy migration [10, 11]:

$$\Delta U(\vec{r}) = U_{hyd}(\vec{r}) = -(1-f)\Omega\sigma_{Hyd}(\vec{r}). \qquad (7.14)$$

Here it is sufficiently accurate to disregard the flux induced by the temperature gradient, which is usually small relative to the total flux. To discriminate GBs and interfaces from the grain interior, we introduce the diffusivity $D = D_0 \exp\left\{-(E_{VD}^G - f\Omega\sigma_{Hyd})/k_B T\right\}$ and valence Z for the grain interior and $D_{int/GB} = D_0 \exp\left\{-(E_{VD}^{int/GB} - f\Omega\sigma_{Hyd})/k_B T\right\}$ and $Z_{int/GB}$ for the GBs and interfaces. The sink/source term R in (7.13), which was first introduced by Rosenberg and Ohring, [12], describes the generation/annihilation rate of vacancy–plated atom pair at the interfaces and GBs. We ignore the contribution from the dislocations in the grain, which can be accounted for, if needed, by introducing a scaling factor converting the grain size to the interdislocation distance.

Due to the large difference in mobility of plated atoms and vacancies, we expect the former to stay where they were generated inside the interfaces and GBs, while the latter can migrate subjected to various driving forces. Therefore, the evolution of the plated atom concentration M can be described by an equation similar to (7.13) without the diffusion-convection term as follows:

$$\frac{\partial M}{\partial t} + R = 0. \qquad (7.15)$$

This indicates that the evolution of the plated atom concentration is governed mainly by the generation/annihilation rate of the vacancy-plated atom pairs. Equations (7.13) and (7.15) show that, similar to the 1D case, we have two mechanisms to change the vacancy concentrations at local GB/interface due to the vacancy flux divergence or generation/annihilation of the vacancy-plated atom pairs. With the vacancy mechanism for atomic diffusion along the GBs/interfaces, the flux divergence can change the vacancy concentration by redistributing them along the GB network. To make this clearer, let us consider a small segment of the line, located near the cathode end of line, as shown in Figure 4.9. It can be seen that EM-induced atomic fluxes are the same at both ends of the segment, but the fluxes caused by the hydrostatic stress gradient are not equal, and instead increase toward the cathode. Thus, the atomic flux leaving the anode end of the segment is larger than that entering the segment at the opposite end, which causes vacancy to accumulate in this segment. This is accompanied in turn by change in the state of stress due to the difference in the volumes occupied by the atom and by the vacancy. Since $\Omega_V < \Omega$, the interaction between the shrinking volume of the segment and the rigid confinement induces a tensile stress in this segment. This destroys the equilibrium between stress and vacancy concentration in both the GB and grain interior. As a result, a transfer of lattice atoms between the grain interior and GB/interface will take place at the GB/interface, accompanying a simultaneous generation or disappearance of the vacancies in the grain interior. In order to avoid the numerical complications associated with the sharp interface between the grain interior and GB and the corresponding boundary conditions, we assume that the vacancy and plated atoms are distributed at the GBs/interface and in the grain interior. This allows us to have vacancies migrating everywhere in the simulation domain but to be generated/annihilated only at the GBs/interfaces. Due to the immobile character of the plated atoms, they will be located just in the GB/interface regions where they are generated. Each generated/annihilated vacancy introduces or removes a volume Ω_V per vacancy, and the generation/annihilation of plated atoms will add or remove an additional volume Ω per atom at the GB/interface. It should be mentioned that the volume change due to vacancy generation/annihilation and diffusion was first considered by Sarychev and co-workers [10] by introducing an additional kinetic equation.

As it was discussed already, under thermodynamical equilibrium, the concentration of vacancies at the interfaces, GBs, and inside grains is in equilibrium with the stress [3]. If any outside load causes a change in the state of stress and hence disturbs the equilibrium in vacancy concentration, then generation/annihilation of the vacancy/plated atom pairs accompanying by vacancy diffusion will be initiated. Depending on the stress, vacancies can be either scavenged or generated at GBs/interfaces. Thus, when a tensile stress is generated under an electric current in a line segment near the cathode, additional vacancies should be generated in the grain. It can be achieved by GB/interface diffusion of vacancies from other parts of the line characterized by less tensile stress and by migration of the interior atoms toward the GB/interface and plating there. In the latter, vacancy-plated atom pairs are generated, and each pair accompanied with a volume increase of $\Delta V = f\Omega$. In the opposite case of a compression stress, vacancies will migrate toward the stressed part of the GB/interface to annihilate the plated atoms there. This reduces the concentration of the vacancies/plated atom pairs with a corresponding volume reduction of $\Delta V = -f\Omega$ per each pair. In combination, we can describe the vacancy generation/annihilation process as a reaction of the first order with a rate of

$$w = k_1\left(T, \sigma_{Hyd}\right) - k_2(T)N, \tag{7.16}$$

where k_1 and k_2 are the constants for the forward (generation of the pairs) and reverse (annihilation of the pairs) reaction rates. At equilibrium, we have $k_2(T)N^{eq} = k_1\left(T, \sigma_{Hyd}\right)$, and so

$$w = -k_2(N - N^{eq}). \tag{7.17}$$

Here,

$$N^{eq} = \Omega^{-1}\exp\left\{-\frac{E_A - f\Omega\sigma_{Hyd}}{k_BT}\right\} \tag{7.18}$$

is the equilibrium concentration of vacancies at temperature T and hydrostatic stress σ_{Hyd}. The rate constant $k_2(T)$ for the generation/annihilation of vacancy/plated atom pairs applies to two sequential processes: diffusion of vacancies through the grain interior and their reaction with the plated atoms. These processes can be expressed as follows:

$$\frac{1}{R_{comb}} = \frac{1}{R_{dif}} + \frac{1}{R_{react}}, \tag{7.19}$$

where R_{comb} is the rate of the combined process, while R_{dif} and R_{react} refer to the rates of diffusion and reaction, respectively. Replacing the rate with the corresponding time constant, we have

$$\tau^* \approx \tau_{dif} + \tau_{react}. \tag{7.20}$$

Here $\tau_{dif} \approx R_g^2/D_{VL}$ is the diffusion time constant for the grain interior, and $\tau_{react} = k_{react}^{-1}$ with k_{react} being the annihilation rate of the vacancy–plated atom pair, R_g is the grain size and D_{VL} the lattice diffusivity of vacancies. Assuming the vacancy–plated atom annihilation occurs much faster than the transport of vacancies through the grain bulk, so the combined process becomes diffusion controlled, then $k_2(T)^{-1} \approx \tau_{dif} \approx R_g^2/D_{VL}$. This formalism can be used for analyzing the effect of grain interior on the kinetics of vacancy generation/annihilation in 1D approximation. It should be mentioned that Kirchheim has implemented a similar approach for accounting an effect of GB diffusion of vacancies on the rate of volume generation [11]. In subsequent 2D and 3D analyses, we will separately solve the kinetic equations for reactions between vacancies and plated atoms and the diffusion of vacancies through the GB/interface and grain bulk. The rate constant $k_2(T)$ then represents the frequency of the plated atom jumping into a vacant site at the GB/interface, or $k_2(T) = \tau_{react}^{-1}$, [13]. The generation/annihilation term in (7.13) and (7.15) can then be presented as follows:

$$G = \frac{N - N^{eq}}{\tau_{react}}, \tag{7.21}$$

which is commonly used in EM and SM modeling [10–12, 14].

7.4 Volumetric Stress Generated by Vacancies and Plated Atoms in a Confined Metal Line

In the previous section, we show that the change in the distributions of vacancies and plated atoms in a metal segment with rigid confinement under an electric current is the primary cause of stress evolution. In this section, we analyze the stress distribution based on the distribution of vacancies and plated atoms as local volumetric strains following the general formulation for thermal stresses developed by Landau and Lifshitz [15].

Traditionally a stress-free state of a solid is assumed to be a state that exists when (i) there are no external forces acting on the body, and (ii) a temperature equals to the stress free temperature T_{ZS} everywhere inside the solid and the surrounded confinement-thermostat. This implies that if $u_{ik} = 0$, then $\sigma_{ik} = 0$, where $u_{ik} = \frac{1}{2}\left(\frac{\partial u_i}{\partial x_k} + \frac{\partial u_k}{\partial x_i}\right)$ are the components of the deformation tensor with u_i the displacement components, and σ_{ik} the components of the stress tensor. Since the stress components are related to the free energy F as $\sigma_{ik} = \frac{\partial F}{\partial u_{ik}}$, the expansion of F in a power series of u_{ik} should not contain the linear terms: $F = F_0 + \frac{1}{2}u_{ll}^2 + Gu_{ik}^2$. The relation between tensors of stress and deformation for an isotropic body can be expressed as follows [15]:

$$\sigma_{ik} = Ku_{ll}\delta_{ik} + 2G\left(u_{ik} - \frac{1}{3}\delta_{ik}u_{ll}\right). \tag{7.22}$$

Here, K and G are the bulk modulus and the shear modulus [16]. Employing this formalism, we derive a description of the deformation caused by change in the vacancy concentration. We assume that the solid is stress-free if it isn't loaded with external forces, the temperature equals to T_{ZS}, and in addition the concentration of vacancies is uniform and equal to $N(T = T_{ZS}) = N^{ZS}$, and the concentration of the plated atoms is zero, $M = 0$. If these concentrations change in any way, then even in the case of absence of external forces and the uniform temperature, the solid will be deformed due to the dilatation effect. This means that in the expansion of the free energy $F(N, M, T)$ in a series of powers of $(u_{ik})^m$ there should be the liner terms u_{ik} in addition to the quadratic terms $\frac{1}{2}u_{ll}^2$ and Gu_{ik}^2, discussed earlier.

So far, we have shown that a change in the vacancy concentration occurs as a result of generation/annihilation of vacancy–plated atom pairs and vacancy diffusion, so the generation of the vacancy–plated atom pair will result in a local volume change. Because of the difference between the vacancy volume Ω_v and the atomic volume Ω, a vacancy generation/annihilation will generate a dilatational volume strain $e_v = \pm\frac{\Omega_v - \Omega}{\Omega}\Delta N = \pm(1 - f)\Delta N$. Here $\Delta N = N - N^{ZS}$ is the change in the vacancy concentration, the minus sign corresponds to the positive ΔN, and $f = \Omega_v/\Omega$. Thus, the vacancy induces a strain of $\varepsilon_v = e_v/3 = -(1-f)\Delta N/3$, and the plated atom corresponds to the generation or removal of an extra volume Ω in the GBs and/or interfaces. Together the total volume strain produced by the generation/annihilation of the vacancy–plated atom pairs ΔM and vacancy diffusion is

$$e_v = -(1-f)\Delta N + \Delta M. \tag{7.23}$$

It should be noted that ΔN and ΔM in (7.23) are not necessarily equal since the dilatational effect is accompanied by vacancy diffusion. Here the concentrations of vacancies and plating atoms are expressed in $1/\Omega$ units, representing the fractions of lattice sites occupied by the vacancies or plating atoms.

The only linear scalar that can be formed from the components of the strain tensor u_{ik} is a sum of the diagonal components u_{ii} [15]. We assume that the deformation due to the changes in the vacancy concentration ($\Delta N = N - N^{ZS}$) and the concentration of plated atoms $\Delta M = M$ is small. In this case, we can assume that the coefficient of the linear term u_{ii} in the F expansion is proportional to $\Delta = M - (1-f)(N - N^{ZS})$ and becomes zero when $N = N^{ZS}$ and $M = 0$, so

$$F(N,M) = F(N^{ZS},0) - K\phi\left(M - (1-f)(N - N^{ZS})\right)u_{ll} + G\left(u_{ik} - \frac{1}{3}\delta_{ik}u_{ll}\right)^2 + \frac{K}{2}u_{ll}^2, \tag{7.24}$$

where the coefficient $K\phi$ is a constant. Differentiating $F(N,M)$ by u_{ik}, we obtain the stress tensor components as follows:

$$\sigma_{ik} = -K\phi\left(M - (1-f)(N - N^{ZS})\right)\delta_{ik} + Ku_{ll}\delta_{ik} + 2G\left(u_{ik} - \frac{1}{3}\delta_{ik}u_{ll}\right). \tag{7.25}$$

Here the first term is the additional stress due to the change in the vacancy–plated atom concentrations. Other terms are the same as in (7.22) relating to the action of the outside forces, which can be originated, for instance, by a rigid confinement in a body where the vacancy-plated atoms are embedded. When vacancy and plated atoms are uniformly distributed through the solid with no outside forces acting, i.e., with free external boundaries, then the internal stress should be zero. By equating σ_{ik} to zero and employing Einstein's summation convention, we have

$$\phi\left(M - (1-f)(N - N^{ZS})\right) = u_{ll}. \tag{7.26}$$

Since u_{ii} is the relative volume change, then $\phi = 1$ and

$$\sigma_{ik} = -K\left(M - (1-f)\Delta N\right)\delta_{ik} + Ku_{ll}\delta_{ik} + 2G\left(u_{ik} - \frac{1}{3}\delta_{ik}u_{ll}\right). \tag{7.27}$$

We employ the following standard relation among bulk, shear, and Young's modulus [16]:

$$K - \frac{2}{3}G = \frac{\nu E}{(1+\nu)(1-2\nu)} = \lambda. \tag{7.28}$$

We can rewrite (7.27) as

$$\sigma_{ik} = -K(M - (1-f)\Delta N)\delta_{ik} + 2Gu_{ik} + \lambda\delta_{ik}u_{ll}. \tag{7.29}$$

Changing the notations in such way as $\sigma_{xx} \equiv \sigma_x$ and $u_{xx} \equiv \varepsilon_x$, we deduce

$$\sigma_x = -K(M - (1-f)\Delta N) + 2G\varepsilon_x + \lambda e, \tag{7.30}$$
$$e = \varepsilon_x + \varepsilon_y + \varepsilon_z.$$

From these expressions, we can derive a relation between the volume strain e and the sum of normal stresses $\theta = \sigma_x + \sigma_y + \sigma_z$ as follows:

$$e = \frac{\theta}{3K} + (M - (1-f)\Delta N). \tag{7.31}$$

Replacing the shear and bulk modules by Young's modulus E and Poisson's ratio ν as: $G = \frac{E}{2(1+\nu)}$, and $K = \frac{E}{3(1-2\nu)}$, we derive the hydrostatic stress as

$$\sigma_{Hyd} = \frac{\theta}{3} = \frac{E}{3(1-2\nu)}[e - (M - (1-f)\Delta N)]. \tag{7.32}$$

Substituting the volume strain e from (7.31) to (7.30) yields the following expression for the strain components:

$$\varepsilon_x - (M - (1-f)\Delta N) = \frac{1}{E}[\sigma_x - \nu(\sigma_y + \sigma_z)]. \tag{7.33}$$

Here, as in (7.30), the second term on the left-hand side of the equation is the strain related to the change in the vacancy–plated atom concentrations. Similar expressions can be written for the two remaining normal strain components ε_y and ε_z. In general, the shear strains are not affected by changes in the vacancy-plated atom concentrations because free dilatation does not produce angular distortion in an isotropic material. From the set of equations (7.33) written for the three strain components, we obtain the following:

$$\sigma_x = \lambda e + 2G\varepsilon_x - \frac{E}{3(1-2\nu)}(M - (1-f)\Delta N). \tag{7.34}$$

Substituting σ_x, σ_y, and σ_z from (7.34) into the differential equations of equilibrium, we derive

$$\frac{\partial\sigma_x}{\partial x} + \frac{\partial\tau_{xy}}{\partial y} + \frac{\partial\tau_{xz}}{\partial z} + X = 0$$
$$\frac{\partial\sigma_y}{\partial y} + \frac{\partial\tau_{xy}}{\partial x} + \frac{\partial\tau_{xz}}{\partial z} + Y = 0 \tag{7.35}$$
$$\frac{\partial\sigma_z}{\partial z} + \frac{\partial\tau_{xz}}{\partial x} + \frac{\partial\tau_{yz}}{\partial y} + X = 0.$$

By assuming that there are no body forces ($X = 0$, $Y = 0$, and $Z = 0$), we obtain

$$(\lambda + G)\frac{\partial e}{\partial x} + G\Delta u - \frac{E}{3(1-2\nu)}\left(\frac{\partial M}{\partial x} - (1-f)\frac{\partial N}{\partial x}\right) = 0 \tag{7.36}$$

Again, similar expressions can be written for the y- and z-components of the displacement. Comparing the derived equilibrium equations (7.36) with the conventional equations describing the equilibrium with the distributed body forces, $(\lambda + G)\frac{\partial e}{\partial x} + G\Delta u + X = 0$ [16], we can identify $-\frac{E}{3(1-2\nu)}\left(\frac{\partial M}{\partial x_i} - (1-f)\frac{\partial N}{\partial x_i}\right)$ as the effective body force components.

Conditions of the equilibrium at the surface of a solid loaded with the surface forces can be written (see, for example, [16]) as follows:

$$
\begin{aligned}
\tilde{X} &= \sigma_x l + \tau_{xy} m + \tau_{xz} n \\
\tilde{Y} &= \sigma_y m + \tau_{yz} n + \tau_{xy} l \\
\tilde{Z} &= \sigma_z n + \tau_{xz} l + \tau_{yz} m.
\end{aligned}
\tag{7.37}
$$

Here, $\tilde{X}, \tilde{Y}$, and $\tilde{Z}$ are the main components of the surface forces per unit area; and l, m, and n are the direction cosines of the external normal to the surface of the body at the point of consideration. This yields the following surface equilibrium condition for the solid with vacancies and plated atoms:

$$
\frac{E}{3(1-2\nu)}(M - (1-f)\Delta N)l = \lambda e l + G\left(\frac{\partial u}{\partial x} l + \frac{\partial u}{\partial y} m + \frac{\partial u}{\partial z} n\right) + G\left(\frac{\partial u}{\partial x} l + \frac{\partial v}{\partial x} m + \frac{\partial w}{\partial x} n\right).
\tag{7.38}
$$

The latter equilibrium condition and the similar written for two other directions were derived for the case when no surface forces were loaded. Thus, the terms $\frac{E}{3(1-2\nu)}(M - (1-f)\Delta N)l$, $\frac{E}{3(1-2\nu)}(M - (1-f)\Delta N)n$ and $\frac{E}{3(1-2\nu)}(M - (1-f)\Delta N)m$, effectively replacing $\tilde{X}, \tilde{Y}$, and $\tilde{Z}$ components of the surface forces for the case of the solid loaded with the surface forces.

Thus, the displacement components u, v, and w, generated by vacancies and plated atoms in the grain interior and GB/interfaces, are equal to the displacements induced by the body forces

$$
X_i = -\frac{E}{3(1-2\nu)}\left(\frac{\partial M}{\partial x_i} - (1-f)\frac{\partial N}{\partial x_i}\right)
\tag{7.39}
$$

distributed inside GB/interfaces and grain interior, and by the tension

$$
\frac{E}{3(1-2\nu)}(M - (1-f)\Delta N)
\tag{7.40}
$$

distributed over the surface. From (7.34) we can see that the normal stress components consist of two parts: one a "traditional" part, which is derived by using elastic strain components; and the hydrostatic part describing the dilatational induced pressure as follows:

$$
-\frac{E}{3(1-2\nu)}(M - (1-f)\Delta N),
\tag{7.41}
$$

which is proportional to the concentrations of vacancies and plated atoms. Hence, the total stress produced in the metal segment embedded into the rigid confinement by nonuniformly distributed vacancies and plated atoms can be obtained by superposing hydrostatic pressure (7.41) on the stress produced by body forces (7.39) and surface forces (7.40).

7.5 Evolution of Stress and Vacancy–Plated Atom Concentrations in a Prevoiding State

After deriving the correlation between the concentrations of vacancies–plated atoms and stresses, we can determine the kinetics of relaxation of the vacancy concentration and stress inside the solid driven by an electric field. To do so, we have to solve a system of PDEs describing the evolution of the concentrations of vacancies $N(\vec{r},t)$ and plated atoms $M(\vec{r},t)$ coupled with the equations describing the evolution of the strain–stress and current density. For the current density, we have the Laplace's equation

$$\nabla\left(\frac{1}{\rho}\nabla V\right) = 0. \tag{7.42}$$

Here $\rho = \rho_0[1 + \beta(T - T_0)]$ is the temperature-dependent resistivity of the metal, and β is thermal coefficient of resistivity.

We can now construct the system of the PDE governing the evolution of the stress tensor components under electromigration. First, we have the following PDE describing the evolutions of the vacancy and plating atom concentrations at the GBs and interfaces as follows:

$$\frac{\partial N}{\partial t} + \vec{\nabla}\left(-D_{int/GB}\vec{\nabla}N - \frac{D_{int/GB}N}{k_BT}\left((1-f)\Omega\vec{\nabla}\sigma_{Hyd} + eZ_{int/GB}\vec{\nabla}V\right)\right) + R = 0 \tag{7.43}$$

$$\frac{\partial M}{\partial t} + R = 0. \tag{7.44}$$

The evolution of the vacancy and plating atom concentrations in the grain interiors is described by the PDE:

$$\frac{\partial N}{\partial t} + \vec{\nabla}\left(-D\vec{\nabla}N - \frac{DN}{k_BT}\left((1-f)\Omega\vec{\nabla}\sigma_{Hyd} + eZ\vec{\nabla}V\right)\right) = 0 \tag{7.45}$$

$$\frac{\partial M}{\partial t} = 0. \tag{7.46}$$

The strain evolution inside the interconnect segment is described by the PDE:

$$(\lambda + G)\frac{\partial e}{\partial x_i} + G\Delta u_i - \frac{E}{3(1-2v)}\left(\frac{\partial M}{\partial x_i} - (1-f)\frac{\partial N}{\partial x_i}\right) = 0. \tag{7.47}$$

Here $x_i = \{x, y, z\}$, and $u_i = \{u, v, w\}$.

Due to constraints for migration of vacancies and plating atoms introduced by barriers and liners, the zero-flux boundary conditions (BCs) are implemented at all outer boundaries of the domains where PDEs (7.43)–(7.46) are solved. BCs for the stress evolution PDE depend on the analyzed problem. It might be specified displacements or tractions at different parts of the outer boundaries. The coordinate dependent tractions, which depend on the boundary concentrations of the vacancies and plating atoms are loaded at the outer borders of the N and M domain:

$$\frac{E}{3(1-2v)}(M-(1-f)\Delta N)\hat{l} = \lambda e\hat{l} + G\left(\frac{\partial u}{\partial x}\hat{l} + \frac{\partial u}{\partial y}\hat{m} + \frac{\partial u}{\partial z}\hat{n}\right) + G\left(\frac{\partial u}{\partial x}\hat{l} + \frac{\partial v}{\partial x}\hat{m} + \frac{\partial w}{\partial x}\hat{n}\right)$$

$$\frac{E}{3(1-2v)}(M-(1-f)\Delta N)\hat{n} = \lambda e\hat{n} + G\left(\frac{\partial v}{\partial x}\hat{l} + \frac{\partial v}{\partial y}\hat{m} + \frac{\partial v}{\partial z}\hat{n}\right) + G\left(\frac{\partial u}{\partial y}\hat{l} + \frac{\partial v}{\partial y}\hat{m} + \frac{\partial w}{\partial y}\hat{n}\right)$$

$$\frac{E}{3(1-2v)}(M-(1-f)\Delta N)\hat{m} = \lambda e\hat{m} + G\left(\frac{\partial w}{\partial x}\hat{l} + \frac{\partial w}{\partial y}\hat{m} + \frac{\partial w}{\partial z}\hat{n}\right) + G\left(\frac{\partial u}{\partial z}\hat{l} + \frac{\partial v}{\partial z}\hat{m} + \frac{\partial w}{\partial z}\hat{n}\right).$$

$$(7.48)$$

Depending on the type of the electrical load, a voltage or current density boundary condition for the Laplace's equation (7.42) is employed. For all other boundaries between conductors and dielectrics, the electric insulating boundary conditions are used.

The initial conditions for all variables in (7.43)–(7.47) can be found as the steady-state values of stress and concentrations of vacancies and plated atoms, which are achieved as the system relaxes after cooling from the anneal temperature down to the shelf temperature. The zero current density should be used everywhere inside the metal line. A solution of this problem will be presented in Section 7.7.

The generation–annihilation term for the vacancy–plated atom pairs is as follows:

$$R = \tau_{react}^{-1}\left(N - \exp\left\{-\frac{E_A - f\Omega\sigma_{Hyd}}{k_BT}\right\}\right). \tag{7.49}$$

The vacancy diffusivities for the grain interior and at the GBs and interfaces are correspondingly the following:

$$D = D_0 \exp\left\{-\frac{E_{VD}^G - f\Omega\sigma_{Hyd}}{k_BT}\right\}, \quad \text{and} \quad D_{int/GB} = D_0 \exp\left\{-\frac{E_{VD}^{int/GB} - f\Omega\sigma_{Hyd}}{k_BT}\right\}.$$

$$(7.50)$$

The hydrostatic stress everywhere inside the simulation domain is deduced as the spherical part of the mechanical stress tensor:

$$\sigma_{Hyd} = (\sigma_x + \sigma_y + \sigma_z)/3. \tag{7.51}$$

Here $\sigma_x, \sigma_y, \sigma_z$ are the normal stress components:

$$\sigma_{x_i} = \lambda e + 2G\varepsilon_{x_i} - \frac{E}{3(1-2v)}(M-(1-f)\Delta N), \tag{7.52}$$

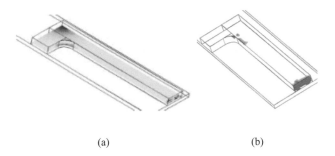

(a)   (b)

Figure 7.1 Current density distribution in a meander segment (a) and corresponding voiding at the cathode edge and current crowding location (b). © 2002 IEEE. Reprinted, with permission, from [17].

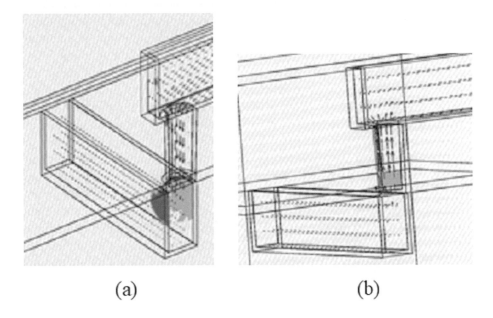

(a)   (b)

Figure 7.2 Void nucleation in the dual-damascene structure caused by the downstream (a) and upstream (b) electron flow.

where, as before, $x_i = \{x, y, z\}$. Figures 7.1 and 7.2 show the results of 3D simulations performed with the preceding discussed methodology.

The 3D model formulated to describe the EM-induced evolution of stress components and concentrations of vacancies and plated atoms in the polycrystalline metal segment embedded into a rigid confinement can be easily reduced to an 1D formulation by introducing the vacancy effective diffusivity and the effective generation/annihilation rate of the vacancy–plated atom pairs averaged on line cross section similarly to (4.46) in Section 4.7. This provides the following system of 1D PDEs:

$$\frac{\partial N}{\partial t} + \frac{\partial}{\partial x}\left[-\frac{D_{eff}N}{k_BT}\left((1-f)\Omega\frac{\partial \sigma}{\partial x} + eZ^*_{eff}\rho j\right)\right] = -G_{eff}$$

$$\frac{\partial M}{\partial t} = -G_{eff} \tag{7.53}$$

$$\sigma = B((1-f)\Delta N - M).$$

Last equation in the system (7.53) describing the relation between hydrostatic stress and the total volumetric strain, which is (4.8) from Section 4.3, replaces the 3D PDEs (7.47) and (7.48) for 3D stress and strain evolution caused by EM. By accepting Korhonen's assumption of immediate equilibration of the vacancy concentration with stress, the system (7.53) can be reduced to the 1D Korhonen equation (4.18), as it was shown in [18].

As an example of simulation of the EM-induced stress distribution in a particular interconnect segment, we consider a standard double-link EM test structure consisting of dual-damascene M1-V1-M2-V1-M1 Cu structure encapsulated by Ta sidewall and bottom liners and by a SiN$_x$ etch stop layer at the top, which was analyzed in [14, 19]. The wiring structure is confined by a thick dielectric (SiO$_2$), so the zero displacement BC are accepted for all interfaces with the rigid confinement. In Figure 7.3a, we show the distribution of the hydrostatic stress σ_{Hyd} along the center line of the M2 copper wire characterized by a particular grain structure under an electric current stressing. Results were obtained with commercial FEA tool Comsol Multiphysics [20]. The distributions of σ_{Hyd} at different times are shown in Figure 7.3b. It is clear that the stress evolution is an extremely nonlinear process, which is characterized by a significant initial nonuniformity, then subsequently evolves to a linear distribution of stress in the grains along the line. The evolution is caused by the generation/annihilation of vacancies and plated atoms and diffusion of vacancies. The distributions of the vacancies and plated atoms are shown in Figure 7.3c and d. Simulations were performed by assuming that (i) Young's modulus of the Cu grains depends on the crystallographic orientation as E(111) = 191.1 GPa, E(110) = 130.3 GPa, E(100) = 66.7 GPa, and (ii) interfaces and GBs are characterized by fast diffusivities of the vacancies $D_{Cu-SiNx} = 10^5 D_{Cu}$, $D_{Cu-Ta} = 10^2 D_{Cu}$, $D_{GB} = 10^4 D_{Cu}$, where D_{Cu} is the vacancy diffusivity in the copper grain interior.

7.6   Major Approaches to Modeling EM-Induced Stress Evolution in Interconnects

Before further discussing the results obtained with the methodology presented here, we discuss first other major approaches to 2D and 3D modeling for simulating the electromigration-induced stress evolution in confined interconnect segments.

Several general models, which in contrast to the 1D approach are not based on the assumption of a local vacancy equilibrium, have been developed by Kirchheim [11], Sarychev et al. [10], Bhate et al. [21] and Sukharev et al. [14, 18]. Kirchheim was the first to formulate two coupled PDEs, one to describe the evolution of vacancy concentration under conventional driving forces and another to resolve stress evolution governed by vacancy generation/annihilation. He proceeded to analyze a case where the vacancy generation/annihilation sites are located at adjacent boundaries and

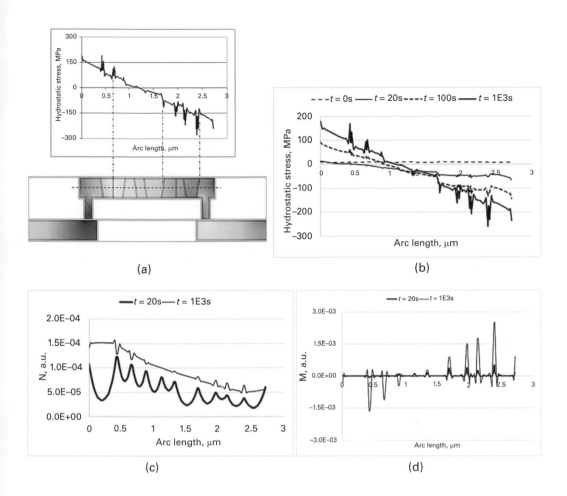

Figure 7.3 (a) Distribution of σ_{Hyd} along central the line in M2 segment of the test structure characterized by different grain orientations; (b) distributions of σ_{Hyd}; (c) vacancies; (d) and plated atoms along central the line in M2 segment in different moments of time. © 2009 IEEE. Reprinted, with permission, from [19].

the vacancy transport was accounted. The coupled PDEs were solved numerically as well as analytically for some limiting cases. However, the employed effective bulk modulus approximation (see (4.8) in Section 4.3) rendered this model in the framework of an 1D approach. A total volumetric strain calculated by Kirchheim was derived from the generation/annihilation of the vacancy–plating atom pairs while the additional volume evolution caused by vacancy diffusion was neglected.

A more general volumetric strain-based model for EM-induced stress evolution was developed by Sarychev et al. [10]. They formulated the general equations for the evolution of the stress tensor components taking into account the elastic and inelastic strain components due to vacancy migration and generation. They have linked the vacancy transport equation containing the vacancy fluxes generated by the electric load and gradients of stress and vacancy concentration, with the PDE describing the kinetics of the inelastic volume deformation. This allowed the determination of the field of displacements and, finally, the full stress tensor.

The kinetics of vacancy transport was evaluated using the standard continuity equation (7.11):

$$\frac{\partial c(\vec{r}, t)}{\partial t} = -\nabla J + R(\vec{r}, t). \tag{7.54}$$

The vacancy flux and their generation and annihilation rates are respectively

$$J = -D\left(\nabla c - c\frac{eZ\rho j}{k_B T} + c\frac{(1-f)\Omega}{k_B T}\nabla\sigma\right) \tag{7.55}$$

$$R = L_v k_B T \ln \frac{c}{c_{Eq}}, \tag{7.56}$$

where $c_{Eq} = c_0 \exp\{f\Omega\sigma/k_B T\}$ is the equilibrium vacancy concentration and L_v is a rate parameter, which is related to the vacancy generation/annihilation time τ_0 as $\tau_0 = -c_{Eq}/k_B T L_v$. The variation of the inelastic volume deformation ε_V induced by vacancy migration and generation/annihilation followed this equation:

$$\frac{\partial \varepsilon_V}{\partial t} = \Omega((1-f)\nabla J + fR). \tag{7.57}$$

The field of elastic displacements $\vec{u} = \{u_i\}$ generated by the inelastic volume deformation ε_V was obtained from the force balance equation:

$$\mu_{el}\nabla^2 u_i + (\lambda_{el} + \mu_{el})\frac{\partial}{\partial x_i}\left(\nabla\vec{u}\right) = B_{el}\frac{\partial \varepsilon_V}{\partial x_i} \tag{7.58}$$

Here λ_{el} and μ_{el} are the Lame's constants, and B_{el} is the material bulk modulus, [16]. Known displacements allowed the calculation of the strain and stress tensors as follows:

$$\varepsilon_{ij} = \frac{1}{2}\left(\frac{\partial u_i}{\partial x_j} + \frac{\partial u_j}{\partial x_i}\right) \tag{7.59}$$

$$\sigma_{ij} = -B_{el}\varepsilon_V \delta_{ij} + \lambda_{el}\varepsilon_{kk}\delta_{ij} + 2\mu_{el}\varepsilon_{ij}, \tag{7.60}$$

where δ_{ij} is Kronecker's symbol, and the Einstein notation is employed. The hydrostatic stress was, as usual, $\sigma = \sigma_{kk}/3$. Finally, the current density distribution was obtained by solving the Laplace equation. These equations were linked with the specific boundary conditions to obtain the analytical solutions for a number of applications. It was mentioned that in a general case a numerical technique such as finite element analysis should be employed.

While acknowledging the accurate mathematical formulation of the analyzed problem, some drawbacks of physical implementation should be mentioned. The most important detail missed in this model is the polycrystalline structure of the

interconnect metals. Sarychev et al. have considered a homogeneous material specified by uniform mechanical properties, uniform vacancy diffusivity, and uniformly
distributed vacancy generation/annihilation sites. While the formalism developed in
[10] allows one to derive the EM-induced evolution of the stress tensor in a 3D space,
the missed discrimination between the grain interior and grain boundaries doesn't
allow one to perform of the analysis of many crucial details such as the effect of grain
boundaries and grain orientation on the stress evolution and voiding.

This drawback was properly corrected by Bower et al. [21–23] and later by
Sukharev et al. [14, 18, 19]. Bower and co-workers have considered the effect of
grain boundaries and interfaces on the failure development in interconnects. Their
approach can be considered a rigorous extension of the classical Fisher model of grain
boundary diffusion [24] by assuming the grain boundaries as venues for fast atomic
migration. Atoms from the grain interior are transferred into the grain boundaries and
diffuse along them in a direction specified by decreasing chemical potential. Bower
and Shankar treated surfaces and interfaces as perfectly sharp boundaries and computed their shape evolution with a front tracking technique [23]. This model has taken
into account the effects of grain boundary sliding and migration, grain boundary
diffusion, as well as elastic deformation within the grain interior. Atom diffusion at
grain boundaries was treated by detaching a surface atom from one of the grains to
migrate under the chemical potential gradient and the electric current along the
boundary, then reattaching to one of the adjacent grains. Following Herring and
Mullins [2, 25], Bower and Shankar took the chemical potential of atoms adjacent
to a grain to be dependent on the normal component of stress σ_n and the local
curvature of the grain surface K:

$$\mu = \Omega\left(\phi \pm \frac{\gamma_{GB}}{2}K - \sigma_n\right), \tag{7.61}$$

where ϕ is the elastic strain energy density in the adjacent material, γ_{GB} is the surface
energy per unit area of the grain boundary in the stress-free configuration, and the $\pm$ signs
refer to the grains above and below the interface. In this case, the atomic flux tangent to
each of the grain boundary interfaces with thickness δ_{GB} can be expressed as follows:

$$J_\pm = -\frac{D_{GBt}\delta_{GB}}{2k_BT}\left(\frac{\partial\mu_\pm}{\partial s} - eZ_{GBt}\frac{\partial V_\pm}{\partial s}\right). \tag{7.62}$$

Here $V_\pm$ is the voltage distribution along the grain surfaces forming GB, and s is the arc
length along the surface; all other notations are shown in Figure 7.4. The total flux tangent
to the grain boundary becomes the sum of the material fluxes tangent to each adjacent
grain surface. The flux of atoms from grain (−) to grain (+) was taken as follows:

$$J_n = -\frac{D_{GBn}}{k_BT\Delta_n}\left(\left[\mu_+ - \mu_-\right] + eZ_{GBn}[V_+ - V_-]\right), \tag{7.63}$$

where Δ_n represents the effective diffusion length for the atom transfer normal to the
boundaries, and the diffusivity D_{GBn} and effective charge Z_{GBn} are attributed to atoms
transferring between grains.

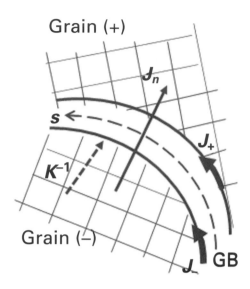

Figure 7.4 Schematic of GB with notations used in modeling grain boundary diffusion.

If a portion of the grain boundary has more atoms flowing in than out, i.e., the flux divergence takes place, the atoms will be incorporating into the lattice of the adjacent grains and cause the grains to move apart. At the same time, there will be a net flux of atoms out of the other side of the grain boundary, and the grains move together. Thus, the divergence of the flux J_+ or J_- will lead to insertion/dissolution of atoms on the corresponding surface element to induce the element to move. The atoms can be added to either one of the two adjacent grains to determine their relative motion with respect to the other. By considering the material transfer between neighboring grain boundary surfaces, Bower and Shankar [23] derived the following equation for the velocity of each side of the grain boundary relative to the underlying material:

$$v_{n\pm} = \pm \frac{\partial J_\pm}{\partial s} - J_n. \tag{7.64}$$

This yielded the relative velocity of the two adjacent grain surfaces, leading to GB narrowing or widening as follows:

$$[v_n] = v_{n+} - v_{n-} = -\frac{\partial}{\partial s}[J_{t+} + J_{t-}]. \tag{7.65}$$

This velocity describes the rate of the normal displacement across the GB, which generates the deformation in the surrounding bulk material, since the boundary separation is incompatible with the surrounding material. When grains are stressed, they tend to slide relative to one another until their surfaces become locked due to their nonplanarity, resulting in a relative tangential displacement [26]. The relative sliding velocity of the adjacent grains can be expressed as follows:

$$[v_t] = \frac{\Omega \eta_0 \sigma_t}{k_B T} \exp\left\{-\frac{E_{GBs}}{k_B T}\right\}. \tag{7.66}$$

Here η_0 is a characteristic sliding velocity, E_{GBs} the activation energy for GB sliding, and σ_t the shear stress [23]. The sliding generates incompatibilities, which must be accommodated by the surrounding grains. Ashby has mentioned three possible accommodation mechanisms: (i) elastic deformation of the surrounding grains that occurs at low stress and temperature and stops when the externally stress is balanced by the internal traction of the GB; (ii) at low stress and high temperature where GB tractions generate a chemical potential gradient to induce diffusion flow along the boundary; and (iii) at higher stress where dislocation glide and climb occur on either side of the boundary to provide the necessary accommodation [26]. Regardless of the mechanism of accommodation of the diffusion-induced deformation, the final stress tensor can be derived from the standard equations of mechanical equilibrium. In this way, Bower et al. have developed a 2D finite element method to model the nucleation, growth, and evolution of voids in polycrystalline interconnects and account for the role of grain boundaries and interfaces on EM-induced void evolution. In this model, void nucleation was based on a critical stress criterion where voids are nucleated at the GB/interfaces when the stress normal to the interface exceeds a critical value that is a material parameter. The model was able to account for effects such as grain boundary and interface diffusion, grain boundary migration and sliding, as well as elastic deformation and electric current flow in the grains.

Obviously, the system of the complex PDEs in such a model cannot be solved analytically, so advanced numerical techniques were required. For the previously described case, the so-called sharp boundary approach was employed by Bower and co-workers. It models the evolving boundaries by tracking and meshing the reference configurations, while the evolution of GB shape is modeled by integrating the corresponding mass balance equations. Stress, current density, and atomic fluxes are computed using FEA methods with the continuously regenerated mesh following the time-dependent geometry changes. Most numerical approaches used the explicit tracking the boundary to study such moving boundary problems. In such sharp interface approaches, the interface is located by specifying points on it. With the duration of time, as the interface evolves, even more points are required to accurately describe the interface. Furthermore, appropriate boundary conditions must be implemented at the increasing number of points at each time step. Thus, such techniques get very complicated and tend to have a rather poor numerical stability [27]. As mentioned in [28], the problem is complex with many different phenomena coupled together in a nonlocal, nonlinear way. Growth rates depend on derivatives of stress components taken along grain boundaries and on curvatures along surfaces. Boundary conditions specify the gradient of the normal stress at junctions where the stress field is singular. Many of the equations couple the displacement field to the stress field, and it is difficult to visualize how this constrains the evolution of the system. These difficulties and, in some cases, incomplete physical reasoning can lead to mathematically ill-posed problems [28]. Introduction of a diffuse interface or phase field models for modeling the interconnect failure [21, 27, 29] has allowed researchers to overcome some problems of the sharp interface method. In this approach, a time-dependent order parameter was introduced to characterize a state of each point in the analyzed domain.

In the case of void evolution, the order parameter takes the same value everywhere within the solid material and a different value inside the void. It rapidly varies from one value to another over a narrow region associated with the void–metal interface. The main advantage of the diffuse interface model is the implicit description of the void surface, which makes it unnecessary to use the surface tracking procedure. We will describe the details of the phase field method in Section 7.10.2 when the void evolution kinetics will be discussed.

Acknowledging the comprehensiveness of the EM model and advancements of numerical technique developed by Bower and co-workers, there are some concerns regarding the physics used to evaluate EM-induced microstructure evolution in polycrystalline interconnects. Mostly, it is about the inelastic deformation in grain interiors and equilibration of vacancy concentration with stresses. The tendency of vacancy concentration to be in equilibrium with local stresses takes place everywhere in metal interconnects, in grain interiors, grain boundaries, and interfaces. It results in additional matter transfer between these regions, which causes additional deformation and stress. Nevertheless, this and some other important details associated with vacancies have been overlooked in the model developed by Bower et al. The problem has since been partly corrected by Bhate et al. [21] and by Sukharev and co-workers [14, 18, 19, 30, 31]. The latter approach, presented in Sections 7.3–7.5, has allowed researchers to address the missing parts in the Sarychev and Bower models and to develop an improved model capable of investigating the effects of the interconnect segment geometry and grain structure of the metal line on stress evolution under electric stressing. This enabled the model to analyze such effects as vacancy-plated atom pair generation/annihilation in grain boundaries and interfaces, variation in vacancy diffusivities in different migration venues, and grain-orientation-dependent mechanical properties' effect on stress evolution. It also allows researchers to analyze the effect of the segment geometry and grain morphology on void migration and void shape evolution. Some results derived by this model on the microstructure effect on EM-induced degradation are discussed in the following section.

7.7 Effect of Microstructure on EM-Induced Degradation in Dual-Damascene Copper Interconnects

Electromigration- and stress-induced degradation are key reliability concerns for copper interconnects with low-k dielectrics [32–34]. Numerous experimental studies have revealed that EM-induced degradation and interconnect failure depend on both the interface bonding and the microstructure of copper interconnect structures [35–37]. The problem has been studied by in situ scanning electron microscopy (SEM), electron backscatter diffraction (EBSD), and numerical simulations, and the results showed that EM-induced degradation depends strongly on the bonding strength and the atomic transport at interfaces [30, 34, 37]. In Chapter 5, we review the experimental studies and the approaches to improve the EM reliability; here we analyze the effect of microstructure and interfaces on EM-induced degradation in Cu low-k interconnects.

The microstructure can affect the EM and stress voiding reliability of Cu low-k interconnects through the dependence of GB diffusivity and the elastic constants on grain structure and orientations. The orientation dependence of grain boundary diffusivity has been investigated for small-angle grain boundaries and found to be proportional to the misorientation angle for angles up to 25° [38], while for large-angle boundaries, it depended strongly on reciprocal density of coincidence sites [39]. The grain misorientation dependency of diffusivity can affect voiding kinetics, especially when GB diffusion is the major path for atom migration. This dependency can also influence local stress evolution under EM in a voidless regime. Due to the high elastic anisotropy of copper, the orientation dependence of the elastic constants is important for stress evolution. This dependence can have a crucial impact and has been included in modeling void nucleation [40] and found that certain local grain orientation can yield high tensile stress to cause delamination and to account for the low value of the critical Blech product found for Cu low-k interconnects.

In this section, a physics-based simulation is done based on the analysis on EM-induced stress evolution described in Section 7.5 and used to analyze the stress evolution in polycrystalline copper interconnect segments. The model takes into account the variations of the GB diffusivities and elastic properties of individual grains to analyze the texture effect leading to copper interconnect degradation. In particular, the stress distributions are simulated for two copper interconnect segments based on experimentally extracted grain structures [19, 41–45]. In addition, we demonstrate a possible approach to detect void nucleation sites and discuss possible mechanisms of void nucleation.

7.7.1   EM-Induced Stress Evolution in Polycrystalline Copper Interconnect Segments

In Cu damascene interconnects, preferential void formation under EM is known to occur at the Cu/capping layer interface, a topic that was discussed in Chapter 5 as well as in many publications; see, for example, [33, 34, 37, 43, 44]. This failure mode has been attributed to the "weakness" of the top interface, which serves as a path for fast atomic transport and with the presence of triple GB junctions to serve as vacancy sinks to induce void nucleation. Experiments have demonstrated that some small voids often arose at the top interface under EM and moved toward the cathode and then agglomerated with other voids, leading eventually to an intrametal-level via failure [37, 43–50]. In this process, the tensile stress developed during EM in the cathode area of the line segment plays an important role for void nucleation when a critical stress level is developed somewhere in the line segment to induce local material yielding and/or fracture of the interface serving as the precursor for void nucleation. Once a void is nucleated, it will evolve as the stress continues to build under EM. Since Cu is highly elastic anisotropic, the stress evolution and plastic deformation are highly inhomogeneous, depending on the local grain structure and the quality of the Cu interface. Depending on the stress level, plastic deformation can be a major mechanism for stress relaxation if the yield stress for dislocation movement is low [51, 52].

However, in polycrystalline lines, the yielding stress can be high due to the overall structural confinement in the damascene line. As a result, the stress buildup can first cause the delamination from the capping layer, and then stress relaxation evolves the flaw to an interfacial failure. In general, both the yielding and fracture phenomena depend on the stress state developed; in particular, it requires significant shear stress components. Such mechanistic details are difficult to observe in experiments, so in the following, a simulation is carried out to model the stress evolution in Cu lines under EM test based on actual grain structures to show the correlation between the shear stress distribution and void formation in Cu damascene lines.

The two Cu damascene lines used in the test and simulation are similar to that described in Section 7.5, where the crystallographic orientations of the individual grains are provided by EBSD analysis, which can be used to determine the elastic constants as a function of grain orientation. In general, the stress and strain components of the stress and strain tensors are connected, according to Hook's law:

$$\sigma_{kl} = C_{ij,kl}\varepsilon_{ij}^{el}.\tag{7.67}$$

Here, $\{C_{ij,kl}\}$ is the stiffness matrix. As was shown in Section 7.4, the strain tensor ε_{ij} includes the elastic components ε_{ij}^{el} and inelastic volumetric strain caused by variations in local concentrations of vacancies and plated atoms ε_{ij}^{inel}:

$$\varepsilon_{ij} = \varepsilon_{ij}^{el} + \varepsilon_{ij}^{inel},$$
$$\varepsilon_{ij}^{inel} = \frac{\Omega}{3}[-(1-f)\Delta N + \Delta M]\delta_{ij}.\tag{7.68}$$

Hence, for the hydrostatic stress we have (see (7.33))

$$\sigma_{Hyd} = C_{ij,kk}\left[\varepsilon_{ij} + \frac{\Omega}{3}((1-f)\Delta N - \Delta M)\delta_{ij}\right].\tag{7.69}$$

Cu is an fcc metal with elastic constants $C_{11} = 168 \cdot 10^9$ Pa, $C_{12} = 121 \cdot 10^9$ Pa, and $C_{44} = 75 \cdot 10^9$ Pa, and with the ratio of the shear modulus in the (100) direction $\mu = C_{44}$ and that in the (110) direction $\mu' = (C_{11} - C_{12})/2$, i.e., $\mu/\mu' \approx 3$. The elastic anisotropy is an important characteristic of Cu, which is reflected in the employment of the stiffness matrix $\{C_{ij,kl}\}$ in our modeling a polycrystalline Cu line.

We prescribed the aforementioned values of elastic constants to a grain oriented along the line. The transformation of the elasticity matrix due to the rotation of the grain can be done by the following expression [53]:

$$C'_{ij,kl} = T_{ig}T_{jh}C_{gh,mn}T_{km}T_{nl}.\tag{7.70}$$

Here $\{T\}$ is a unitary orthogonal matrix describing the transformation of the coordinate systems as $x'_i = T_{ij}x_j$. If the grain orientation relative to the line is described according to a set of Euler angles (ϕ, θ, κ) in Bunge's form [54], the grain rotation can be described by the product of three matrices, $T = A_z(\phi)A_x(\theta)A_z(\kappa)$, where

$$A_z(\phi) = \begin{pmatrix} \cos\phi & -\sin\phi & 0 \\ \sin\phi & \cos\phi & 0 \\ 0 & 0 & 1 \end{pmatrix}, \; A_x(\theta) = \begin{pmatrix} 1 & 0 & 0 \\ 0 & \cos\theta & -\sin\theta \\ 0 & \sin\theta & \cos\theta \end{pmatrix},$$

$$A_z(\kappa) = \begin{pmatrix} \cos\kappa & -\sin\kappa & 0 \\ \sin\kappa & \cos\kappa & 0 \\ 0 & 0 & 1 \end{pmatrix}.$$

$$(7.71)$$

Using these transformations and (7.67), the elastic constant can be determined for grains with orientations measured by EBSD analysis.

Two test structures were used for in situ EM studies: Sample I, with a line length 800 μm, height 400 nm, and width 450 nm; and Sample II, with the same length, height 410 nm, and width 350 nm. The samples were cut to parallel test structures, each separated by a distance of about 20 μm, and then mounted and wire-bonded to modified 24-pin test chip packages. The cross section of the sample was finished for the in situ EM investigation using focused ion beam (FIB) milling. Special care was taken to avoid cutting of the metal line with the ion beam, so approximately 5–100 nm of the dielectric material surrounding the metal line was left over in the sample front. With this procedure, all interfaces were kept in the as-manufactured state, so void formation, movement, and growth can be continuously monitored by SEM in fully embedded Cu line test structures in in situ EM degradation experiments. During tests, up to several hundreds of SEM images of the test structure were recorded as video sequences, making it possible to track the details of the time-dependent EM-induced degradation processes.

The stress state generated by EM was evaluated using the Comsol Multiphysics FEA software [19] for modeling of anisotropic materials for the Cu line segments based on the grain structures observed by EBSD. For the predominantly "bamboo" and "near-bamboo" structures in the line, we assumed that the grain sizes transverse to the line direction were equal to the linewidth. In this case, a uniform strain state can be established across the line [55], so we can use the plane–strain approximation in the 2D modeling.

The simulated structure of Sample I is presented in Figure 7.5a. Under electric stressing, the maximal tensile stress (near the cathode) is about 130 MPa. Since this segment is a small part of the long line, the gradient of the hydrostatic stress along the segment is small, but there are stress differences between the grains due to their crystallographic orientation. Sharp peaks in the hydrostatic stress can be observed at intersections of grain boundaries with the interface as shown in Figure 7.5b. Comparing Figure 7.5b with the experimental results in Figure 7.6a, we note that these peaks are somewhat higher at voiding sites. However, it is difficult to define a criterion for void initiation by the peaks of the hydrostatic stress. As will be shown, the shear stress distribution appears to be more predictive for void nucleation.

The EM video of Sample I revealed several sites of particular interest (A–C in Figure 7.6a). Positions A and B in Figure 7.6a were identified as void nucleation sites where continuous void formation was observed at position B, in contrast to position C acting as a void-trapping site as voids moved toward the cathode end of

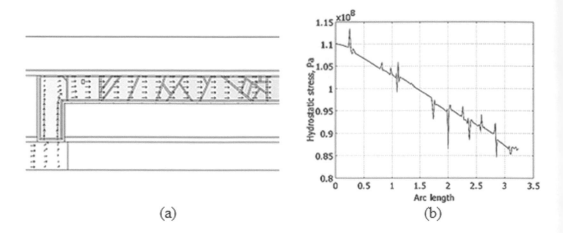

(a)   (b)

Figure 7.5 (a) The cathode-side region of the simulated structure, Sample I. The arrows indicate direction of electrons flow; vacancies migrate in the opposite direction. (b) Distribution of the hydrostatic stress under the top interface. © 2009 IEEE. Reprinted, with permission, from [19].

the test structure, where significant material redeposition was observed at the line end above the via.

The first two voids appeared almost simultaneously after about 15 minutes at position A (0.5 μm away from the line end) and position B (2.2 μm away from the line end). Both voids remained fixed at their positions for some time. After 50 minutes, another void appeared at position C (3.5 μm away from the line end), then became trapped at this position and did not move away. Toward the end of the experiment, this void continued to grow. Immediately after a void appearing at position C, the void at position B started to move along the line toward the via and, after several shape and speed changes, finally merged into the void at position A. The time for the void to travel from position B to position A was approximately 25 minutes, where the voids merged and then moved toward the line end starting to grow in size, and eventually reached the via. Meanwhile, at least one more void appeared at position C, moved toward the via, then merged into a large void near the via. The large void underwent several shape changes as it grew into the via while depositing material into the upper region near the via. The sample failed due to the large via void after 3:15 hours.

To investigate the effect of microstructure on void formation, EBSD analysis was carried out at the completion of the EM experiment. Figure 7.6 shows an SEM image of the cross section of Cu Sample A together with an inverse pole figure (IPF) map showing the crystallographic orientation of the individual grains relative to the wafer surface. The void nucleation site B is located above a cluster of small grains. On both the cathode side on the left and the anode side on the right of B, the microstructure is dominated by large grains extending across the entire height of the line to yield a bamboo-like structure. At position C, the microstructure is dominated by a cluster of small grains above which the void was trapped. Again, this cluster is enclosed by

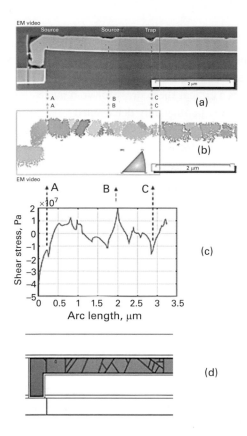

Figure 7.6 Postmortem SEM/EBSD analysis of Sample I and simulation results: (a) SEM image of the cross-section of Cu-sample together with (b) an inverse pole figure (IPF) from EBSD measurement of the crystallographic orientation of individual grains relative to the wafer surface; (c) simulated distribution of the shear stress at the top copper–dielectric barrier interface for copper line with the grain structure shown in (d). © 2009 IEEE. Reprinted, with permission, from [19].

regions of bamboo-like microstructure. At the cathode side of the cluster (left), a twinned region can be seen, consisting of four grains. The grain-to-grain misorientation of 60° together with periodic arrangement of the grains is characteristic of recrystallization twins in fcc materials. The microstructure of the via at the end of the interconnect line is dominated by a single large grain. No additional grain boundaries were observed in the region where the redeposition of material was observed during the final stages of the in situ EM experiment.

The distribution of the shear stresses derived from simulation under the top interface is shown in Figure 7.6c. A large negative shear stress was found to exist at the edge of the line, due to the geometry of the segment. In addition, local shear stress peaks appeared in the line, most probably due to large differences in the elastic constants of neighboring anisotropic grains. Voids were observed at the A–C sites in Figure 7.6a, indicating that the local shear stress exceeded some critical value $|\tau_{cr}|$, which was taken to be $|\tau_{cr}| \sim 30$ MPa together with a hydrostatic tensile stress of 130 MPa. At Sites A and C, the shear stress is negative, while positive at Site B. One can assume that excessive shear stress is a precursor for void generation at the top interface. In our simulation, τ_{cr} is a critical shear stress for void nucleation, depending on the interface strength and to be experimentally determined.

The top interface is assumed to be the fastest path for vacancy migration in the simulation with an interface diffusivity $D_{int} \sim 10^3 D_{bulk}$. The grain boundary diffusivity is taken to be $D_{GB} \sim 10 \times (misorientation\ angle) \times D_{bulk}$. We found that the variations of diffusivities in the bulk, in the GB, and along the interface can affect the magnitude of the stresses although not the locations of the shear stress peaks. Results from the simulation clearly show voiding kinetics depend on the diffusivities of vacancies at the A–C sites, which can serve either as a source or trap of voids. This is manifested in Figure 7.6, where a cluster of fine grains located at Site C results in a greater density of GBs than in the regions at Sites A and B. As a result, the higher vacancy diffusivity provides an outflow of vacancies from the top interface, making Site C an effective void pinning location.

It is worth noting that the condition $|\tau| > |\tau_{cr}|$ is not sufficient to determine the location of void nucleation. It is possible that voids will not nucleate at locations where the shear stress is rather high but the grains are large. Instead, voids in the line occur at the sites where the grains are small, making the grain size an additional parameter that can influence void formation. This effect can be related to plastic deformation of the metal. According to the Hall–Petch relation, the yielding stress of the polycrystalline metal is $\tau^{yield} \sim d^{-1/2}$, where d is the grain diameter [56], so the region with fine grains may fracture in a relatively brittle manner, while at a region with large grains the shear stress can nucleate dislocation glide instead.

Similar to Sample I, void formation in Sample II also occurred at the capping layer interface, where the void characteristics at two sites, A and B in Figure 7.7a, are of particular interest. After 10 minutes, the first void appeared at A Site, 2.5 μm away from the line end. From there, the void began to migrate relatively fast toward Site B located 1.3 μm away from the line end in 16 minutes. On the way toward Site B, the void experienced several shape and speed changes. On one occasion, it became divided into two voids, with the smaller void following the larger one for some time, then remerged into one larger void before reaching position B. There it became trapped and remained stationary for a relatively long period of time of 28 minutes. Then another void formed at the line end directly above the via and started to grow rather slowly. After 35 minutes, more voids appeared near Site A and traveled quickly toward Site B and merged into the existing void. Finally, after 54 minutes, the void at Site B began traveling again, then after some shape changes, it moved slowly toward the line end, thereby stopping briefly at position C, 0.75 μm away from the line end. After 76 minutes, all voids had merged into the void at the line end. This void then continued to grow in size and to extend into the via. Again, material redeposition into the line end region occurred, causing the sample to fail due to the via void after 2:56 hours.

The EBSD analysis of this sample, shown in Figure 7.7b, showed that there are only $\langle 111 \rangle$-oriented grains present in the segment between Sites A and B. This

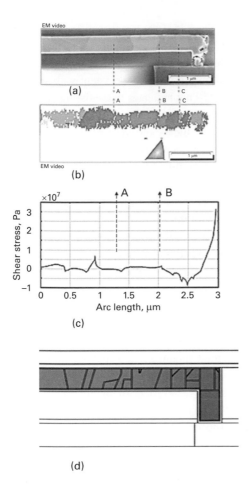

Figure 7.7 Postmortem SEM/EBSD analysis of Sample II and simulation results:
(a) SEM image of the cross section of Cu sample together with (b) an IPF map from
EBSD showing the crystallographic orientation of the individual grains relative to the
wafer surface; (c) simulated shear stress distribution at the top copper–dielectric
barrier interface for copper line with the grain structure shown in (d). © 2009 IEEE.
Reprinted, with permission, from [19].

highly textured $\langle 111 \rangle$ region extends even beyond Site A on the anode side, the left
side of the image. On the cathode side of Site B, right in the image, a $\langle 511 \rangle$-oriented
grain was found. Such grains are known to represent coherent twin grains to neigh-
boring $\langle 111 \rangle$-oriented grains. From position B toward the line end, such $\langle 511 \rangle$-
oriented grains are repeatedly found alternating with $\langle 111 \rangle$-oriented grains. One of
the $\langle 511 \rangle$-oriented grains is located at position C, where the void had briefly stopped
on its way toward the line end. It is worth noting that no voids or signs of delamination
were seen along the capping layer interface during the postmortem SEM/EBSD
analysis. All voids were completely refilled during the EM experiment.

The particular Sites A and B of Sample II are not that different in stress characteristics. Since the simulation did not reveal any region with high shear stresses except at the edge of the line, the shear stress peaks are quite small at both Site A and B in Figure 7.7c. The low shear stress along the line is consistent with why void initiation was not observed at the top interface. This indicates that existing defects at Site A can initiate void nucleation, while high vacancy diffusivity along grain boundaries at Site B can contribute to the outflow of vacancies and void pinning. However, without a precursor for void generation, such as a microscopic delamination or cracking at these sites, the voiding process becomes reversible, leading to restoration of the copper line during further migration of atoms.

The experimental data together with results from numerical simulations provide detailed information on stress evolution during EM in dual-inlaid copper interconnect and can be used to predict the specific sites for void nucleation. For example, the triple junctions at the intersection of grain boundaries with the interface can serve as voiding sites if copper delamination occurs at these points. By introducing the copper grain orientations obtained by EBSD analysis into the EM model and comparing the simulations results with the SEM experiment, we conclude that the high shear stresses are indeed responsible for the delamination. Such high shear stress can arise as a result of the different elastic moduli of the neighboring grains. Overall, the model can account for the stress evolution in the Cu damascene structures during EM and predict the potential voiding sites in the copper lines with observed microstructure.

The next step toward understanding the effect of microstructure on interconnect degradation is to analyze the voiding dynamics and verify the simulation results with in situ SEM experimental data. The orientation dependence of activation energy of atomic transport along the GBs is expected to have significant influence on the void growth and migration processes as evidenced in the preliminary results from simulations presented in Figure 7.8. Here, when the diffusion along the top interface prevails, i.e., $D_{int} \gg D_{GB}$, voids were found to be nucleated at delamination sites as depicted by the semispherical cutouts of the copper line, and the subsequent movement toward the cathode is not affected by the grain boundaries (Figure 7.8a). In the opposite limit, when $D_{int} \ll D_{GB}$, the void moved along the grain boundary and can be transformed into a slit, as shown in Figure 7.8b. In both cases, the void nucleated at the line edge grew preferentially toward the bottom of the via. A detailed description of the different modeling approaches to analysis of the void evolution under EM action is given in Section 7.10.

We found that an accurate estimate of the GB diffusivity is required in order to analyze the microstructure texture effect in modeling void evolution leading to interconnect degradation, particularly for lines with strengthened interfaces. Results from the void kinetics model in turn can provide useful recommendations for interfacial engineering in copper deposition, and good estimate of the EM lifetime for the interconnect lines. We will discuss modeling of void kinetics in Section 7.9. In the following section, we discuss first the initial conditions in a line segment passivated by ILD/IMD oxides, focusing on the stress distribution and concentrations of vacancies and plated atoms in a polycrystalline metal segment in the prevoiding state before electric stressing is applied.

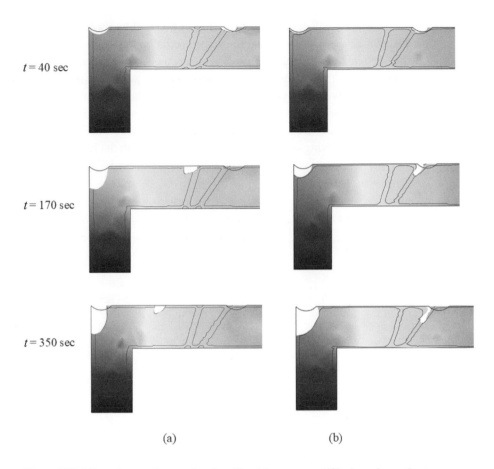

Figure 7.8 Migration and growth of voids: (a) vacancy diffusion along the top interface is the dominating mechanism of atomic transport; and (b) diffusivity along grain boundaries dominates. © 2009 IEEE. Reprinted, with permission, from [19].

7.7.2 Effect of Microstructure on Stress Evolution Induced by Thermal Processing

To analyze EM-induced void formation, it is relevant to quantify the stress state in the chip after it is manufactured where a number of stress relaxation process can occur [55]. Among them there are self-anneals, which can modify the grain size distribution while the chip is stored, where atom migration can be induced by nonuniform stresses developed during temperature ramps and stress-induced voiding, etc. Such processes can create a new equilibrium state under a steady stress state with the vacancy concentration in equilibrium. In some cases, voids can be formed by high tensile stresses developed in interconnect segments, as discussed in Section 4.6. We can apply the distributions of stress and vacancy–plated atoms existing in the new equilibrium state as an initial condition to derive the system evolution under the electric load. This is demonstrated for a confined polycrystalline metal segment under cooling from thermal processing to a use or a shelf temperature.

The system will evolve under the stress generated by cooling in the confined metal segment as the equilibrium vacancy concentration is perturbed by stress generation due to temperature change. Both factors, the change of the temperature T and generation of the thermal stress σ_T, destroy the equilibrium in the vacancy concentration. It activates the generation/annihilation of the vacancy–plated atom pairs and the vacancy migration. Plated atoms generate a compression at the GBs/interfaces, while the tensile stress is generated everywhere by vacancies. The kinetics can be extracted by solving simultaneously (7.43)–(7.46) with the EM terms omitted to yield vacancy and plating atom concentrations at the grain boundaries and interfaces as follows:

$$\frac{\partial N}{\partial t} + \vec{\nabla}\left(-D_{int/GB}\vec{\nabla}N - \frac{D_{int/GB}N}{k_B T}(1-f)\Omega\vec{\nabla}\sigma_{Hyd}\right) + R = 0 \tag{7.72}$$

$$\frac{\partial M}{\partial t} + R = 0. \tag{7.73}$$

While the evolution kinetics of the vacancy and plating atom concentrations in the grain interior are described by:

$$\frac{\partial N}{\partial t} + \vec{\nabla}\left(-D\vec{\nabla}N - \frac{DN}{k_B T}(1-f)\Omega\vec{\nabla}\sigma_{Hyd}\right) = 0 \tag{7.74}$$

$$\frac{\partial M}{\partial t} = 0 \tag{7.75}$$

Strain evolution everywhere in the confined interconnect segment is described by the modified equation (7.47) as follows:

$$(\lambda + G)\frac{\partial e}{\partial x_i} + G\Delta u_i - \frac{E}{3(1-2\nu)}\left(\frac{\partial M}{\partial x_i} - (1-f)\frac{\partial N}{\partial x_i} + 3\alpha\frac{\partial T}{\partial x_i}\right) = 0. \tag{7.76}$$

Here the additional body force components $3\alpha\frac{\partial T}{\partial x_i}$ are generated due to the nonuniform temperature distribution $T(x, y, z, t)$ [16], with α for the corresponding coefficient of thermal expansion. The temperature distribution can be obtained by solving the heat equation:

$$\rho C \frac{\partial T}{\partial t} - \nabla(k\nabla T) = Q, \tag{7.77}$$

where ρ, k and C are the density, thermal conductivity, and heat capacity of the corresponding materials, and Q is the heat source. This results in the following modified boundary equations for the surface tractions when the temperature deviates from the zero-stress temperature T_{ZS}:

$$\frac{E}{3(1-2\nu)}(M - (1-f)\Delta N + 3\alpha\Delta T)\hat{l} = \lambda e\hat{l} + G\left(\frac{\partial u}{\partial x}\hat{l} + \frac{\partial u}{\partial y}\hat{m} + \frac{\partial u}{\partial z}\hat{n}\right)$$
$$+ G\left(\frac{\partial u}{\partial x}\hat{l} + \frac{\partial v}{\partial x}\hat{m} + \frac{\partial w}{\partial x}\hat{n}\right)$$

$$\frac{E}{3(1-2\nu)}(M-(1-f)\Delta N + 3a\Delta T)\hat{n} = \lambda e\hat{n} + G\left(\frac{\partial v}{\partial x}\hat{l} + \frac{\partial v}{\partial y}\hat{m} + \frac{\partial v}{\partial z}\hat{n}\right)$$

$$+ G\left(\frac{\partial u}{\partial y}\hat{l} + \frac{\partial v}{\partial y}\hat{m} + \frac{\partial w}{\partial y}\hat{n}\right) \qquad (7.78)$$

$$\frac{E}{3(1-2\nu)}(M-(1-f)\Delta N + 3a\Delta T)\hat{m} = \lambda e\hat{m} + G\left(\frac{\partial w}{\partial x}\hat{l} + \frac{\partial w}{\partial y}\hat{m} + \frac{\partial w}{\partial z}\hat{n}\right)$$

$$+ G\left(\frac{\partial u}{\partial z}\hat{l} + \frac{\partial v}{\partial z}\hat{m} + \frac{\partial w}{\partial z}\hat{n}\right),$$

where $\Delta T = T - T_{ZS}$. Here, in contrast to the volumetric stresses generated by the vacancy and plated atoms, the thermal stress is generated everywhere in the confined line segment, where the boundary conditions are applied to the outmost surfaces.

In this way, we can obtain the hydrostatic stress generated by both the thermal and vacancy–plated atom–induced dilatations as shown in Figure 7.9. Figure 7.9a shows the steady-state distribution of the hydrostatic stress developed immediately after reaching the test temperature T_{test}, and Figure 7.9b shows the final hydrostatic stress distribution achieved after stress relaxation caused by vacancy diffusion and vacancy–plated atom pair generation was completed. In Figure 7.10, we show the evolution of the hydrostatic stress, and the concentrations of vacancies and plated atoms along the central line in the M2 layer (the second metal layer in the interconnect structure in Figure 7.9). The dashed lines show the distributions at the beginning of the relaxation process immediately after a new temperature T_{test} was reached, while the solid lines show the corresponding distributions at the final stage of relaxation.

The results in Figures 7.9 and 7.10 show that the cooling and the tensile stress generated upon cooling affect the equilibrium concentration of vacancies in opposite directions. The details, however, depend on other factors such as grain size and orientation distributions. So, a simple approach to study such relaxation processes is to model a single copper grain embedded in a rigid confinement. Following Herring [2], we model a grain as a spherical object of a radius R with a subsurface region of thickness δ to represent a grain boundary, where vacancy–plated atom pairs are generated or annihilated. In

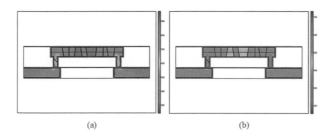

(a)   (b)

Figure 7.9 Distributions of the hydrostatic stress generated by an abrupt temperature change (a) and at the steady state due to generation/annihilation of the vacancy–plated atom pairs and vacancy redistribution (b).

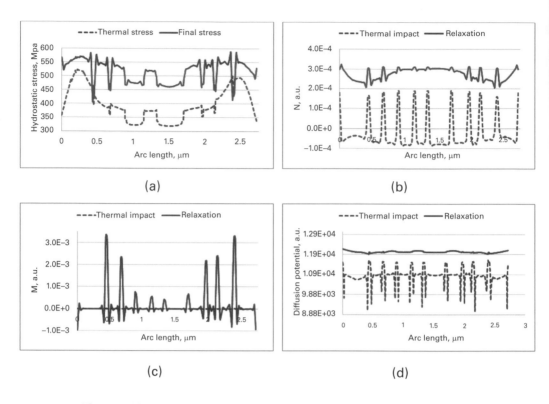

Figure 7.10 Distributions of the hydrostatic stress (a), vacancy concentration (b), concentration of the plated atoms (c), and the diffusion potential $\mu_{dif}(90)$ (d) along the central line of an M2 interconnect segment. Dashed lines show these distributions when the test temperature was reached, and solid lines indicate the postrelaxation distributions.

Figure 7.11, we show the changes in the vacancy and plated atom concentrations together with the relaxation of the hydrostatic stress at GBs and at two positions in the grain interior: one at the grain center, and other halfway between the GB and the grain center. These changes are induced by cooling from the zero-stress temperature, taken as $T_{ZS} = 650$ K to 498 K. Here the grain has a radius $R \approx 1$ μm and GB thickness of $\delta \approx 10^{-3}$ μm or 10 Angstroms and embedded in a rigid confinement, i.e., with zero-displacement boundary conditions at the outer grain interface. An activation energy for the vacancy formation was taken to be $E_A = 1.3 \cdot 10^{-19} J$, the same as for vacancy diffusion in the grain interior E_{VL}, to yield $D_V = D_0 \exp\{-[E_{VL} - (1-f)\Omega\sigma_{Hyd}]/k_BT_{test}\}$, where $D_0 = 5.2 \cdot 10^{-5} \, m^2/s$. The activation energy for the GB diffusion was taken to be $E_{VGB} = 0.9 \cdot 10^{-19}$ J, corresponding to an acceleration factor of 500 compared with diffusion in the grain interior. For τ_{react}, which was discussed in Section 7.3 as a reverse value of the reaction rate constant for vacancy annihilation by a plated atom $\tau_{react} = k_2^{-1} = \tau_0 \exp\{E_R/k_BT\}$, so at $T = T_{test}$, we have $\tau_{react} = 8.4 \cdot 10^{-3}$ s. A time scale for the diffusion-induced vacancy transfer from the grain interior to the GB can be approximated as $\tau_D \approx R^2/2D_V \approx 0.5$ s.

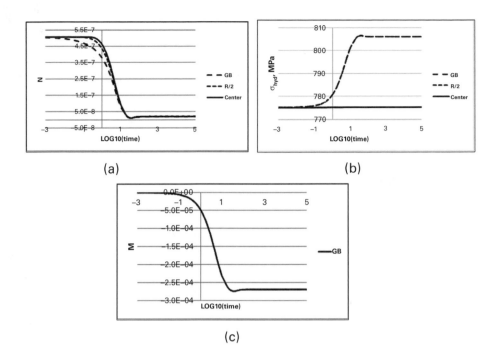

(a) (b)

(c)

Figure 7.11 Evolution of vacancy concentration (a), hydrostatic stress (b) at the GB and in the grain interior, and plated atom concentration (c) at the GB induced by cooling from $T_{ZS} = 650$ K to $T_{test} = 498$ K.

The analysis of relaxation kinetics as shown in Figure 7.11a indicates that a reduction in the vacancy concentration at the GB begins almost immediately after generation of the thermal stress. The reduction in vacancy concentrations at the points located half-radius distance from the GB and at the grain center is delayed to begin at $t \approx 0.2$ s and $t \approx 0.3$ s correspondingly. Figure 7.12a shows clearly that the change of the vacancy distribution inside the grain interior is diffusion controlled. Based on this result and an estimate of the time constants for the reaction and diffusion kinetics, we found that the overall defect relaxation is controlled by two processes induced by temperature change: diffusion of excessive vacancies to the GB, and annihilation with the plated atoms there.

Since the rate of vacancy annihilation is much higher than the rate of vacancy supply by diffusion, the overall kinetics is controlled by vacancy diffusion, so the GB hydrostatic stress will increase following the reduction in the plated atom concentration, which is governed by vacancy annihilation as shown in Figures 7.11 and 7.12. Simulations show that the vacancy concentration at the GB evolves from the initial N_0^{ZS} at zero-stress temperature T_{ZS} to the equilibrium concentration of $N_{GB}^{eq} =$ $\exp\left\{-\left(E_A - f\Omega\sigma_{Hyd}^{GB}\right)\Big/k_B T_{test}\right\}$ with the corresponding equilibrium vacancy concen-tration in the grain interior of $N_{GR}^{eq} = \exp\left\{-\left(E_A - f\Omega\sigma_{Hyd}^{GR}\right)\Big/k_B T_{test}\right\}$. The hydrostatic stress in the grain interior becomes equal to the thermal stress after cooling from T_{ZS} down

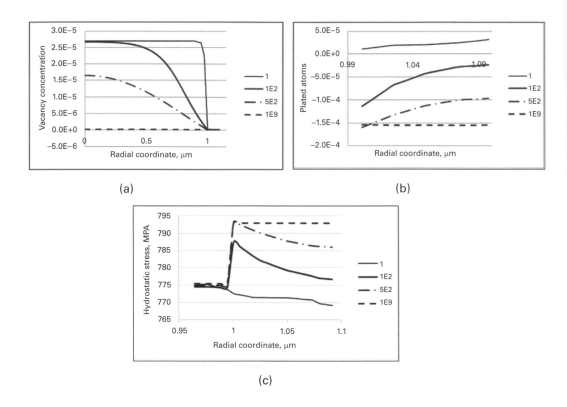

Figure 7.12 Evolution of the radial distributions of (a) vacancies, (b) plated atoms, and (c) hydrostatic stress upon cooling.

to T_{test} with an increase at the GB of about 30 MPa (see Figure 7.12c). Simulation shows further that the difference between the hydrostatic stresses at GB and in the grain interior increases with increasing ratio of R/δ, indicating an increase in the volumetric deformation of the GB layer due to the annihilation of the excessive vacancies in the large grain.

7.7.3   Effect of Microstructure on Stress Relaxation Induced by Thermal Processing: Analytical Formulation

In this section, we further analyze the effect of microstructure of a damascene line on stress relaxation induced by thermal processes. The hydrostatic stresses and vacancy concentrations obtained from numerical simulations at the steady state for GB and grain interior can be checked by comparing with results from analytical solutions derived for the same single metal-grain model of radius R and GB thickness $\delta \ll R$. The grain is embedded in a rigid confinement and subjected to abrupt temperature changes by rapid cooling from the zero-stress temperature T_{ZS} to the test temperature T_{test}. In this model, we solve for the equilibrium stress and concentrations of vacancies and plated atoms everywhere inside grain.

This model has a central symmetry, so the force balance equation describing the mechanical equilibrium takes the following form:

$$\frac{d\sigma_r}{dr} + \frac{2}{r}(\sigma_r - \sigma_t) = 0. \tag{7.79}$$

The equation is then subjected to the following stress–strain relations (7.52):

$$\sigma_r = \frac{E}{(1+v)(1-2v)}[(1-v)\varepsilon_r + 2v\varepsilon_t - (1+v)V] + \frac{\sigma_T}{3} \tag{7.80}$$

$$\sigma_t = \frac{E}{(1+v)(1-2v)}[\varepsilon_t + v\varepsilon_r - (1+v)V] + \frac{\sigma_T}{3}. \tag{7.81}$$

The hydrostatic stress is

$$\sigma_{Hyd} = \sigma_T + \frac{E}{3(1-2v)}[\varepsilon_r + 2\varepsilon_t - 3V]. \tag{7.82}$$

Here $V(r) = M - (1-f)\Delta N$ describes a dilatational strain, and other notations are defined in Figure 7.13. Employing the relations between the radial displacement $u(r)$ and the strain components ε_r, ε_θ, and ε_φ, [16]: $\varepsilon_r = \frac{du}{dr}$, $\varepsilon_\theta = \varepsilon_\varphi = \varepsilon_t = \frac{u}{r}$, we obtain the following solution to (7.79) as follows:

$$u(r) = -\frac{1+v}{1-v}\left[\frac{1}{r^2}\int_0^r Vr^2 dr - \frac{r}{(R+\delta)^3}\int_0^{R+\delta} Vr^2 dr\right]. \tag{7.83}$$

This solution was derived by applying the central symmetry condition of $u(r=0) = 0$ and zero displacement at the interface under the rigid confinement: $u(r=R+\delta) = 0$.

Since vacancies and plated atoms are generated and annihilated as pairs at GB and plated atoms stay inside the GB, the dilatational strain can be presented as

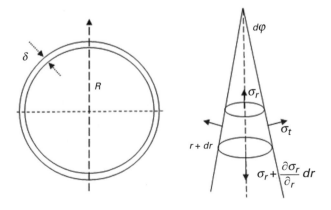

Figure 7.13 Schematics for the geometry of the analyzed sample and the employed notations.

$$V(r) = M - (1-f)\Delta N = \begin{cases} -(1-f)\left(N(r) - N_0^{ZS}\right), & 0 \le r \le R \\ M(r) - (1-f)\left(N(r) - N_0^{ZS}\right), & R \le r \le R + \delta. \end{cases}$$

(7.84)

Here, as before, $N_0^{ZS} = \exp\{-E_A/k_B T_{ZS}\}$ is the equilibrium vacancy concentration at zero stress. This leads to the following relation, which is valid at any time:

$$M = \frac{1}{3}\left(\frac{R}{\delta}\right)\left(N_{GR} - N_0^{ZS}\right) + \left(N_{GB} - N_0^{ZS}\right).$$

(7.85)

Here we denote the vacancy concentration at the GB as N_{GB}, and inside the grain interior as N_{GR}. Since we are interested in the steady state when all relaxation processes are completed, we assume that vacancies and plated atoms at the end will be uniformly distributed across the grain interior and GB. This assumption is supported by the results of the FEA simulation (Figure 7.12). In this case, the hydrostatic stress σ_{Hyd} in the grain interior, σ_{Hyd}^{GR}, and at the GB, σ_{Hyd}^{GB} takes the form

$$\sigma_{Hyd} = \begin{cases} \sigma_{Hyd}^{GR} \approx \sigma_T + \dfrac{2}{3}\dfrac{E}{(1-v)}\left(N_0^{ZS} - N_{GR}\right); & 0 \le r \le R \\[2mm] \sigma_{Hyd}^{GB} \approx \sigma_T + \dfrac{2}{9}\dfrac{E}{(1-v)}\left(\dfrac{R}{\delta}\right)\left(N_0^{ZS} - N_{GB}\right); & R \le r \le R + \delta. \end{cases}$$

(7.86)

This confirms that the hydrostatic stress in the grain interior is mainly determined by the thermal stress, so additional work is required against the stress in order to put the plated atom on the GB and vacancy in the grain interior.

This follows with the equation for the equilibrium vacancy concentration inside the GB as follows:

$$N_{GB} = N_0 \exp\left\{\frac{f\Omega\sigma_T}{k_B T} + \frac{2}{9}\frac{\Omega}{k_B T}\frac{Ef}{1-v}\left(\frac{R}{\delta}\right)\left(N_0^{ZS} - N_{GB}\right)\right\}.$$

(7.87)

An approximate solution to this equation can be derived since under typical operating conditions, $\frac{2}{9}\frac{E}{\sigma_T(1-v)}\left(\frac{R}{\delta}\right)\left(N_0^{ZS} - N_{GB}\right) < 1$, so

$$N_{GB} \approx N_0 e^{\frac{f\Omega\sigma_T}{k_B T}}\left(1 + \frac{2}{9}\frac{\Omega}{k_B T}\frac{Ef}{1-v}\left(\frac{R}{\delta}\right)\left(N_0^{ZS} - N_0 e^{\frac{f\Omega\sigma_T}{kT}}\right)\right) \approx N_0 e^{\frac{f\Omega\sigma_T}{k_B T}}\left(1 + \frac{2}{9}\frac{\Omega}{k_B T}\frac{Ef}{1-v}\left(\frac{R}{\delta}\right)N_0^{ZS}\right).$$

(7.88)

For the corresponding vacancy concentration in the grain interior, we have the following:

$$N_{GR} \approx N_0 e^{\frac{f\Omega\sigma_T}{k_B T}}\left(1 + \frac{2}{9}\frac{\Omega}{k_B T}\frac{E}{1-v}\left(\frac{R}{\delta}\right)N_0^{ZS}\right).$$

(7.89)

Substituting N_{GB} and N_{GR} from (7.88) and (7.89) to (7.85), we obtain the plated atom concentration as follows:

$$M \approx -\frac{1}{3}\left(\frac{R}{\delta}\right)\left(N_0^{ZS} - N_0 e^{\frac{f\Omega\sigma_T}{k_B T}}\right).$$

(7.90)

Comparing the solutions (7.88) and (7.89), we find that the postrelaxation vacancy concentrations inside the grain interior and at the GB are almost the same. This is in agreement with the results from the FEA simulations shown in Figure 7.12a. The same condition, $\frac{2}{9}\frac{E}{\sigma_T(1-v)}\left(\frac{R}{\delta}\right)\left(N_0^{ZS} - N_{GB}\right) < 1$, indicates that the hydrostatic stress in the grain interior is the same as the thermal stress but smaller than that at the GB, which is consistent with the numerical results shown in Figures 7.11b and 7.12c. Radial σ_r and tangential σ_t stresses calculated with the obtained N_{GR}, N_{GB}, and M are satisfied the force balance equation (7.79) everywhere inside grain and GB. Uniform distributions of the steady-state hydrostatic stresses and concentration of vacancies at GB and inside the grain lead to zero vacancy fluxes. Fluxes of vacancies through the interface between GB and grain interior, caused by differences in the hydrostatic stress and vacancy concentration there, compensate each other, providing

$$-D\frac{\partial N}{\partial r} - \frac{DN}{k_B T}\Omega\frac{\partial \sigma_{Hyd}}{\partial r} = 0. \tag{7.91}$$

This condition is equivalent to that of a uniform chemical potential everywhere within the system [3], as shown in Figure 7.10d:

$$\mu_{diff} = \mu_0 + k_B T \ln\left(\frac{N}{N_0}\right) + \Omega\sigma_{Hyd} = const. \tag{7.92}$$

The preceding analytical solutions allow us to analyze the hydrostatic stresses and concentrations of plated atoms and vacancies as a function of the grain size, crystallography, temperature, etc. We found that the hydrostatic stress and plated atom concentration inside GB linearly increase with the grain size, while the vacancy concentration and hydrostatic stress in the grain interior are almost independent on this parameter, as well as vacancy concentration in the GB. The amount of the changes in the hydrostatic stress and vacancy concentration have been evaluated with the following material parameters: $E_{Cu} = 1.2 \cdot 10^{11}$ Pa, $v_{Cu} = 0.3, f \approx 0.6$, $\Omega \approx 1.66 \cdot 10^{-29}$ m^3. The residual stress σ_T developed by cooling down is $\sigma_T \approx \frac{E_{Cu}}{3(1-2v)}\Delta\alpha\Delta T$. Taking $\Delta T \sim 300-500$ K and $\Delta\alpha \approx 1.7 \cdot 10^{-5} 1/$ K, we get σ_T in the range of 500–800 MPa. In addition, by taking $N_{ZS} = 5 \cdot 10^{-7}$ corresponding to $N_0 \approx 10^{-11}$ and $T_{ZS} = 800$ K [57], we find the hydrostatic stress relaxes by $\Delta\sigma_{Hyd}^{GB} \approx \frac{2}{9}\frac{E}{1-v}\left(\frac{R}{\delta}\right)N_0^{ZS} \approx 20 \div 200$ MPa, in a range of R/δ ratio of $10^3 - 10^4$. In comparison, the stress relaxation in the grain is very small, reaching only about 100 kPa. This result indicates that the excessive tensile stress at the GBs compared with the grain interior is due to the large number of vacancies being annihilated with plated atoms, which is much larger than the number of vacancies generated by tensile stress developed during cooling down from T_{ZS} to T_{test}. Interestingly, the change in the GB energy induced by the hydrostatic stress, $\frac{2}{3}\frac{E\Omega}{1-v}\left(\frac{R}{\delta}\right)N_0^{ZS}$, is about one to two orders of magnitude larger than that induced by surface tension $\gamma_{GB}\Omega/2R$. Indeed, for typical values of the surface energy density $\gamma_{GB} \approx 1$ N/m^2, $\Omega = 1.6 \cdot 10^{-29}$ m^3, and $R \approx 2$ μm, the change in GB energy due to surface tension is 10^{-4} eV compared to 10^{-2} eV due to the hydrostatic stress.

Let us consider the stress distribution developed in a polycrystalline metal line with a wide range of grain size being cooled down from annealing to a test temperature.

Based on the analysis presented earlier, all grain interiors should have the same stress σ_T, while the GB network is characterized by a nonuniform distribution of the hydrostatic stress, depending on the grain size distribution before fast vacancy diffusion starting along grain boundaries. This follows with a redistribution of the vacancy concentration along the GB network to reach a final stress distribution with uniform diffusion potential, which is governed by the diffusion kinetics driven by the hydrostatic stress gradients. The distribution of the diffusion potential μ_{diff} (7.92) along the center of the M2 line segment is shown in Figure 7.10d, with the dashed line for the initial time and the solid line at $t = 10^3$ s and when the steady state was almost achieved. For the latter, there is a slight deviation from the straight line at $t = 10^3$ s because the relaxation process is not yet completed.

Hence to simulate the evolution of the EM-induced stresses, a set of initial conditions consisting of postannealing relaxation of stress and the vacancy and plated atom concentrations generated has to be considered. The results would be quite different if the stress generated in the line segment during cooling exceeds the critical stress and a saturated void is created. The steady-state stress distribution generated by the saturated void and the kinetics of stress evolution will be quite different in comparison with the voidless case, as discussed in Section 4.8. The general approach to simulation of stress-induced voiding we will present in Section 7.11.

7.7.4   Effect of Mass Exchange between Grain Interior and Grain Boundaries on EM-Induced Stress Evolution

To demonstrate the predictive capability of developed methodology, we consider a real situation characterized by low concentrations of vacancies that exist in the interconnect line at the chip operation temperatures. The realistic relations between grain sizes and grain boundaries and interfaces' thicknesses should be used. To keep a reasonable simulation time, we take a grain of 1 μm × 4 μm in size and a GB/interface thickness of 5 nm, corresponding to $R/\delta \approx 10^2 - 10^3$. This ratio allows us to explore the influence of the grain interior on evolution of stress and vacancy concentration. We choose the zero-stress condition at $T_{ZS} = 650$ K with $N_0^{ZS} \approx 2.7 \cdot 10^{-7}$ followed by cooling down to a test temperature of $T_{test} = 498$ K to drop in the vacancy concentration from N_0^{ZS} to $N_0 \approx 6.5 \cdot 10^{-10}$. Other parameters are current density $j \approx 4.2 \cdot 10^{-10}$ A/m², the energy of generating the vacancy-plated atom pair $E_V = 1.77 \cdot 10^{-19}$ J, the same as the activation energy for vacancy diffusion with an preexponential factor of $D_0 = 5.2 \cdot 10^{-5}$ m²/s, and thermal mechanical properties of Cu: Young's modulus $E_{Cu} = 1.2 \cdot 10^{11}$ Pa, Poisson ratio $v_{Cu} = 0.3$, and CTE $\alpha_{Cu} = 1.7 \cdot 10^{-5}$ 1/K. We assume that vacancy diffusivity at the GBs and interfaces is 500× faster than in the grain interior, corresponding to an activation energy of GB/interface diffusion of $E_{VD}^{GB} = 1.34 \cdot 10^{-19}$ J. The 2D distribution of steady-state vacancy concentration and hydrostatic stress caused by a combined impact of cooling and electric stressing is shown in Figure 7.14a and b. Figure 7.15a shows the calculated hydrostatic stress distribution at a sidewall interface with rigid confinement, and along the central line of the grain interiors in Figure 7.15b. Figure 7.15c–d shows

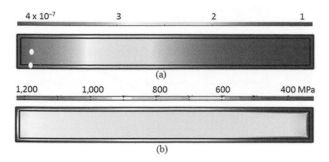

Figure 7.14 The steady-state distribution of (a) the vacancy concentration and (b) the hydrostatic stress in a rigid confined line induced by an electric current and cooling. The deformed shape scale factor is 150 for (b).

the vacancy concentration distributions along the same cross-section lines, and finally Figure 7.15e shows the distribution of plated atoms at the interface. All shown distributions were obtained at different moments in time.

These results confirm that the maximum hydrostatic stress is developed at GBs and interfaces due to accumulation or annihilation of the layers of plated atoms inserted between grains or between the surrounding dielectric. Much smaller stress variations are generated in the grain, so that the steady-state hydrostatic stress near the cathode differs from that at the anode edge by just tens of megapascals. Figure 7.16 shows the evolution of hydrostatic stresses in two points located at the grain interior and at the GB near the cathode shown in Figure 7.14a. The kinetics of stress evolution at these two points are almost identical, as shown in Figure 7.16a and b indicating that the stress modification occurring in the grain is mainly caused by the dilatation at the GBs due to generation of the plated atoms. Regarding the vacancy concentration, as can be seen in Figure 7.17a, the initial change of the vacancy concentration inside GB and interface is governed by the vacancy generation/annihilation with plated atoms (Figure 7.17b). Delayed reduction of the vacancy concentration in the grain interior, which is shown in Figure 7.17a, can be explained by the slow vacancy diffusion toward the GB or interface.

There is one more factor that can affect the evolution kinetics of the stress and vacancy concentrations under EM in a confined line, which can be attributed to the exchange of vacancies between the GB and in the grain interior. The role of this vacancy exchange on evolution kinetics can be analyzed by simulating the case when the vacancies and plated atoms are confined to the GBs and interfaces only. Evolution of stress and concentrations of vacancies and plated atoms in the segment shown in Figure 7.14 is simulated with the same parameters, BC, and loads, except vacancies cannot diffuse through the grain interior. Comparison of the results show that for the case where vacancies are allowed in the grain interior, the kinetics is delayed by vacancy exchanges between GB and grain interior, as shown in Figure 7.18a, and the GB steady-state hydrostatic stress is increased by 30–50 MPa as shown in Figure 7.18b.

All simulation results discussed so far are obtained with stress-dependent vacancy diffusivities everywhere in the simulated domain. We can now find out the error in the

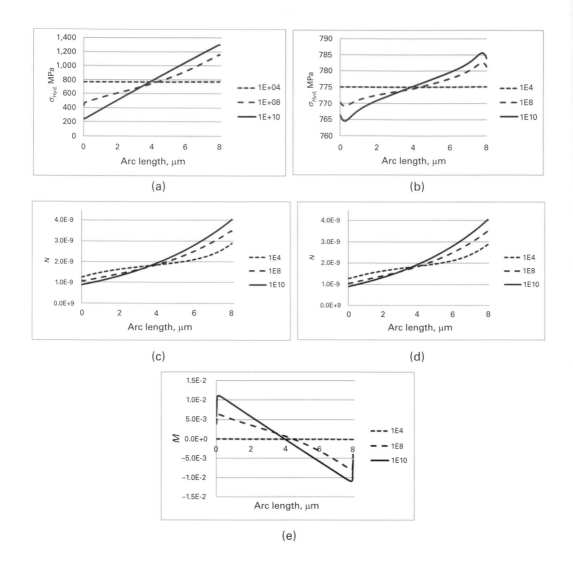

Figure 7.15 Calculated distributions of the σ_{Hyd} along the long (sidewall) interface (a) and along the central line of the grain interior (b); distributions of the vacancy concentration N along the long interface (c) and the central line of the grain interior (d), and plated atom concentration M (e).

simulation if the stress dependency is ignored. We will simulate the same system as discussed in Figure 7.14 with a metal line segment rigidly confined and driven by the same electric current but with a stress-independent vacancy diffusivity in the grain interior of $D_V = D_0 \exp\{-E_V/k_B T_{test}\}$ and at the GB/interfaces $D_V = D_0 \exp\{-E_{VD}^{GB}/k_B T_{test}\}$. The results demonstrate the same steady-state distributions of the hydrostatic stress and concentrations of vacancies and plated atoms as shown in Figure 7.15. A difference between these two cases of stress-dependent and stress-independent vacancy diffusivities is in the kinetics of evolution of these variables (Figure 7.19).

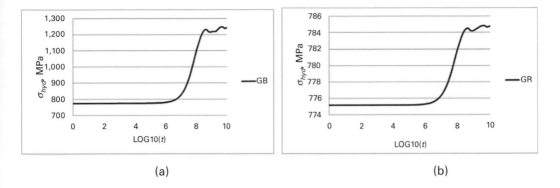

(a)   (b)

Figure 7.16 (a) Evolution of the hydrostatic stress at GBs and (b) in the grain interior at the two locations near the cathode shown in Figure 7.14 by small circles.

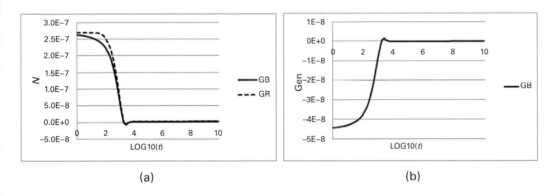

(a)   (b)

Figure 7.17 (a) Evolution of the vacancy concentrations at the GB and in the grain interior and (b) evolution of the generation/annihilation rate of vacancy–plated atom pairs as a function of time.

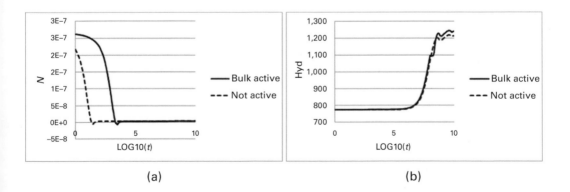

(a)   (b)

Figure 7.18 Effects of vacancies exchange between GB and grain interior on (a) the evolution of vacancy concentration N and (b) the steady-state hydrostatic stress σ_{Hyd} under EM.

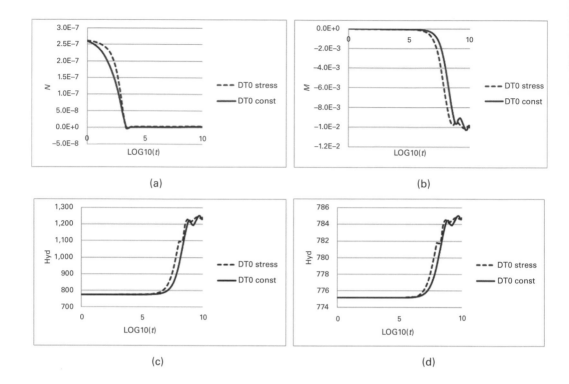

Figure 7.19 Effects of stress-independent vacancy diffusivity on (a) evolution of vacancy concentration at GBs, (b) concentration of plated atoms, (c) evolution of the hydrostatics stress at GBs, and (d) in the grain interior.

As can be seen from Figure 7.19, the stress dependency of the vacancy diffusivity accelerates the rate of hydrostatic stress increase, which can result in a shorter void nucleation time. Assuming, for example, $\sigma_{crit} \approx 1\,\text{GPa}$, the void nucleation time for a stress-dependent vacancy diffusivity is $t_{nuc} \approx 5.5 \cdot 10^{7}\,\text{s}$, which is more than $2\times$ faster than in the case of stress-independent diffusivity $t_{nuc} \approx 1.5 \cdot 10^{8}\,\text{s}$, or one year and nine months versus four years and nine months. The difference is quite significant and would not be obtained if the stress dependency of the vacancy diffusivity is ignored as in most of the 1D modeling simulations discussed in Chapter 4.

The last two examples clearly demonstrate the importance of including the stress-dependent vacancy diffusivity and the GB–grain interior interaction; both are often ignored in EM modeling.

7.8   Experimental Studies of EM-Induced Voiding in Interconnects

In the mid-1990s, Arzt et al. pointed out that the modeling of void formation and growth was a long-standing scientific problem [58]. To be useful for applications, the modeling should be able to predict the EM-induced time to failure in interconnects

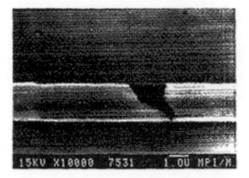

Figure 7.20 SEM micrograph of the top view of the slit-like void in an Al line. Reprinted from [58], with the permission of AIP Publishing.

and help to understand the underlying physics of such processes. This practical demand has initiated enormous research activities targeting development and employment of novel experimental methodologies to track and record the voiding phenomena in metal films and stripes caused by electric stressing, as well as to develop new theoretical models and numerical techniques to explain and predict the phenomenon. The big interest in the voiding phenomenon was heated up particularly by a crucial role played by a special type of voids, so-called slit-like voids (Figure 7.20), in developing an early failure of the aluminum-based on-chip interconnects accepted at that time by the semiconductor industry. An architecture of the Al-based metallization was responsible for this type of early failure resulting in an open circuitry. As was discussed in Chapter 3 and illustrated in Figure 3.4a, Al interconnects are surrounded by ILD materials, most commonly SiO_2. Al reacts chemically with SiO_2 to form alumina, which eliminates atomic diffusion of Al into the surrounding ILD and along the interface between Al and SiO_2. As a result, the patterned metal lines are formed by etching the deposited blanket Al film. Architecturally, Al interconnects have thick, highly EM-resistant refractory metal layers, which are usually made of titanium nitride (TiN), at the top and bottom of the lines. Tungsten (W)–filled vias are used to connect layers of Al metallization. Under and over TiN layers serve as shunts for electron flow, and W-filled vias serve as fully blocking boundaries for electromigration.

The void shown in Figure 7.20, when is grown over the entire linewidth and extended to the line bottom, undercuts the electric current path through Al and forces the current to pass through the highly resistive TiN shunts. It results an enormous increase in the line resistance, which is accompanied by a current crowding and, as a result, by a huge heat release provoking further line degradation.

Currently accepted Cu interconnect technology also demonstrates a presence of the slit-like voids cutting the entire line, usually along the grain boundary (Figure 7.21), though not so frequent as was observed in the case of Al interconnects. More important is that slit-like voids do not generate such abrupt change in the Cu line resistance, as was the case in Al metallization. It is related to the interconnect architecture of the Cu dual-damascene technology shown in Figure 3.4b in

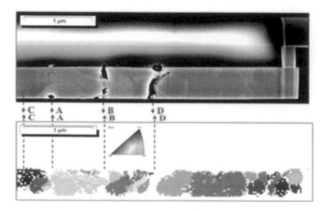

Figure 7.21 SEM/EBSD analysis of a Cu/CoWP interconnect sample after an EM experiment. Reprinted from [60], with the permission of AIP Publishing.

Chapter 3. Since Cu does not chemically reduce SiO_2 as Al does, Cu interconnects are fabricated by the damascene method, where a trench is first etched into a blanket layer of ILD before it is filled with Cu by electroplating. Since Cu easily diffuses in most dielectric materials, including SiO_2 and carbon-doped silica, in order to prevent such diffusion, thin refractory metal layers consisting of Ta, TaN, or their combination are deposited on the sidewalls and bottom of the Cu trenches and vias (Figure 3.4b). Then, Cu lines are capped with a dielectric diffusion barrier, which is usually made of Si_3N_4 or SiC_xN_y. Cu-filled vias are used to connect multiple layers of metallization. Presence of the sidewall metal liners is responsible for lesser increase of the line resistance, which observed when voids undercut the entire Cu line, in comparison with Al line. In addition, these types of voids are less frequent in Cu lines, if compared with Al interconnects, partly due to a difference in the surface energies of Cu and Al interfaces with the Si_3N_4 and TiN correspondingly. Cu atoms diffuse from the Cu–Si_3N_4 interface to the void surface to minimize the total energy [60]. Nevertheless, EM-induced voiding remains the major reliability problem for the Cu metallization.

For the case of Al metallization, experimental observations of the damage development, performed on lines with the optimized near-bamboo grain structure, have demonstrated a nontrivial void evolution showing a complex sequence of void motion, growth, and shape changes [58]. First, voids were not static but rather moved in the direction opposing the electron wind. This has been confirmed by in situ SEM studies on unpassivated Al lines [61, 62] and by field-emission SEM or scanning transmission electron microscopy (STEM) on passivated Al lines [63–65]. It was also observed that voids can break up into smaller fragments or grow by coalescing with other voids [65]. Second, it was observed that besides conventional grain boundary diffusion, surface and interface diffusion can contribute to damage development. This was suggested by in situ transmission electron microscopy (TEM) studies on large-grained Al stripes and films revealing voids inside the grains [62]. These voids had grown in the direction of current flow lines, sometimes without apparently interacting with grain boundaries. Thinning of large regions within a single grain in Al film during electromigration testing was also observed [66]. Both observations cannot be explained if grain boundaries are the only diffusion paths. Third, voids did not grow in a self-similar manner, but could show significant shape changes. This point has been demonstrated in electromigration tests that were interrupted several times for damage characterization in SEM [67]. The resulting fatal void often had a slit-like morphology, which gave the appearance of a crack perpendicular to the line. Following detailed study by SEM and focused ion beam techniques [68, 69] and by TEM [70, 71], it was found that these slits frequently do not follow grain boundaries, as might be expected, but are transgranular. These observations have clearly indicated the necessity to consider mass transport mechanisms other than grain boundary diffusion.

Copper interconnects technology, which was adopted at the end of 1990s, with new manufacturing process steps and changed combinations of thin film materials, has resulted in a changed microstructure of the metal interconnects, new types of interfaces, and new degradation phenomena observed during accelerated reliability tests of high-performance microelectronic products. The major difference between Al- and Cu-based architectures, which affects interconnect EM resistance, is the presence of interfaces between copper and sidewall metal liners and, more importantly, between copper and capping dielectric diffusion barrier. These interfaces provide additional venues for atom diffusion and hence modify the dynamics of failure development. Numerous experimental studies have indicated that EM-induced degradation and eventually interconnect failure depend on both the interface bonding and the microstructure of the copper interconnect structures [36, 37].

Insitu SEM interconnect degradation studies, EBSD-based copper microstructure studies, and numerical simulations, based on a physical model, have shown that EM-induced degradation mechanisms depend strongly on the bonding strength of interfaces [37, 41]. Particularly, different degradation mechanisms have been described for weak and strengthened top interfaces, i.e., for interfaces with different bonding strength and/or activation energies for atomic transport along interfaces [30, 43]. When the activation energy for interface diffusion along strengthened interfaces becomes comparable to the activation energy for copper grain boundary diffusion [38, 72, 73], mass transport along grain boundaries must be taken into consideration for interconnects with polycrystalline microstructures.

In the early age of the copper interconnect technology, the top copper interfaces were relatively weak, and consequently these interfaces provided the fastest pathways for mass transport. Therefore, it can be understood that only a few papers at that time reported that grain boundary diffusion can be the dominant mechanism for mass transport in copper [74, 75] or can be comparable [76] to the mass transport along interfaces. The more physics-based explanation is that the EM-induced mass transport along the copper interconnect depends on competing activation energies for atomic migrations along the interfaces and the grain boundaries. As long as the activation energy for mass transport along at least one interface is smaller than along the grain boundaries, which is usually the case for the Cu/SiN_x interface in inlaid Cu interconnect structures, the mass transport along the weakest interface dominates the

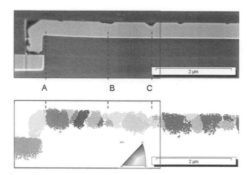

Figure 7.22 EBSD investigation of a cross-sectioned via/line dual-inlaid Cu interconnect structure. Reprinted from [42], with the permission of AIP Publishing.

EM-induced degradation process. In other words, the degradation of copper interconnect is a function of the bonding strength of the weakest interface [77, 78]. In this case, the Cu microstructure is a second-order effect. The strong dependence of the degradation mechanisms on the Cu microstructure occurring in Cu lines with strengthened interfaces was proven experimentally by in situ SEM studies of void evolution in dual-inlaid Cu interconnects [44].

Void formation, movement, and growth in a copper interconnect during an EM test of fully embedded test structures were continuously monitored in an in situ SEM degradation experiment [41]. Typical sequences of interconnect degradation during such an experiment for a standard Cu–SiN$_x$ capping layer interface, and for a strengthened top interface (for a summary of several approaches how to stress the interface, see [43]), show significantly different degradation mechanisms. In case of the standard top interface, the voids were formed initially at the Cu–SiN$_x$ interface (often away from the cathode end), then they moved discontinuously to the cathode end of the interconnect test structure (see Figure 7.22). Eventually a large void grew at the top corner of the via, as shown in Figures 7.23 and 7.24 [79, 80]. CoWP-coated Cu lines are an example of interconnects with strengthened top interfaces, and show void formation at and void movement along the copper–liner interfaces, as shown in Figures 7.25 and 7.26 [35, 80], which was similar to the case of standard unstrengthen top interface but with a lower velocity. In some cases, the mass transport along this interface was reduced to an extent that the copper–liner interface became the dominant diffusion path (see Figure 7.27). Again, discontinuous step-like void movement, void shape changes, void growth, and sometimes void coalescence processes were observed. Eventually, failures occurred either as a result of the dominant void growth perpendicular to the direction of electric current flow or the coalescence of voids moved along the upper and the lower interface, or as a result of the void growth at the bottom of the via (see Figure 7.28) [80]. That means the interface strengthening slows down the mass transport along the top interface of the interconnect line, which indicates a stronger interface bonding. As a result of the previously described interface strengthening, the mass transport along the copper–capping layer interface is not faster than those along the copper–liner interface. This observation suggests that the bonding

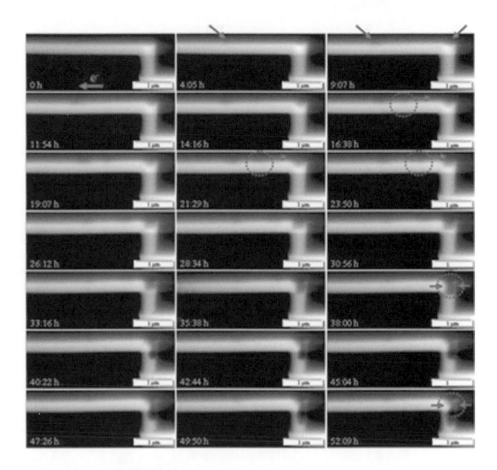

Figure 7.23 In situ SEM image sequence of the cathode region of Cu sample. Reproduced from [80] with permission of M. A. Meyer.

strength of the copper–capping layer interface is at least as high as that of the copper–liner interfaces.

Since the activation energy for mass transport along strengthened top interfaces (e.g., applying an additional CoWP coating) can be similar to the activation energy for mass transport along grain boundaries in copper [35, 81], the Cu microstructure has to be considered in the explanation of EM-induced degradation process in Cu interconnects.

From these experimental studies, it was concluded that electromigration failure is the result of a complicated competition between growth, shape change, and motion of voids. The interaction among these mechanisms was not well understood at the time when these experiments were done. Since the 1970s, several important attempts have been made to model such events. Void motion has been treated, for example by Ho [82], with the result that small voids migrate more rapidly. Nix and Arzt [83] have suggested that larger voids catch up with smaller ones, moving more rapidly, which resulted in a catastrophic mechanism of void growth and failure. As described by Borgesen et al. [84], grain boundaries can trap voids until they reach a critical size.

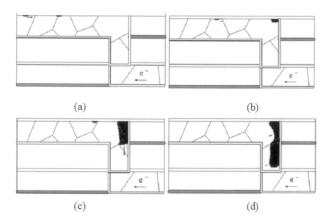

Figure 7.24 Degradation sequence in copper via/line structures with a Cu–SiN$_x$ top interface. (a) Interface diffusion: initial void formation at the copper/capping layer interface; (b) void movement and agglomeration at the line end, inner surface diffusion; (c) void growth into the via from top to bottom, discontinuous process; (d) material redeposition in the upper region of the via. Failure occurs when the remaining cross section is reduced to a critical size. Reprinted from [30], with the permission of AIP Publishing.

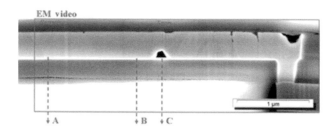

Figure 7.25 SEM image of the cathode region. Reproduced from [80] with permission of M. A. Meyer.

    The first successful attempt to understand the physics underlying the void shape change was by Arzt et al. [58]. These authors, based on the observation that fatal voids have a typical slit or wedge shape with a pronounced asymmetry, as shown in Figure 7.29, with a straight cathode boundary normally oriented to the line, explained that the fast propagation through the linewidth and thickness is driven by EM to move atoms on a void surface, depending on the angle between the surface and the direction of the electron flow. A void with the shape shown in Figure 7.29a can become fatal, because the EM-induced atomic flux from the tip 2 to tip 1 on the anode side surface is larger than from tip 3 to tip 2 on the void cathode side. Due to normal orientation of the cathode side of the void to the direction of the electron flux, there is no EM-induced mass flux from tip 3 to tip 2, which is not the case for the mass flux from tip

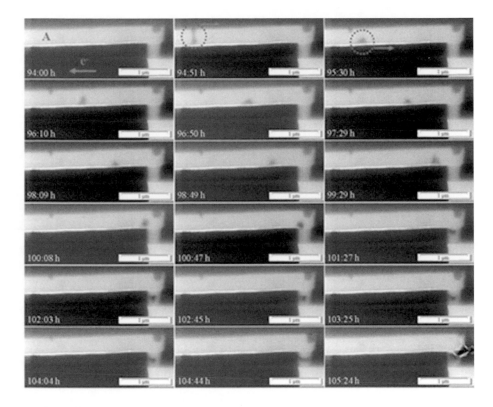

Figure 7.26 In situ SEM image sequence for a CuAl sample. Reproduced from [80] with permission of M. A. Meyer.

2 to tip 1. The tangential component of the electric current is a function of the angle between the void side and direction of the electron flux.

Consequently, mass is removed from tip 2, and the void tends to become fatal, as was observed in the experiment shown in Figure 7.30a. The void with a shape shown in Figure 7.29b is characterized by the zero EM-induced mass flux on the anode void surface from tip 2 to tip 1, but by nonzero mass flux from tip 3 to tip 2. As a result, mass is removed from tip 3 and deposited at tip 2, so the void grows along the line, which corresponds to the experimental observations (Figure 7.30b). An analytical model describing the initial change of the void shape was developed in [58]. For both phenomena, the atomic diffusion on the void surfaces was assumed to be the dominant transport mechanism, and other diffusion paths such as grain boundaries and interfaces have been ignored. The experimental observations that the sidewalls of the voids remained perpendicular to the substrate have justified the validity of the 2D model.

In the next section, we will present a general model describing the void motion and void shape change caused by the applied electric current and outline different numerical techniques that were developed for solving a set of governing equations. A major part of the presented material is based on results reported by Arzt et al. [58, 67–69], Bower et al. [21–23, 85, 86], as well as Mahadevan and Bradley [27, 87].

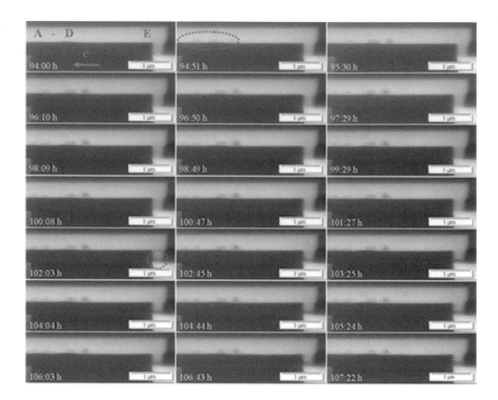

Figure 7.27 In situ SEM image sequence for a CuAl sample. Reproduced from [80] with permission of M. A. Meyer.

7.9   Modeling of EM-Induced Void Motion and Shape Evolution

EM-induced void motion along the metal line as well as change of the void shape is caused by redistribution of atoms along the void surface by means of surface diffusion. Additional impact on the void evolution is caused by atomic exchange between metal and void surface. Indeed, as was shown in Figure 4.20 in Section 4.10, a redistribution of atoms on the surface of the circular void driven by the electron wind force causes the void drift against the electron flow but with the void shape unchanged. Same electron wind force might be responsible for the change of the large circular void shape to the slit-like shape in the narrow lines with width comparable with the initial void size [88]. In both cases, the surface diffusion of atoms is a major mechanism responsible for void evolution. In the absence of electric current, surface diffusion is driven by a variation in chemical potential, which causes atoms to migrate from regions of high chemical potential to those where chemical potential is lower. There are two contributions to the chemical potential of an atom on a free surface. The first is related to the free energy of the surface, while the second is due to the strain energy density of material adjacent to the void surface [89]:

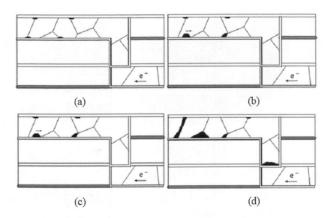

Figure 7.28 Degradation sequence in a copper via/line structures with strengthened Cu/SiN$_x$ top interface: CoWP coating or local alloying with Al. (a) Void formation at the Cu–liner and the Cu–capping layer interfaces, interface diffusion; (b) void movement and agglomeration, void growth, and interface diffusion; (c) void shape modification, pinning at grain boundaries and triple points, void growth, and interface and grain boundary diffusion; (d) failure occurs if a pinned void grows across the entire cross section or if a void reaches the via, thereby reducing the cross section to a critical size. Reprinted from [30], with the permission of AIP Publishing.

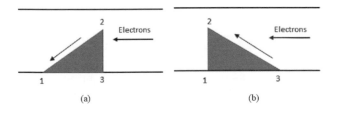

Figure 7.29 Schematic illustration of the interaction between the void shape and the electron wind.

$$\mu_S = \Omega(W - \gamma_S K).\tag{7.93}$$

Here, $W = \sigma_{ij}\varepsilon_{ij}/2$ is the elastic strain energy density; γ_S is the surface energy, which in a general case can be anisotropic; K is the principal curvature of the void surface in the 2D case, and a mean of the sum of the principal curvatures in the 3D case, defined in a way that a concave surface has a negative curvature; and Ω is the atomic volume. In the following discussion, we will assume small elastic deformation, which allows to approximate the chemical potential of a surface atom as $\mu_S = -\Omega\gamma_S K$. Similarly to the bulk atomic flux, we assume that the surface mass transport is driven by a gradient of the atomic chemical potential

$$F_{SD} = \nabla_S\mu_S = -\Omega\nabla_S(-W + \gamma_S K),\tag{7.94}$$

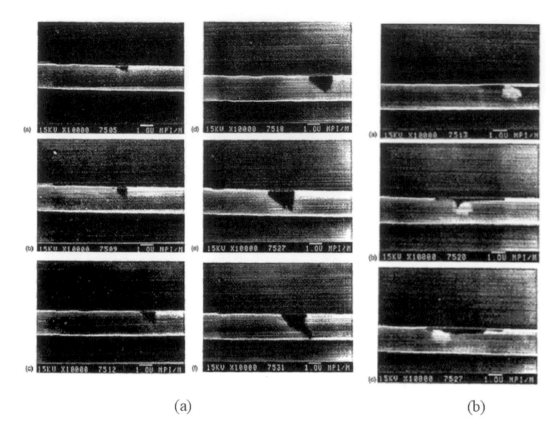

Figure 7.30 Kinetics of evolution of (a) the fatal void (critical void shape) and (b) the void growing along the line with uncritical shape. Electron flow is directed from right to left. Reprinted from [58], with the permission of AIP Publishing.

where s is the arc length along the void surface. Electric current flow gives rise to the additional driving force for diffusion, which is proportional to the component of the current density tangential to the void surface j_S:

$$F_{EM} = -eZ_S^* \rho j_S, \tag{7.95}$$

where Z_S^* is the effective charge number of a surface atom. The atomic flux on the surface, similarly to the bulk atomic flux (see Section 7.2, (7.1)), is to be linearly proportional to the driving force:

$$\Gamma_S = N_S M_S F. \tag{7.96}$$

Here, Γ_S is the volume of material crossing a line of unit length normal to the void surface per unit time; and M_S is the mobility, which relates to the coefficient of self-diffusion on the surface D_S through the Einstein relation $M_S = D_S/k_B T$, where k_B, as before, is the Boltzmann's constant and T is the absolute temperature. The surface atomic concentration N_S is approximated as $N_S = \delta_S/\Omega$, where the parameter δ_S, which was introduced to fix the dimension, is associated with the effective thickness of the diffusion layer [21, 23, 27]. Thus, the total flux of atoms along the void surface may be expressed as

$$\Gamma_S = -\frac{\delta_S D_S}{\Omega k_B T}\left(\nabla_S \mu_S - eZ_S^* \rho j_S\right). \tag{7.97}$$

The rate of atom deposition on or removal from the void surface element is governed by the surface flux divergence. Based on the kinetic law describing the mass conservation condition, the normal velocity of the void surface at any surface point is described as

$$v_n^\nabla = -\Omega \nabla_S \Gamma_S. \tag{7.98}$$

Combining (7.97) and (7.98) provides

$$v_n^\nabla = \frac{\delta_S D_S}{k_B T}\left(-\Omega \gamma_S \frac{\partial^2 K}{\partial s^2} + eZ_S^* \frac{\partial^2 V}{\partial s^2}\right), \tag{7.99}$$

where V is the electric potential. The rate of surface evolution should be adjusted further by introducing a flux of vacancies $\Gamma_{V,B \to S}$, which the void surface is exchanging with the bulk of metal as well as with grain boundaries and interfaces. Vacancies coming to the void surface annihilate lattice sites. In this case, the velocity of the void surface is

$$v_n = -\Omega \nabla_S \Gamma_S - \Gamma_{V,B \to S}. \tag{7.100}$$

Following Suo [56, 89], the equilibrium vacancy concentration on the void surface is

$$N_{VS} = N_0 \exp\left\{\frac{\Omega \gamma_S K}{k_B T}\right\}, \tag{7.101}$$

where, as before, $N_0(T) = \Omega^{-1} \exp\left\{-E_V/k_B T\right\}$. The equilibrium vacancy concentration far away from the void region with the hydrostatic stress σ is

$$N_V = N_0 \exp\left\{\frac{\Omega \sigma}{k_B T}\right\}. \tag{7.102}$$

Comparing formulas (7.101) and (7.102) shows that in the case of

$$\sigma > \gamma_S K, \tag{7.103}$$

the vacancies from the metal regions with the hydrostatic stress satisfying the condition (7.103) diffuse to the void surface, and the void enlarges. This is the same condition that we derived in Section 4.6 for the case of a circular void shape.

Flux of vacancies coming to or from the void surface can be found from the solution to the vacancy diffusion boundary value problem. If the vacancy concentration satisfies the steady-state condition $\partial N_V/\partial t = 0$, then its spatial distribution is a solution of the Laplace equation $\Delta N_V = 0$. We consider an infinite crystal with a circular void of the radius R. As the boundary condition, we take the concentrations of vacancies at the void surface and at locations, which are far from the void, as described by (7.101) and (7.102). We introduce the spherical symmetric coordinate system with the center of the void. The solution to the Laplace equation in the spherical coordinate is $N_V(r) = A + B/r$, where r is the distance from the void center, and constants A and B are found from the boundary conditions (7.101) and (7.102):

$$A = \frac{1}{\Omega} e^{-\frac{E_V}{k_B T}} \exp\left\{\frac{\Omega\sigma}{k_B T}\right\},$$

$$B(s) = R\frac{1}{\Omega} e^{-\frac{E_V}{k_B T}}\left(\exp\left\{\frac{2\gamma_S\Omega}{Rk_B T}\right\} - \exp\left\{\frac{\Omega\sigma}{k_B T}\right\}\right) \approx Re^{-\frac{E_V}{k_B T}}\frac{1}{k_B T}\left[\frac{2\gamma_S}{R} - \sigma\right]. \quad (7.104)$$

Thus, the generalized vacancy flux per unit area from the bulk phase to the surface of an arbitrarily shaped void is taken to be

$$\Gamma_{V,B\rightarrow S} = -D_v\frac{\partial N_V}{\partial r} = -\frac{D_v^i}{k_B T}e^{-\frac{E_V}{k_B T}}K(s)[\sigma - \gamma_S K(s)], \quad (7.105)$$

where D_v^i is the corresponding coefficient of self-diffusion of vacancies in the grain bulk or in the grain boundaries and interfaces, and the curvature K depends on position on the surface, which is determined in 2D geometry by the arc length s along the void surface. Combining equations (7.99) and (7.105) results in the velocity of void surface evolution caused by electromigration:

$$v_n = \frac{\delta_S D_S}{k_B T}\left(-\Omega\gamma_S\frac{\partial^2 K}{\partial s^2} + eZ_S^*\frac{\partial^2 V}{\partial s^2}\right) - \frac{\Omega D_v^i}{k_B T}e^{-\frac{E_V}{k_B T}}K[\sigma - \gamma_S K]. \quad (7.106)$$

It should be mentioned that the employed distribution of hydrostatic stress comes from the solution of coupled equations (7.42)–(7.52). Equation (7.106) provides an additional to (7.49) generation/annihilation term for the vacancies, describing the action of the void surface as a sink/source of the vacancies, which should be accounted for in (7.43). We can see that the formalism described here is similar to one that was accepted by Bower et al.; see, for example, [21, 23] for description of EM-induced evolution of the void and GB.

Equation (7.106) should be combined with the system of PDEs representing the general model of EM-induced stress evolution, which was described in the Section 7.5. By solving them together with the proper boundary conditions at the void surface, such us a zero normal stress, a zero normal electric current component, and zero values for all material and kinetic parameters inside the void region, we can derive the kinetics of void movement and the void shape evolution. Details of the solution procedure will be demonstrated in the following sections.

7.10   Numerical Techniques

One of the first models describing the void growth and change of void shape was developed by Arzt et al. [58]. The analytical model describes the atomic diffusion along the void surface as a function of the current density component tangential to that surface. The diffusion caused by the curvature gradient was ignored as well as the exchange of atoms between void surface and surrounding metal. Analytical expressions for the tangential component of the current density along the perimeter of 2D circular and elliptic voids were employed to solve the void evolution equation (7.106). The solution has confirmed the classical Ho result [82], indicating that the circle void under

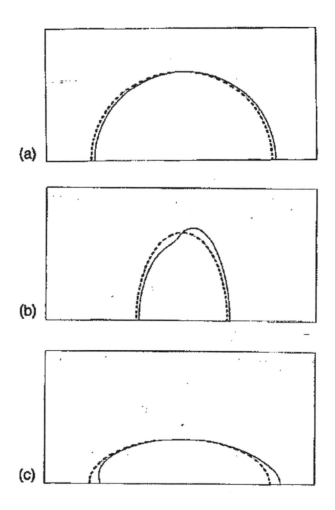

Figure 7.31 Calculated void shape changes for three different initial void shapes (dotted line: initial shape, electron flow from the right to the left): (a) circular void: the void moves without shape change; (b) elliptical void with $a < b$: the void tends to form a flat facet at the cathode end and shows a slight lateral growth; (c) elliptical void with $a > b$: the void tends to elongate and shows no lateral growth. Reprinted from [58], with the permission of AIP Publishing.

the action of DC current moves against the electron flow while keeping the original circle shape (Figure 7.31a). Elliptic voids with longer axes normal to the current lines (transverse ellipse) have demonstrated a shape distortion, which produced a flatter face at the cathode edge, as shown in Figure 7.31b, while a longitudinal ellipse has elongated toward the cathode (Figure 7.31c). All these results have demonstrated the influence of the initial void shape on its evolution. However, the model by itself, due to the conservation of the total number of surface atoms, was not able to explain the void volume growth observed in the experiment. Also, the semi-infinite width of the metal line has yielded a less intense current crowding at the void surface than in a finite width line.

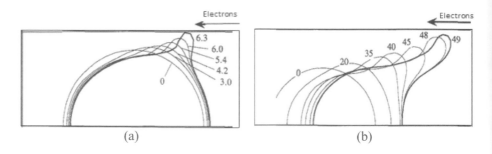

Figure 7.32 Simulation of the development of initially semicircular voids with radii of (a) 0.8 mm and (b) 0.6 mm, in a 1 mm wide line. Applied current density was 2 MA/cm^2 and the time is given in units of 100 s. Reprinted from [88], with permission from Elsevier.

Despite the preliminary nature of the results, the model has demonstrated that diffusion on the void surface is the definite mechanism for void shape change.

Many of the simplified assumptions accepted in [58] have been elaborated in subsequent models described in [87–92]. In [88], Kraft et al. modeled 2D void evolution by accounting for the mass transfer on the void surface induced by the electromigration and curvature gradient. A finite width of the metal line was used in the FEA simulation of distributions of electric current density and temperature distribution in the vicinity of the void. Current density component tangential to the void surface and temperature resolved in the FEA simulation were introduced to each surface node of the developed finite-difference (FD) scheme analysis to calculate the void surface propagation. At each time step, the evolved void surface was remeshed and FEA/FD simulations were iterated until either a fatal void was formed or noncritical void development was established. Since the model did not include the atom exchange between void surface and metal, i.e., the surface atom conservation was assumed, the void volume (the void area in the 2D simulation) was conserved. To account for the experimentally observed effect of the void size on shape change and void migration, a void growth with similar shape was artificially modeled by enlarging the void at each time step to fit the experimentally measured growth rate. It was found that the shape of an initial large semicircular transgranular void would evolve to a wedge-shaped void (Figure 7.32a), while the smaller void had a more pronounced shape change, leading to a slit-like failure (Figure 7.32b).

So far, the simulation results have shown that growth, motion, and shape change of voids are competing processes. In general, fast growth tends to suppress the other two mechanisms and as a result, an extended wedge void with a typical asymmetric shape is obtained. By contrast, a slowly growing void can accommodate more motion and shape change and becomes slit-like, rather than wedge-shaped. Another important result of these simulations has demonstrated that void shape changes tend to reduce the lifetime of the line, which can explain why the slit-like voids are attributed to the early failure. Indeed, a mass exchange between the void and the surrounding metal for void growth begins when the EM-induced stress in the void vicinity exceeds a threshold, which depends on the original void size. This can take a long time, months if not years. Thus, before the threshold stress is developed, the void evolves through the drift or shape change. In this case, the narrowest lines loaded with large current densities could fail in days or even hours. It was argued in [87], however, due to the requirement of void area conservation [88], the growth of a slit-shaped protuberance assumes the original void would take up mass. The protuberance could not become too narrow because the surface energy cost would be prohibitive. Thus, the size of the original void in general limits the length of slit-shaped voids that could be formed, and only large voids could create slits long enough to produce line failure. Therefore, if large interface cavities are difficult to form in a mature interconnect processing, the origin of slit-shaped voids remains unclear, unless large stress-induced voids are formed to relax the thermal stress [93]. It was suggested that if the transport of mass along the undisturbed line edge (interfaces) is accounted for, the growth of a fatal slit can occur from a small perturbation. In addition, as was mentioned in [88], the numerical FD scheme had a problem with the stability of the void shape calculation, so the time step used in the iteration scheme had to be very small in order to link with FEA simulations of the current density and temperature.

To address these simulation problems, more advanced numerical techniques were employed to simulate the void evolution kinetics. As was discussed in the previous section, the principal challenge in modeling interconnect failures is that void growth occurs as a result of mass transport, driven by the electric current and stress in the lines. Consequently, it is necessary to track the evolution of the reference configuration as part of the solution. Three methods were employed to do this: in one approach, the surfaces and interfaces in the solid are idealized as perfectly sharp boundaries, which are computed and updated using some front tracking algorithms [23, 85, 86, 94–102]. In these studies, the moving interface is represented and tracked explicitly using markers distributed on the interface. The second approach, the so-called phase field method, tracks the evolving reference configuration by solving for an order parameter field, or phase field, being constructed so that zero contours of the phase field represent moving interface [21, 27, 29, 103]. The third approach, the so-called level-set method, introduces a continuous additional function, the level-set function, over the whole simulation domain, and embedded the interface as the zero level set of the higher-dimensional function [104–107]. In the two latter approaches, the interface is implicitly captured by a certain globally defined scalar function. All these approaches have their own advantages. Thus, the sharp interface models are relatively straightforward to implement, at least in two dimensions, and are computationally efficient. In contrast, implicit models have the advantage of easily accounting for boundary topology changes, such as merging curves, and can be more easily implemented in 3D than the sharp interface models. Despite some differences, both the phase-field and level-set methods are similar in defining the interface in an implicit manner [108]. The following discussion will be focused on the use of the sharp interface approximation and phase-field method to analyze void evolution.

7.10.1   Sharp Interface Approximation

A numerical procedure to describe the void evolution subjected to the perfectly sharp boundary approximation was explained in details by Xia et al. in [85]. A finite element procedure should be used to solve the equation for the void evolution presented in the previous section together with the equations from Section 7.5 describing the current density distribution and evolution of strain and stress everywhere inside simulation domain. Additional boundary conditions for void surface reflecting the zero traction and vanishing normal component of the current density are implemented calculating the distributions of the voltage and stress fields. The equations describing surface diffusion and the vacancy exchange between the void surface and surrounding material are solved in order to get the change in the void shape. The initial shape of the void in stress-free condition and absence of electrical load, the so-called reference configuration, is given. Surface velocity $v_n(t)$, for the case of conservation of the total mass of solid, can be calculated by writing a weak form of (7.106):

$$\int_{S_V} v_n \delta v_n ds = \int_{S_V} \left(\frac{\delta_S D_S}{k_B T} \left(-\Omega \gamma_S K + e Z_S^* V \frac{\partial^2 V}{\partial s^2} \right) \frac{\partial^2 \delta v_n}{\partial s^2} - \frac{\Omega D_v^i}{k_B T} e^{-\frac{E_V}{k_B T}} K [\sigma_S - \gamma_S K] \delta v_n \right) ds.$$

(7.107)

Here S_V is the void surface. To compute the change of the void shape, (7.107) is integrated with respect to time. Details of this computation, for the case when the vacancy exchange between void surface and material, were omitted can be found in [85], including an estimate of the evolution kinetics of void surface curvature by using a semi-implicit Euler scheme for a 2D geometry, and in [86] for a 3D geometry. Detailed description of the mesh generation procedure can be found in [23, 85, 109]. Reduction of (7.107) to a system of linear equations, by using a specific interpolation technique, allowed one to solve it for the nodal displacements. The coordinates of nodes can be updated on the void surface and used as new control points to specify the void shape at time Δt. Then a new mesh can be generated with the procedure repeated to provide the progressive change in the void shape.

The developed simulation scheme was applied to analyze several interesting cases [85]. In the first case, the solid was free of stress and electric current loads, and thus the surface atomic diffusion was driven by the surface free energy only. It was shown that a cavity with initial elliptical profile would evolve due to diffusion driven by the surface curvature gradient into the circle-shaped void characterized by a minimal surface energy. In the second case, the void evolution was analyzed in a stress-free electrically conducting strip of length L and height H with the voltage drop of V_0 applied to the strip ends and a cavity with radius R_0 that lies on the symmetry axis of the strip. The result as shown in Figure 7.33 is in agreement with previous studies of small circular voids with radii much smaller than the strip dimensions drifting in the current direction with its shape unchanged [82, 88]. Similar to [88], an increase of the initial void size R_0 was found to generate a void shape change (Figure 7.34). The results confirmed that the void shape is governed by two dimensionless numbers: the ratio of void size to linewidth R_0/H and the ratio of the electromigration driving force

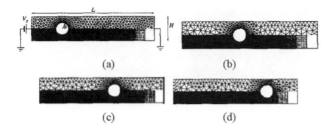

Figure 7.33 Stable migration of a void in a stress-free, electrically conducting strip:
(a) $t/t_0 = 0$, (b) $t/t_0 = 0.1581$, (c) $t/t_0 = 0.3084$, and (d) $t/t_0 = 0.4544$,
$\chi = 10, R_0/H = 0.25, L/H = 5$. The characteristic time is $t_0 = a^4/D_S\Omega\gamma_S$. Reprinted
from [85], with permission from Elsevier.

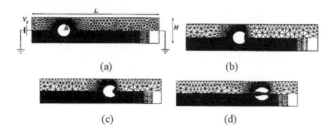

Figure 7.34 A void in a stress-free, electrically conducting strip, as it collapses into
slits: (a) $t/t_0 = 0$, (b) $t/t_0 = 0.0467$, (c) $t/t_0 = 0.0936$, and (d) $t/t_0 = 0.1362$,
$\chi = 30$, $R_0/H = 0.25, L/H = 5$. Reprinted from [85], with permission from Elsevier.

(7.95) to the driving force caused by variation in the surface energy (7.94) [89, 110].
Indeed, the former force prevails over the later one, when

$$eZ_S^* E_S > -\Omega\gamma_S \frac{\partial K}{\partial s} = \Omega\gamma_S \frac{1}{R^2}\frac{\partial R}{\partial s}. \tag{7.108}$$

Here, $R = 1/K$ is the void radius, and $E_S = -\rho j_S$ is the tangential component of the
electric field, which are functions of the surface coordinates. It is clear from (7.108)
that the void with initial smooth shape will collapse into slit-like shaped void if

$$\frac{\partial R}{\partial s} = \chi = \frac{eZ_S^* E_S R^2}{\Omega\gamma_S} > 1. \tag{7.109}$$

In the case of initially circular void, the dimensionless number χ takes the form
derived in [89, 110]

$$\chi = \frac{eZ_S^* E_S R_0^2}{\Omega\gamma_S}. \tag{7.110}$$

Effect of stress distribution in the line on the void shape was analyzed in [85],
where the outer boundaries of the electric current free segment were loaded with
normal tractions to generate uniform normal stress components far from the void

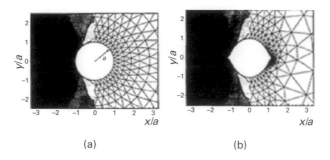

(a)   (b)

Figure 7.35 Evolution of cracks from an initially circular void subjected to remote stress: $\Lambda = 0.6, \sigma_{xx} = 0, \sigma_{yy} = \sigma_0, t/t_0 = 0.00513$. Reprinted from [85], with permission from Elsevier.

(Figure 7.35). Variations in strain energy around the void surface generated by nonuniform stress–strain distribution cause surface atoms to migrate along the gradient of the chemical potential (7.93). A competition between the elastic strain energy and surface energy driving forces causes the change of the void shape. A ratio of these forces, as shown in [85], can be written as

$$\Lambda = \frac{\sigma_\infty^2 R}{E\gamma_S}. \tag{7.111}$$

Here E is Young's modulus and σ_∞ is the principal stress remote from the void. It was demonstrated that in the case of a small Λ, the surface energy is dominated and the initially circular void evolves to an elliptical shape. In the case when Λ exceeds some critical value, the void shape evolves to a wedge shape oriented normally to the maximal stress, as shown in Figure 7.35 [85, 111, 112].

When both the stress and electric current are loaded, the evolution of the void shape shows even more complex behavior [85]. One such evolution was caused by the voltage drop of V_0 applied on strip ends, and normal traction σ_{xx} is shown in Figure 7.36, where the shape evolution is governed by both parameters χ and Λ as well as the geometries R_0/H and L/H. For large values of χ and Λ, both the electromigration and surface energy driving forces generate the slit or crack development (Figure 7.36c and d). A variety of cases characterized by different relations among these parameters were analyzed in [85] and demonstrated different kinds of shape evolution, from wedge and slit-shaped to crack-like voids propagating along the current flow.

The preceding results were obtained with the advanced sharp boundary approximation, which demonstrated a good predictive capability of the simulation model for void shape evolution. However, a number of simplifying assumptions used in this model have limited its capability to predict EM-induced failure in certain interconnect segments. These included ignoring the inelastic deformation and the mass transfer between void surface and surrounding material, as well as the role of grain boundaries and interfaces in atomic exchanges with the void surface. Ignoring the self-consistent

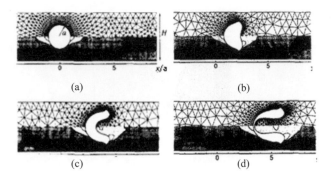

Figure 7.36 A void collapsing into cracks under combined stress and electric field:
(a) $t/t_0 = 0$, (b) $t/t_0 = 0.0805$, (c) $t/t_0 = 0.1497$, and (d) $t/t_0 = 0.2079$, $\chi = 10$,
$\Lambda = 0.2$, $\sigma_{xx} = \sigma_0$, $\sigma_{yy} = 0$. Reprinted from [85], with permission from Elsevier.

modeling of the stress evolution due to applied electric current and stress relaxation
due to void evolution is major drawback for employing this model for EM assessment.

All of these modeling drawbacks have been addressed later by Bower and Shankar
in [23]. Evolution of EM-induced stress was described (as discussed in Section 7.6) by
generation of the inelastic deformations inside grain boundaries due to divergences of
atomic fluxes driven by a combination of a gradient in chemical potential and an
electron wind force. An atom exchange between grain boundaries and void surface
has been accounted for, thus removing a limitation on the void volume conservation.
The grain boundary migration and shape change were also computed taking into
account constraints at the triple junctions and junctions between grain boundaries,
void surfaces, and interfaces. Several improvements of the numerical technique were
implemented to improve the existing problems with solving the diffusion equations
and tracking the boundary topology changes [23].

The model and the simulation technique were tested on a simple case of a single-
level interconnect consisting of two grains separated by a grain boundary, with
additional pathways for mass transfer represented by immobile interfaces between
interconnect and passivation. This is shown in Figure 7.37, where an initial semicircular void with the junction between the void surface and the grain boundary is
located at one of the interfaces. The simulation showed that different void evolutions
can be observed depending on model parameters: in Figure 7.37, the void migrates
together with the grain boundary while remaining connected to each other, and in
Figure 7.38 the void migrates and detaches from the grain boundary. It was demonstrated also that the void can grow or shrink while migrating.

The phenomenon of void trapping by grain boundary and then detaching has been
observed in a number of experimental studies; see, for example, [64, 67]. In one case,
a void was observed to break away from the grain boundary when the electric field
exceeds a critical value depending on the void size [110, 113], and the critical product
of the electric field and void size was estimated for the circular void in [89]. Numerical
calculations showed that voids with evolving shapes can break away from the grain

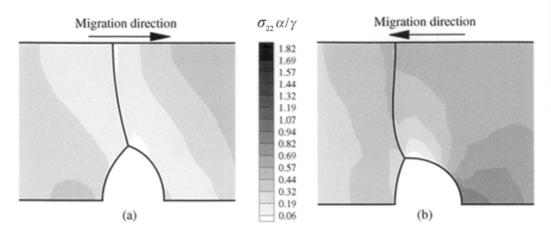

Figure 7.37 The steady-state profile of a void migrating together with a grain boundary. In each case, the electric current flows from right to left. In case (a), the grain boundary is mobile, and so matter can cross the grain boundary. In case (b), the boundary is immobile, and matter flows across the triple junction only in response to curvature differences of the two surface segments. In this case, the migration direction is reversed [23]. © IOP Publishing. Reproduced with permission of Allan F. Bower. All rights reserved.

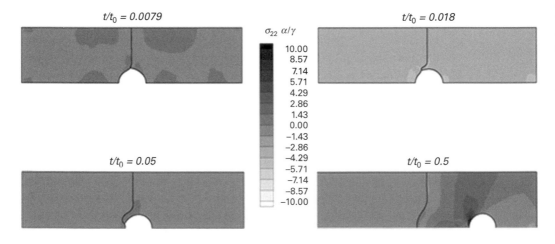

Figure 7.38 A void escaping from a grain boundary [23]. © IOP Publishing. Reproduced with permission of Allan F. Bower. All rights reserved.

boundary, as reported in [23]. In [113], the void detachment time was estimated based on consideration of the energy that should be consumed for restoration/creation of the grain boundary area (Figure 7.39). The analysis can be further improved, however, by accounting for the anisotropy in surface energy and atomic diffusivity, the inelastic deformation inside the grain bulk, and, as we mentioned before, an equilibration of the vacancy concentration with stress everywhere inside metal interconnects, including

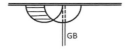

Figure 7.39 When a void has drifted from the original location, the GB of the area $\pi R^2/2$ will be restored at the cost of $\Delta E_{GB} = \gamma_{GB}\pi R^2/2$.

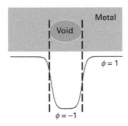

Figure 7.40 Phase-field function in voided metal.

the grain interiors, grain boundaries, and interfaces, should be accounted for. In general, the most serious limitation of the sharp boundary approximation methodology is that it is restricted mainly to analysis of 2D cases, although some attempts to extend the computations to 3D have been made [86].

7.10.2 Phase-Field Method

As we discussed in the previous section, the sharp interface model and surface tracking technique, when applied for analyzing an electromigration void evolution, can provide realistic results to account for experimental observations. However, this approach is characterized by certain complexities due to the need of mesh adaptation around the void surface and implementation of the proper boundary conditions at the propagating front nodes at each iteration step. This limits the capability of the model in treating the topological changes caused by a coalescence of two or more voids or a fragmentation of a large void into smaller voids. These factors have inspired the development of new, technically simpler, but still accurate approaches to tracking the void shape evolution and void migration. The phase-field method is one of them. The method tracks the time-space void evolution by solving the governing PDEs for new field variable, called an order parameter $\phi(\vec{r}, t)$, tracking the changes in the phase or the state of material of the voided interconnect metal. Since the voided interconnect is a two-component system containing one region occupied by the voids and another region occupied by a metal, it is customary to set $\phi = -1$ inside the void and $\phi = 1$ everywhere in the metal (Figure 7.40). The order parameter varies smoothly between -1 and $+1$ in the narrow interfacial layer, which represents the void surface. The same order parameter $\phi(\vec{r}, t)$ describes the spatial evolution of the mechanical and electrical properties of the voided interconnect. Together with setting the electrical conductivity η, Young's modulus E, and Poisson's factor ν as

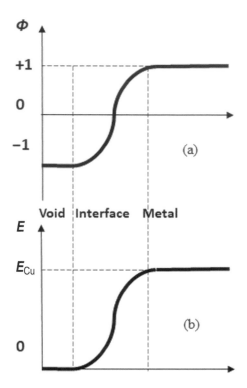

Figure 7.41 Diffuse interface between the void region and metal: (a) variations of the order parameter ϕ, and (b) Young's modulus E.

$$\eta(\phi) = \eta \frac{1 + \phi}{2}, \quad E(\phi) = E \frac{1 + \phi}{2}, \quad v(\phi) = v \frac{1 + \phi}{2}, \tag{7.112}$$

one can assign all original parameters to the metal regions and vanish inside void regions, as shown in Figure 7.41. In the case when a thermal load is considered, similar settings should be applied for the thermal properties of the voided interconnect. Then the motion of the void interface $\phi(\vec{r}, t)$ can be deduced by solving the appropriate governing equation for the order parameter, either the Cahn–Hilliard equation originally proposed for spinodal decomposition [114, 115] or the Allen–Cahn equation [116] (a modified time-dependent Ginsburg–Landau equation [117, 118]), which is coupled with the electrical and mechanical PDEs through (7.112). In this way, by solving the coupled PDEs for the order parameter, vacancy concentration, electric current and stress, allows the void evolution to be computed with the fixed finite element mesh while avoiding the discretization and explicit tracking of the void surface. Obviously, this phase-field formulation of void evolution provides the same results as those obtained in the sharp interface approximation in the limit of zero diffuse interface width. A detailed asymptotic analysis of recovering the sharp interface model in the limit of a zero-width diffuse interface was published in [21, 27].

The phase-field technique has been used with success in obtaining accurate results in a variety of problems involving mobile interfaces, such as the aforementioned spinodal decomposition [114, 119, 120], dendritic solidification [121, 122], two-phase fluid flow [123], and motion of the activated reaction front [124]. A major development to implement the phase-field model for tracking the electromigration-induced void evolution has been performed in [21, 27, 87].

For electromigration-induced void evolution, the order parameter ϕ is governed by a combination of the extended Cahn–Hilliard and Allen–Cahn equations describing the diffuse motion of the void surface due to both the surface and bulk diffusions. To understand the underlying physics, we briefly review the nature of the formulation starting with the Cahn–Hilliard equation. Following [114, 115], let us consider an isotropic solid solution with the nonuniform composition C. The local free energy per atom in a region of nonuniform composition depends on the local composition and its spatial derivatives $f(C, \nabla C, \nabla^2 C, \ldots)$. Assuming that C and its derivatives are independent variables, we can expand f in a Taylor series about f_0, which is the free energy per atom of a solution of uniform composition C:

$$f(C, \nabla C, \nabla^2 C, \ldots) = f_0(C) + \sum_i L_i \frac{\partial C}{\partial x_i} + \sum_{ij} K_{ij}^{(1)} \frac{\partial^2 C}{\partial x_i \partial x_j} + \frac{1}{2} \sum_{ij} K_{ij}^{(2)} \frac{\partial C}{\partial x_i} \frac{\partial C}{\partial x_j} + \cdots,$$

(7.113)

where

$$L_i = \left(\frac{\partial f}{\partial (\partial C / \partial x_i)} \right)_0, \quad K_{ij}^{(1)} = \left(\frac{\partial f}{\partial (\partial^2 C / \partial x_i \partial x_j)} \right)_0, \quad K_{ij}^{(2)} = \left(\frac{\partial^2 f}{\partial (\partial C / \partial x_i) \partial (\partial C / \partial x_j)} \right)_0.$$

(7.114)

By assuming that the local free energy f is a function only of f_0 and the composition derivatives, and since f is a scalar, it must be invariant with respect to the direction of the gradient, thus only terms in even powers of the gradient can appear. Hence, we have

$$f(C, \nabla C, \nabla^2 C, \ldots) = f_0(C) + k_1 \nabla^2 C + k_2 (\nabla C)^2 + \cdots. \tag{7.115}$$

The total free energy of the solid solution can be obtained by integration (7.115) over its volume

$$F = \int_V \left[f_0(C) + k_1 \nabla^2 C + k_2 (\nabla C)^2 + \cdots \right] dV. \tag{7.116}$$

Further, integrating the term $k_1 \nabla^2 C$ by parts, we obtain

$$\int_V k_1 \nabla^2 C \, dV = - \int_V \frac{\partial k_1}{\partial C} (\nabla C)^2 dV + \int_S \left(k_1 \nabla C \cdot \vec{n} \right) dS, \tag{7.117}$$

where the last integral is taken over an external surface, and $\vec{n}$ is the unit vector normal to the surface. Since we are not interested in the effects at the surface, we can choose the boundary of integration in a way to vanish the surface integral; then the total free energy can be written as

$$F = \int_V \left[f_0(C) + \frac{k}{2}(\nabla C)^2 + \cdots \right] dV, \tag{7.118}$$

where $f_0(C)$ is the free energy of a unit volume of the homogeneous material with composition C, and $k(\nabla C)^2$ represents the increase in free energy due to composition gradient, giving rise to the surface tension, $k/2 = -\partial k_1/\partial C + k_2$. Since the analysis performed in [114, 115] was dealing with the effect of infinitesimal composition fluctuations on the stability of an initially homogeneous solution, the gradients will also be infinitesimal, so keeping the second term would be sufficient to describe the contribution from the transition layers between the regions different in composition. We can then expand $f_0(C)$ near the average composition C_0 as

$$f_0(C) = f_0(C_0) + (C - C_0)\left(\frac{\partial f_0}{\partial C}\right)_{C=C_0} + \frac{1}{2}(C - C_0)^2 \left(\frac{\partial^2 f_0}{\partial C^2}\right)_{C=C_0}. \tag{7.119}$$

In a homogeneous binary solution, $\partial f_0/\partial C$ is proportional to the difference in the chemical potentials of the two components, corresponding to the change in free energy caused by replacement of some amount of one component by the other. The composition change will also change the local gradient. Let us consider a variation in free energy δF caused by a composition variation δC

$$\delta F = \int_V \left\{ \frac{k}{2}\left[(\nabla(C + \delta C))^2 - (\nabla C)^2 \right] + [f_0(C + \delta C) - f_0(C)] \right\} dV$$

$$\approx \int_V \left[\frac{k}{2}(2\nabla C \cdot \nabla \delta C) + \frac{\partial f_0}{\partial C}\delta C \right]. \tag{7.120}$$

After integrating the first term by parts and setting the integral taken over an external surface to vanish, it takes the form

$$\delta F = \int_V \left(\frac{\partial f_0}{\partial C} - k\nabla^2 C \right) \delta C \, dV. \tag{7.121}$$

The quantity in the brackets is the change in local free energy due to a local change of composition δC. When it is negative, the solution becomes unstable to infinitesimal fluctuations in composition, and the decomposition will happen. For more accurate estimation of the free energy change, the strain energy associated with the composition changes should be included. Then the free energy variational derivative takes the following form [114]:

$$\frac{\delta F}{\delta C} = \frac{\partial f_0}{\partial C} + \frac{2\eta^2 E}{1 - \nu}(C - C_0) - k\nabla^2 C. \tag{7.122}$$

Here, η is the linear expansion per unit composition change, and E and ν are Young's modulus and Poisson's ratio for the isotropic material. By introducing an effective mobility M, we derive the flux of any atomic component in the solid solution as

$$\vec{\Gamma} = -M\vec{\nabla}\left[\frac{\partial f_0}{\partial C} + \frac{2\eta^2 E}{1-\nu}(C-C_0) - k\nabla^2 C\right]. \tag{7.123}$$

For a conservative system, a divergence of the component flux results in the change of the local composition as described by the continuity equation, i.e., the Cahn–Hilliard equation:

$$\frac{\partial C}{\partial t} = -\text{div}\vec{\Gamma} = \nabla M\nabla\left(\frac{\partial f_0}{\partial C} + \frac{2\eta^2 E}{1-\nu}(C-C_0) - 2k\nabla^2 C\right)$$

$$= M\left(\frac{\partial^2 f_0}{\partial C^2} + \frac{2\eta^2 E}{1-\nu}\right)\nabla^2 C - 2Mk\nabla^4 C. \tag{7.124}$$

Here the term $M\left(\frac{\partial^2 f_0}{\partial C^2} + \frac{2\eta^2 E}{1-\nu}\right)$ is the stress-dependent diffusion coefficient, while Mk is the thermodynamic correction factor for the evolving surfaces [114]. If we know how the free energy depends on the solution composition $f_0(C)$, then (7.124) can be used to track the evolution of the incipient surface governed by volume diffusion of the components.

As discussed in [115], in the case of two isotropic phases α and β of compositions C_α and C_β separated by a flat interface of area S, the free energy of the nonequilibrium material with composition between C_A and C_B can be represented by a continuous function $f_0(C)$. The analysis in [116] showed that the composition profile across the interface is sigmoidal in shape with the interface thickness w as

$$w = 2(C_\alpha - C_\beta)\left(\frac{k}{\Delta f_{\text{max}}}\right)^{1/2}, \tag{7.125}$$

where Δf_{max} is the local maximum value of $\Delta f(C) = f_0(C) - [C\mu_B(e) + (1-C)\mu_A(e)]$. Here, $\mu_A(e)$ and $\mu_B(e)$ are the chemical potentials per molecule of the species A and B in the α and β phase respectively.

When the composition C is not a conserved, as for example in the case of adsorption of atoms on the solid surface, then instead of the Cahn–Hilliard equation (7.124), the change of the local composition is described by the Allen–Cahn equation [116]:

$$\frac{\partial C}{\partial t} = -\alpha\frac{\delta F}{\delta C} = -\alpha\frac{\partial f_0}{\partial C} + 2\alpha k\nabla^2 C. \tag{7.126}$$

In the general case, the order parameter $\phi(\vec{r}, t)$ indicating the phase or the state of material is used instead of the composition C. As shown in [125], this formalism can be applied to the phase-field approximation by introducing the following relation between C and ϕ: $2C = (C_\alpha - C_\beta)\phi + (C_\alpha + C_\beta)$ or

$$\phi = \frac{2(C - C_\beta)}{C_\alpha - C_\beta} - 1. \tag{7.127}$$

It can be seen from (7.127) that $\phi = -1$ when $C = C_\beta$ and $\phi = 1$ when $C = C_\alpha$, as shown in Figure 7.42.

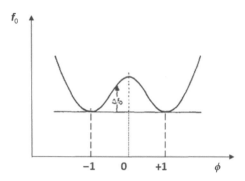

Figure 7.42 Free energy per unit volume f_0 is an even function of long-range order parameter ϕ.

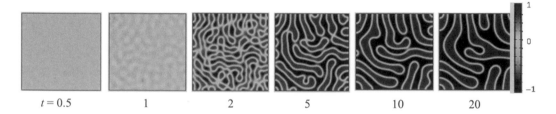

$t = 0.5$ 1   2   5   10   20

Figure 7.43 Phase separation. The evolution of the order parameter calculated with Comsol Multiphysics [20], t is time in arbitrary units.

The Cahn–Hilliard and Allen–Cahn equations take the corresponding forms of (7.128) and (7.129):

$$\frac{\partial \phi}{\partial t} = \nabla M \nabla \left(-\varepsilon^2 \nabla^2 \phi + \frac{\partial \Delta f_0}{\partial \phi} \right) \tag{7.128}$$

$$\frac{\partial \phi}{\partial t} = -\alpha \frac{\partial \Delta f_0}{\partial \phi} + 2\alpha k \nabla^2 \phi. \tag{7.129}$$

Here, Δf_0 is the free energy difference between a homogeneous state of an arbitrary order parameter and that with $\phi = \pm 1$.

After deriving the equations tracking the evolution of the order parameter, we should discuss how they can be employed to analyze the void evolution induced by electromigration. However, the implementation of the Cahn–Hilliard equation is not straightforward, since it describes the continuous transformation of a metastable phase to a more stable phase induced by a small composition fluctuation spread over a large volume, as shown in Figure 7.43. In our case, this metastable phase is a homogeneous solution of vacancies in the metal subjected to EM-induced mechanical stress, which generates large-scale variations in the vacancy concentration accompanied by small-scale variations induced by sharp stress variations between grains with different crystallographic orientations. Such fluctuations in vacancy–metal composition could initiate a transformation to a more stable phase consisting of voids inside the metal if not an extremely small vacancy concentration. As was estimated in [114], the solution became unstable with respect to composition fluctuations when $\partial^2 f_0 / \partial C^2 = -8\pi^2 k / L^2$, where L is the largest linear dimension of the system. Since the parameter k is proportional to the vacancy concentration, there is a minimal concentration limit required for initiating a decomposition. In real cases, the vacancy concentration is much smaller than this limit, and even in the case of high concentration of vacancies near the metal melting point the phase decomposition will not necessarily proceed by a spinodal mechanism. As discussed in [114], such a mechanism is slower than the nucleation and growth mechanism. As we already discussed, the formation of a multivacancy agglomeration at normal conditions is highly unlikely. Also, the Cahn–Hilliard equation (7.128) does not describe nucleation but rather the monoto-nical decrease in the free energy due to a diffusion process and doesn't permit the small but finite fluctuations necessary for nucleation [114]. The Cahn–Hilliard and Allen–Cahn equations describe the interfacial motion caused by different physical phenomena and thus can be in principle employed to describe electromigration-induced postnucleation void evolution. It was argued in [125] that while the original derivation of the Cahn–Hilliard equation (7.128) was based on the volume diffusion of the order parameter, the phase-field approach can be applied to interface motion induced by surface diffusion by solving the Cahn–Hilliard equation with an appropri-ate ϕ-dependent diffusional mobility $M(\phi)$. It was proposed to approximate surface diffusion by introducing a narrow interfacial band characterized by finite diffusional mobility and zero mobility everywhere outside. This idea was employed by Mahadevan et al. [27, 87] and Bhate et al. [21] to describe the shape evolution of electromigration-induced voids. We will discuss their approaches and the major results obtained with the phase-field method later in this section.

Another approach to capture implicitly the EM-induced void evolution was pro-posed in [126, 127] with the atomic concentration used as the order parameter with a specific value, N_C, to determine the void surface in the concentration space. All the physical parameters used in this approach, such as mechanical properties, electrical conductivity, and atomic diffusivity, were assumed to vanish inside the void regions where $N < N_C$. A narrow band surrounding the void surface with the relative atomic concentration $N_C \leq N \leq N_C + \delta N$ with $\delta N \approx 10^{-5}$ was characterized by a much higher atomic diffusivity than the surrounding metal. In this model, the tangential electron wind force was the only driving force for the atomic surface migration. The tangential component of the electric field E_τ was obtained by solving the Laplace's equation everywhere in the simulation domain as $\vec{E}_\tau = E_x \vec{\tau}_x + E_y \vec{\tau}_y$, where $\vec{\tau}_x$ and $\vec{\tau}_y$ are components of the unit tangential to the void surface as $\tau_x = N_y \theta / \sqrt{N_x + N_y}$ and $\tau_y = -N_y \theta / \sqrt{N_x + N_y}$, and $\theta(N)$ is a smoothed step function of N, discrimin-ating the void surface region $N_C \leq N \leq N_C + \delta N$ from the surrounding metal: $\theta(N_C) = 1$, $\theta(N \geq N_C + \delta) = 0$.

Figure 7.44 shows the simulated line edge drift taking place when the electron wind and hydrostatic stress gradient were taken to be the driving forces for atomic volume migration. This picture fits well with the experimental data, demonstrating that material removal was initiated at the upper-left corner and then propagated to the right and down to completely deplete the cathode and caused the strip edge to drift [8].

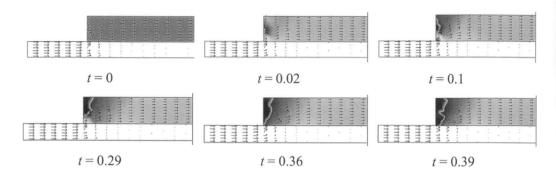

$t = 0$ $t = 0.02$ $t = 0.1$

$t = 0.29$ $t = 0.36$ $t = 0.39$

Figure 7.44 Simulated an EM-induced drift of the edge of the Blech strip test structure (Al strip on a top of TiN line) with the simulation setup described in [127]. Time is in the arbitrary units.

Figure 7.45 Simulated void growth when stress gradient–induced flux was ignored.

Figure 7.46 Simulated void evolution in the via-line test structure with the weak top interface in the presented grain boundaries, with the simulation setup described in [126].

Interestingly, a completely different picture of the degradation was obtained when the stress gradient–induced atomic flux was ignored, as shown in Figure 7.45, where a simulated void growth was found to replace the edge drift. Here the atomic flux driven by the electron wind force has resulted in a void nucleated at the lower-left corner, where the electron flux entered the Al strip, and then the void growth continued along the Al–TiN interface. This example demonstrates the importance of including all the forces driving atom migration in predicting voiding-induced failure.

Figures 7.46 and 7.47 provide additional examples of void drift and shape evolution obtained for the via-line test structures with a weak top interface between metal and the dielectric barrier. Figure 7.46 shows the void nucleated at the triple points between grain boundaries and the top interface and then drifted toward the cathode end of the test structure. Figure 7.47 shows evolution of the void nucleated at the top-right corner of the cathode end of the line where no grain boundaries were presented. The void grew along the metal line and down toward the via bottom.

More accurate techniques for analyzing the EM-induced void evolution have been described in [21, 27, 87, 128, 129], where a diffuse interface model was employed. As

Figure 7.47 Simulated void evolution in the same test structure as shown in Figure 7.46 when no grain boundaries were presented. Electrons' flow is from right to left in both cases. © 2005 IEEE. Reprinted, with permission, from [126].

shown in [20, 127], the free energy functional depends on the order parameter, and its gradient and strain are as follows:

$$F = \int_V \left[\frac{2\gamma}{\pi\varepsilon} \left(f_0 + \frac{1}{2}\varepsilon^2 |\nabla\phi|^2 \right) + W(\varepsilon, \phi) \right] dV. \tag{7.130}$$

Here, $f_0(\phi)$, as before, is the free energy of a unit volume of a homogeneous, unstrained material with an order parameter ϕ; γ is the surface energy; ε is used to replace the original parameter κ to control the thickness or width of the interfacial layer associated for the void surface; and $W(\varepsilon, \phi)$ is the strain energy density replacing the strain energy associated with the composition variation in (7.122). Different mathematical representations used for the free energy function were accepted by different authors. Here we use a quartic double well function $f_0(\phi) = \left(1 - \phi^2\right)^2/2$ with equal minima at $\phi = \pm 1$ as in [21, 27, 129]. Similar to (7.122), the chemical potential of an atom in the interfacial region has the form

$$\mu_s = 2\Omega \frac{\delta F}{\delta \phi} = 2\Omega \left[\frac{2\gamma}{\pi\varepsilon} \left(\frac{\partial f_0}{\partial \phi} - \varepsilon^2 \nabla^2 \phi \right) + \frac{\partial W(\varepsilon, \phi)}{\partial \phi} \right]. \tag{7.131}$$

As in the sharp interface model (see (7.97)), the surface atomic flux is taken to be proportional to the chemical potential gradient and the electromigration driving force, but with surface diffusion confined to the interfacial layer $|\phi| < 1$:

$$\Gamma_S = -\frac{\delta_S D_S}{\Omega k_B T \pi \varepsilon} \left(\nabla_S \mu_S - eZ_S^* \rho j_S \right) \tag{7.132}$$

and $\Gamma_S = 0$ for $|\phi| \geq 1$. It should be mentioned that the surface energy associated with the curvature of the void surface in the sharp interface model is represented in the phase-field model by the term $\frac{2\gamma}{\pi\varepsilon} \left(\frac{\partial f_0}{\partial \phi} - \varepsilon^2 \nabla^2 \phi \right)$, which, as discussed before, describes the free energy increase associated with the formation of interfaces or surfaces.

In addition to the surface flux divergence, the void surface can evolve due to vacancy exchange between the void surface and surrounded metal. Similar to the sharp interface formalism used to calculate vacancy exchange flux as described in (7.105) in Section 7.9, Bhate et al. in [21] expressed the vacancy flux per unit volume from the bulk phase to the surface as

$$\Gamma_{VS} = \frac{2A}{\pi\varepsilon}\left(1 - \phi^2\right)(\mu_S + \mu_V),\tag{7.133}$$

where μ_V is the chemical potential of a vacancy in the metal bulk (7.92), μ_S is the chemical potential of an atom on the void surface, and A is a rate parameter. It should be mentioned that the parameters $2/\pi\varepsilon$ and $\left(1 - \phi^2\right)$ were used in this approach in order to match the results of the sharp interface model and to specify the region with $|\phi| < 1$ where void evolution occurs. The order parameter evolves due to divergence of the surface flux and the exchange between void interfacial layer and metal bulk as described by the equation

$$\frac{1}{2}\frac{\partial\phi}{\partial t} = -\nabla\Gamma_S - \Gamma_{VS}.\tag{7.134}$$

The vacancy concentration in the metal evolves in accordance with the formalism described in Section 7.3 with the additional generation-recombination term given by the vacancy exchange flux at the void surface Γ_{VS}, serving as a sink/source of vacancies [130].

Evolution of the order parameter given by (7.134) is represented by a combination of the Cahn–Hilliard and Allen–Cahn formalisms describing the change of the local order parameter in the conservative and open systems. Similar formalism was adopted in [129], and in [27, 87] where the surface flux divergence was the only cause of evolution in the $|\phi| < 1$ region.

The phase-field model was tested on several practical cases [21, 128]. First, it was demonstrated that a simulated migration of the circular void caused by applied electric field has corresponded to the exact Ho solution [82]. A failure caused by a cathode edge void undercutting the tungsten via was simulated as shown in Figure 7.48a. Here the anode end of the line was represented as an infinite source of vacancies. It was

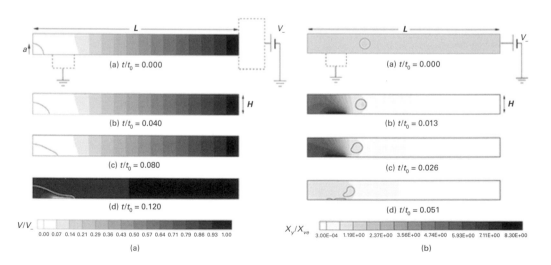

Figure 7.48 (a) A simulation of an open circuit by an evolving void initiated at the line cathode edge recess; and (b) a formation of an open circuit due to shape instability of a void near a via. Reprinted from [21], with permission from Elsevier.

shown that a diffusional exchange of vacancies between voids can explain phenomena such as the coarsening of voids and the break of large void into a number of smaller voids. Figure 7.48b demonstrates that a circular void migrated toward the cathode tungsten via, then developed a shape instability to cause its breakup into three separate voids, with two of them cutting the via further.

A diffuse interface model developed by Mahadevan and Bradley [27] has been simultaneously applied to evaluate the evolution of a void surface and an edge of the metal line, where the initial void precursor was located, to simulate the growth of a flaw to become a slit-shaped void spanning the line [87]. They demonstrated that ignoring the mass transport along the line edges can conserve the void area [88] and, if the line is too narrow and the surface energy too high, it would be impossible for the original protuberance to grow long. Thus, since the size of the original flaw cannot be very large, it will limit the length of the slit-shaped void. However, an atomic exchange between void surface and line edge of a confined interface can effectively remove this model limitation and permits the growth of a fatal slit from a small precursor as shown in Figure 7.49. All the examples cited here and many others available in the literature have demonstrated the advantage of the phase-field method over the methods employing the moving boundary technique, especially for cases characterized by changes in the void topology since the latter require explicit tracking of the boundary and continuous remeshing of the domain.

7.11   EM-Induced Evolution of Stress-Induced Voids (SIV)

To complete the discussion of the advantages of the phase-field method, we present in this section the results of the simulation of the stress-induced voiding (SIV) caused by postanneal cooling and the further void evolution caused by electrical stressing in the polycrystalline segment discussed in Sections 7.5 and 7.72. For this analysis, an additional PDE similar to (7.134) is added to the system of PDEs (7.72)–(7.78) and (4.42)–(4.53) to track the evolution of the order parameter ϕ that enables us to analyze the evolution of stress and vacancy–plated atom concentrations induced by a cooling from the zero stress temperature to the test temperature and by applied electric current [130]. In Figure 7.50, we show the steady-state distribution of the hydrostatic stress along the top interface between the copper and dielectric barrier in a dual-damascene interconnect line, shown in Figure 7.51. The stress shown by the dashed line was developed immediately after cooling down from the zero-stress temperature $T_{ZS} = 700$ K to the target temperature $T_{test} = 498$ K, while the solid line shows the final hydrostatic stress distribution achieved when stress relaxation induced by vacancy diffusion and vacancy–plated atom pair generation was completed. Here, as the critical stress of $\sigma_{crit} = 600$ MPa was developed at the middle of the top interface, the growth of the small semispherical void precursor was initiated at the same location. The growth of the void precursor, as shown in Figure 7.51, was due to the vacancy exchange between the void surface and surrounding metal and accompanied by a reduction of the tensile stress in the metal line as expected. In Figure 7.52 the evolution of the hydrostatic stress at the top interface is presented as a function of

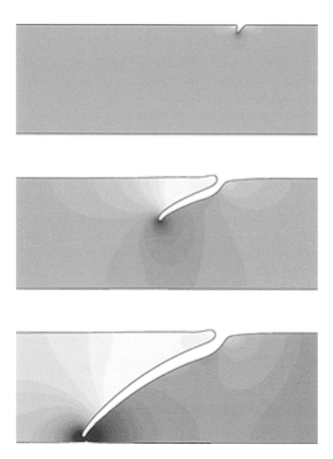

Figure 7.49 Snapshots of a portion of the line showing the development of an edge perturbation into a slit-like void. Reprinted with permission from M. Mahadevan and R. M. Bradley, *Phys. Rev. B*, 59, 11037(1999), [87]. Copyright (1999) by the American Physical Society.

time. It provides the scales of the hydrostatic stress reduction with time and the rate of void growth. It should be noted that the time intervals shown in Figures 7.51 and 7.52 are counted from the instance when the σ_{crit} was developed, which is around $t \approx 10^4$s after a rapid cooling was performed.

In Figure 7.53, we show the effect of current stressing on evolution of the previously discussed stress-induced void. The applied voltage of $\Delta V = 0.5$ V between two M1 line segments generates a current density of $j = 6 \cdot 10^{10}$ A/m^2 in the M2 line. Here the electric stressing changes the stress and vacancy distributions across the M2 segment, providing an additional driving force for atoms to migrate along the void surface. The snapshots show the evolution of the shape of the preexisting stress-induced void and the evolution of distribution of hydrostatic stress in the test structure due to electromigration. In Figure 7.54, we show the evolution of the hydrostatic stress along the top interface during the same duration of time. Here the growth of

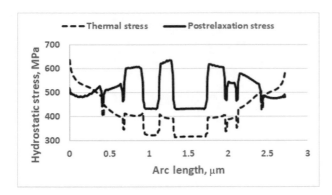

Figure 7.50 Distributions of the hydrostatic stress along the top interface of the M2 segment. Distributions were taken when the cooling from anneal temperature 700 K down to test temperature 498 K was done (the dashed line) and when the postrelaxation steady state was achieved, $t \approx 5 \cdot 10^3$ s (the solid line).

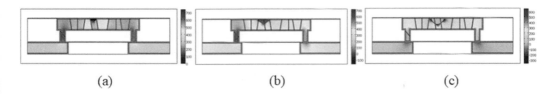

(a)   (b)   (c)

Figure 7.51 A simulation of a stress-induced void evolution at the top interface of the M2 segment. (a) $t = 10$ s, (b) $t = 50$ s, and (c) $t = 250$ s.

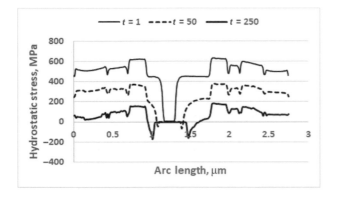

Figure 7.52 Simulated stress-induced void enlargement and accompanying reduction of the tensile stress in surrounding metal.

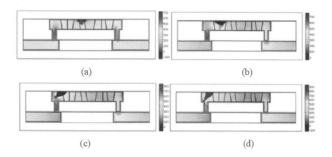

Figure 7.53 Snapshots of void shape evolution and modification of the hydrostatic stress distribution caused by the applied electric current: (a) $t = 0$ s, (b) $t = 27$ s, (c) $t = 32$ s, and (d) $t = 40$ s.

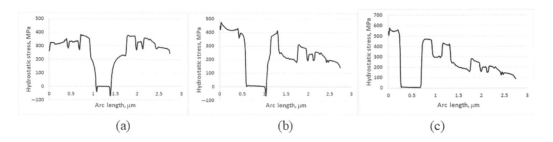

Figure 7.54 Simulated stress-induced void drift accompanied by EM-induced growth of tensile stress in the near-cathode region at the top interface.

hydrostatic tension with time at the cathode region clearly accompanies with the drift of the void (zero-stress regions) toward the cathode at a speed of about 0.028 μm/s. Time intervals shown in Figures 7.53 and 7.54 are counted from the instance when the electric stressing was applied, which is 50 s after the σ_{crit} was developed. It is worth noting that such short time intervals and fast void drift can be accounted for by the high test temperature of 225°C employed for these simulations.

The picture of the electromigration-induced void evolution reported here is quite different from the result predicted by 1D modeling. It clearly demonstrates that the stress relaxation taking place during a chip shelf-life can and should affect the kinetics of electromigration-induced failure for an on-chip interconnect. Such a correspondence between the predicted by simulation and observed by SEM [80] voiding evolution in the dual-damascene interconnect line is shown in Figure 7.55.

The examples discussed here show that electric current stressing can indeed drive existing stress-induced voids to further evolve, and the results clearly demonstrate the predictive power of the phase-field method. Such physics-based simulations can replace costly real-silicon experiments to address reliability concerns and develop approaches to optimizing the properties of interconnect materials and the physical design of the interconnect layouts in order to achieve electromigration-resistive on-chip interconnects.

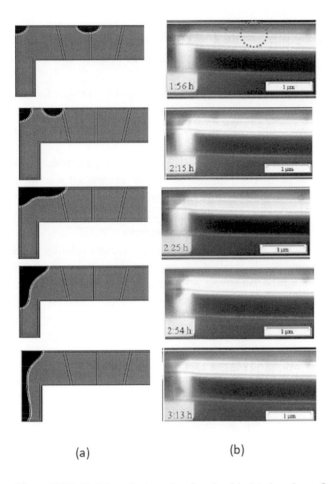

(a)   (b)

Figure 7.55 Void evolution simulated with (a) the phase-field model vs. (b) the SEM image sequence in Cu dual-damascene sample. Figure 7.55b reproduced from [80] with permission of M. A. Meyer.

7.12   Summary

In this chapter, we summarize the major results that have been achieved with the 2D and 3D multiphysics models of EM. The physical models and the respective numerical techniques that were developed for simulation of the EM-induced mass transport processes have allowed us to describe and to understand the degradation mechanisms and to estimate the EM lifetimes based on microstructural data and/or activation energies for atomic transport along grain boundaries and interfaces. When the activation energy for interface diffusion along strengthened interfaces becomes comparable to the activation energy for copper grain boundary diffusion, mass transport along grain boundaries must be taken into consideration for interconnects with polycrystalline microstructures. Then the characterization and monitoring of microstructural parameters such as grain size, texture, and stress for copper interconnects become even more important for back end of the line (BEOL) process control. The implementation of the copper microstructure into the numerical simulation makes the model more universal and precise. This model helps to provide a closer link between microstructural data and EM lifetime.

Different numerical techniques that were developed to simulate failure in interconnect lines due to electromigration and stress-induced voiding are discussed. Sharp interface models that have been applied extensively in the past demonstrate the considerable success of the approach, but the complexity of the required surface tracking has restricted its application to 2D single-voiding phenomena. An advanced diffuse interface model for electromigration and stress-induced voiding has extended simulation capabilities to realistic three-dimensional interconnects, allowing researchers to simulate void splitting and coalescence as a result of electromigration.

These simulation efforts are in progress. But even now, the predicted locations of the void nucleation site as well as the predicted stress evolution for a variety of interconnect segments fits the available experimental data quite well. The capability of this model for optimization of the physical and electrical design rules is emphasized. By varying the interconnect architecture, segment geometry, material properties, and some of the process parameters, users will be in a position to generate on-chip interconnect systems with a high immunity to EM-induced failures.

References

1. A. Einstein, Über die von der molekularkinetischen Theorie der Wärme geforderte Bewegung von in ruhenden Flüssigkeiten suspendierten Teilchen, *Annalen der Physik* (in German) **322** (1905), 549–560.
2. C. Herring, Diffusional viscosity of a polycrystalline solid, *Journal of Applied Physics* **21** (1950), 437–445.
3. F. C. Larche and J. W. Cahn, Overview No. 41: the interactions of composition and stress in crystalline solids, *Acta Metallurgica* **33** (1985), 331–357.
4. C. Jones, P. J. Grout, and A. B. Lidiard, The heat of transport of vacancies in model fcc solids, *Philosophical Magazine A* **79** (1999), 2051–2070.
5. P. S. Ho and T. Kwok, Electromigration in metals, *Reports on Progress in Physics* **52** (1989), 301–348.
6. M. J. Aziz, Thermodynamics of diffusion under pressure and stress: relation to point defect mechanisms, *Applied Physics Letters* **70** (1997), 2810–2812.
7. P. Shewmon, *Diffusion in Solids*, 2nd ed. (Switzerland: Pergamon International Publishers, 2016).
8. I. A. Blech, Electromigration in thin aluminum films on titanium nitride, *Journal of Applied Physics* **47** (1976), 1203–1208.
9. K. F. Riley, M. P. Hobson, and S. J. Bence, *Mathematical Methods for Physics and Engineering*, 3rd ed. (New York: Cambridge University Press, 2010).
10. M. E. Sarychev, Y. V. Zhitnikov, L. Borucki, C. L. Liu, and T. M. Makhviladze, General model for mechanical stress evolution during electromigration, *Journal of Applied Physics* **86** (1999), 3068–3075.

11. R. Kirchheim, Stress and electromigration in Al-lines of integrated circuits, *Acta Metallurgica Materialia* **40** (1992), 309–323.

12. R. Rosenberg and M. Ohring, Void formation and growth during electromigration in thin films, *Journal of Applied Physics* **42** (1971), 5671–5679.

13. R. W. Balluffi and A. V. Granato, Dislocations, vacancies and interstitials. F. *Dislocations in Solids*, vol. 4, ed. R. N. Nabarro (Amsterdam, New York, Oxford: North-Holland Publishing Company, 1979), 1.

14. V. Sukharev, E. Zschech, and W. D. Nix, A model for electromigration-induced degradation mechanism in dual-inlaid copper interconnects: effect of microstructure, *Journal of Applied Physics* **102** (2007), 053 505, 1–14.

15. L. D. Landau and E. M. Lifshitz, *Theory of Elasticity* (Oxford: Pergamon Press, 1970).

16. V. Timoshenko and J. N. Goodier, *Theory of Elasticity* (New York: McGraw-Hill, 1951).

17. V. Sukharev, R. Choudhury, and C. W. Park, Electromigration simulation in Cu-low-*K* multilevel interconnect segments, *IEEE International Integrated Reliability Workshop Final Report, 2002* (Lake Tahoe: IEEE, 2002), 55–61, doi: 10.1109/IRWS.2002.1194233.

18. V. Sukharev, A. Kteyan, J.-H. Choy, S. Chatterjee, and F. N. Najm, Theoretical predictions of EM-induced degradation in test-structures and on-chip power grids with analytical and numerical analysis, *Proceedings of the 2017 IEEE International Reliability Physics Symposium (IRPS)*. (Monterey: IEEE, 2017), 6B-5.1-10.

19. V. Sukharev, A. Kteyan, E. Zschech, and W. D. Nix, Microstructure effect on EM-induced degradation in dual-inlaid copper interconnects, *IEEE Transactions on Device and Materials Reliability* **9** (2009), 87–97.

20. COMSOL, Inc. Burlington, MA. 8 New England Executive Park.

21. D. N. Bhate, A. F. Bower, and A. Kumar, A phase field model for failure in interconnect lines due to coupled diffusion mechanics, *Journal of the Mechanics and Physics of Solids* **50** (2002), 2057–2083.

22. A. F. Bower and L. B. Freund, Analysis of stress-induced void growth mechanism in passivated interconnect lines, *Journal of Applied Physics* **74** (1993), 3855.

23. A. F. Bower and S. Shankar, A finite element model of electromigration induced void nucleation, growth and evolution in interconnects, *Modelling and Simulation of Materials Science and Engineering* **15** (2007), 923–940.

24. J. C. Fisher, Calculation of diffusion penetration curves for surface and grain boundary diffusion, *Journal of Applied Physics* **22** (1951), 74–77.

25. W. W. Mullins, Mass transport at interfaces in single component systems, *Metallurgical and Materials Transactions A* **26** (1995), 1917–1929.

26. M. Ashby, Boundary defects, and atomistic aspects of boundary sliding and diffusion creep, *Surface Science* **31** (1972), 498–542.

27. M. Mahadevan and R. Bradley, Phase field models of surface electromigration in thin metal films, *Physica D* **126** (1999), 201–213.

28. J. Wilkening, L. Borucki, and J. A. Sethian, Analysis of stress-driven grain boundary diffusion. Part I, *SIAM Journal of Applied Mathematics* **64** (2004), 1839–1886.

29. H. Ceric and S. Selberherr, An adaptive grid approach for the simulation of electromigration induced void migration, *IEICE Transactions on Electronics* **E86–C(3)** (2002), 421–426.

30. V. Sukharev and E. Zschech, A model for electromigration-induced degradation mechanisms in dual-inlaid copper interconnects: effect of interface bonding strength, *Journal of Applied Physics* **96** (2004), 6337–6343.

31. V. Sukharev, Beyond Black's equation: full-chip EM/SM assessment in 3D IC stack, *Microelectronic Engineering* **120** (2014), 99–105.

32. P. S. Ho, K. D. Lee, E. T. Ogawa, S. Yoon, and X. Lu, Impact of low-*k* dielectrics on electromigration reliability for Cu interconnects, *Characterization and Metrology. for ULSI Technology*, ed. D. G. Seiler, A. C. Diebold, T. J. Shaffner, et al. (Austin: AIP Conference Proceedings, 2003), 533–539.

33. C. K. Hu, S. G. Malhotra, and L. Gignac, Electromigration in submicron copper lines deposited by chemical vapor deposition and physical vapor deposition techniques, *Proceedings of the International Symposium on Interconnect and Contact Metallization for ULSI, Vol. PV 99–31, Electrochemical Society Proceedings*, ed. G. S. Mathad, H. S. Rathore, and Y. Arita (Pennington: Electrochemical Society, 1999), 206–213.

34. E. T. Ogawa, K. D. Lee, V. A. Blaschke, and P. S. Ho, Electromigration reliability issues in dual-damascene Cu interconnects, *IEEE Transactions on Reliability* **51** (2002), 403–419.

35. M. W. Lane, E. G. Liniger, and J. R. Lloyd, Relationship between interfacial adhesion and electromigration in Cu metallization, *Journal of Applied Physics* **93** (2003), 1417–1423.

36. E. Zschech and V. Sukharev, Microstructure effect on EM-induced copper interconnect degradation: Experiment and simulation, *Microelectronics Engineering* **82** (2005), 629–638.

37. E. Zschech, M. A. Meyer, S. G. Mhaisalkar, et al., Effect of interface modification on EM-induced degradation mechanisms in copper interconnects, *Thin Solid Films* **504** (2006), 279–283.

38. I. Kaur and W. Gust, *Handbook of Grain and Interphase Boundary Diffusion Data* (Stuttgart: Ziegler Press, 1989).

39. A. Suzuki, Y. Mishin, Atomistic modeling of point defects and diffusion in copper grain boundaries, *Interface Science* **11** (2003), 131–148.

40. J. R. Lloyd, C. E. Murray, T. M. Shaw, M. W. Lane, X.-H. Liu, and E. G. Liniger, Theory for electromigration failure in Cu conductors, *Proceedings of the 8th International Workshop on Stress-Induced Phenomena in Metallization*, ed. E. Zschech, K. Maex, P. S. Ho, H. Kawasaki, and T. Nakamura (Dresden: AIP Conference Proceedings 817, 2006), 23–33.

41. M. A. Meyer, M. Herrmann, E. Langer, and E. Zschech, In situ SEM observation of electromigration phenomena in fully embedded copper interconnect structures, *Microelectronics Engineering* **64** (2002), 375–382.

42. A. Kteyan, V. Sukharev, M. A. Meyer, E. Zschech, and W. D. Nix, Microstructure effect On EM-induced degradations in dual-inlaid copper interconnects, *Proceedings of the 9th International Workshop on Stress-Induced Phenomena in Metallization*, ed. S. Ogawa, P. S. Ho, and E. Zschech. (Kyoto: AIP Conference Proceedings 945, 2007), 42–55.

43. E. Zschech, H.-J. Engelmann, M. A. Meyer, et al., Effect of interface strength on electromigration-induced inlaid copper interconnect degradation: experiment and simulation, *Z. Metallkunde* **96** (2005), 966–971.

44. E. Zschech, M. A. Meyer, and E. Langer, Effect of mass transport along interfaces and grain boundaries on copper interconnect degradation, *Materials, Technology and Reliability Advanced Interconnects and Low-k Dielectrics, vol. 812, Materials Research Society Symposium Proceedings*, ed. R. J. Carter, Proc. C. S. Hau-Riege, G. M., Kloster, T.-M. Lu, and S. E. Schulz (Warrendale: Materials Research Society, 2004), 361–372.

45. M. A. Meyer, M. Grafe, H.-J. Engelmann, E. Langer, and E. Zschech, Investigation of void formation and evolution during electromigration testing, *Proceedings of the 8th International Workshop on Stress-Induced Phenomena in Metallization*, ed. E. Zschech, K. Maex, P. S. Ho, H. Kawasaki, and T. Nakamura (Dresden: AIP Conference Proceedings 817, 2006), 175–184.

46. E. Zschech, H. Geisler, I. Zienert, et al., Reliability of cooper inlaid structures: geometry and microstructure effects, *Proceedings of the Advanced Metallization Conference (AMC)*, ed. B. M. Melnick, T. S. Cale. S. Zaima, and T. Ohta (San Diego: Materials Research Society, 2002), 305–312.

47. K. D. Lee, E. T. Ogawa, H. Matsuhashi, et al., Electromigration critical length effect in Cu/oxide dual-damascene interconnects, *Applied Physics Letters* **79** (2001), 3236–3238.

48. C. K. Hu, L. Gignac, S. G. Malhotra, and R. Rosenberg, Mechanism for very long electromigration lifetime in dual-damascene Cu interconnects, *Applied Physics Letters* **78** (2001), 904–906.

49. S. Yokogawa, N. Okada, Y. Kakuhara, and H. Takizawa, Electromigration performance of multi-level damascene copper interconnects, *Microelectronics Reliability* **41** (2001), 1409–1416.

50. Q. Guo, A. Krishnamoorthy, N. Y. Huang, and P. D. Foo, Resistance degradation profile in electromigration of dual-damascene Cu interconnects, *Proceedings of the Advanced Metallization Conference (AMC)*, ed. B. M. Melnick, T. S. Cale. S. Zaima, and T. Ohta. (San Diego: Materials Research Society, 2002), 191–195.

51. W. D. Nix, Mechanical properties of thin films, *Metallurgical and Materials. Transactions A* **20** (1989), 2217–2245.

52. C. V. Thompson, The yield strength of polycrystalline films, *Journal of Materials Research* **8** (1993), 237–238.

53. J. Hirth and J. Lothe, *Theory of Dislocations* (New York: Wiley, 1982).

54. K. Hayashi, M. Osada, Y. Kurosu, Y. Miyajima, and S. Onaka, Log angles: characteristic angles of crystal orientation given by the logarithm of rotation matrix, *Materials Transactions* **57** (2016), 507–512.

55. R. P. Vinci, E. M. Zielinski, and J. C. Bravman, Thermal strain and stress in copper thin films, *Thin Solid Films* **262** (1995), 142–153.

56. Z. Suo, *Reliability of Interconnect Structures" in Volume 8: Interfacial and Nanoscale Failure, Comprehensive Structural Integrity*, eds. W. Gerberich and W. Yang (Amsterdam: Elsevier, 2003), 265–324.

57. D. R. Askeland, P. P. Fulay, and W. J. Wright, *The Science and Engineering of Materials*, 6th ed. (Stamford: Cengage Learning, 2011).

58. E. Arzt, O. Kraft, W. D. Nix, and J. E. Sanchez, Electromigration failure by shape change of voids in bamboo lines, *Journal of Applied Physics* **76** (1994), 1563–1571.

59. N. L. Michael, C. U. Kim, Q. T. Jiang, R. A. Augur, and P. Gillespie, Mechanism of electromigration failure in submicron Cu interconnects, *Journal of Electronic Materials* **31** (2002), 1004–1008.

60. M. A. Meyer and E. Zschech, New microstructure-related EM degradation and failure mechanisms in Cu interconnects with CoWP coating, *Proceedings of the 9th International Workshop on Stress-Induced Phenomena in Metallization*, ed. S. Ogawa, P. S. Ho, and E. Zschech. (Kyoto: AIP Conference Proceedings 945, 2007), 107–114.

61. S. Shingubara and Y. Nakasaki, Electromigration in a single crystalline submicron width aluminum interconnection, *Applied Physics Letters* **58** (1991), 42–44.

62. I. Vavra and P. Lobotka, TEM in-situ observation of electromigration in A1 stripes with quasi-bamboo structure, *Physica Status Solidi A*, **65** (1981), K107–K108.

63. E. Castano, J. Maiz, P. Flinn, and M. Madden, In situ observations of dc and ac electromigration in passivated Al lines, *Applied Physics Letters* **59** (1991), 129–131.

64. P. R. Besser, M. C. Madden, and P. A. Flinn, In situ scanning electron microscopy observation of the dynamic behavior of electromigration voids in passivated aluminum lines, *Journal of Applied Physics* **72** (1992), 3792–3797.

65. M. C. Madden, E. V. Abratowski; T. Marieb, and P. A. Flinn, High resolution observation of void motion in passivated metal lines under electromigration stress, *Materials Reliability in Microelectronics II, Vol. 265, Materials Research Society Symposium Proceedings*, ed. C. V. Thompson and J. R. Lloyd (Pittsburgh: Materials Research Society 1992), 33–38.

66. R. W. Vook and C. Y. Chang, The catastrophic phase of electromigration: surface diffusion, island formation, and film thinning, *Applied Surface Science* **60–61** (1992), 71-78.

67. 0. Kraft, S. Bader, J. E. Sanchez, Jr., and E. Arzt, Observation and modeling of electromigration-induced void growth in Al-based interconnects, *Materials Reliability in Microelectronics III, Vol. 309, Materials Research Society Symposium Proceedings*, ed. W. F. Filter, H. J. Frost, P. S. Ho, and K. P. Rodbell (Pittsburgh: Materials Research Society, 1993) 199–204.

68. J. E. Sanchez, Jr. and E. Arzt, Microstructural aspects of interconnect failure, *Materials Reliability in Microelectronics II, Vol. 265, Materials Research Society Symposium Proceedings*, ed. C. V. Thompson and J. R. Lloyd (Pittsburgh: Materials Research Society, 1992), 131–142.

69. J. E. Sanchez, Jr., 0. Kraft, and E. Arzt, Electromigration induced transgranular slit failures in near bamboo Al and Al-2% Cu thin-film interconnects, *Applied Physics Letters* **61** (1992), 3121–3123.

70. J. H. Rose, Fatal electromigration voids in narrow aluminum-copper interconnect, *Applied Physics Letters* **61** (1992), 2170–2172.

71. J. H. Rose and T. Spooner, The microstructural nature of electromigration and mechanical stress voids in integrated circuit interconnect, *Materials Reliability in Microelectronics III, Vol. 309, Materials Research Society Symposium Proceedings*, ed. W. F. Filter, H. J. Frost, P. S. Ho, and K. P. Rodbell (Pittsburgh: Materials Research Society, 1993), 409–416.

72. N. I. Peterson, Self-diffusion in pure metals, *Journal of Nuclear Materials* **69–70** (1978), 3–37.

73. D. Gupta, C. K. Hu, and K. L. Lee, Grain boundary diffusion and electromigration in Cu-Sn alloy thin films and their VLSI interconnects, *Defect Diffusion Forum* **143–147** (1997), 1397–1406.

74. O. V. Kononenko, V. N. Matveev, Yu. I. Koval, S. V. Dubonos, and V. T. Volkov, Electromigration in submicron wide copper lines, *Advanced Metallization for Future ULSI, Vol. 427, Materials Research Society Symposium Proceedings*, eds. L. J. Chen, J. W. Mayer, J. M. Poate, and K. N. Tu (Pittsburgh: Materials Research Society, 1996), 127–133.

75. R. Gonella, Key reliability issues for copper integration in damascene architecture, *Microelectronics Engineering* **55** (2001), 245–255.

76. E. Glickman and M. Nathan, On the unusual electromigration behavior of copper interconnects, *Journal of Applied Physics* **80** (1996), 3782–3791.

77. C. K. Hu, R. Rosenberg, and K. L. Lee, Electromigration path in Cu thin-film lines, *Applied Physics Letters* **74** (1999), 2945–2947.

78. D. Edelstein, C. Uzoh, C. Cabral Jr., et al., A high performance liner for copper damascene interconnects, *Proceedings of the International Interconnect Technology Conference (IITC)* (Piscataway: IEEE, 2001), 9–11.

79. E. Zschech, H. Geisler, I. Zienert, et al., Reliability of cooper inlaid structures: geometry and microstructure effects, *Proceedings of the Advanced Metallization Conference (AMC)*, ed. B. M. Melnick, T. S. Cale, S. Zaima, and T. Ohta (San Diego: Materials Research Society, 2002), 305–312.

80. M. A. Meyer, Effects of advanced process approaches on electromigration degradation of Cu on-chip interconnects, PhD thesis, Brandenburgischen Technischen Universitat Cottbus (2007). https://opus4.kobv.de/opus4-btu/frontdoor/index/index/docId/343.

81. M. Lane, R. Rosenberg, Interfacial relationship in microelectronic devices, *Materials, Technology and Reliability for Advanced Interconnects and Low-k Dielectrics, Vol. 766, Materials Research Society Symposium Proceedings*, ed. A. McKerrow, J. Leu, O. Kraft, and T. Kikkawa (Pittsburgh: Materials Research Society, 2003), E9.1.

82. P. S. Ho, Motion of inclusion induced by a direct current and a temperature gradient, *Journal of Applied Physics* **41** (1970), 64–68.

83. W. D. Nix and E. Arzt, On void nucleation and growth in metal interconnect lines and electromigration conditions, *Metallurgical and Materials Transactions A* **23** (1992), 2007–2013.

84. P. Borgesen, M. A. Korhonen, T. D. Sullivan, D. D. Brown, and C.-Y. Li, Electromigration damage by current induced coalescence of thermal stress voids, *Thin Films: Stresses and Mechanical Properties III, Vol. 239, Materials Research Society Symposium Proceedings*, ed. E. Arzt, J. C. Bravman, L. B. Freund, and W. D. Nix (Pittsburgh: Materials Research Society, 1991), 683–689.

85. L. Xia, A. F. Bower, Z. Suo, and C. F. Shih, A finite element analysis of the motion and evolution of voids due to strain and electromigration induced surface diffusion, *Journal of the Mechanics and Physics of Solids* **45** (1997), 1473–1493.

86. Y. W. Zhang, A. F. Bower, L. Xia, and C. F. Shih, Three-dimensional finite element analysis of the evolution of voids and thin films by strain and electromigration induced surface diffusion, *Journal of the Mechanics and Physics of Solids* **47** (1999), 173–199.

87. M. Mahadevan and R. M. Bradley, Simulation and theory of electromigration-induced slit formation in unpassivated single-crystal metal lines. *Physical Review B* **59** (1999), 11037–11046.

88. O. Kraft and E. Arzt, Electromigration mechanism in conductor lines: void shape changes and slit-like failure, *Acta Materialia* **45** (1997), 1599–1611.

89. Z. Suo, Motion of microscopic surfaces in materials, *Advances in Applied Mechanics* **33** (1997), 193–294.

90. W. Yang, W. Wang, and Z. Suo, Cavity and dislocation instability due to electric current, *Journal of the Mechanics and Physics of Solids* **42** (1994), 897–911.

91. D. Maroudas, Dynamics of transgranular voids in metallic thin films under electromigration conditions, *Applied Physics Letters* **67** (1995), 798–800.

92. O. Kraft and E. Arzt, Numerical simulation of electromigration-induced shape changes of voids in bamboo lines, *Applied Physics Letters* **66** (1995), 2063–2065.

93. J. He, Z. Suo, T. N. Marieb, and J. A. Maiz, Electromigration lifetime and critical void volume, *Applied Physics Letters* **85** (2004), 4639–4641.

94. M. R. Gungor and D. Maroudas, Electromigration-induced failure of metallic thin films due to transgranular void propagation, *Applied Physics Letters* **72** (1998), 3452–3454.

95. M. R. Gungor and D. Maroudas, Theoretical analysis of electromigration-induced failure of metallic thin films due to transgranular void propagation, *Journal of Applied Physics* **85** (1999), 2233–2246.

96. D. Fridline and A. Bower, Influence of anisotropic surface diffusivity on electromigration induced void migration and evolution, *Journal of Applied Physics* **85** (1999), 3168–3174.

97. M. R. Gungor and D. Maroudas, Modeling of electromechanically induced failure of passivated metallic thin films used in device interconnections, *International Journal of Fracture* **109** (2001) 47–68.

98. M. Schimschak and J. Krug, Electromigration-induced breakup of two-dimensional voids, *Physical Review Letters* **80** (1998), 1674–1677.

99. M. Schimschak and J. Krug, Electromigration-driven shape evolution of two-dimensional voids, *Journal of Applied Physics* **87** (2000), 695–703.

100. T. O. Ogurtani and E. E. Oren, Computer simulation of void growth dynamics under the action of electromigration and capillary forces in narrow thin interconnects, *Journal of Applied Physics* **90** (2001), 1564–1572.

101. T. O. Ogurtani and E. E. Oren, Electromigration-induced void grain-boundary interactions: the mean time to failure for copper interconnects with bamboo and near-bamboo structures, *Journal of Applied Physics* **96** (2004), 7246–7253.

102. T. O. Ogurtani and O. Akyildiz, Grain boundary grooving and cathode voiding in bamboo-like metallic interconnects by surface drift diffusion under the capillary and electromigration forces, *Journal of Applied Physics* **97** (2005), 093520.

103. J. W. Barrett, R. Nürnberg, and V. Styles, Finite element approximation of a phase field model for void electromigration, *SIAM Journal on Numerical Analysis* **42** (2004), 738–772.

104. P. Smereka, Semi-implicit level set methods for curvature and surface diffusion motion, *Journal of Scientific Computing* **19** (2003), 439–456.

105. Z. Li, H. Zhao, and H. Gao, A numerical study of electromigration voiding by evolving level set functions on a fixed Cartesian grid, *Journal of Computational Physics* **152** (1999), 281–304.

106. S. Osher and J. A. Sethian, Fronts propagating with curvature-dependent speed: algorithms based on Hamilton–Jacobi formulations, *Journal of Computational Physics* **79** (1988), 12–49.

107. M. Khenner, A. Averbuch, M. Israeli, M. Nathan, and E. Glickman, Level set modeling of transient electromigration grooving, *Computational Materials Science* **20** (2001), 235–250.

108. E. Maitre, C. Misbah, P. Peyla, and A. Raoult, Comparison between advected-field and level-set methods in the study of vesicle dynamics. https://hal.archives-ouvertes.fr/hal-00460668v3.

109. A. F. Bower and E. Wininger, A two-dimensional finite element method for simulating the constitutive response and microstructure of polycrystals during high temperature plastic deformation, *Journal of the Mechanics and Physics of Solids* **52** (2004), 1289–1317.

110. W. Q. Wang, Z. Suo, and T. H. Hao, A simulation of electromigration induced transgranular slits, *Journal of Applied Physics* **79** (1996), 2394–2403.

111. Z. Suo and W. Wang, Diffusive void bifurcation in stressed solid, *Journal of Applied Physics* **76** (1994), 3410–3421.

112. W. Q. Wang and Z. Suo, Shape change of a pore in a stressed solid via surface diffusion motivated by surface and elastic energy variation, *Journal of the Mechanics and Physics of Solids* **45** (1997), 709–729.

113. C.-Y. Li, P. Borgesen, and M. A. Korhonen, Electromigration-induced failure in passivated aluminum-based metallization. The dependence on temperature and current density, *Applied Physics Letters* **61** (1992), 411–413.

114. J. W. Cahn, On spinodal decomposition, *Acta Metallurgica* **9** (1961), 795–801.

115. J. W. Cahn and J. E. Hilliard, Free energy of nonuniform system. I. Interfacial free energy, *Journal of Chemical Physics* **28** (1958), 258–267.

116. S. M. Allen and J. W. Cahn, A microscopic theory for antiphase boundary motion and its application to antiphase domain coarsening, *Acta Metallurgica* **27** (1979), 1085–1095.

117. L. D. Landau and E. M. Lifshitz, *Statistical Physics, Part I, Course of Theoretical Physics, Vol. 5* (Moscow: Institute of Physical Problems, USSR Academy of Sciences; Oxford: Pergamon Press, 1980).

118. E. M. Lifshitz and L. P. Pitaevskii, *Statistical Physics, Part II, Course of Theoretical Physics, Vol. 9* (Moscow: Institute of Physical Problems, USSR Academy of Sciences; Oxford: Pergamon Press, 1980).

119. J. Cahn, Phase separation by spinodal decomposition in isotropic systems, *Journal of Chemical Physics* **42** (1965), 93–99.

120. M. Copetti and C. Elliott, Kinetics of phase decomposition processes: numerical solutions to Cahn–Hilliard equation, *Materials Science and Technology* **6** (1990), 273–283.

121. R. Kobayashi, A numerical approach to three-dimensional dendritic solidification, *Experimental Mathematics* **3** (1994), 59–81.

122. E. Fried and M. Gurtin, A phase field theory for solidification based on general anisotropic sharp interface theory with interfacial energy and entropy, *Physica D* **91** (1996), 143–181.

123. D. Jacqmin, Calculation of two-phase Navier–Stokes flows using phase-field modeling, *Journal of Compuational Physics* 155 (1999), 96–127.

124. N. Provatas, M. Grant, and K. R. Elder, Phase-field model for activated reaction fronts, *Physical Review B* **53** (1996), 6263–6272.

125. J. W. Cahn and J. E. Taylor, Overview NO. 113. Surface motion by surface diffusion, *Acta Metallurgica Materialia* **42** (1994), 1045–1063.

126. V. Sukharev, Physically based simulation of electromigration-induced degradation mechanisms in dual-inlaid copper interconnects, *IEEE Transactions on Computer-Aided Design of Integrated Circuits and Systems* **24** (2005), 1326–1335.

127. V. Sukharev, Physically-based simulation of electromigration induced failures in copper dual-damascene interconnect, *Proceedings of the 7th International Workshop on Stress-Induced Phenomena in Metallization*, eds. P. S. Ho, S. P. Baker, and C. Volkert (Austin: AIP Conference Proceedings 741, 2004), 85–96.

128. D. N. Bhate, A. Kumar, and A.F. Bower, Diffuse interface model for electromigration and stress-voiding, *Journal of Applied Physics* **87** (2000), 1712–1721.

129. S. Sadasiva, G. Subbarayan, L. Jiang, and D. Pantuso, Numerical simulations of electromigration and stress migration driven void evolution in solder interconnects, *Journal of Electronic Packaging* 134 (2012), 020907-1-9.

130. A. Kteyan and V. Sukharev, Physics-based simulation of stress-induced and electromigration-induced voiding and their interactions in on-chip interconnects, *Microelectronic Engineering* 247 (2021) 111585-1-7.

8 Massive-Scale Statistical Studies for Electromigration

In this chapter, the statistical nature of electromigration will be described. Along with understanding the basic physical degradation mechanisms, this is an important area of research due to the need for extrapolations from simple test structures to the product level. One must keep in mind that electromigration testing is usually done on single-link structures, just encompassing one metal line with a contact and/or a via connected at each end. Sometimes, a few links are stitched together in a series or parallel fashion, but massive-scale studies with large interconnect arrays have not been implemented yet as a standard testing methodology. Only the application of very large test structures with an extended amount of interconnect links and contacts/vias can lead to the detection of "early" failures, which are the limiting factor in the extrapolation to product-level interconnect systems. The detection of these early failures in electromigration and the complicated statistical nature of this important reliability phenomenon have been difficult issues to treat for decades in the past. A satisfactory experimental approach for the detection and the statistical analysis of these early failures has only been proposed by about the year 1999. This is mainly due to the rare occurrence of early failures and difficulties in testing of large sample populations. In this chapter, a technique utilizing large interconnect arrays in conjunction with the well-known Wheatstone Bridge will be discussed. Both Al- and Cu-based interconnect technologies will be addressed.

8.1 Requirement for Massive Statistical EM Tests

The term "early" or "extrinsic" failure in electromigration describes the occurrence of a failure that is not consistent with a monomodal distribution. The general assumption during the acquisition of EM test data is the fact that only a single failure mode is operative throughout the full temperature and current density range between accelerated test and device operation conditions. Furthermore, and most importantly, it is assumed that with increasing test sample size all failure times will coincide on one single failure distribution, which is usually assumed to be log-normal. An experimental test run typically contains far less than 100 samples; only about 30 are used as a standard. However, a full-scale device such as a state-of-the-art microprocessor or memory chip can contain up to several billion possible failure links. For the sake of simplicity, let us assume that about one million interconnects provide the necessary wiring between the transistors on a chip, e.g., a relatively small microcontroller application. Modern state-of-the-art chips can employ billions of interconnects, as

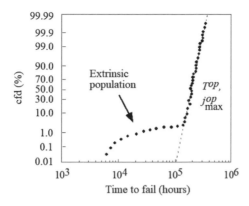

Figure 8.1 Illustration of how an extrinsic failure mode may lead to undesired consequences in the field. The extrinsic failure population occurs much earlier than expected and not according to the extrapolation from accelerated test data to a given operating temperature T^{op} and a maximum sustainable current density j^{op}_{max}. Reprinted from [6] with permission from AIP Publishing.

previously mentioned, such as in high-performance central processing units (CPUs) or graphic processing units (GPUs). If one assumes that 10^4 devices operate in the field, a relatively low number, one must account for 10^4 times one million, or 10 billion interconnects, which all can possibly fail due to EM phenomena. It has to be kept in mind that the failure of the first out of those 10 billion interconnects can be responsible for the fallout of the first full-scale device. Usually, due to high-performance and high-density requirements, only a very limited amount of redundancy is built into the design of a chip. A conservative model of a full-scale device is therefore a chain of many interconnects with a weakest-link approach. The first fail out of the many possible links then determines the lifetime of the whole chain. It is of utmost importance to prove that the very first device failures are part of the same intrinsic population that is characterized during accelerated testing, using very limited sample sizes. Figure 8.1 depicts a case where an extrinsic population interferes with the extrapolation of an intrinsic population to a reliability goal of, e.g., one failure in time (FIT). The intrinsic population, assumed at a given operating temperature T^{op} and a maximum sustainable current density j^{op}_{max}, extrapolates to a cumulative failure distribution (cfd) value of 0.01% at 10^5 hours. This satisfies the reliability goal of one FIT, or one failure in 10^9 device hours. However, the present extrinsic population leads to much shorter lifetimes at low failure rates. This can lead to a large amount of undesirable field failures. Hence, a massive-scale statistical approach to EM testing needs to be applied.

8.2   Wheatstone Bridge Technique for Al Interconnects

The previously described considerations have led to the need of early failure detection in EM. It is impossible to test several billion interconnect samples in an experimental approach. Only few studies have been performed on Al-based metallization that extend the test sample size beyond the typical number of several tens of failure units [1–3]. In this section, a different approach is described to gain information about the statistical behavior of many thousand interconnects and to investigate possible deviations from perfect log-normal statistics. A Wheatstone Bridge arrangement and large arrays of 480 Al-based interconnects are used to prove that deviations do not occur down to a cumulative failure rate of about one out of 20,000. Shorter reports on this technique have been published first in 1999 and 2000, respectively [4, 5], and an extended study was published in 2001 [6]. The goal of these three publications, investigating Al-interconnect-based metallization, was the characterization of the statistical EM behavior as a function of temperature, as well as a detailed description of the complex statistical analysis (deep censoring). Over a temperature range from 155–200°C, more than 75,000 interconnects were tested without any early failure occurrences. This number is still far away from several million or billion interconnects, but is the largest test sample size reached to date on realistic, Al-based multilevel test structures known to the authors.

8.2.1   Description of Test Structures

The test structures used for massive-scale EM studies were first presented in the description of experiments aimed at the statistical behavior of EM failures [5–8]. A basic unit of five interconnects in parallel serves as the smallest building block. Figure 8.2 depicts a top and side view of this basic unit.

The Metal1/Via/Metal2 (M1/Via/M2) chains were chosen to be 5 μm long on the lower level and 100 μm long on the upper level. The metallization scheme employed was a multilayer stack of Ti/TiN/Al(1 wt% Cu)/TiN. The M2 linewidth and W (tungsten) via size were 0.6 μm. By keeping the lower-level interconnects well below the critical Blech length, EM failure is induced in the upper level only [9]. This basic unit was repeated 96 times in series to build a large interconnect array of 480 M2 segments as possible failure links. Figure 8.3 shows a schematic drawing of this structure.

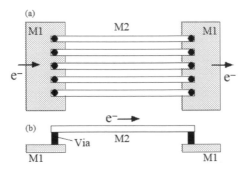

Figure 8.2 Basic unit of five interconnects at the upper metal level (M2) in parallel. By keeping the lower-level interconnects short, EM failure is induced in the upper level only: (a) represents a top view, (b) a side view. Arrows indicate the electron path. Reprinted from [6] with the permission from AIP Publishing.

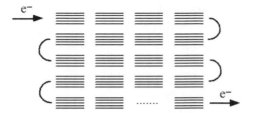

Figure 8.3 By repeating the basic unit from Figure 8.2 96 times, a large array of 480 interconnects was created. This array was used as a resistor in the Wheatstone Bridge device described later in this section. Reprinted from [6] with the permission from AIP Publishing.

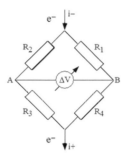

Figure 8.4 Wiring scheme according to the well-known Wheatstone Bridge technique. Instead of measuring the current between points A and B as in the original purpose of this technique, the voltage imbalance is monitored during an EM experiment while the current (i–, i+) is passing through the two branches. Each resistor, R_1 through R_4, consists of 480 interconnects; therefore, 1,920 possible failure sites are tested simultaneously. Reprinted from [6] with the permission from AIP Publishing.

The increase in interconnect links per test device sets limits to the resolution of small resistance changes during an EM test. With an increasing number of links, the measured voltage across each unit increases. Therefore, the resolution decreases. A possible remedy for this problem is the use of the well-known Wheatstone Bridge technique. Originally intended to measure the resistance of an unknown device, it consists of four resistors wired in parallel and series, as shown in Figure 8.4.

For EM testing purposes, it is not necessary to adjust any of the resistors. Instead of measuring the current passing between points A and B in the diagram depicted in Figure 8.4, the voltage imbalance between these points is monitored during an EM experiment. The initial imbalance can be estimated to be very small, therefore not leading to improper current settings in the two branches of the bridge. The initially measured resistance values for each single array, resistors R_1 through R_4, differ by only a few percent at the one sigma level. One can easily calculate that the imbalance in the two branches leads to a difference of only a few percent in stressing current density.

8.2.2   Characterization of Al(Cu) Microstructure

Transmission electron microscopy (TEM) was used to characterize the microstructure of the interconnect lines tested in the EM experiments. Figure 8.5 depicts a TEM micrograph of a part of an interconnect array with 0.6 μm wide lines at 0.6 μm spacing.

The lines are of a "near-bamboo" microstructure. The gray features at the sides of the interconnects are remainders of the passivation encapsulation that is present prior to the sample preparation. A total of 100 grains were measured to obtain the statistical distribution of grain sizes. Figure 8.6 shows a log-normal cumulative probability plot for the grain sizes obtained from Figure 8.5.

The median grain size was determined to 0.43 μm with a log-normal sigma of 0.39. Along with the grain size data for the 0.6 μm interconnect lines, grain size distributions for 0.4 and 1.0 μm wide interconnects and for films are given (the TEM micrographs are not shown here). The thin film data were obtained at contact pads of 100 μm × 100 μm size. All data were obtained using samples from the same wafer. Interestingly, the median grain sizes in the interconnects are almost identical and range from 0.43–0.47 μm. All log-normal sigma values are on the order of about 0.40–0.50. The thin film data yielded a median grain size of 1.3 μm and a log-normal sigma of 0.46. Given the data represented in Figure 8.6, we can conclude that 1 μm wide interconnects are of a polycrystalline microstructure. At 0.6 μm linewidth, the

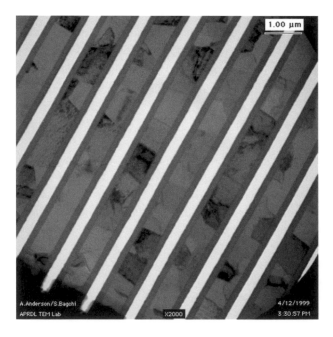

Figure 8.5 TEM of 0.6 μm wide Al(Cu) interconnects used in this study. At this linewidth, the microstructure approaches a "near-bamboo" arrangement (magnification 2,000×). Reprinted from [6] with the permission from AIP Publishing.

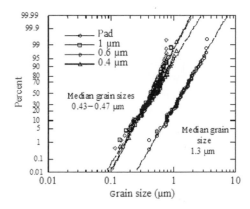

Figure 8.6 Cumulative probability plots of grain sizes, obtained from TEM micrographs, as a function of linewidth. The median grain sizes in interconnects of 0.4–1.0 μm linewidth were measured to range from 0.43–0.47 μm with log-normal sigma values of around 0.40–0.50. Along with the interconnect grain size data, a thin film grain distribution from the same samples is plotted (measured at contact pads). The median grain size for films was measured to 1.3 μm with a log-normal sigma of 0.46. Reprinted from [6] with the permission from AIP Publishing.

microstructure approaches a "bamboo-type" arrangement. Very fine lines of 0.4 μm width were found to be almost perfectly "bamboo type."

8.2.3   Experimental Data for Al(Cu) Metallization

8.2.3.1   Electromigration Data as a Function of Test Structure

The three types of devices previously described were tested in an EM experiment conducted at an ambient temperature of $T = 170°C$ and at a moderate current density of $j = 8.3 \times 10^5$ A/cm^2. A total of 32 basic units with five interconnects each (first type), 13 samples with 480-interconnect arrays (second type), and eight Wheatstone Bridge devices (third type) were tested initially. The total number of interconnects for this single test run was 21,760. The failure criterion used in this study is the time to first discernible resistance increase for the basic units and array structures, and the time to first discernible voltage imbalance change for the Wheatstone Bridge devices. The choice of this criterion reflects the fact that the incubation time during which Cu diffuses past the critical length is the dominating failure mechanism at operating conditions [10, 11]. A relatively low current density was chosen to shift the failure mechanism toward a Cu incubation time-controlled mode. The onset of Al drift, concurrent with void formation and resistance/voltage changes, signals the end of the incubation time. The choice of this failure criterion also reflects the fact that the focus of of this study is the very early failure distribution, i.e., the characterization of the very first void formation phenomena leading to possible circuit failure. In addition to the detection of the earliest failures, another advantage is the fact that this criterion

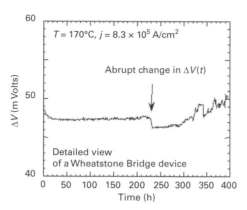

Figure 8.7 Detailed view of the change in $\Delta V(t)$ in one of the Wheatstone Bridge devices tested. The abrupt change in the voltage imbalance correlates with the end of the incubation time and signals the onset of the very first, early void formation. Due to the specific wiring scheme and the randomness of the initial void formation location, abrupt changes in the upward and downward direction are equally likely. Reprinted from [6] with the permission from AIP Publishing.

is independent of the length of the interconnect test stucture. Usually, failure criteria such as 10% or 20% resistance increase are chosen, or other values relating to a certain percentage of the initial resistance at the beginning of the experiment. A certain, constant void size creates different relative resistance increase values for different test structures, which can be an issue in extrapolations to actual interconnect design cases. The here chosen criterion is independent of these layout specifics.

A detailed view of the voltage imbalance in a Wheatstone Bridge device as a function of time, $\Delta V(t)$, is depicted in Figure 8.7. The initial voltage decrease is due to commonly encountered annealing effects and coarsening of Al_2Cu precipitates, reducing the resistance of each interconnect and hence also the small voltage imbalance of less than 50 mV.

During the incubation time, the voltage imbalance remains constant. As soon as void formation occurs anywhere in the four arrays, the imbalance changes and an abrupt behavior, as shown in Figure 8.7, can be detected. The abrupt change can occur both to higher or lower voltage imbalances, depending on the location of the first void to form. In fact, about half of the devices showed decreases during the test. It is important to note that the voltage resolution and therefore the sensitivity to detect electromigration-induced void formation processes are greatly increased by the Wheatstone Bridge technique. Due to the fact that not absolute but relative changes are measured, the monitored voltage is minimized. The typical resolution limit of commercial testing systems available in the late 1990s was on the order of 1 mV in the lowest voltage range. Due to the relative imbalance measurement, full advantage of this resolution limit was taken. The smallest corresponding resistance changes in the interconnect arrays can be estimated to be on the order of about 0.2 Ω. The abrupt behavior of the voltage imbalance can be explained by the fact that the end of the

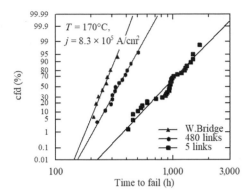

Figure 8.8 Cumulative failure distributions for Wheatstone Bridge devices (W.Bridge), 480-interconnect arrays (480 links), and basic units of five interconnects (five links) at testing conditions of $T = 170°C$ and $j = 8.3 \times 10^5$ A/cm^2. The lifetimes are shortened with increasing number of failure links. The spread in the failure distribution also continues to decrease, as expected. The straight lines are for illustrative purposes only. Reprinted from [6] with the permission from AIP Publishing.

incubation time in the first interconnect to fail signals a maximum tensile stress at the cathode end and a maximum compressive stress at the moving Cu front, which has reached the critical length. At this point in time, the stress-induced backflow cannot completely counteract the electron wind force, and considerable void formation commences [9]. The size of the voltage imbalance change will depend on the microstructure at the cathode via, which influences the void location, size, and shape.

Figure 8.8 represents a plot of cumulative failure distributions (cfds) for Wheatstone Bridge devices, 480-interconnect arrays, and basic units of five intercon-nects (links) at the aforementioned stressing conditions. As presented in earlier studies [7, 12], a general trend of decreasing lifetimes and spread in the distributions as a function of increasing failure links within each device can be discerned.

This is in accordance to the weakest link approach. The failure of the weakest link determines the lifetime of the whole assembly of multiple links. The straight lines through the failure data in Figure 8.8 are for illustrative purposes only to visualize the trend in decreasing lifetimes and spread. A log-normal fit could only be applied to a single-interconnect population where the failure mechanism is typically assumed to follow log-normal statistics. As soon as more than one failure link is tested in a chain or array, the behavior cannot be log-normal.

It must further be noted that the weakest link approach is only applicable if the incubation time is used as the failure criterion for EM. After the end of the incubation time in the first link to fail, all other links start to fail consecutively and contribute to the total resistance increase or voltage imbalance change. This part of the $R(t)$ or $\Delta V(t)$ curves cannot be used for further analysis. For example, if the first link out of 1,920 interconnects within a Wheatstone Bridge device has failed, the information about the remaining 1,919 links must be discarded. This is important in the following statistical deconvolution to the single-interconnect level.

8.2.3.2   Deconvolution to the Single-Interconnect Level

In order to assess alternate EM failure mechanisms (or "early" failures), the failure data presented in Figure 8.8 must be deconvoluted using conditional reliabilities. The goal is to characterize the EM failure distribution with respect to the single-interconnect (or single failure link) level. Most EM test data have traditionally been treated with log-normal statistics. The probability density function $pdf(t)$ then can be expressed as

$$pdf(t) = \frac{1}{\sqrt{2\pi}\sigma t} \exp\left[-\frac{1}{2}\left\{\frac{\ln(t) - \ln(t_{50})}{\sigma}\right\}^2\right], \tag{8.1}$$

where t is the time, σ the standard deviation or width of the distribution, and t_{50} the median time to fail. At each given time t, the product $pdf(t)\ dt$ gives the probability of a failure between t and $t + dt$. The integral over time gives the cumulative failure distribution $cfd(t)$, or $F(t)$:

$$F(t) = cfd(t) = \int dt\, pdf(t). \tag{8.2}$$

The fact that most test data appear to follow log-normal statistics is most probably a consequence of microstructural thin film parameters. As was shown earlier, the grain size distributions of test samples used in this study seem to obey log-normal statistics. However, for good statistical confidence, the sample size (here the number of measured grain sizes) would have to be increased considerably.

The aforementioned lack of redundancy in large-scale devices such as microprocessors allows a statistical treatment of a full-scale device as a long chain of contact/via-metal segments whose overall lifetime is defined by the time to failure of the weakest link. If we assume that each single link is statistically independent and obeys a cumulative failure distribution $cfd_1(t)$ or $F_1(t)$, the statistical behavior of a chain or an array of N links can be described as

$$F_N(t) = 1 - [1 - F_1(t)]^N, \tag{8.3}$$

where $F_N(t)$ is the cumulative failure distribution of the whole chain, parallel unit, array, or Wheatstone Bridge device. The statistical deconvolution to the single failure link level can be accomplished using conditional reliabilities in conjunction with censored data analysis. This procedure is commonly used for reliability tests where a certain number of test devices are removed after previously set readout times [13]. Table 8.1 shows the data for a calculation of conditional and unconditional reliabilities entering the procedure to deconvolute the failure data acquired for the eight Wheatstone Bridge devices containing 1,920 interconnects each.

The first column in Table 8.1 gives the number of failed Wheatstone Bridge devices, i. The second column lists the failure times t_i for each device in hours, as determined from the experiment. The third column gives the total number of interconnects n_i entering the test at time t_{i-1}, e.g., at time t_0 all eight devices with 1,920 interconnects each entered the test; therefore, there were 15,360 links at the beginning of the test. This number decreases in multiples of 1,920 since after each failure of just one interconnect

Table 8.1. Calculation of conditional and unconditional reliabilities for the statistical deconvolution of a Wheatstone Bridge device population to the single-interconnect level.

Col. 1	Col. 2	Col. 3	Col. 4	Col. 5	Col. 6	Col. 7	Col. 8
$i = \#\ of$ arrays failed	$t_i = time$ array failed (h)	$n_i = \#\ of$ links at t_{i-1}	$1/n_i$	$R_i' = 1-1/n_i$ cond. reliab.	$R_i = R_1'$ $R_2'...R_i'$ uncond. reliab.	$F_i = 1-R_i$ uncond. failure	NSD_i
1	209	15,360	0.00007	0.99993	0.99993	0.00007	−3.825
2	219	13,440	0.00007	0.99993	0.99986	0.00014	−3.635
3	229	11,520	0.00009	0.99991	0.99977	0.00023	−3.508
4	251	9,600	0.00010	0.99990	0.99967	0.00033	−3.405
5	271	7,680	0.00013	0.99987	0.99954	0.00046	−3.314
6	274	5,760	0.00017	0.99983	0.99937	0.00063	−3.223
7	292	3,840	0.00026	0.99974	0.99911	0.00089	−3.123
8	334	1,920	0.00052	0.99948	0.99859	0.00141	−2.986

Source: Reprinted from [6] with the permission from AIP Publishing.

in an array the whole four-array assembly has to be taken out from the overall test population. This procedure ensures that the stressing current density for all intercon-nects is constant over the testing time, except small deviations due to the initial voltage imbalance. A more detailed analysis, not shown here [14], enables the acquisition of several consecutive voltage imbalance changes due to their small size. Column four gives the amount of interconnects failed at time t_i (which is always only one), divided by the amount of all interconnects in the running sample population. Therefore, column four is just the reciprocal of column three. Column five gives the conditional reliability, which is $R_i' = 1 - 1/n_i$. This number gives the fraction of the population unfailed at time t_{i-1}, which also reaches the time t_i unfailed. Column six shows the recursive calculation of the unconditional reliability at time t_i. This number is given by $R_i = R_i' \times R_{i-1}$, where $R_0 = 1$. Column seven gives the fraction of the population that has failed by the time t_i, which is simply given by $F_i = 1-R_i$. Column eight yields the number of standard deviations NSD_i, which directly correlates to F_i, assuming log-normal statistics.

In the following cumulative failure plots, the number of standard deviations will be used as the ordinate instead of a percentage axis as used previously. The use of NSD_i is more convenient when analyzing multi-interconnect device failure data in terms of a single-interconnect population. The same procedure as described for Wheatstone Bridge devices was applied to the 480- and five-interconnect data shown in Figure 8.8. Due to the large amount of data, the details of this calculation are not shown here. The deconvoluted failure data for the three different device types are shown in Figure 8.9.

Note that the failure times in Figure 8.9 are identical to the ones depicted in Figure 8.8, only the probabilities change according to the preceding calculations. It is evident from Figure 8.9 that no alternate EM failure mechanisms are acting down to the four sigma level. All data coincide on one log-normal distribution, correctly represented by the straight line fit through the data. It must be noted that with the choice of five interconnects as the smallest unit and a sample size of 32 for this device type, the maximum NSD value barely reaches past zero, i.e., the intrinsic population is

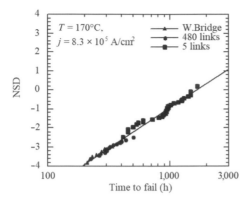

Figure 8.9 Deconvoluted failure distributions of the cumulative plots shown in Figure 8.8. The deconvolution, employing conditional reliabilities, is performed with reference to the single-interconnect link level. As evident from these data, all distributions coincide on one probability density function, therefore ruling out any alternate EM failure mechanisms down to the four sigma level (NSD = number of standard deviations). Reprinted from [6] with the permission from AIP Publishing.

not fully characterized toward higher *NSD* values. In an ideal case, it is desirable to additionally run a single-interconnect population with a large sample size. However, time constraints and testing capacity issues may interfere. Possibly a second distribution will be missed, but the main goal here is the characterization of the early failure distribution, which limits the reliability of the device population. The choice of five links as the smallest unit also helps to map out the intermediate range between −2 < *NSD* < −1, as seen in Figure 8.9. In general, test structure sizes spanning multiple orders of magnitude, such as failure link numbers of $N = 1, 10, 100, 1,000$, and 10,000 are advisable. Any number above about $N = 100$ would then have to be wired using the Wheatstone Bridge technique to avoid resolution limit issues.

8.2.3.3   Activation Energy Characterization

In order to verify the preceding results as a function of temperature and to substantiate that the failure mechanism studied here under a statistical approach correctly represents the Cu incubation time, experiments were run in the temperature range between 155–200°C. Three additional temperature splits were chosen, leading to a total sample size of 75,860 interconnects. Figure 8.10 shows deconvoluted cumulative failure plots as a function of temperature. The statistical analysis works remarkably well at all temperatures.

Figure 8.11 represents Arrhenius plots for both the five-interconnect and Wheatstone Bridge populations. The activation energy (Q or E_a) values extracted from these plots are $Q = 1.2 \pm 0.1$ eV for the five-interconnect structures and $Q = 1.08 \pm 0.05$ eV for the Wheatstone Bridge devices.

Due to the tight distributions for the latter devices, the error bars (using 90% confidence) decrease and a determination of activation energies to the second decimal

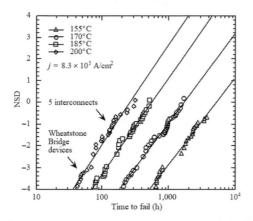

Figure 8.10 Deconvoluted data for five-interconnect basic units and Wheatstone Bridge devices as a function of temperature. The statistical analysis works remarkably well at all temperatures. Data for 480-interconnect arrays at 170°C were omitted for clarity (NSD = number of standard deviations). Reprinted from [6] with the permission from AIP Publishing.

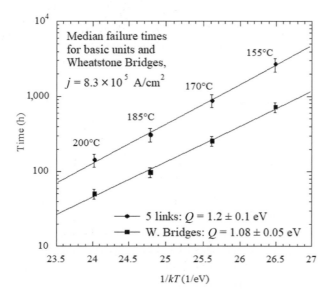

Figure 8.11 Arrhenius plots for basic units with five interconnects (links) and Wheatstone Bridge devices. The activation energies are in excellent agreement with values obtained in previous studies, using the same metallization. Reprinted from [6] with the permission from AIP Publishing.

can be performed. The activation energies obtained here are in excellent agreement with values found in earlier experiments, using near-bamboo or bamboo interconnects [11, 15], and in good agreement with generally reported values ranging from 0.87–1.0 eV for Cu diffusion in Al(Cu) metallization [15, 16]. We surmise that interface diffusion of Cu along the Al(Cu) sidewalls and along the top and bottom refractory layers, coupled with grain boundary diffusion within the interconnects, constitutes the Cu incubation mechanism. A slightly higher activation energy above $Q = 1.0$ eV, as found here, may be mainly due to the fact that the diffusion process incorporates paths along the interfaces of the Al(Cu) interconnects with the surrounding materials. This is in contrast to EM data on wide interconnects, where the Cu diffusion mainly occurs along grain boundaries.

8.2.4   Discussion of Experimental Data for Al(Cu) Metallization

Most electromigration test data have traditionally been treated with log-normal statistics, although test sample sizes are typically too small to decisively determine the true nature of experimentally acquired failure distributions. In that respect, it is important to address the issue of scaling. Assuming that the electromigration failure at the single-link level obeys a certain statistical distribution function, it is generally the case that a chain of N links does not obey this function. This is due to the fact that distribution functions usually do not scale. For example, if the single failure link obeys a log-normal distribution function, the whole chain cannot be described by log-normal behavior. The distribution function then assumes "multi-log-normal" behavior. In fact, the only distribution function that scales is the Weibull distribution function.

Lloyd and Kitchin have elucidated the conceptual problem of scaling as follows [17]: a typical electromigration experiment yields failure data that seem to follow log-normal statistics. The number of failure times generated is usually not large enough to distinguish a log-normal distribution from a different statistical behavior. The experiment was conducted with interconnects of a certain length l. If we now were to test two samples of length l as a pair, the failure of each test device would be determined by the failure of the first interconnect to fail. If the single-interconnect population had been of log-normal nature, the devices consisting of pairs could not behave according to log-normal statistics. If in turn one would cut each interconnect in half, the experimentally acquired distribution could not obey log-normal statistics. One is then left with the concern that it is very unlikely to have chosen the exact interconnect length, which leads to log-normal behavior. Furthermore, many research labs are testing interconnects of various lengths, and almost each group seems to report log-normal behavior.

Within the scope of this study, it is possible to rule out Weibull statistics as the underlying failure distribution if the spread in the distribution of electromigration data is a function of the test structure. The Weibull probability density function can be expressed as

$$pdf(t) = \left(\frac{\beta}{\alpha^{\beta}}\right) t^{\beta-1} \exp\left[-\left(\frac{t}{\alpha}\right)^{\beta}\right], \tag{8.4}$$

where α is the characteristic lifetime (reached at a 63.2 percentile) and β is the shape parameter. A high value for β corresponds to a small spread in the failure distribution. This distribution is mentioned here due to its uniqueness in terms of scaling, as mentioned earlier. If a single-interconnect population obeys Weibull failure statistics, then a chain of independent links from the same pool as the single-interconnect population will obey Weibull failure statistics as well. The Weibull shape parameter β does not change as a function of the number of possible failure links. Only the failure times are reduced with an increase in tested links per device. The failure distributions in Figure 8.8 clearly show a decrease in the spread with an increasing number of possible failure links. We can therefore rule out Weibull statistics for the EM data acquired here.

For a broader statistical analysis of electromigration data as a function of possible failure links, the traditionally used log-normal distribution function was chosen as a starting point for numerical simulations. A previous study describes this statistical approach and its implications in detail [12]. An experiment was simulated where several groups of 60 test devices were used. The difference between each group of devices is the number of via–interconnect links in the chain, where this number varies from 1, 10, 100, and up to 10^6. A maximum of 60 million failure times, obeying a log-normal distribution function, were obtained by a random number generator. The failure times were randomly grouped to form 60 devices in each case. The shortest lifetime of all via–interconnect links within each device was used as the failure time of that particular device, according to the weakest link approach. Figure 8.12 shows the simulated data in a log-normal probability plot, where each data point represents a device with a chain of a specific number of links.

The underlying single-failure link population was generated around an arbitrarily chosen t_{50} value of 100 hours and a log-normal standard deviation of 0.5. The two important features of this plot are that the t_{50} values and the widths of the distributions

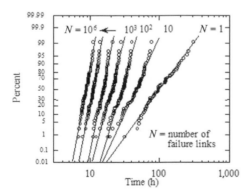

Figure 8.12 Simulated failure distributions of chains with various numbers of links. Millions of failure times were randomly generated along the one-via log-normal distribution, and groups of 10, 100, etc., used to determine the behavior of multivia chains. The weakest link determines the lifetime of the whole chain. Reprinted from [6] with the permission from AIP Publishing.

decrease with increasing number of vias or interconnects in the chain, as found experimentally in this study. The behavior of a population of full-scale devices is therefore expected to show very different characteristics than a population representative of typically used test structures with only one or a few via–interconnect links. The device population will be characterized by a much tighter distribution and shorter lifetimes. However, a correct use of one-via–interconnect data predicts the first fail within the device population correctly. It is important to note that the straight line fits in Figure 8.12 are for illustrative purposes only, just as in Figure 8.8. The straight lines suggest a log-normal fit to the failure data; however, the multilink failure distributions cannot obey log-normal statistics, as mentioned previously. The lines are drawn to highlight the change in the spread of the failure distributions. This is to be expected if the distribution at the single-failure link level is of a log-normal type.

A reliability goal of, e.g., 1 FIT for a full-scale device population can be represented in two ways: keeping in mind that one has to account for the first failure out of a population of 10^4 full-scale devices at 10^5 hours, one has to ensure a cumulative failure of $F = 10^{-4}$, or 0.01%, at that time. Under the weakest link approach, this time coincides exactly with the failure of the first link out of a population of 10^{10} single links if one assumes 1 million links per full-scale device, all of them being exposed to a certain maximum current density limit (which needs to be checked through chip-level design simulations). Therefore, a cumulative failure of $F = 10^{-4}$ for the device population with 1 million links per device is identical to a cumulative failure of $F = 10^{-10}$ for a single-link population. This fact is often overlooked in the extrapolation of EM data acquired on test structures containing only one or a few vias or interconnects. It is important to distinguish between a one-FIT reliability goal for a device population or a test structure population.

We are still left with the question whether the "core" of an experimentally acquired EM failure distribution, namely the single-link population, obeys log-normal statistics. The smallest unit tested in this study contained five via–interconnect links. We therefore can be certain that this distribution is not of a log-normal type. If we had acquired data on a single-interconnect test structure, we would not have been able to distinguish between a log-normal and a multi-log-normal distribution function due to the vast number of experimental failure times needed. The failure distribution of the single-interconnect test structure may already have been of a multi-log-normal type. The length of the interconnect test structure was 100 µm, and we cannot be certain that this is the "correct" length for a log-normal distribution, as discussed by Lloyd and Kitchin [17]. However, a closer look at the EM failure mechanism studied here will prove helpful. The unique choice of the Cu incubation time as the EM failure criterion, rather than a customarily chosen relative resistance increase value, simplifies the argument considerably. In addition, the test structures are not single-level metal stripes as often used in EM experiments. The failure mechanism therefore is the depletion of Cu at the cathode end of each interconnect, caused by the W via, which acts as a flux divergence for Cu diffusion [18–20]. The incubation time is reached when the Cu concentration over a length equal to the critical Blech length, starting from the cathode via, has fallen below a critical concentration to inhibit Al drift. This concentration may be on the order of several 100 ppm [21]. The critical length can be estimated to be about 60 μm for the current density of $j = 8.3 \times 10^5$ A/cm^2 used in our experiment. A previous study using the same metallization scheme yielded a critical current density-length product $(jl)^*$ of about 4,800 A/cm [22]. The incubation time is therefore the time that is necessary to deplete about 1 wt% of Cu over a length of approximately 60 μm. The depletion mechanism is a combination of diffusion along interfaces and through grain boundaries. Assuming that the microstructure is characterized by a grain size distribution that obeys log-normal statistics, it is not surprising that the depletion times obey the same behavior. If the Cu diffusion paths are distributed according to a log-normal distribution and the EM driving force is constant in time at each point in the diffusion path, the failure times will obey log-normal statistics as well. The specific length l leading to log-normal behavior is therefore the Blech length, which is set by the current density chosen in the experiment.

We now have shifted the problem from EM failure time distributions to microstructural parameters. As in the case of experimentally acquired EM failure distributions, the amount of grain sizes measured is typically too low to determine the true statistical nature of their distribution. In this study, only a few hundred grain sizes were measured (see Figure 8.6). However, theoretical considerations can be used to explain the fact that thin film grain distributions typically obey log-normal statistics [23]. Within the scope of this work, log-normal behavior was found to be valid down to a cumulative failure of about one out of 20,000 per testing temperature. This corresponds to an *NSD* value of almost four. More data collection and optimization of test structures will lead to cumulative failures of about one out of several million, reaching the five sigma level. However, the goal of one out of 10^{10}, as mentioned earlier in the discussion of theoretical FIT values, will most probably never be reached. In the later section on Cu-based metallization, failures of about one out of several 100,000 links will be presented, coming close to a 1 ppm goal that is typically required for automotive applications.

The question remains why early failures have been reported in accelerated EM testing in the past. Keeping in mind that the dominating failure mechanism at operating conditions of around 100°C and about 1×10^5 A/cm^2 current density for Al(Cu) metallization is the Cu incubation time, it is difficult to explain how the diffusion process of Cu past a critical length of a few 100 μm could be of a bimodal nature. Again, a closer look at the EM failure criterion will prove useful. If a relative resistance increase criterion is used and the experiments are conducted at relatively high temperatures and stressing currents, the failure mode is shifted toward Al diffusion-controlled processes. Due to the high activation energy of Cu depletion mechanisms and due to the high current exponent of $n = 2$, this first stage of the EM damage process will be masked completely [10, 11]. The EM failure times will then become very dependent on the local microstructure of the Al(Cu) interconnect close to the via. For example, a grain boundary triple point right on top of the via would lead to very fast diffusion and manifest itself in an early failure data point. However, the preceding Cu incubation time at low temperatures and current densities would have rendered this data point irrelevant.

8.3   Statistical Tests and Analysis for Cu Interconnects

The drive to ever higher current densities due to device scaling has prompted the switch from Al(Cu)-based metallization toward Cu-based interconnects by the end of the last millennium. In August 1997, and soon thereafter at the 1997 IEEE IEDM, IBM announced a schedule for the first implementation of Cu interconnect technology on ICs, in this case for logic products in its 0.22 μm CMOS generation [24]. At the same International Electron Devices Meeting (IEDM) meeting, Motorola announced its entry into Cu technology [25]. By the end of 1998 and into 1999, both IBM and Motorola started shipping ICs with Cu metallization in high volume, while other companies followed after a few years of delay. The introduction of Cu technology enabled a significant performance and reliability improvement [26]. However, the introduction of the dual-damascene process brought with it a profound change in EM statistics, leading to distinct bimodality in both up- and downflow current directions. This is in strong contrast to the previously described EM statistics for Al(Cu), which has not shown any deviation from monomodality down to the four sigma level. This is mainly due to the fact that Al(Cu) metalliation typically applies a refractory, Ti/TiN-based barrier layer at both the bottom and top interfaces, in addition to the use of W vias, which are very stable and do not take part in the overall diffusional processes leading to electromigration-induced voiding. The following subsections will discuss these implications in detail.

8.3.1   Specific Considerations for Cu Interconnects

Electromigration failures in Cu interconnects are mainly the result of void growth at the cathode end of the line, just as for Al-based metallization. Also, the corresponding times to failure (*TTF*) usually follow a log-normal distribution with the median lifetime (*MTTF*) defined by the quality of the top interface, which controls the mass transport [27–30]. This is one of the main differences to Al(Cu)-based metallization where grain boundary diffusional processes are the main cause for void formation. Since electromigration experiments are performed at much higher current densities and temperatures compared to operating conditions, extrapolations are necessary to assess reliability at operating conditions using the following equation:

$$TTF_{oper} = MTTF_{stress} \left({}^{I_{stress}}\!/\!_{I_{oper}} \right)^n \exp\left[\frac{E_a}{k_B} \left(\frac{1}{T_{oper}} - \frac{1}{T_{stress}} \right) + NSD*\sigma \right], \quad (8.5)$$

where I is the current, n the current exponent, E_a (or Q) the activation energy, k_B Boltzmann's constant, T the temperature, NSD the number of standard deviations, and σ the log-normal standard deviation. The subscripts $_{stress}$ and $_{oper}$ refer to EM testing and actual operating conditions, respectively. Equation (8.5) indicates that the main factors requiring attention are the activation energy related to the dominating diffusion mechanism, the current exponent, as well as the median lifetimes and log-normal standard deviation values of experimentally acquired failure time distributions.

Whereas the origins of EM activation energy and current exponent as well as the behavior of median lifetimes with continuing device scaling are relatively well understood [26, 29], the origin and scaling behavior of the log-normal standard deviation is more complex, with a few studies focusing on a detailed analysis [31–35]. Strong bimodality for the electron upflow direction in dual damascene (or "dual-inlaid") Cu interconnects has added complexity to the analysis of the standard deviation [36–38]. Contrary to single-inlaid technology such a Al(Cu)-based metallization and very early Cu integration schemes, the failure voids can now occur both within the via ("early" mode) or within the trench ("late" mode). Subsequently, bimodality has been reported also in downflow EM, leading to very short lifetimes due to small, slit-shaped voids under vias [39, 40]. In addition to shorter *MTTFs*, these early failure modes can have different activation energy, current exponent, and sigma values compared to the late mode depending on the diffusion and void formation mechanisms. Hence, a more thorough investigation of these early failure phenomena is needed. To accomplish this task, again specific test structures were designed based on the Wheatstone Bridge (WSB) technique, as described above for Al(Cu) metallization.

8.3.2 Experimental Results: Standard Electromigration Testing Methodology

Before an overview of the statistical EM testing results, the standard EM testing methodology as well as typical lifetime data will be reviewed. Standard EM testing is performed on a Cu line connected by vias on both ends of the structure to a lower- or upper-metal level. The current flows from a wide metal line through a via into the test line. If the test line is on a lower-metal level compared to the current supply line, the test is called "downflow," whereas it is called "upflow" for test lines on a higher-metal level. Only one via–line interface is being tested per sample, as shown in Figure 8.13.

The length of the 90 nm technology test lines used in this study is 250 µm and the width is 0.12 µm for downflow and 0.14 µm for upflow EM, respectively. The Cu interconnect is surrounded by a Ta-based barrier layer and an SiCN cap layer. SiCOH is used as the dielectric at metal and via levels. EM testing was performed at temperatures between 225–325°C. A 10 % resistance increase was used as the failure criterion. The EM lifetimes were analyzed using a log-normal distribution function. Additionally, scanning electron microscope (SEM) images were obtained to illustrate the failure mechanism. Lifetime distributions for downflow EM tests performed at five different temperatures using a current density of 1.5 MA/cm^2 are shown in Figure 8.14 [41].

In general, the distributions appear to be well behaved, monomodal, and with longer lifetimes at lower temperatures, as expected. However, increased sigma values at some temperatures indicate an early failure distribution, especially at 275°C. Since a distinct separation of an early and a late failure mode does not exist, the possible early failure population appears to be small and thus not definable given the total sample size of 281. Failure analysis using focused ion beam cutting and SEM imaging was performed on samples with short and median lifetimes. Two basic void locations and shapes were identified as shown in Figure 8.15a and b, consistent with previous studies [39, 40].

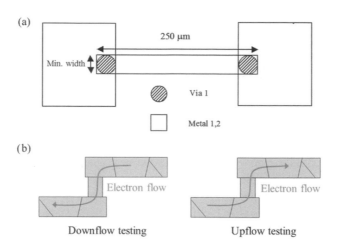

Figure 8.13 Standard electromigration test structures: (a) top view, and (b) cross-sectional view for downflow and upflow testing. The test line (M1 or M2) is kept at the minimum width as allowed by design guidelines.

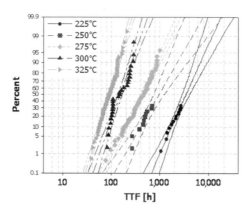

Figure 8.14 Lifetime distributions for downflow electromigration in single-link devices. Reprinted from [41] with the permission from AIP Publishing.

Early failure samples show slit voids under the via in the metal 1 line. In contrast, samples with median lifetime have faceted voids in the metal 1 line under the via or adjacent to the via. This observation supports the existence of two distinct failure mechanisms. Void formation either starts under the via possibly at a defect location leading to faster failure, or voids form at the interface away from the via and evolve toward the line end.

Figure 8.16 shows lifetime distributions of upflow EM experiments obtained at four different temperatures and a current density of 1.5 MA/cm^2. A monomodal log-normal distribution was used to fit the data.

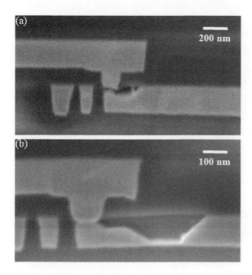

Figure 8.15 SEM images of (a) an early failure void and (b) a late failure void for downflow electromigration. The lateral via size at midheight is about 140 nm. Reprinted from [41] with the permission from AIP Publishing.

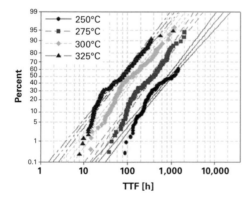

Figure 8.16 Lifetime distributions for upflow electromigration in single-link devices. Reprinted from [41] with the permission from AIP Publishing.

However, significant deviations between the fit and the data points were observed, clearly indicating the existence of multiple failure modes. Approximately 35% of the samples contribute to an early mode. Failure analysis was performed on samples with short and long lifetimes to examine the different failure mechanisms. Two different failure locations were identified in agreement with earlier work [36–38]. Voiding in the via as indicated in Figure 8.17a corresponds to smaller lifetimes, while the formation of a larger failure void in the metal 2 trench results in longer lifetimes, as shown in Figure 8.17b.

Causes for the different failure void location might be process-induced defects in the via and/or grain boundary orientation in the via/line area. For both up- and

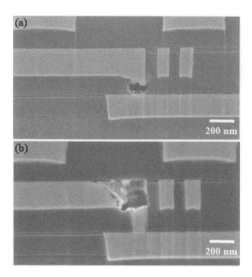

Figure 8.17 SEM images of (a) an early failure void and (b) a late failure void for upflow electromigration. The lateral via size is about 140 nm at midheight (compare to Figure 8.16). Reprinted from [41] with the permission from AIP Publishing.

downflow void formation mechanisms, the initial nucleation process will be the limiting factor. Besides "extrinsic" effects such as imperfect cleaning processes at the bottom of the via or insufficient TaN/Ta barrier and/or Cu liner coverage, local stress effects may play a role. Due to the high mechanical anisotropy of Cu, misorientation between grains affects the state of local stresses that in turn could affect stress migration or stress-induced void formation [42, 43]. These effects could influence the local electromigration mechansims, in the sense of a stress migration–assisted electromigration phenomenon. Especially in the case of adjacent $\langle 100 \rangle$ and $\langle 111 \rangle$ oriented grains, local stress–strain mismatch could lead to premature, i.e., early nucleation, which then leads to an early failure due to fast void formation at critical sites. For example, if such a critical juncture occurs directly under or in a via, a stress-assisted nucleation process could induce fast failure.

8.3.3   Experimental Results: Statistical Electromigration Testing Methodology

As mentioned earlier, extrapolations from EM stressing conditions to operating conditions require correct values for *MTTF* and sigma, obtained through distribution fitting routines of the experimental data. Due to the rather high percentage of early failures for the upflow case, it is possible to easily use a bimodal distribution fit to obtain separate sets of *MTTF* and sigma values for the early and late populations. These parameters can be used for separate, adequate extrapolations for both modes. However, the downflow EM case is much more challenging. Due to the apparently small percentage of early failures for this mechanism, the identification of separate parameters for the early and late modes is not possible with a typical dataset. Even a few hundred or thousands of samples will not be sufficient to extract the necessary parameters. However, for an extrapolation to use conditions, the assumption of a monomodal distribution will not be adequate. Since it is not feasible to test enough parts using standard, single-link structures, the statistical testing methodology based on the WSB technique as described for Al(Cu) metallization was used for a detailed examination of these failures.

8.3.3.1 Experimental Procedure Using Wheatstone Bridges

Several studies employed statistical test structure designs to examine the void formation behavior in interconnects [6, 14, 29, 36, 40, 41, 44]. The test structures used in this part of the analysis consist of an arrangement of parallel and serial interconnects, wired into a Wheatstone Bride design. This particular test design was successfully employed for early failure studies in Al(Cu) interconnects [6, 14] and later for Cu metallization [45–49]. While additional details can be found in these references, the specifics of the first Wheatstone Bridge design in Cu metallization will be described in the following. The smallest unit is composed of 10 interconnects in parallel, as shown schematically in Figure 8.18a.

Note that the number of interconnects in parallel was increased from five to 10 when compared to the initial studies with Al(Cu) metallization; see Figure 8.2. The ability of void detection decreases with the number of increasing links in parallel. However, the resolution of a typical electromigration test system is usually adequate and enables an increase to about 10. With better resolution limits, this number can be increased to about 20. Each trench is terminated by vias on both line ends. The metallization scheme and most trench geometries were identical to the standard single-link samples. Only the line length was chosen differently to be 50 μm as a result of test system limitations (voltage compliance). However, at the current densities used in this study, no differences were observed between 50–250 μm long standard structures validating a direct comparison between the 50 μm long statistical structures and the 250 μm single-link samples. The basic unit of 10 parallel interconnects was repeated 23 times in series to build a large interconnect array of 230 segments as possible failure links.

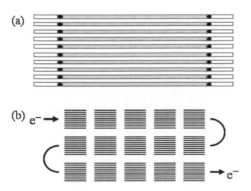

Figure 8.18 Building blocks for large-scale Cu electromigration testing: (a) a basic unit of 10 interconnects in parallel, and (b) an interconnect array formed of basic units in series. Reprinted from [41] with the permission from AIP Publishing.

Figure 8.18b displays a schematic of this structure. With an increasing number of interconnect links, the resistance of each test device increases, resulting in a decrease in resolution of small resistance changes during an EM test. Again, the solution to this problem is the use of the Wheatstone Bridge. Four such arrays are being used as shown in Figure 8.4 for Al(Cu) metallization. This way, the bridge technique enables an increase in the number of tested links per device without the loss of detection of small resistance changes. In this study, each bridge structure then consisted of $4 \times 230 = 920$ interconnects instead of just one in standard test designs.

Downflow EM experiments were performed at seven temperatures between 150–325°C and a current density of 1.5 MA/cm^2. Upflow tests were limited to 325°C and 1.5 MA/cm^2, since bimodality is readily seen in standard, single-link EM testing. The total number of initially examined interconnects (here for the 90 nm technology node, the first Cu node evaluated with WSB devices) was 115,920 and 686,400 for upflow and downflow directions, respectively. Figure 8.19a and b shows resistance imbalances as a function of time for 275°C downflow tests at 1.5 MA/cm^2 and 0.15 MA/cm^2, respectively.

In both graphs, the resistance imbalances are initially close to zero and remain constant. However, in the higher current density experiment, changes in the imbalance are clearly observable for all samples. In contrast, no such shifts are found in the test using the significantly reduced EM driving force. This observation supports the expectation that a change correlates with failure void formation in one of the 920 interconnects. Additionally, the steadiness of the resistance imbalances over time, as seen in Figure 8.19b, indicates the stability of the WSB technique.

The observed changes can occur both to higher or lower resistance imbalances, depending on the location of the first failure void within the four interconnect arrays. In agreement with expectation, about half of the devices showed upward trends and the other half downward trends. The failure criterion used in this study is the time to first discernible resistance imbalance change. A detailed view of resistance imbalances in selected WSB devices as a function of time is depicted in Figure 8.20.

The first steps in the traces are clearly visible. It is important to note that the resistance resolution and therefore the sensitivity to detect EM-induced void formation are greatly increased by the WSB technique. Due to the fact that not absolute but relative changes are measured, the monitored resistance is minimized. As expected, the technique works very well for both Al(Cu) and Cu metallization. Figure 8.21 represents a plot of lifetime distributions for downflow WSB devices and the corresponding single-link samples tested at 325°C and 1.5 MA/cm^2.

The lines through the failure data using a monomodal log-normal fit are for illustrative purposes only. A log-normal fit could only be applied to a single-interconnect population where the failure mechanism is typically assumed to follow log-normal statistics. As soon as more than one failure link is tested, the behavior cannot be log-normal [14, 17]. A significant decrease in lifetimes can be discerned for the WSB samples compared to standard single-link devices. This is in accordance to the weakest link approach, i.e., the failure of the weakest link determines the lifetime of the whole assembly of multiple links. It needs to be remembered that one data point

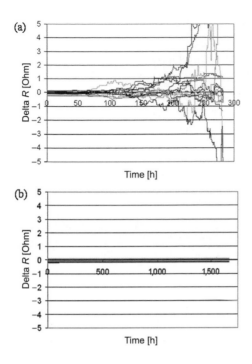

Figure 8.19 The resistance imbalance as a function of time for downflow testing at 275°C and a current density of (a) 1.5 MA/cm^2 and (b) 0.15 MA/cm^2. Reprinted from [41] with the permission from AIP Publishing.

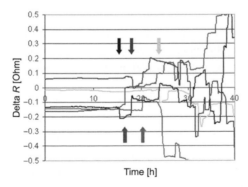

Figure 8.20 Detailed view of selected resistance imbalances as a function of time. Arrows indicate the first distinctive steps. Reprinted from [41] with the permission from AIP Publishing.

in the WSB distribution represents the first failed line out of 920 interconnects. Thus, while showing only 112 data points in this experimental subset, the Wheatstone distribution contains information about 103,040 interconnects. However, exact failure times for these trenches besides the first link to fail usually cannot be obtained, since all other links start to fail consecutively and contribute to the total resistance

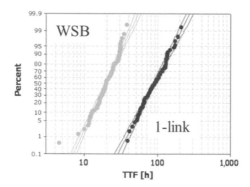

Figure 8.21 Lifetime distributions for downflow Wheatstone Bridge devices and the corresponding single-link samples at 325°C and 1.5 MA/cm^2. Reprinted from [41] with the permission from AIP Publishing.

imbalance change. Then this part of the $\Delta R(t)$ curves cannot be used for further analysis. Hence, once the first link out of 920 interconnects within a Wheatstone Bridge device has failed, generally no detailed information about the remaining 919 links can be extracted. This is important for the statistical deconvolution to the single-interconnect level discussed further in Section 8.3.3.3.

8.3.3.2   Characterization of Joule Heat

It is well known that interconnects under certain EM testing conditions exhibit Joule heat effects resulting in a higher effective temperature of the test lines compared to the actual oven temperature. This effect is expected to be more pronounced in the WSB arrangement, since current is applied simultaneously to a large number of adjacent interconnects (the number of interconnects in parallel was increased to 10, as compared to 5 for Al(Cu) metallization). To examine the resulting Joule heat, the resistances of individual 230 line arrays were measured as a function of current. The lowest current value represented a current density of 0.15 MA/cm^2. Using this value as a reference, the temperature change as a function of applied current was obtained. For comparison, the Joule heat was assessed in single-link structures using a similar method. Both upflow and downflow structures were examined. Figure 8.22 shows a plot of temperature change as function of applied current density for multi- and single-link downflow samples.

Clearly, considerably more Joule heat is observed in WSB structures compared to single-link samples due to the denser arrangement of lines and vias. It is important to note that at standard testing conditions of 1.5 MA/cm^2, Joule heat remains at reasonable levels of 2.3°C for the WSBs and 0.2°C for the single-link structures. Thus, the lifetime data are not influenced significantly. In contrast, at higher current densities, the temperature increase in the WSBs as a result of Joule heating can exceed 20°C. Since EM is exponentially dependent on temperature, this Joule heat increase can significantly influence the lifetime data by inducing failure earlier than expected. Thus, to enable

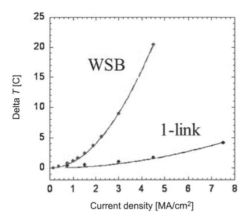

Figure 8.22 Temperature change as a function of the applied current density for Joule heat analysis in downflow samples. Reprinted from [41] with the permission from AIP Publishing.

an accurate comparison, the failure times reported in this study were adjusted to account for the difference in oven temperature and effective temperature close to the test lines.

8.3.3.3 Deconvolution to the Single-Interconnect Level

In order to assess the EM early failure mechanism, the WSB failure data presented in Figure 8.21 must be deconvoluted using conditional reliabilities. The goal is to include the limited information about the nonfailed links in the EM failure distribution, and thus enable a direct comparison between Wheatstone and single-link structures. In other words, one characterizes the entire distribution with respect to the single-interconnect level. The statistical deconvolution can be accomplished using conditional reliabilities in conjunction with censored data analysis. This procedure is commonly used for reliability tests where a certain number of test devices are removed after previously set readout times; see Section 8.2.3.2 and [13]. Table 8.2 shows the data for a calculation of conditional and unconditional reliabilities illustrating the procedure to deconvolute the failure data acquired for 10 WSB devices containing 920 interconnects each.

This small subpopulation was tested at 275°C using 3 MA/cm^2 as a portion of a current exponent evaluation. This subpopulation was chosen here due to its small size and simple representation of the deconvolution technique. The first column in Table 8.2 contains the number i of failed WSB devices. The second column lists the failure times for each device as determined from the experiment. The third column gives the total number of interconnects n_i running at time t_{i-1}. For instance, at time t_0, all 10 devices with 920 interconnects each are in test, resulting in 9,200 links at the beginning of the experiment. This number decreases in multiples of 920, since after the failure of just one interconnect in a WSB structure the entire four-array assembly has to be removed from the overall test population. This procedure ensures that the stressing current density for all interconnects is constant over the testing time, except

Table 8.2. Calculation of conditional and unconditional reliabilities for 10 Wheatstone Bridge devices tested at 275°C using 3 MA/cm².

i: # of failed samples	t_i = failure time [h]	n_i = # of links running at t_{i-1}	$R_i' = n_i/(n_i + 1)$	$R_i = R_i^* R_{i-1}$	$F_i = 1-R_i$	NSD_i
1	12.4	9,200	0.999891	0.999891	0.000109	−3.697927
2	13.6	8,280	0.999879	0.999771	0.000229	−3.503691
3	13.8	7,360	0.999864	0.999635	0.000365	−3.377869
4	14.5	6,440	0.999845	0.999480	0.000520	−3.279233
5	14.9	5,520	0.999819	0.999299	0.000701	−3.194041
6	15.8	4,600	0.999783	0.999081	0.000919	−3.115338
7	17.3	3,680	0.999728	0.998810	0.001190	−3.038173
8	20.35	2,760	0.999638	0.998448	0.001552	−2.957277
9	23.55	1,840	0.999457	0.997906	0.002094	−2.863615
10	24	920	0.998914	0.996822	0.003178	−2.728860

Source: Reprinted from [41] with the permission from AIP Publishing.

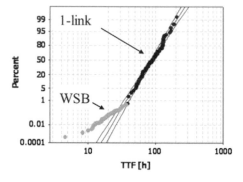

Figure 8.23 Deconvoluted lifetime data for downflow Wheatstone Bridge devices and the corresponding single-link data at 325°C. Reprinted from [41] with the permission from AIP Publishing.

small deviations due to the initial resistance imbalance. Column four contains the conditional reliability calculated using the Herd–Johnson method, i.e., it is defined as $R_i' = n_i/(n_i + 1)$. This number provides the fraction of the population not failed at time t_{i-1}. Column five shows the recursive calculation of the unconditional reliability at time t_i. This number is given by $R_i = R_i' \times R_{i-1}$, where $R_0 = 1$. Column six lists the fraction of the population that has failed by the time t_i, which is given by $F_i = 1-R_i$. Finally, the number of standard deviations NSD_i is displayed in column seven, which directly correlates to F_i assuming log-normal statistics. Cumulative distribution plots can now be represented using either the failure percentage or the corresponding NSD values as the ordinate. For plotting convenience, most graphs display the latter.

In addition to single-link data, deconvoluted lifetime data for downflow WSBs tested at 325°C and 1.5 MA/cm² are shown in Figure 8.23.

Note that the lifetimes are identical to the ones depicted in Figure 8.21, only the probabilities are different according to the preceding calculation. The deconvoluted WSB data and the single-link results align well at the transition point. However, it is evident that the majority of the WSB data deviates from the monomodal behavior, which is represented by the straight line fit through the single-link data. This observation clearly indicates the existence of an early failure mechanism for approximately 0.1% of downflow samples. It is important to note that a distinct separation of an early and a late failure mode was not possible using only the single-link data due to the rather small number of early failures. However, a detailed characterization of these failures is essential, since these devices limit the reliability of the entire population and hence the product in the field.

8.3.3.4 Characterization of Early Failure Modes

As seen in Figure 8.23, WSB structures are necessary to characterize the early failure mechanism in downflow devices. Thus, EM experiments using these structures were performed at seven different temperatures to determine the activation energy corresponding to the early mode, extending the temperature range to 150°C, which is very low compared to standard testing temperatures. In contrast to the downflow case, results from single-link tests on upflow structures clearly showed an early failure distribution. Thus, activation energy and current exponent values of both early and late modes can be determined from the single-link population.

8.3.3.4.1 Downflow EM Testing

Lifetime distributions of downflow WSB devices corresponding to seven temperatures in the range of 150–325°C are shown in Figure 8.24. All experiments were performed at a current density of 1.5 MA/cm^2.

The data in this figure are not deconvoluted, i.e., one data point represents 920 interconnects. The lifetime distributions show consistent *MTTF* and sigma values, with increasing *MTTF* as the temperature decreases and comparable sigma values over

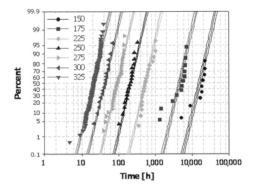

Figure 8.24 Lifetime distributions of downflow Wheatstone Bridge devices as a function of temperature. Reprinted from [41] with the permission from AIP Publishing.

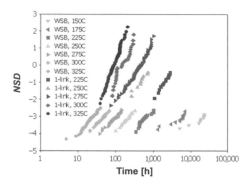

Figure 8.25 Deconvoluted lifetime data for downflow Wheatstone Bridge devices and single-link data. Reprinted from [41] with the permission from AIP Publishing.

the entire temperature range. Figure 8.25 displays the deconvoluted WSB data as well as the corresponding single-link results.

The distributions originating from multilink and single-link testing align reasonably well at all temperatures with only slight disconnects at 275°C and 250°C. Due to the long testing times, no single-link data at 175°C and 150°C is available. The WSB distributions show consistently higher sigma values, indicating the existence of the early failure mode at all temperatures. According to the weakest link approach, it is expected that the majority of the multilink samples fail according to the early failure mechanism. Thus, they are thought to be caused by slit-like failure voids under the via. The fact that the sigma value is larger for the early failure tail compared to the late failure mode can have several reasons. It might be partly attributed to the void formation being defined by the quality of the interface between via bottom and metal line. The sigma values of the strong mode are rather well defined by the line dimensions and the top interface. However, the location of the early failure slit voids – the via bottom area – is rather sensitive to processing issues, e.g., etch damage or residue might lead to preexisting voids. Depending on the type of defect, the time to form a failure void can be significantly impacted, resulting in a broader lifetime distribution of the early mode. In addition, the initial stages of EM-induced void formation have been shown to display a larger log-normal sigma value when compared to the later stages [31–35]. This observation can be explained by the initially lesser effect of averaging over multiple diffusional processes, since only the area directly underneath the via significantly contributes to void formation. At later stages and larger void sizes, the diffusion is averaged over a larger area of the top interface, leading to a "sharpening" of the effective diffusion path. This sharpening effect in turn leads to tighter EM lifetimes at later stages. Another factor possibly contributing to the larger sigma value of the early mode might be an elevated tensile stress level in the Cu line (directly under the via). This stress level is impacted by the exact shape of the via and the quality of the surrounding materials, and might have an influence on the void formation time of a slit void. It appears that the percentage of these fails is small. However, it needs thorough investigation since this failure population defines the actual chip lifetime.

A few very early fails were found at 175°C and 150°C, possibly indicating a "third" failure mode. It has been observed that stress-induced voiding in Cu interconnects occurs predominantly in the 150–200°C range [50]. To examine whether stress-induced voiding could have caused the observed very early failures under EM testing conditions, WSB testing was performed at 0.15 MA/cm^2, 10× lower than regular testing conditions, to rule out any EM-induced effects. Thirteen and 9 samples were tested at 175°C and at 150°C, respectively, resulting in 11,960 and 8,280 examined interconnects. Resistance imbalances as a function of time for both temperatures were stable just as represented in Figure 8.19b. No failure was observed in more than 1.9 years of testing at 175°C and 2.7 years of testing at 150°C. As the very early fails in the EM tests occurred significantly earlier, "classical" stress-induced voiding is unlikely to be the cause of these fails. However, "stress-assisted" EM is a possibility, as the Cu lines are subject to higher tensile stresses at these testing temperatures, possibly reducing the critical stress level for EM void formation and/or facilitating EM mass transport. Ideally, to determine activation energy values for both failure modes, a fit to the experimental lifetime distribution needs to be obtained using two log-normal distributions, one for each failure mechanism. It is reasonable to assume that both mechanisms follow log-normal distribution characteristics, since their kinetics are both controlled by the top interface of the metal 1 line. However, due to the large size of the dataset, this is a rather challenging task. Since the majority of the multilink devices appear to fail according to the early mode and most of the single-link samples display late mode characteristics, it seems to be a reasonable approximation to determine the characteristics of the early mode from the WSB devices and the parameters of the late mode from the single-link samples. Figure 8.26 shows Arrhenius plots for both device types.

The single-link samples yield an activation energy of 0.91 ± 0.03 eV. This value represents diffusion at the Cu–SiCN interface on top of the metal 1 line, in good agreement with data reported in literature [26, 29, 35, 44]. In contrast, a slight decrease

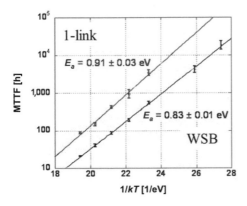

Figure 8.26 Arrhenius plots for downflow Wheatstone bridge structures and standard single-link samples indicating different activation energies. Reprinted from [41] with the permission from AIP Publishing.

in activation energy is observed for the WSB structures, namely a value of 0.83 ± 0.01 eV. Due to the large amount of data and resulting small error bars, this difference is deemed to be significant. Thus, the early failure mode in downflow samples shows a decrease in activation energy. This change is possibly the result of void formation being defined by the quality of the interface between via bottom and metal line, and not solely by the diffusion along the top metal 1 interface. In addition, the elevated tensile stress level in the Cu line (directly under the via) may lead to a reduction in the apparent activation energy for the early failure mechanism. This slight decrease in activation energy will have a significant impact on the extrapolations to use conditions, as will be discussed further later in this section.

8.3.3.4.2 *Upflow EM Testing*

Only limited data using upflow WSB structures are needed, since single-link results already show a clear distinction between an early and a late failure mode. Thus, only the validity of the WSB technique will be shown, followed by the characterization of the different failure modes from single-link data. Figure 8.27 displays the lifetime distribution of upflow WSB samples obtained at 325°C and 1.5 MA/cm^2 in addition to single-link data already shown in Figure 8.16.

The deconvoluted WSB and the single-link data align well at the transition point, providing additional support for the validity of the measurement technique. The multi-link results confirm the bimodality in the upflow failure distribution already observed with single-link testing. Thus, even though no failure analysis could be performed on WSB samples, it is expected that they fail by voiding in the via as observed in early failed single-link samples. The determination of the activation energy value for the early and late failure modes in upflow samples was accomplished using single-link results. The lifetime distributions were divided into early and late failure times and analyzed separately. Figure 8.28 shows the resulting Arrhenius plots for both modes.

In contrast to the downflow structures, comparable activation energy values of 0.91 ± 0.05 eV and 0.93 ± 0.03 eV were obtained for the late and early failure modes of upflow tested samples. This result indicates that both failure mechanisms – void

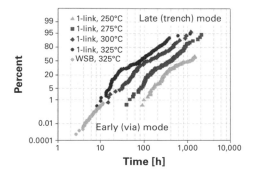

Figure 8.27 Deconvoluted lifetime data for upflow Wheatstone Bridge devices and the corresponding single-link data. Reprinted from [41] with the permission from AIP Publishing.

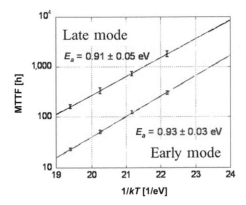

Figure 8.28 Arrhenius plots for early and late failure modes of upflow standard single-link samples. Reprinted from [41] with the permission from AIP Publishing.

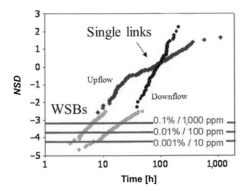

Figure 8.29 Lifetime distributions for upflow and downflow Wheatstone Bridges and single-link devices at 325°C and 1.5 MA/cm^2. Various reliability targets are represented by the horizontal lines (10–1,000 ppm). Reprinted from [41] with the permission from AIP Publishing.

formation in the via as well as in the metal 2 line – are controlled by diffusion at the Cu–SiCN interface on top of the metal 2 line. It seems that even the diffusion of the Cu ions that leave the via and induce void formation in the via is defined by the top interface of the metal line and not by diffusion paths within the via. Even if a void nucleates within the via volume, the rate limiting step must be the top interface, which would indicate that the grain boundary diffusion within the via microstructure most likely has a slightly higher activation energy than that top interface diffusion mechanism.

8.3.3.4.3 *Comparison between Downflow and Upflow EM Data*

Figure 8.29 shows the lifetime data for upflow and downflow WSB devices as well as single-link structures at 325°C and 1.5 MA/cm^2. All acquired data, namely 115,920 and 686,400 for upflow and downflow directions, respectively, are represented here.

Clearly, the *TTF* values of the early mode in upflow samples originating from WSB samples as well as single-link structures are significantly smaller compared to the early mode data in downflow devices. Furthermore, it seems that the upflow samples show a third failure mode exhibiting very long lifetimes possibly due to barrier leakage. However, the regular late mode failures in up- and downflow samples appear to show comparable *MTTF* and sigma values. These results indicate that at accelerated testing conditions, the upflow interface is more critical than the downflow direction in defining the chip lifetime. From these accelerated testing conditions, the lifetimes need to be extrapolated to use conditions employing (8.5). The target operating conditions, namely temperature and current density, as well as the failure rate are defined differently by each company depending on the customer's needs. While generally all of these parameters need to be considered, in the following only the temperature was taken into account when extrapolating from stress conditions to operating conditions. Since the WSB data provides actual lifetime values for low ppm levels, the extrapolation was performed at 1,000, 100, and 10 ppm for the sake of simplicity instead of employing the *MTTF* and sigma values of the respective early modes. Thus, (8.5) simplifies to

$$TTF_{oper} = TTF_{stress}\exp\left[\frac{E_a}{k_B}\left(\frac{1}{T_{oper}} - \frac{1}{T_{stress}}\right)\right], \tag{8.6}$$

where TTF_{stress} corresponds to the measured failure time at a chosen cumulative percent. While it is important to examine the extrapolated values of all failure modes, the early mode is expected to be the limiting factor in determining chip lifetime. Thus, the activation energy values determined for the early failure modes were used for the calculations here, namely 0.83 eV and 0.93 eV for the downflow and upflow interfaces, respectively. The chosen stress and operating temperatures were 325°C and 105°C. Table 8.3 lists the measured lifetimes at various failure levels at 325°C and the calculated values at an operating temperature of 105°C.

Table 8.3. Extrapolation to use temperatures at 1.5 MA/cm^2 using measured activation energy values for up- and downflow EM. Bold values represent the respective shorter lifetimes for each condition.

Downflow	325°C	105°C
1,000 ppm	22.1 h	29.5 yrs
100 ppm	12 h	**16 yrs**
10 ppm	6.5 h	**8.7 yrs**
Upflow	325°C	105°C
1,000 ppm	**6.8 h**	**281 yrs**
100 ppm	**4.2 h**	17.4 yrs
10 ppm	**3 h**	12.4 yrs

Source: Reprinted from [41] with the permission from AIP Publishing.

Performing the extrapolation at a 1,000 ppm level, the early mode for the upflow direction yields a smaller lifetime compared to downflow EM in agreement with the trend of the measured values. However, at 10 ppm and at 105°C, downflow EM appears to yield smaller lifetimes than upflow EM as a result of the smaller activation energy value and the broader distribution, namely 8.7 years versus 12.4 years. Lifetimes are comparable at 100 ppm. Overall, all extrapolated values are rather high, providing passing data for typical reliability specifications. However, it is possible that a different interface determines the chip lifetime at accelerated testing conditions and selected use conditions. Further specific extrapolations will be necessary depending on each company's individual current target using experimentally determined values for the current exponent. In general, this result clearly shows the importance of a thorough characterization of all possible failure modes, especially at the small percentage level. The WSB technique provides a valuable path to obtain this critical data, especially when setting maximum EM current density targets for IC design.

8.3.4   Discussion of Experimental Data for Cu Metallization

The main result found in the experimental evaluation of the downflow, early EM mode is the lowered activation energy. Within the very small error bars of the massive-scale statistical study described here, a drop by about 0.07 eV from the expected ~0.9 eV for Cu–SiCN interface diffusion is a significant amount. As previously mentioned, this drop may have various reasons. Some of the vias (in this study, about every one out of 1,000) may not have been processed perfectly, i.e., the via bottom shape may be irregular, etch residue may be present at the interface to the underlying metal line, a very small initial void may be present at the via location even before testing, etc. This imperfection could be the cause for initial void formation and growth directly under the via, as opposed to further away from the via. The strong or "trench" mode usually occurs through initial void formation away from the via and subsequent movement and growth toward the via. Both upflow and downflow late mode EM mechanisms should follow this behavior, an expectation that is supported by the late mode lifetime distributions in Figure 8.29. Apart from the very late failures in the upflow mode (indicating a slight amount of barrier leakage), the strong mode distributions are very similar. In the case of a "nonperfect" bottom interface of the via to the metal line, it is not surprising that the activation energy is lowered by a certain amount. The energy barrier for void formation may be lowered through any type of imperfection at this critical interface.

A second possibility is the influence of the stress state in the Cu lines on the void formation behavior [43]. It is well known that the Cu interconnects are typically at a zero-stress state close to about 300°C, while the room temperature stress is on the order of several 100 MPa [51]. Therefore, the increased tensile stress with lowered testing temperature may reduce the required energy for void formation, especially under a via. The indication of the appearance of a "third" mode (see Figures 8.24 and 8.25) suggests that this is a possibility. While it seems that only a few samples might be significantly affected by this phenomenon, a certain influence of the stress state on the void formation under the via would be expected in all samples, possibly impacting the *MTTF* value. The additional tensile stress below the via could very well lead to a reduction of about 0.07 eV in the required energy for void formation. Since the measured EM activation energy, based on the *MTTF* values of the WSB populations, appears stable over the entire evaluated temperature range, the stress state could be influencing the *MTTF* values proportionally at all temperatures. In this case, the lowered activation energy would be "artificial," but nonetheless important. The lowest testing temperature shown in this study reached down to 150°C, quite close to typically used, highest operating temperatures of about 125°C. While the activation energy appears stable down to 150°C, the observation of the few very early failed samples raises some concern. However, even with the highly efficient WSB technique presented here, testing times are rather long. Thus, an adequate experimental evaluation of the activation energy between about 100–150°C remains very challenging.

Further studies were conducted to extend the WSB-based technique described here to more advanced technology nodes [52–54]. The main work outlined in this chapter was performed on 90 nm metallization employing pure Cu seed and standard Cu electrochemical deposition. With advancing technology nodes, the need for ever higher current densities during operation neccessiated the development of alloyed Cu seeds. One widely implemented solution to address reduced electromigration lifetime margins (and also concurrent concerns with respect to stress-induced voiding) was the introduction of Cu(Al) [55, 56] and Cu(Mn) [56–62] as sputter targets during physical vapor deposition (PVD). After Cu plating and anneal, the solute diffuses into the interconnect and suppresses Cu diffusion, leading to significant electromigration lifetime enhancements. Using standard single-link testing, the improvement factor can be up to about 6×, depending on the solute concentration [63]. A clear improvement using Cu(Mn), as well as electroless CoWP top surface plating, was also found in the very early failure distribution using 40 and 28 nm technology nodes; see Figure 8.30 [63].

Figure 8.30 shows the expected decrease in electromigration lifetime as a function of feature size scaling using standard Cu integration processes. Assuming a constant current density and not changing the main metallization integration flow, which leads to a Cu–SiCN top interface as the main diffusion path, the decrease without process improvements would have been drastic. The implementation of Cu(Mn) as a seed option leads to a lifetime increase of about 9× for 40 nm technology and even 18× for 28 nm technology, which is even higher than for standard single-link testing [63]. For both technologies, WSB devices were used to get a direct readout at the *NSD* = –3 level. The strong improvement effect for the early failure population can be explained by the fact that Mn also has a tendency to accumulate at the top surface, therefore strengthening the critical via bottom interface and reducing the tendency for early void nucleation. Electroless CoWP top surface plating can therefore achieve an even more significant improvement, e.g., about 40× in the 28 nm technology node. More recently, the testing technique was extended to 12 nm technology. During the writing of this chapter, a total of 12,800 interconnects have been tested and plans are under way to evaluate the activation energy and current density exponent using approximately 480,000 interconnects wired in WSB devices (both down- and upflow).

Table 8.4. Massive-scale statistical database encompassing WSB devices from multiple technology nodes, reaching down to 12 nm. The total sample size surpasses 2 million Cu-based interconnects.

Technology and electron direction	Sample size
90 nm downflow	686,400
90 nm upflow	115,920
65 nm downflow	408,816
45 nm downflow	14,440
40 nm downflow	496,800
28 nm downflow	266,400
12 nm downflow	12,800
Sum	2,001,576

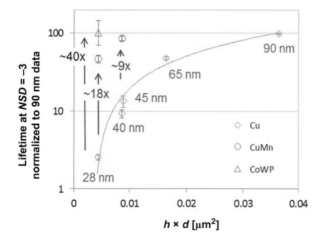

Figure 8.30 Experimentally obtained EM lifetimes at the $NSD = -3$ level as a function of the product of line height (h) and via size (d). This NSD level corresponds to a cumulative failure of about 0.1%. Thus, the plot shows the scaling behavior of the EM early failure distribution from 90 nm to 28 nm technology. Various process options are included. © IOP Publishing. Reprinted from [63] with the permission, all rights reserved.

Table 8.4 summarizes the total amount of Cu interconnects evaluated using this technique, leading to a massive-scale statistical database encompassing more than 2 million interconnects across 90, 65, 45, 40, 28, and 12 nm technology nodes.

8.4   Implications for EM Reliability

The massive-scale EM study described in this chapter investigated in detail the early failure phenomenon in Al- and Cu-based metallization. No deviations from monomodal behavior were found for Al(Cu) technology down to the four sigma level

($NSD = -4$). The advantage of the Al(Cu) integration scheme is the fact that the W-based vias are very stable and for all practical purposes inert in terms of diffusional processes. Furthermore, the Al(Cu) portion of the interconnect is typically clad between stable Ti, TiN, or a combination thereof, and hence has a type of "built-in redundancy." The consequence is the absence of early failure mechanisms. However, the maximum sustainable current densities in Al(Cu) were not able to keep pace with the ever-increasing device performance needs, and Cu-based metallization had to be introduced at the turn of the last millennium. In contrast to Al(Cu)/W integration, Cu metallization has an inherent weakness in terms of early failure mechanisms due to the dual-damascene (or "dual-inlaid") integration process. The vias are not inert in terms of diffusional mechanisms, as is the case for W-based processes. Strong bimodality for the electron upflow direction was found initially, with failure voids occuring both within the via ("early" mode) or within the trench ("late" mode). Subsequently, bimodality has also been reported in downflow EM, leading to very short lifetimes due to small, slit-shaped voids under the Cu via. In addition to shorter $MTTFs$, these early failure modes can have different activation energy, current exponent, and sigma values compared to the late mode depending on the diffusion and void formation mechanisms. Hence, a very thorough investigation of these early failure phenomena is needed. The WSB technique presented in this chapter was successfully employed to increase the examined interconnect size well past standard single-link testing capabilities. Downflow EM can exhibit bimodality at very small percentage levels around 0.1%, not readily identifiable with single-link test structures. The activation energy corresponding to the early mode in downflow interfaces can be slightly lower than expected values, which has important implications for extrapolations to operating conditions, especially when targeting high-reliability applications such as the automotive field. Using the WSB technique, several 100,000 interconnects can be tested in a relatively efficient way, enabling direct extrapolations into the single-digit ppm regime, which is essential when high reliability is in demand.

A downside of WSB-type testing is the fact that the structure design, verification, data inspection, and analysis are much more complex than with standard single-link testing. Especially the evaluation of the resistance steps can be cumbersome, and automated, software-based readout routines are not easily implemented. A compromise between mostly insufficient single-link testing and large-scale WSB-type evaluations is the use of mutilink structures, e.g., a combination of single and repeated ($N = 1$, 10, 50, and 100) serial chains of nominally identical interconnects that can be used in conjunction with statistical analyses based on the weakest link concept [36, 64]. The first fail out of N interconnects can readily be seen in $R(t)$-curves as long as N is kept at a reasonable number, as noted previously, and visual inspection is not needed as long as the failure criterion is adjusted for a very low percentage number. If the weakest link within the chain shows a resistance increase of the usually used 10–20%, the chain will have a much smaller resistance increase.

Alternatively, a relatively small repetition of the previously described parallel five-link structure (Figure 8.2) can help considerably to extend the statistical sample size toward a cumulative failure number of about 0.1%. Stitching four of these units using

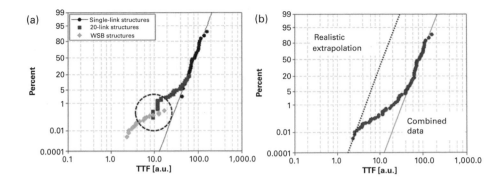

Figure 8.31 (a) Comparison of single links, 20-link structures, and WSBs; (b) combined distribution of all data, creating a clearly bimodal distribution plot.

a series wiring and obtaining a 20-link structure can create an intermediate step between single links and WSBs, as shown in Figure 8.31.

Figure 8.31a shows the three structure types separately, where the 20-link data bridges well between the single links and the WSBs. Figure 8.31b replots the same data as one distribution, and this leads exactly to the predicted bimodal shape as proposed in Figure 8.1 at the beginning of this chapter. Standard data analysis routines can now separate the two distributions and determine the median lifetimes and log-normal sigma values for both the early and late modes.

Especially the setting of EM current density design guidelines should involve some type of larger-scale, statistical testing. If the straight extrapolation from the single-link data as shown in Figure 8.31b (indicated by the solid line) is used, the lifetime during operation in the field can be overestimated by almost a factor of $10\times$. The dashed line represents the correct extrapolation using the early failure population obtained by a combination of 20-link structures and WSB devices. If no intermediate, large-, or massive-scale statistical EM data are acquired during the early stages of technology development, a sufficient lifetime margin needs to be built in to protect against early failures in the field. This is a typical procedure followed by most semiconductor companies until today. Standard testing methodologies still propose only single-link testing, and it will take time to implement large-scale testing procedures as a general practice.

References

1. L. P. Muray, L. C. Rathbun, and E. D. Wolf, New technique and analysis of accelerated electromigration life testing in multilevel metallizations, *Applied Physics Letters* **53** (1988), 1414–1416.
2. H. H. Hoang, E. L. Nikkel, J. M. McDavid, and R. B. Macnaughton, Electromigration early-failure distribution, *Journal of Applied Physics* **65** (1989), 1044–1047.

3. T. Nogami, T. Nemoto, N. Nakano, and Y. Kaneko, The electromigration lifetime determined from the minimum time to failure in an acceleration test, *Semiconductor Science and Technology* **10** (1995), 391–394.

4. M. Gall, C. Capasso, D. Jawarani, R. Hernandez, H. Kawasaki, and P. S. Ho, Electromigration early failure distribution in submicron, *Interconnects. IEEE International Interconnect Technology Conference* (1999), 270–272.

5. M. Gall, C. Capasso, D. Jawarani, R. Hernandez, H. Kawasaki, and P. S. Ho, Detection and analysis of early failures in electromigration, *Applied Physics Letters* **76** (2000), 843–845.

6. M. Gall, C. Capasso, D. Jawarani, R. Hernandez, H. Kawasaki, and P. S. Ho, Statistical analysis of early failures in electromigration, *Journal of Applied Physics* **90 (2)** (2001), 732–740.

7. M. Gall, C. Capasso, D. Jawarani, R. Hernandez, H. Kawasaki, and P. S. Ho, Statistical evaluation of device-level electromigration reliability, *AIP Conference Proceedings* **418** (1998), 483–494.

8. M. Gall, C. Capasso, D. Jawarani, R. Hernandez, H. Kawasaki, and P. S. Ho, Electromigration early failure distribution in submicron interconnects, *AIP Conference Proceedings* **491** (1999), 3–14.

9. I. A. Blech, Electromigration in thin aluminum films on titanium nitride, *Journal of Applied Physics* **47** (1976), 1203–1208.

10. H. Kawasaki and C.-K. Hu, An electromigration failure model of tungsten plug contacts/vias for realistic lifetime prediction, *IEEE Symposium VLSI Technology* (1996), 192–193.

11. D. Jawarani, M. Gall, R. Hernandez, C. Capasso, and H. Kawasaki, New insight on electromigration failure mechanism and its impact on design guidelines, *IEEE Symposium on VLSI Technology* (1997), 39–40.

12. C. Capasso, M. Gall, S. G. H. Anderson, D. Jawarani, R. Hernandez and H. Kawasaki, Statistical modeling and experimental verification of electromigration time to fail distribution in metal interconnects, *Electrochemical Society Proceedings* **97** (31) (1998), 196–202.

13. W. Nelson, *Accelerated Testing* (New York: Wiley, 1990), 145.

14. M. Gall, PhD dissertation, University of Texas at Austin, 1999.

15. M. Gall, D. Jawarani, and H. Kawasaki, Characterization of electromigration failures using a novel test structure, *Materials Research Society Symposium Proceedings* **428** (1996), 81–86.

16. F. d'Heurle and P. S. Ho, Electromigration in thin films, *Thin Films: Interdiffusion and Reactions*, eds. J. Poate, K. N. Tu, and J. Mayer (New York: Wiley, 1978), 265.

17. J. R. Lloyd and J. Kitchin, The electromigration failure distribution: the fine-line case, *Journal of Applied Physics* **69** (1991), 2117–2127.

18. C.-K. Hu, P. S. Ho, and M. B. Small, Electromigration in Al(Cu) two-level structures: effect of Cu and kinetics of damage formation, *Journal of Applied Physics* **74** (1993), 969–978.

19. C.-K. Hu, R. Rosenberg, and K. N. Tu, Stages of damage formation by electromigration in line/stud structures, *AIP Conference Proceedings* **305** (1994), 195–210.

20. C.-K. Hu, K. P. Rodbell, T. D. Sullivan, K. Y. Lee, and D. P. Bouldin, Electromigration and stress-induced voiding in fine Al and Al-alloy thin-film lines, *IBM Journal of Research and Development* **39** (1995), 465–497.

21. C. M. Su, H. G. Bohn, K.-H. Robrock, and W. Schilling, Chemical, microstructural, and internal friction characterization of Al/Cu thin-film reactions, *Journal of Applied Physics* **70** (1991), 2086–2093.

22. M. Gall, J. Müller, D. Jawarani, C. Capasso, R. Hernandez, and H. Kawasaki, Critical length and resistance saturation effects in Al(Cu) interconnects, *Materials Research Society Symposium Proceedings* **516** (1998), 231–236.

23. S. K. Kurtz and F. M. A. Carpay, Microstructure and normal grain growth in metals and ceramics. Part I. Theory, *Journal of Applied Physics* **51** (1980), 5725–5744.

24. D. C. Edelstein et al., Full copper wiring in a sub-0.25 μm CMOS ULSI technology, *Tech. Dig. IEEE International Electron Devices Meeting* (1997), 773–776.

25. S. Venkatesan et al., A high performance 1.8 V, 0.20 μm CMOS technology with copper metallization, *Tech Dig. IEEE International Electron Devices Meeting* (1997), 769–772.

26. R. Rosenberg, D. C. Edelstein, C.-K. Hu, and K. P. Rodbell, Copper metallization for high performance silicon technology, *Annual Review of Materials Science* **30** (2000), 229–262.

27. C.-K. Hu, R. Rosenberg, H. S. Rathore, D. B. Nguyen, and B. Agarwala, Scaling effect on electromigration in on-chip Cu wiring, *Proceedings of the IEEE International Interconnect Technology Conference* (1999), 267–269.

28. C. S. Hau-Riege and C. V. Thompson, Electromigration in Cu interconnects with very different grain structures, *Applied Physics Letters* **78** (22) (2001), 3451–3453.

29. E. T. Ogawa, K.-D. Lee, V. A. Blaschke, and P. S. Ho, Electromigration reliability issues in dual-damascene Cu interconnections, *IEEE Transactions on Reliability* **51** (4) (2002), 403–419.

30. M. A. Meyer, M. Herrmann, E. Langer, and E. Zschech, In situ SEM observation of electromigration phenomena in fully embedded copper interconnect structures, *Microelectronics Engineering* **64** (2002), 375–382.

31. M. Hauschildt, PhD dissertation, University of Texas at Austin, 2005.

32. M. Hauschildt, M. Gall, S. Thrasher, et al., Statistical analysis of electromigration lifetimes for Cu interconnects, *AIP Conference Proceedings* **817** (2006), 164–174.

33. M. Hauschildt, M. Gall, S. Thrasher, et al. Analysis of electromigration statistics for Cu interconnects, *Applied Physics Letters* **88** (2006), 211907-1-3.

34. M. Hauschildt, M. Gall, S. Thrasher, et al., Statistical analysis of electromigration lifetimes and void evolution, *Journal of Applied Physics* **101** (2007), 043523-1-9.

35. M. Gall, M. Hauschildt, P. Justison, et al., Scaling of statistical and physical electromigration characteristics in Cu interconnects, *Materials Research Society Symposium Proceedings* **914** (2006), 305–316.

36. E. T. Ogawa, K.-D. Lee, H. Matsuhashi, et al., Statistics of electromigration early failures in Cu/oxide dual-damascene interconnects, *Proceedings of the IEEE International Reliability Physics Symposium* (2001), 341–349.

37. J. Gill, T. Sullivan, S. Yankee, H. Barth, and A. v. Glasow, Investigation of via-dominated multi-modal electromigration failure distributions in dual damascene Cu interconnects with a discussion of the statistical implications. *Proceedings of the IEEE International Reliability Physics Symposium* (2002), 298–304.

38. J. B. Lai, J. L. Yang, Y. P. Wang, et al., A study of bimodal distributions of time-to-failure of copper via electromigration. *Proceedings of the International Symposium on VLSI Technology* (2001), 271–274.

39. B. Li, C. Christiansen, J. Gill, R. Filippi, T. Sullivan and E. Yashchin, Minimum void size and 3-parameter lognormal distribution for EM failures in Cu interconnects, *Proceedings of the IEEE International Reliability Physics Symposium* (2006), 115–122.

40. S.-C. Lee and A. S. Oates, Identification and analysis of dominant electromigration failure modes in copper/low-*k* dual damascene interconnects, *Proceedings of the International Reliability Physics Symposium* (2006), 107–114.

41. M. Hauschildt, M. Gall, and R. Hernandez, Large-scale statistical analysis of early failures in Cu electromigration, Part I: Dominating mechanisms, *Journal of Applied Physics* **108** (2010), 013523-1-10.

42. J. H. An, PhD dissertation, University of Texas at Austin, 2007.

43. J. R. Lloyd, Black's law revisited: nucleation and growth in electromigration failure, *Microelectronics Reliability* **47** (9–11) (2007), 1468–1472.

44. K.-D. Lee and P. S. Ho, Statistical study for electromigration reliability in dual-damascene Cu interconnects, *IEEE Transactions on Device and Materials Reliability* **4** (2) (2004), 237–245.

45. H. Tsuchiya and S. Yokogawa, Electromigration lifetimes and void growth at low cumulative failure probability, *Microelectronics Reliability* **46** (9–11) (2006), 1415–1420.

46. M. Hauschildt, M. Gall, P. Justison, R. Hernandez, and M. Herrick, Large-scale statistical study of electromigration early failure for Cu/low-k interconnects, *AIP Conference Proceedings* **945** (2007), 66–81.

47. M. Hauschildt, M. Gall, and R. Hernandez, Large-scale statistics for Cu electromigration, *AIP Conference Proceedings* **1143** (2009), 31–46.

48. M. Hauschildt, M. Gall, and R. Hernandez, Large-scale electromigration statistics for Cu interconnects, *Materials Research Society Symposium Proceedings* **1156** (2009), 121–132.

49. M. Gall, M. Hauschildt, and R. Hernandez, Large-scale statistical analysis of early failures in Cu electromigration, Part II: Scaling behavior and short-length effect, *Journal of Applied Physics* **108** (2010), 013524-1-6.

50. E. T. Ogawa, J. W. McPherson, J. A. Rosal, et al., Stress-induced voiding under vias connected to wide Cu metal leads. *Proceedings of the IEEE International Reliability Physics Symposium* (2002), 312–321.

51. S.-H. Rhee, PhD dissertation, University of Texas at Austin, 2001.

52. M. Hauschildt et al., Electromigration early failure void nucleation and growth phenomena in Cu and Cu(Mn) interconnects, *Proceedings of the IEEE International Reliability Physics Symposium* (2013), 2C.1.1–2C.1.6.

53. M. Hauschildt et al., Electromigration void nucleation and growth analysis using large-scale early failure statistics, *AIP Conference Proceedings* **1601** (2014), 89–98.

54. M. Gall, Large-scale statistical analysis of early failures in Cu electromigration, presentation at the 2020 IEEE North Atlantic Test Workshop (unpublished work) (2020).

55. S. Yokogawa and H. Tsuchiya, Effects of Al doping on the electromigration performance of damascene Cu interconnects, *Journal of Applied Physics* **101** (2007), 013513.

56. T. Nogami et al., High reliability 32 nm Cu/ULK BEOL based on PVD CuMn seed, and its extendibility, *Proceedings of the IEEE International Electron Devices Meeting* (2010), 33.5.1–33.5.4.

57. T. Usui et al., Low resistive and highly reliable Cu dual-damascene interconnect technology using self-formed MnSixOy barrier layer, *Proceedings of the IEE International Interconnect Technology Conference* (2005), 188–190.

58. J. Koike and M. Wada, Self-forming diffusion barrier layer in Cu–Mn alloy metallization, *Applied Physics Letters* **87** (2005), 041911.

59. J. Koike, M. Haneda, J. Iijima, and M. Wada, Cu alloy metallization for self-forming barrier process, *Proceedings of the IEEE International Interconnect. Technology Conference* (2006), 161–163.

60. J. Koike, M. Haneda, J. Iijima, Y. Otsuka, H. Sako and K. Neishi, Growth kinetics and thermal stability of a self-formed barrier layer at Cu-Mn/SiO$_2$ interface, *Journal of Applied Physics* **102** (2007), 043527.

61. C. Christiansen, B. Li, M. Angyal, et al., Electromigration-resistance enhancement with CoWP or CuMn for advanced Cu interconnects, *Proceedings of the IEEE International Reliability Physics Symposium IEEE IRPS* (2011), 3E.3.1–3E.3.5.

62. C.-K. Hu et al., Electromigration in Cu(Al) and Cu(Mn) damascene lines, *Journal of Applied Physics* **111** (2012), 093722.

63. M. Hauschildt et al., Advanced metallization concepts and impact on reliability, *Japanese Journal of Applied Physics* **53** (2014), 05GA11.

64. P. R. Justison, PhD dissertation, University of Texas at Austin, 2003.

9 Assessment of Electromigration Damage in Large On-Chip Power Grids

9.1 Introduction

Due to technology scaling, electromigration (EM) signoff has become increasingly difficult, mainly due to the use of inaccurate methods for EM assessment, such as the empirical Black model. Results of recent measurements performed on power grid-like structures isolated in the power grid environment have demonstrated that the weak link approach cannot accurately predict the lifetime of the power grids. It calls for significant overdesign, while, today, there is very little margin left for electromigration. Numerous observations clearly indicate that there is a need for a new approach that accurately assesses EM degradation based on reliability physics combined with a mesh model to account for redundancy, while being fast enough to be practically useful. This chapter presents a novel approach for power grid EM assessment using physics-based models that can account for process, voltage, and temperature variations across the die. Existing physical models for EM in metal branches are extended to track EM degradation in multibranch interconnect trees. Results are presented for a number of power grid benchmarks to confirm that the Black model is unacceptably inaccurate. The lifetimes found using the described physics-based approach are on average $2.75\times$ longer than those based on the Black model when extended to mesh power grids. With a maximum runtime of 6.5 hours among all assessed test grids, this method appears to be suitable for large VLSI circuits.

9.2 Problems with the Standard Weakest Link Approximation for On-Chip Power Grid EM Assessment

As a result of continued scaling of integrated circuits (IC) technology, electromigration has become a major reliability concern for design of on-die power grids in large digital integrated circuits [1]. The ongoing IC component miniaturization results in a reduction of the wiring cross sections and hence an increase of the current densities because the required current cannot be reduced to the same extent, even by reducing the supply voltage and gate capacitance. As was discussed in previous chapters, EM-induced voiding is responsible for the interconnect line resistance increase, while hillock formation can generate the electrical short-induced failure. In this chapter, we concentrate on void-induced degradation because voids occur much more frequently than hillocks in practice. In general, there are two major functions of the on-chip interconnect: (i) as signal lines to provide intra- and intercell connectivity for proper signal propagation, and (ii) as power (and ground) lines to deliver a well-regulated supply voltage to every cell. While the resistance degradation of individual metal line segments can destroy both these functions, the time scales of EM-induced degradation in the power supply versus the signal lines are quite different. As discussed in Chapters 4 and 5, EM-induced voiding is due to the growth of preexisting microscopic defects by high tensile stress generated around these defects. The difference in the degradation rate between the power and the signal circuits is in the types of electric currents employed in these two cases. Indeed, the majority of signal lines in digital chips carry bidirectional currents that lead, as it will be demonstrated in this chapter, to a repetitive increase and decrease of the mechanical stress at the line diffusion blocking ends, which results in very long times to the EM-induced failure. In contrast, power lines carry mostly unidirectional currents and so can fail in much shorter times due to the absence of, or negligible, stress relaxation. Thus, we can conclude that, in most cases, EM-induced digital chip failure is due to the failure of the power network to deliver needed voltages to some cell of the circuitry [2]. Hence the focus of this chapter is on EM reliability assessment of power grids.

Today it is becoming harder to sign off on chip designs using traditional EM checking tools, as there is very little margin left between the predicted by circuit simulation electric currents and that allowed by EM design rules [3]. This loss of safety margin can be traced back to the inaccurate and oversimplified EM assessment methodology used by existing tools, which is based on an extension of the traditional single-link test-structure EM criterion to the domain of the power grid.

The EM test of a standard single-link test structure, as it is discussed in Chapter 8, is typically performed on a set of Cu lines as in Figure 9.1a, in which the electric current flows from the wide metal supply lines through the vias and into the narrower test lines. Changes in voltage and resistance over time are measured and recorded. An increase of resistance of individual lines above some threshold value is considered a failure. Details describing this methodology characterized by a variety of measurement techniques can be found elsewhere; see, for example, [4]. Here, we will only give a brief summary of the approach.

The time to failure (TTF) of each metal line, stressed by direct current (DC) of density j at the temperature T, is recorded for hundreds of identical lines in the test structures (see Figure 9.1a). The mean time to failure (MTTF) is extracted from the measured TTF ensembles, and it is known to follow the Black equation [6]:

(a)   (b)

Figure 9.1 (a) Multilink test structure; (b) upstream and downstream EM test structures. Arrows show the electron flow directions. © 2018 IEEE. Reprinted, with permission, from [5].

$$MTTF = \frac{A}{j^n} \exp \left\{ \frac{E_A}{k_B T} \right\}, \tag{9.1}$$

where, k_B is the Boltzmann constant and A is a proportionality coefficient, which depends on line geometry, residual stress, and temperature. Two critical parameters, the current density exponent n and activation energy E_A, are extracted by regression from the measured TTFs. In order for the failure to happen in reasonable time periods, these measurements are carried out at so-called stressed conditions, characterized by elevated temperatures of 200–400°C and high current densities of 3×10^9–5×10^{10} A/m^2. Failure times are customarily fitted to a log-normal distribution:

$$F(t) = \int_0^t \frac{1}{\sqrt{2\pi}\sigma u} \exp \left\{ \frac{[\log(u) - \log(MTTF)]^2}{2\sigma^2} \right\} du \tag{9.2}$$

or a Weibull distribution:

$$F(t) = 1 - \exp \left\{ -0.693 \frac{t^\beta}{MTTF} \right\}. \tag{9.3}$$

Here, $F(t)$ is the cumulative percent failures at time t and σ is the standard deviation. Both $MTTF$ and σ are found by plotting the experimental TTF data on a log-normal plot; the coefficient β is extracted from a Weibull plot. All these parameters are extracted from measurements that are done on a variety of test structures designed for different metal layers and different current directions (upstream and downstream tests, as in Figure 9.1b).

Translation of the $MTTF$ obtained at the stress conditions to the operating conditions, characterized by lower temperatures and current densities, is performed based on the following equation:

$$MTTF_{OPER} = MTTF_{STRESS} \left(\frac{j_{STRESS}}{j_{OPER}} \right)^n \times \exp \left\{ \frac{E_A}{k_B} \left(\frac{1}{T_{OPER}} - \frac{1}{T_{STRESS}} \right) \right\}. \tag{9.4}$$

It is assumed that knowing the $MTTF$ and the failure probabilities for each type of interconnect segment and for all metal layers allows one to calculate the failure times for every segment, with different geometries using the independent element model shown in Figure 9.2. When the failure probability of the ith element, $F_i(t)$, is known, the chip-level failure probability, $F_{Chip}(t)$, is calculated based on the weakest link statistics:

$$F_{Chip}(t) = 1 - \prod_{i=1}^m (1 - F_i(t)), \tag{9.5}$$

where m is the total number of elements with EM concerns in the chip. Thus, EM-induced failure rates of individual segments are considered a measure of EM-induced reliability and, in the extreme end, the MTTF of the weakest segment is accepted as a measure for the chip lifetime.

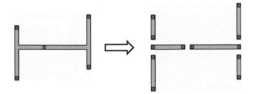

Figure 9.2 An interconnect tree broken into segments. In conventional reliability assessment approaches, the reliability of the segments is independently estimated. Reprinted from [7], with the permission of AIP Publishing.

The scaling equation (9.4) is used also for generating a set of so-called EM design rules, which provide maximum allowed current densities for all interconnect segments characterized by different geometries, current directions, and temperatures. Indeed, accepting the requirement that each segment should survive not less than a required period of time $MTTF_{SPEC}$, which can be any specified amount of years, (9.6) provides the maximum current density satisfying that requirement:

$$ j_{MAX} = j_{STRESS} \left(\frac{MTTF_{STRESS}}{MTTF_{SPEC}} \right)^{\frac{1}{n}} \times \exp\left\{ \frac{E_A}{nk_B} \left(\frac{1}{T_{OPER}} - \frac{1}{T_{STRESS}} \right) \right\} \qquad (9.6) $$

According to this methodology, the standard practice employed in industry is to break up a grid into isolated metal branches, assess the reliability of each branch separately using the Black model [6], and then use the series model (earliest branch failure time) with the weakest link approximation to determine the failure time for the whole grid (see Figure 9.2). This approach, when applied to the on-chip power grid, is highly inaccurate, for at least three reasons. First, the fitting parameters obtained for the Black model under accelerated testing conditions are not necessarily valid at actual operating conditions, and this can lead to significant errors in lifetime extrapolation [4, 8–10]. Second, the Black model ignores the material flow between branches. In today's mesh structured power grids, many branches within the same metal layer are connected (there are no diffusion barriers between them), forming an interconnect tree, and atomic flux can flow freely between them [7]. As a result, the physical analysis of the failure of a single line segment with diffusion barriers at both ends, which is based on the accumulation of stresses at these ends, is not applicable. This was confirmed by results of direct failure time measurements on a variety of trees done by Thompson's group [7, 11]. They performed electromigration tests on simple Al (0.5 wt.%Cu) tree structures, such as straight lines with an additional via in the center of the line (Figures 9.3 and 9.4). It was experimentally demonstrated that the reliability of a segment cannot be predicted without knowledge of the conditions for stress migration and electromigration in connecting segments, thus the reliabilities of the different segments of the tree are not independent of each other. Analogous results have been obtained from the lifetime tests on Cu straight via-to-via lines with an additional via in the middle of the line [11]. Similar analysis was performed by Lin and Oates on Cu interconnect multisegment

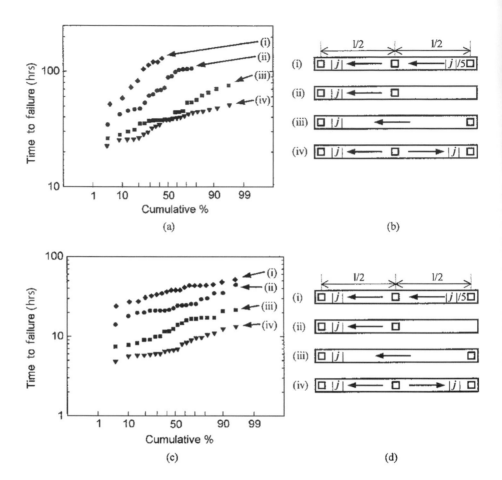

Figure 9.3 (a) Times to failure, defined as a 30% increase in electrical resistance, for 0.27 μm wide lines tested at $T = 250°C$ and $j = 2 \times 10^6$ A/cm^2 for electron current configurations shown in (b). (c) Times to failure for 0.27 μm wide lines tested at $T = 350°C$ and $j = 5 \times 10^5$ A/cm^2 for electron current configurations shown in (d). The line length $l = 500$ μm. Reprinted from [7], with the permission of AIP Publishing.

structures [12]. It was shown that the most highly stressed segment in a Cu interconnect tree is not always the least reliable. Instead, most failures were found to occur in the right limbs of the (i)–(iii) structures in Figure 9.3a, b, and d, even though the current density was much higher in the left limbs. The results obtained in these tests [7, 11–15] demonstrated that the segment-based analysis is inaccurate since current densities from different segments of an interconnect tree can interact with each other, and additionally, the stress evolutions in different segments are coupled.

This conclusion was confirmed by the results of the analytical solution derived by Chen et al. [16] for the EM-induced stress evolution in multisegment interconnect trees. Figure 9.5a and b shows examples of the EM-induced stress evolution along segments in the three-terminal interconnect tree shown in Figure 9.5d. Such

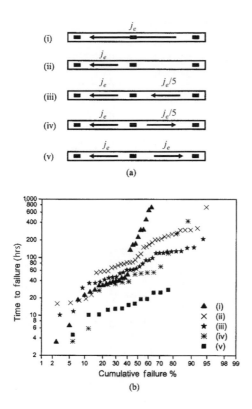

Figure 9.4 (a) Schematic diagrams of dotted-I test structures with the five different current configurations used in the experiments. Tests were carried out at $T = 350°C$ and $j_e = 2.5$ MA/cm^2. The arrows show the direction of electron flow. (b) Times to failure for 500 μm long, 0.28 μm wide dotted-I structures with the different electron current configurations shown in (a). Reprinted from [11], with the permission of AIP Publishing.

time-dependent stress distributions cannot be obtained by decomposing the tree on individual segments as shown in Figure 9.2. In that case, the stress evolution derived for each segment by assuming diffusion barriers at both ends would be the well-known symmetrical curves shown in Figure 9.5c. Therefore, a direct use of the Black model to evaluate the MTTF for each segment based on extracted current density and geometry cannot be justified for multiline interconnect trees. In the case where individual branches happen to be short, they would be immortal due to the Blech short-length effect, [17], so the interconnect tree would also be immortal, which is highly optimistic and can be entirely misleading for design.

Finally, the third problem lies with the series model assumption. A series model applies to the case where a power grid is deemed to have failed as soon as the first of its branches has failed, typically due to an open circuit. However, modern power grids use a mesh structure [18–21] where there are many paths for the current to flow from the flip-chip bumps to the underlying logic, a characteristic we refer as a redundancy

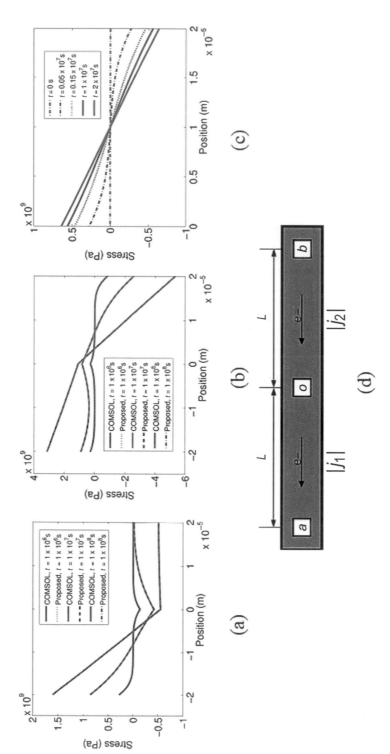

Figure 9.5 EM-induced stress development along the lines 1 and 2 in the three-terminal interconnect tree: (a) $j_1 = 2 \times 10^{10}$ A/m^2, $j_2 = 0$ A/m^2; (b) $j_1 = 2 \times 10^{10}$ A/m^2, $j_2 = 6 \times 10^{10}$ A/m^2; (c) single segment with diffusion blocking ends; (d) the straight-line three-terminal interconnect tree. © 2016 IEEE. Reprinted, with permission, from [16].

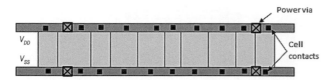

Figure 9.6 An example of a power grid. The standard cells, which are the gray rectangles separated by vertical lines, can be supplied by two paths, from either the left or the right power vias.

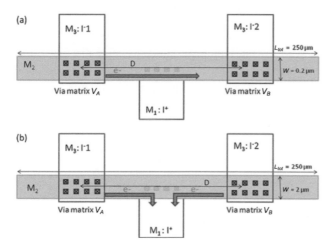

Figure 9.7 Test structures: (a) reference, (b) redundant. Reprinted from [20], with permission from Elsevier.

(see Figure 9.6). Such a mesh power grid is in fact closer in structure to a parallel system. As such, it is highly pessimistic to assume that a single branch failure will always cause the whole grid to fail.

The effect of the structure difference was demonstrated by Ouattara et al. [20] by direct measurements of the resistance change in the two structures shown in Figure 9.7, which were composed at the silicon level of a long wire (250 μm) of 0.2 μm width in the metal 2 layer (M2) connected to supply wires in M3, and a wire in M1 was tapped in the middle of the structure to complete a standard cell connection. The line under test was connected through matrices of multiple vias to the supply layers. In the reference structure shown in Figure 9.7a, the current density j_1 was established between the supply vias V_A and V_B, while the layer M1 was not used. In the redundant structure shown in Figure 9.7b, an electric current of $2\times$ current density was injected from the M1 pad into the M2 wire. Due to the equidistance of vias V_A and V_B from the M1 pad, the same current density j_1 was expected in the left and right legs of the M2 wire. For the redundant structure, two abrupt jumps of resistance were observed, while for the reference structure, only one jump coincident with the first jump of the redundant

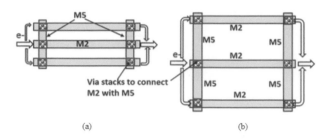

Figure 9.8 Schematics of two EM power grid-like structures isolated in the power grid design. © 2018 IEEE. Reprinted, with permission, from [21].

structure was observed. In the redundant structure, after the first critical void was formed, the current continued to flow through the redundant path, so only after formation of the second critical void in the redundant path this structure would fail. For these two structures, the cumulative failure distributions versus TTF should have the same MTTF, although the total gate current in the redundant structure is $2\times$ higher. It is worth noting that the current EM assessment methodology would consider the redundant test structure as failed immediately after the first critical void was formed. Thus, it is clear that the weak link approximation based on the series model is not applicable for EM reliability assessment for power grids with redundant structures.

Another demonstration of the redundancy effect on EM-induced power grid reliability was recently reported by Li et al. [21]. In this study, the resistance degradation was analyzed for small interconnect power grid-like structures isolated inside a power grid environment, where two different three-leg power grid-like structures were used, as shown in Figure 9.8. The connections between all three parallel M2 legs of 50 μm in length provided an isolated structure with almost equal initial resistances for all three legs (Type I, shown in Figure 9.8a). Given the parallel connection of the three legs and the very short length of the M5 connectors (~0.7 μm), there were almost equal initial currents in the three lines. Another structure (Type II, shown in Figure 9.8b) was used where the center leg had a lower resistance than the other two legs, so that the initial current density in the center leg was higher than in the other two. To compare the EM behavior of these two types of power grid-like structures with a traditional single-link EM structure, a V2/M2 EM test structure, Figure 9.1b, was used as a reference.

The measured distributions of the time to resistance jump for both types of power grid-like structures and the single-link reference structure showed that both power grid-like structures demonstrated much longer time to the first jump in resistance than the single-link reference structure. This was attributed to the current redistribution in the power grid-like structures due to the grid redundancy, which provided not only the alternative current paths, but also allowed the electric current to redistribute when the EM-induced void caused a local resistance increase. The time differences of the first resistance drop observed for the different types of power grid-like structures was accounted for by the larger initial current flowing through the central leg and slower current redistribution caused by the growing void in the Type II structure as compared with Type I.

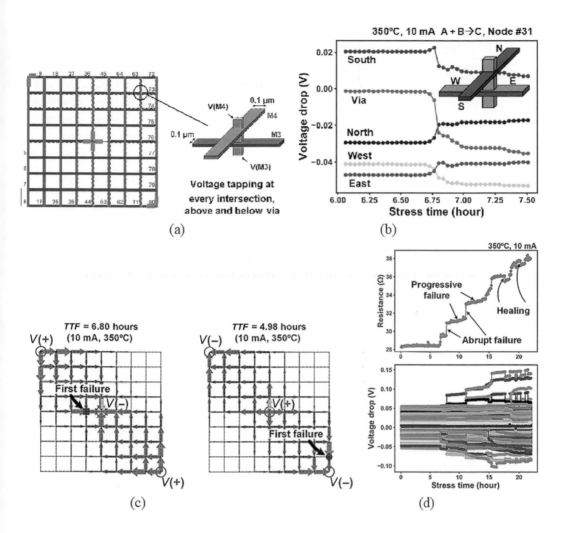

Figure 9.9 (a) A 9 × 9 M3–M4 power grid with three voltage connection points, (b) EM-induced evolution of the resistance in the via and in adjacent metal segments, (c) across-grid voltage drop distributions caused by different current stressing, and (d) power grid resistance and voltage drop traces of all segments and vias. © 2018 IEEE. Reprinted, with permission, from [22].

New, very interesting experimental results on EM-induced voltage drop degradation in the on-chip power grid were reported recently in [22]. The previously developed technique of the local sensing [23–25], which is called a voltage tapping in [22], was employed to analyze the voltage evolution in all 162 nodes formed by 81 intersections of two metal layers M3 and M4 (Figure 9.9a). The V4 vias were used to tap the node voltages of the M4 and the V2 vias to tap the node voltages of the M3, as shown in Figure 9.9a. Measured node voltages were used to calculate the voltage drops across the 20 μm M3 and M4 segments and the V3 vias (Figure 9.9b).

The current pad connection points, shown by the cross marks in Figure 9.9a, were located at two corners and the center of the grid. On-chip poly heaters used to keep the die temperature during experiment at 350°C. The active circuits used for the voltage measurements were placed 400 μm away from the heating region. Figure 9.9c shows measured magnitude and polarity of the voltage across each segment at arbitrary moments of time, where the circle and square show the sign of voltage drop across the via. The kinetics of the voltage drop degradation in a particular cross section due to the EM voiding is shown in Figure 9.9b. It was proposed that the change of the voltage can be attributed to two reasons: resistance increase in the via or wire due to voiding, or the current increase due to voiding at the nearby location. Figure 9.9d shows the evolution of the power grid resistance between the corner pad connections and the grid center, along with the voltage drop across each segment. Self-healing phenomena were observed along with the abrupt and progressive failures. Voltage drop traces provided better insight than resistance traces. Based on the voltage fluctuations, locations of the vias that were undergoing self-healing were found. It was concluded that a presence of the redundant paths in the power/ground grids allows the current to bypass the failure location immediately upon the voiding, thus providing the opportunity for healing and further increasing the grid lifetime.

Thus, on-chip power/ground interconnect grids have distinct EM characteristics due to the parallel network configuration where the standard weakest link approximation used to evaluate EM lifetime is not applicable for evaluating the EM performance and would have been too pessimistic in projecting the EM lifetime. As a result, the Blech short-length effect for a single line segment that requires diffusion barriers at both ends for accumulation of stresses at these ends would not be applicable for redundant structures. Hence, the use of the Black model to project the MTTF for each segment based on extracted current densities and geometries cannot be justified for multiline interconnect trees. Clearly, there is a need for a new EM approach and criterion that can accurately model EM degradation for power grid structures to account for mesh redundancy.

9.3   EM Assessment of Power Grids Based on Physical Modeling

A new approach for EM assessment was recently proposed based on the criterion that EM-induced chip failure is deemed to happen only when the grid interconnect cannot deliver the needed voltage to properly operate certain parts of the underlying circuitry. If the power supply for a circuit is not within the user-defined specification, the result can be a timing violation or a reduction of the noise margin. The loss of performance, which is a *parametric* failure, should be considered a more realistic and practical failure criterion to define EM-induced power grid failures.

This novel approach has recently been developed [18, 19, 26–28], first by Chatterjee et al. [18], who proposed a mesh model as an alternative to the series model, where a grid is deemed to have failed, not when the first line fails, but when enough lines have failed so that the voltage drop at some grid node has exceeded a

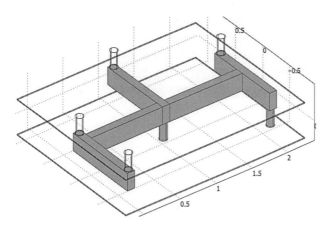

Figure 9.10 Interconnect segment confined by diffusion barriers/liners. © 2016 IEEE. Reprinted, with permission, from [27].

predefined threshold to induce errors in the underlying logic circuit. However, the Black model was still used to evaluate the reliability of individual branches [18]. Subsequently, Huang et al. [19, 27] proposed to adapt the Korhonen physical EM model [29] to analyze the interconnect trees. The use of the Korhonen model can be traced first to Hau-Riege et al. [7] and later Alam et al. [30], who developed a closed-form solution to describe stress evolution at a junction (a point where multiple branches meet) by replacing the connected branches with semi-infinite limbs, which was later used by Li et al. [31] in their EM verification tool. It should be noted since the connected branches were replaced by semi-infinite limbs, the atomic flow across the whole tree was not accounted for.

In [19, 27], the Korhonen EM model was adapted for interconnect trees consisting of continuously connected, highly conductive metal lines terminated by diffusion barriers within one layer of metallization, as shown in Figure 9.10. Later the approach developed in [19, 26, 27, 31] was extended to account for temperature variation [11, 28, 32–34]. For transient source currents, the effective DC has been calculated in a way as proposed in [19, 26, 35]. Here we provide a short description of the effective DC approach.

9.4 EM Induced Stress Evolution under AC and Pulse Current Loads

The EM-induced degradation of the electrical resistance of test structures, as discussed in Chapters 5 and 8, is traditionally monitored by applying DC stressing. A majority of EM models also assume DC stressing to describe the EM phenomenon. However, all semiconductor IC chips are operated with time-dependent, pulse electric currents. Since EM-induced failures continue to be the primary reliability concern for on-chip interconnects, different conversion factors have been introduced to predict electric degradation due to transient current on the basis of well understood DC current–induced

(a)   (b)

Figure 9.11 (a) Schematic illustration of the structure used to test for electromigration-induced damage. The inner line is electrically isolated from the outer line and carries no current. (b) Damage in the middle line has formed without a current. Reprinted from [41], with the permission of AIP Publishing.

degradation. Such factors are used by chip designers to design EM-resilient chips and for the final design sign-off verification.

Several models have been developed to describe the EM phenomenon under AC/pulse current stressing, [36–40] based on the evolution of vacancy concentration in a metal line confined by the rigid dielectric. However, the evolution of the mechanical stress induced by EM, which is the main cause of the void growth leading to resistance degradation, was not properly considered. In the following, we describe the stress evolution by solving the 1D continuity equation in an initially void-free metal line under time-dependent electric current stressing. The analytical solution shows that different waveforms of electric loads can develop a stress exceeding the critical level required for stable void growth from preexisting microflaws, as discussed in Chapter 4. In particular, the formalism predicts immortality of the lines loaded with symmetric bidirectional currents, although the experiments demonstrated that in some cases the lines did fail under AC and symmetrical bidirectional pulse currents. One of the possible answers for this paradox can be found in the paper by Moning et al. [41] based on the test where identical defects were found in three identical lines located in close proximity in the test structure with two outer lines loaded with sinusoidal voltages and the middle line not carrying any current. The temperature oscillations generated by the outer lines have resulted in identical defects formed in all three lines, as shown in Figure 9.11. It was concluded that thermal fatigue was responsible for formation of the defects different from the EM-induced voids, which can be observed by SEM.

9.4.1   Stress Evolution Caused by a Time-Dependent Current

The stress evolution under a time-dependent current load can be derived by solving the continuity equation (4.18) from Chapter 4 by converting it to the following nonhomogeneous form:

$$\frac{\partial\left(\sigma + \frac{eZ\rho j}{\Omega}\left(x - \frac{L}{2}\right) - \sigma_T\right)}{\partial t} - \kappa^2 \frac{\partial^2}{\partial x^2}\left(\sigma + \frac{eZ\rho j}{\Omega}\left(x - \frac{L}{2}\right) - \sigma_T\right) = \frac{eZ\rho j}{\Omega}\left(x - \frac{L}{2}\right)\frac{\partial j}{\partial t},$$

$$(9.7)$$

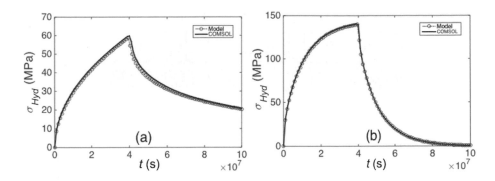

Figure 9.12 Stress relaxation at (a) $T = 373$ K and (b) $T = 400$ K. Reprinted from [35], with the permission of AIP Publishing.

while subjected to the same diffusion blocking BCs: $\partial\sigma/\partial t|_{x=0,l} = -eZ\rho j/\Omega$, and initial condition $\sigma(x, t = 0) = \sigma_T$ used in (4.24). By introducing new variables $u(x,t) = \sigma + \frac{eZ\rho j}{\Omega}\left(x - \frac{L}{2}\right) - \sigma_T$ and $h(x,t) = \frac{eZ\rho j}{\Omega}\left(x - \frac{L}{2}\right)\frac{\partial j}{\partial t}$, the nonhomogeneous BCs can be converted to the homogeneous BCs $\partial u/\partial t|_{x=0,l} = 0$. This allows us to obtain a solution by applying the cosine series expansion for $u(x,t)$ and $h(x,t)$ as

$$\sigma(x,t) = \sigma_T + \frac{4eZ\rho}{\Omega L}\kappa^2 \sum_{n=0}^{\infty} \cos\frac{(2n+1)\pi x}{L} e^{-\kappa^2\frac{(2n+1)^2\pi^2}{l^2}t} \int_0^t j(t)e^{\kappa^2\frac{(2n+1)^2\pi^2}{L^2}\tau}d\tau. \quad (9.8)$$

Here, it is easy to show that by substituting the time-dependent current density $j(t)$ a DC current density j and using the trigonometric identity $4\sum_{n=0}^{\infty}\dfrac{\cos\left(\dfrac{(2n+1)\pi x}{L}\right)}{(2n+1)^2\pi^2} = \frac{1}{2} - \frac{x}{L}$, [42], (9.8) yields the classical stress evolution kinetics described by (4.34) in Section 4.5. If at an instant $t = t^*$ the current is switched off, then the stress will start to relax. The stress relaxation kinetics can be obtained by solving (9.7) with the initial condition given by (9.8) at $t = t^{**}$:

$$\sigma(x, t \geq t^*) = \sigma_T + \frac{4eZ\rho}{\Omega L}\kappa^2 \sum_{n=0}^{\infty} \cos\frac{(2n+1)\pi x}{L} e^{-\kappa^2\frac{(2n+1)^2\pi^2}{L^2}t} \int_0^{t^*} j(t)e^{\kappa^2\frac{(2n+1)^2\pi^2}{L^2}\tau}d\tau. \quad (9.9)$$

Solution (9.9) describes the relaxation of stress accumulated during the time interval $[0, t^*]$ as an exponential decay with the partial time constants $\tau_0^{(2n+1)} = L^2/\kappa^2\pi^2(2n+1)^2 = \tau_0/(2n+1)^2$, determined by the atomic diffusivity. Figure 9.12 shows the relaxation kinetics of stresses accumulated at the cathode end in 4×10^7 s under a DC current of 1×10^9 A/m^2 at $T = 373$ K and 400 K. The kinetics was derived from the analytical expression (9.9) and by numerical solution of the (4.17) using the finite element analysis tool COMSOL [43] for the following parameters: $B = 1 \cdot 10^{11}$ Pa, $\Omega = 1.66 \cdot 10^{-29}$ m^3, $k_B = 1.38 \cdot 10^{-23}$ J/K, $L = 1 \cdot 10^{-4}$ m, $\sigma_T = 0$, $Z = 10$, $e = 1.6 \cdot 10^{-19}$ q, $\rho = 3 \cdot 10^{-8}$ Ohm·m, $D_0 = 7.56 \cdot 10^{-5}$ m^2/s, and $E_a = 1.28 \cdot 10^{-19}$ J.

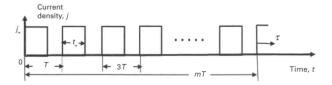

Figure 9.13 Schematics for the unipolar pulse current. Reprinted from [35], with the permission of AIP Publishing.

Stress evolutions induced by a unipolar (unidirectional) pulse current (UPC) with current density j_+, pulse period T and pulse time t_+ can be obtained by solving (9.7) subjected to the initial condition:

$$\sigma(x, t = T) = \sigma_T + \frac{4eZ\rho}{\Omega l}\kappa^2 \sum_{n=0}^{\infty} \cos\frac{(2n+1)\pi x}{l} e^{-\kappa^2\frac{(2n+1)^2\pi^2}{l^2}T} \int_0^{t_+} j_+(t)e^{\kappa^2\frac{(2n+1)^2\pi^2}{l^2}t} dt.$$

The kinetics of stress evolution has the form

$$\sigma(x,t) = \sigma_T + \frac{4eZ\rho}{\Omega l}\kappa^2 \sum_{n=0}^{\infty} \cos\frac{(2n+1)\pi x}{l} e^{-\kappa^2\frac{(2n+1)^2\pi^2}{l^2}\tau}$$

$$\times \left(\int_0^\tau j_+(\tilde{\tau})e^{\kappa^2\frac{(2n+1)^2\pi^2}{l^2}\tilde{\tau}} d\tilde{\tau} + e^{-\kappa^2\frac{(2n+1)^2\pi^2}{l^2}T} \int_0^{t_+} j_+(\tilde{\tau})e^{\kappa^2\frac{(2n+1)^2\pi^2}{l^2}\tilde{\tau}} d\tilde{\tau} \frac{1 - e^{-\kappa^2\frac{(2n+1)^2\pi^2}{l^2}mT}}{1 - e^{-\kappa^2\frac{(2n+1)^2\pi^2}{l^2}T}} \right).$$

$$(9.10)$$

Here, $t = mT + \tau$ is the global time as shown in Figure 9.13.

The solution of (9.7) for the bipolar (bidirectional) pulse current (BPC) with the positive and negative current densities j_+ and j_-, pulse period T, and positive and negative phase durations t_+ and t_- results in the following stress evolution kinetics.

For the positive pulse phase, the global time is $t = mT + \tau$:

$$\sigma(x,t) = \sigma_T + \frac{4eZ\rho}{\Omega l}\kappa^2 \sum_{n=0}^{\infty} \cos\frac{(2n+1)\pi x}{l} e^{-\frac{\tau}{\tau_0^{(2n+1)}}} \left(\int_0^\tau j_+ e^{\frac{\tilde{\tau}}{\tau_0^{(2n+1)}}} d\tilde{\tau} + K_n \frac{1 - e^{-\frac{mT}{\tau_0^{(2n+1)}}}}{1 - e^{-\frac{T}{\tau_0^{(2n+1)}}}} \right),$$

$$(9.11)$$

where $K_n = e^{-\frac{t_+}{\tau_0^{(2n+1)}}} \int_0^{t_+} j_+ e^{\frac{\tilde{\tau}}{\tau_0^{(2n+1)}}} d\tilde{\tau} + \int_0^{t_-} j_- e^{\frac{\tilde{\tau}}{\tau_0^{(2n+1)}}} d\tilde{\tau}$.

For the negative pulse phase, the global time is $t = mT + t_+ + \tau$:

$$\sigma(x,t) = \sigma_T + \frac{4eZ\rho}{\Omega l}\kappa^2 \sum_{n=0}^{\infty} \cos\frac{(2n+1)\pi x}{l} e^{-\frac{\tau}{\tau_0^{(2n+1)}}}$$

$$\times \left(\int_0^\tau j_- e^{\frac{\tilde{\tau}}{\tau_0^{(2n+1)}}} d\tilde{\tau} + e^{-\frac{t_+}{\tau_0^{(2n+1)}}} \int_0^{t_+} j_+ e^{\frac{\tilde{\tau}}{\tau_0^{(2n+1)}}} d\tilde{\tau} + e^{-\frac{T}{\tau_0^{(2n+1)}}} K_n \frac{1 - e^{-\frac{mT}{\tau_0^{(2n+1)}}}}{1 - e^{-\frac{T}{\tau_0^{(2n+1)}}}} \right).$$

$$(9.12)$$

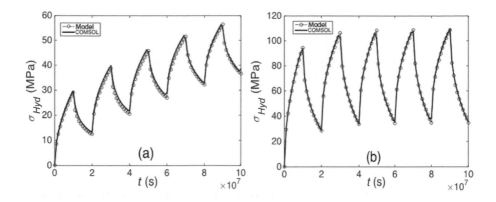

Figure 9.14 Stress evolution caused by the unipolar pulse current at $T = 373$ K (a); and $T = 400$ K (b). Reprinted from [35], with the permission of AIP Publishing.

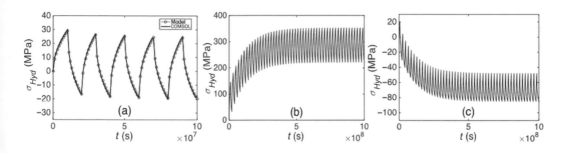

Figure 9.15 Stress evolution caused by bipolar symmetrical pulse ($t_+ = 1 \times 10^7$ s and $t_- = 1 \times 10^7$ s) current density of 1×10^9 A/m² taking place at $T = 373$ K, (a); long-term stress kinetics caused by BPC with $j_+ = 5 \times 10^9$ A/m² and $j_- = -1 \times 10^9$ A/m², $t_+ = 1 \times 10^7$ s and $t_- = 1 \times 10^7$ s, (b), and with $j_+ = 1 \times 10^9$ A/m² and $j_- = -1 \times 10^9$ A/m², $t_+ = 0.5 \times 10^7$ s and $t_- = 1.5 \times 10^7$ s (c) at $T = 373$ K. Reprinted from [35], with the permission of AIP Publishing.

The stress evolution induced by a unipolar pulse ($t_+ = 1 \times 10^7$ s and $T = 2 \times 10^7$ s) current with the density of 1×10^9 A/m² at $T = 373$ K and 400 K is shown in Figure 9.14. Figure 9.14b demonstrates the fast stress relaxation occurring between the pulses, which happens at elevated temperatures. The solutions to (9.7) for the bipolar (bidirectional) BPC with the different positive and negative current densities j_+ and j_-, pulse period T, and positive and negative phase durations t_+ and t_- are shown in Figure 9.15. Figure 9.15a shows the fit between stress evolution curves caused by a bipolar symmetrical pulse ($t_+ = 1 \times 10^7$ s and $t_- = 1 \times 10^7$ s) current density of 1×10^9 A/m² at $T = 373$ K predicted by (9.11) and (9.12) and with COMSOL-based numerical solution of (9.7).

All the results shown here were obtained for slow varying current densities with the positive and negative parts of the pulse applied for about 100 days [35]. The long-time intervals allowed the system to reach a new equilibrium state, corresponding to the

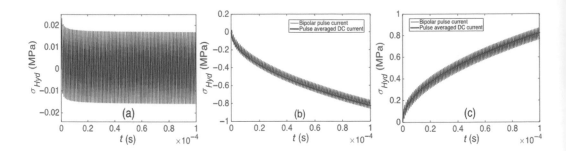

Figure 9.16 Stress evolution in the metal line caused by high-frequency (1 MHz) bipolar pulse currents: with $j_+ = 1 \times 10^9$ A/m^2 and $j_- = -1 \times 10^9$ A/m^2, (a), $j_+ = 1 \times 10^9$ A/m^2 and $j_- = -5 \times 10^9$ A/m^2, (b), $j_+ = 5 \times 10^9$ A/m^2 and $j_- = -1 \times 10^9$ A/m^2, (c) at $T = 500$ K. Reprinted from [35], with the permission of AIP Publishing.

applied current pulses, by means of metal atom diffusion. Contrary to this case, conventional semiconductor chips are operated at the MHz and GHz frequencies. For comparison, the stress evolution induced by applying 1 MHz symmetrical and non-symmetrical BPC, with the characteristic pulse time of 5×10^{-7} s, is shown in Figure 9.16. Here, the stress evolutions induced by high-frequency pulse currents correspond well to the stress evolutions induced by the averaged DC pulse currents $j = (j_+ t_+ + j_- t_-)/T$ as shown by the solid lines in Figure 9.16b and c. This result can be derived from the BPC stress evolution kinetics (9.11) and (9.12). Indeed, remembering that for all the time intervals involved in this case, we have $T, t_+, t_- \ll \tau_0$, so we can expand the exponents in series and keep just the liner terms to yield

$$\sigma = \sigma_T - \frac{eZ\rho L}{\Omega}\left(\frac{j_+ t_+ - |j_-|t_-}{T}\right)\left(\frac{x}{L} - \frac{1}{2} + 4\sum_{n=0}^{\infty}\frac{\cos\left(\frac{(2n+1)\pi x}{L}\right)}{(2n+1)^2\pi^2}e^{-\kappa^2\frac{(2n+1)^2\pi^2}{L^2}t}\right)$$

$$+ O\left(\frac{j_+ t_+ - |j_-|t_-}{\tau_0}\right),$$

$$(9.13)$$

where $O((j_+ t_+ - |j_-|t_-)/\tau_0)$ represents all the terms providing small (~0.01 MPa) fluctuations around the curve describing the stress evolution under the effective DC currents as shown in Figure 9.16.

Hence, for high-frequency pulse currents, which are typical for the standard cells operation in IC chips, the accumulated stress distributions along the intracell metal lines correspond to the stresses caused by the effective DC currents with the values scaled proportionally to the pulse duty factors, which is $(j_+ t_+ - |j_-|t_-)/T$ for bidirectional currents, and $j_+ t_+/T$ for the unidirectional pulse currents. And since the length of the metal lines located in the IC cells is usually so short, of the order of a micron or less, any essential stress buildup should not be expected. It means that EM-induced void nucleation and growth cannot be responsible for the resistance degradation observed in IC standard cells as reported [44].

9.5   On-Chip Power/Ground EM Assessment

A quite different picture of voiding due to time-dependent pulse currents can be developed in the power/ground nets. In this case, the currents distributed across the grid are a superposition of the unipolar pulses generated by switching cells, as shown in Figure 9.17. The intensity of these pulses depends on the activity factors of cells, which are the ratios of the durations of the "on" and "off" states of the cells and different for different workloads. In general, the total current passing between the neighbor power vias is time dependent, so the unidirectional character and the long length of power net segments can provide the conditions for σ_{crit} to accumulate. Thus, the effective-current model as demonstrated previously should be used to estimate the power grid EM degradation, so that the lifetime of a metal line carrying the time-varying transient current becomes the same as one carrying the constant (DC) effective current [45]. Hence, a power grid can be modeled as a DC linear system with constant (effective) currents, voltages, and conductance, with the effective branch currents being obtained directly from effective source currents by performing a simple DC analysis. As voids are being nucleated due to EM, the branch resistances would change fairly quickly, followed by the quick change of the currents and the voltage drops to their new effective values. Thus, the power grid model must consider the branch conductance change over time, and can be expressed as

$$G(t)V(t) = i_S, \tag{9.14}$$

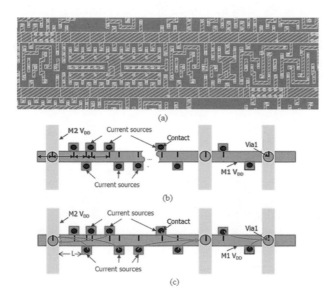

Figure 9.17 Schematics for the first metal layer (M1) layout: (a) power lines are the wide lines connecting power multiple vias, signal lines are the narrow lines; (b) schematics of the current source distribution along the power line; and (c) schematics of source currents in the power line at the arbitrary instance in time.

where $G(t)$ is the piecewise-constant conductance matrix, $V(t)$ is the corresponding time-varying but piecewise-constant node voltage drop vector, and i_S the effective DC source current vector modeling the underlying logic blocks.

9.5.1   Compact Model–Based Methodology for Power Grid EM Assessment

The EM assessment time scale for the power grids can be months or years. Instead of solving a system of the Korhonen equations linked at all the junctions by stress/flux continuity boundary conditions, Huang et al. [19, 27] developed a compact model to provide the void nucleation times at the cathodes (electron flow inlet ports) of tree branches where the steady-state stresses can exceed the critical stress. This model assumed that the distribution of the DC current densities and the current flow directions are given, and the void-free steady state for a given interconnect tree was achieved. At the steady state, the atomic flux in each branch under the electron flow is balanced by the atomic backflow induced by the stress gradient, so all atomic fluxes vanish. This condition is used for a simple estimate of the stresses developed at the cathode and anode ends of each branch under the void-free assumption as follows:

$$\sigma_i^c - \sigma_j^a = \Delta\sigma_{ij} = \frac{eZ\rho\left(j_{ij}L_{ij}\right)}{\Omega}. \tag{9.15}$$

Here, σ_i^c and σ_j^a are the hydrostatic stresses at the cathode (electron flow inlet) and anode (electron flow exit) ends of the ij-branch, respectively. Figure 9.18a shows an example of a multibranch interconnect tree. Equation (9.15) allows one to obtain the stresses at all the branches' ends as a function of the "reference" stress developed at the end of any arbitrarily chosen branch. The uncertainty can be easily eliminated by accounting for the atom conservation condition. Indeed, at the unblocked branch ends, the atom redistribution between branches would determine the final stress distribution while the total number of atoms remain unchanged. This leads to the following equation for determining the missing "reference" stress as

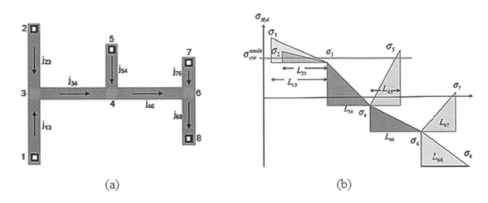

(a)   (b)

Figure 9.18 (a) Example of an interconnect tree; (b) hydrostatic stress distribution in the interconnect tree. © 2016 IEEE. Reprinted, with permission, from [27].

$$\sum_{i=1}^{k} \left(\sigma_i - \left(\sigma_T + \frac{eZ\rho\left(j_{ij}L_{ij}\right)}{2\Omega} \right) \right) L_{ij} = 0. \tag{9.16}$$

Thus, the steady-state stress distribution in the interconnect tree shown in Figure 9.18b can be derived by solving the following matrix equation:

$$M \times \begin{bmatrix} \sigma_1 \\ \sigma_2 \\ \sigma_3 \\ \sigma_4 \\ \sigma_5 \\ \sigma_6 \\ \sigma_7 \\ \sigma_8 \end{bmatrix} = \begin{bmatrix} \chi(j_{13}L_{13}) \\ \chi(j_{13}L_{13}) \\ \chi(j_{13}L_{13}) \\ \chi(j_{13}L_{13}) \\ \chi(j_{13}L_{13}) \\ \chi(j_{13}L_{13}) \\ \chi(j_{13}L_{13}) \\ \Theta \end{bmatrix}, \tag{9.17}$$

where:

$$M = \begin{bmatrix} 1 & 0 & -1 & 0 & 0 & 0 & 0 & 0 \\ 0 & 1 & -1 & 0 & 0 & 0 & 0 & 0 \\ 0 & 0 & 1 & -1 & 0 & 0 & 0 & 0 \\ 0 & 0 & 0 & -1 & 1 & 0 & 0 & 0 \\ 0 & 0 & 0 & 1 & 0 & -1 & 0 & 0 \\ 0 & 0 & 0 & 0 & 0 & -1 & 1 & 0 \\ 0 & 0 & 0 & 0 & 0 & 1 & 0 & -1 \\ L_{13} & L_{23} & L_{34} & L_{54} & L_{46} & L_{76} & L_{68} & 0 \end{bmatrix} \tag{9.18}$$

$\Theta = \sum_{i=1}^{7} \left(\sigma_T + \frac{eZ\rho\left(j_{ij}L_{ij}\right)}{2\Omega} \right) L_{ij} = 0$, and $\chi = \frac{eZ\rho}{\Omega}$.

The calculated stresses should be compared with the critical stress (σ_{crit}) responsible for the void nucleation. If any calculated stress exceeds σ_{crit}, then the time for void nucleation at the cathode, characterized by the biggest stress σ_m, can be calculated as

$$t_{nuc}^{m} \approx \frac{L_{max/min}^{2}}{2D_0} \frac{k_B T}{\Omega B} e^{\frac{E_V+E_D}{k_B T}} \exp\left\{ -\frac{f\Omega\sigma_{m(j_1,j_2,\dots,j_n)}}{k_B T} \right\} \times \ln\left\{ \frac{\sigma_{m(j_1,j_2,\dots,j_n)} - \sigma_T}{\sigma_{m(j_1,j_2,\dots,j_n)} - \sigma_{crit}} \right\}. \tag{9.19}$$

Equation (9.19) can be considered as an extension of (4.44) in Section 4.7 for EM assessment of a multibranch interconnect tree. Here $\sigma_m(j_1, \dots j_n)$ is the steady-state stress at the cathode end of the branch with the highest tensile stress in the interconnect tree, n is the number of branches, and $L_{max/min}$ is the cumulative length of the branches connecting the cathode (maximal tensile stress) to the anode (maximal compressive or minimal tensile stress). In the case of several cathode ends with stresses exceeding σ_{crit}, the nucleation time should be calculated for the cathode with

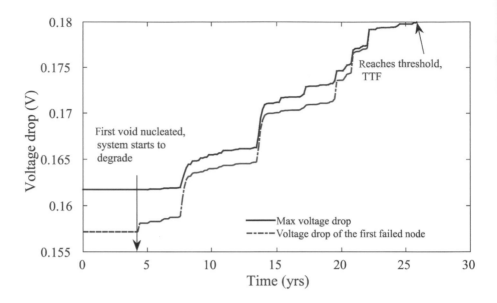

Figure 9.19 Voltage drop of the first failed node and the maximum voltage drop in IBMPGNEW1, showing the evolution over time. © 2016 IEEE. Reprinted, with permission, from [27].

the highest stress. A correct stress distribution in the interconnect tree at a time when the first void was nucleated, and during the time when the void evolved, can be derived by solving a system of equations, similar to (4.18) from Section 4.4, describing the simultaneous stress evolution in all branches of the tree subjected to the boundary conditions representing the continuity of stress and the atomic flux conservation at the branch joints. The resistance increases of individual branches due to void growth are calculated as described in [46]. Because all trees are electrically connected, a void-induced resistance change in one tree can change the currents in all neighboring trees, which can then change the volumes of the already formed saturated voids and hence the resistances of the corresponding branches. That condition would trigger a new calculation iteration, a process that will continue until the voltage drop at some node of the power/ground network reaching a user-specified threshold value, or the total time exceeding the required lifetime as shown in Figure 9.19. Such an EM assessment method was tested on IBM power grid benchmark circuits [47] on a 2.3 GHz Linux server with 132 GB memory.

Huang et al. [19] demonstrated for the first time that EM assessment based on the Black equation for either series or mesh models would lead to overly pessimistic results when compared to the method described here. The results are shown in Table 9.1. In the analysis based on the Black model, (9.4) was used to estimate MTTF of a single metal line, where $T_{STRESS} = 600$ K, $j_{STRESS} = 3 \times 10^{10}$ A/m^2, $E_a = 0.86$ eV, and the $MTTF_{STRESS}$ was obtained from t_{nuc} calculated under these stressed conditions [27] with a critical stress of 500 MPa. The use/operating conditions were characterized by a

Table 9.1. Comparison of power grid MTTF obtained using the Black model and the proposed model.

Power grid		Black equation		Proposed model		
				Time to failure (years)		
Name	Nodes	Series	Mesh	No saturated voids	With saturated voids.	Runtime
IBMPG2	61,797	6.17	12.83	16.85	18.78	6.36 m
IBMPG3	407,279	12.79	17.90	23.56	31.97	5.83 h
IBMPG	474,836	13.23	22.27	26.97	33.41	14.71 h
IBMPG	497,658	4.41	12.34	19.13	25.16	40.64 m
IBMPG	807,825	8.44	10.89	14.62	19.85	1.75 h
IBMPG NEW1	715,022	12.85	13.96	18.84	25.97	16.78 h
IBMPG NEW2	715,022	12.73	13.84	15.60	21.79	15.32 h

Source: © 2016 IEEE. Reprinted, with permission, from [27].

constant temperature of 373 K and the branch current densities distributed between 3.4×10^3 and 4.7×10^{10} A/m^2, which were extracted with SPICE simulations.

The methodology developed by Huang et al. [19, 27] had the difficulty of being very slow, requiring up to 17 hours to estimate the failure time of a grid with 700 thousands of nodes. Thus, there was a need for a new EM checking approach that can accurately model EM degradation using physics-based models, and to combine them with a mesh model to account for branch redundancy, but is still fast enough to be practically useful.

9.5.2   Extended Korhonen Model for Power Grid EM Assessment

This novel approach extending the Korhonen model was developed recently by Chatterjee et al. [24, 28, 48]. While the physical model was similar to that used by Huang et al. in [19, 27], the advanced numerical technique has improved the capability for EM assessment of power grids in large VLSI circuits. The method proceeded by first decomposing the power grid into a number of interconnect trees, solving the set of discretized PDEs (9.20) for all branches of each tree characterized by different current densities and geometries (length and width), and linking the solutions at the segment junctions through the proper BCs given in (9.21) and (9.22):

$$\frac{\partial \sigma_n}{\partial t} = \frac{\partial}{\partial x}\left[\kappa_n^2\left(\frac{\partial \sigma_n}{\partial x} + G_n\right)\right]. \tag{9.20}$$

Here, $\sigma_n(x,t)$ is the time-varying hydrostatic stress at location x in the nth branch of the tree and the standard notations are used as $\kappa_n^2 = D_{eff}^n B_n \Omega / k_B T_n$, and $G_n = eZ\rho j_n / \Omega$, where D_{eff}^n is the effective atomic diffusivity in the nth branch, B_n is the effective bulk modulus, Ω is the atomic volume, k_B is the Boltzmann constant, T_n is the absolute temperature, eZ is the effective charge of migrating atoms, ρ is the metal resistivity, and j_n is the electric current density in the nth branch. The BCs reflecting the continuity of stress and atomic flux at every junction between neighboring branches are the following:

$$\sigma_n(x,t) = \sigma_{n+1}(x,t), \quad x = x_n, t > 0 \tag{9.21}$$

$$\kappa_n^2\left(\frac{\partial\sigma_n}{\partial x} + G_n\right) = \kappa_{n+1}^2\left(\frac{\partial\sigma_{n+1}}{\partial x} + G_{n+1}\right), \quad x = x_n, t > 0. \tag{9.22}$$

The initial conditions for (9.20) determine the stress in interconnect trees at $t = 0$ before the electric stressing was applied. In on-chip interconnects, since the metal lines are embedded in a rigid confinement, the difference in the coefficients of thermal expansion (CTE) of the Cu metal α_m and Si confinement α_{conf} generates stresses as the chip cools down after anneal, which can be expressed as follows:

$$\sigma_n^{T_n}(t) = B_n\left(\alpha_m - \alpha_{conf}\right)(T_{ZS} - T_n(t)), \tag{9.23}$$

where $\sigma_n^{T_n}(t)$ is the thermal stress, $T_n(t)$ is the temperature of branch n, B_n is the effective bulk modulus, and T_{ZS} is the stress-free annealing temperature. Here we assume that the initial stress $\sigma_n(0,t)$ in branch n is equal to its thermal stress at $t = 0$, so that

$$\sigma_n(0,t) = \sigma_n^{T_n}(0). \tag{9.24}$$

The temperature distribution across the metal layers is taken into account by employing a compact thermal model that represents a die as an array of cuboidal thermal cells with effective local thermal properties. Specifically, each thermal block is represented as a thermal node connected to six resistors, a current source, and a capacitor. Thermal resistors are used to account for heat propagation in the lateral and vertical directions, and a thermal capacitor is included for transient thermal analysis. The analysis consists of three steps: (i) extracting the effective thermal properties of each thermal cell, (ii) generating a thermal netlist for the whole chip, and (iii) evaluating the temperature at each thermal node by circuit solvers as in [49].

The effective thermal conductivities are calculated as a function of metal density and routing direction of wires in each metal layer based on the theory of effective thermal properties of anisotropic composite materials [50]. With the extracted thermal resistances, estimated power sources, as well as the thermal boundary conditions, the chip can be represented as a thermal netlist, in which the nodal temperatures correspond to the nodal voltages and the powers corresponds to the current sources [49]. The total power dissipated in a thermal block can be represented as the sum of the average power dissipated by Joule heating of the metal branches within the thermal block and the underlying logic circuits due to active switching and leakage currents [32, 33]. The electric circuit solver is used to obtain the temperature for each thermal node. The thermal grid is first generated at $t = 0$ to calculate the diffusivities and thermal stress in all branches at the initial temperature. After a void is nucleated, the branch currents will change and the average powers dissipated by Joule heating are updated for all thermal nodes in order to find the new temperature distributions, branch diffusivities, and thermal stresses. As an example, we have analyzed the effect of temperature on the estimated lifetime using the extended Korhonen model. For this study, we used the interconnect tree taken from the ibmpg2 grid with 192 branches and a high current density profile corresponding to a maximum branch current density of 5.31×10^9 A/m^2 [28]. We first estimated the MTTF using the actual temperature distribution, varying between 318–333 K. For this case, the first failure happens around 13.2 years. Then, we assumed a constant temperature of 325 K throughout the tree, corresponding to the average of the actual branch temperatures. In this case, the first failure happens around 20.26 years, indicating that a higher nominal temperature would result in a shorter failure time, as expected. Hence, the temperature distribution plays a very important role and should be taken into account for EM assessment of on-chip power grids.

It should be mentioned that the variation in the thermal stress before the electric stressing is applied, can be attributed to the dependency of the effective bulk modulus B on the line geometry: its width, the aspect ratio, and the grain morphology [51]. The additional residual stress developed across the whole interconnect structure after packaging assembly can be assessed using the methodology as described in [52, 53]. Here the stress relaxation is induced mainly by stress-gradient-induced atomic diffusion, i.e., stress migration (SM), which occurs during chip shelf life, resulting in a uniform stress distribution within every interconnect tree [2]. Following He et al. [54], we assume that the residual stresses in some of interconnect trees exceed the critical stress, leading to formation of the saturated voids, whose sizes can be estimated from the initial strain and zero stress everywhere else in the line. Here the possibility of multivoid generation in the interconnect trees should also be considered and the actual stress distribution, which is used as an initial stress in the EM analysis, depends on the duration of the chip shelf life, which introduces another uncertainty factor. The kinetics of stress evolution under EM are evaluated for two cases: one initially void free, and the other with a saturated void. Results are different only in the beginning, which disappear at long times, showing the same steady-state stress distributions [55]. To be in a more pessimistic side in predicting the EM-induced MTTF, we should use the SM-induced steady-state stress and saturated voids located in trees with residual stresses exceeding the critical one as an initial condition in all interconnect trees. The BC at the void edges, which is used for calculating the postvoiding stress evolution, takes the following form [55]:

$$\frac{\partial \sigma \left(x_{edge}, t \right)}{\partial x} = \frac{\sigma \left(x_{edge}, t \right)}{\delta}. \tag{9.25}$$

Here δ is the thickness of the void interface, $\sigma \left(x_{edge}, t \right)$ is the stress near the void edge equal to the critical stress at the time of void nucleation. Resolving the postvoiding stress evolution kinetics allows us to calculate the void volume evolution from the volume of atoms transferred from the void surface into the metal $V_{void}(t) = A \int_0^L \left(\sigma(x,t)/B \right) dx$, where A is the cross-sectional area and L is the length of the segment as shown in Figure 9.20. The void evolution kinetics used to calculate the resistance evolution has replaced the approximation employed in [26, 28, 48], leading to very fast void growth and thus very short times (relative to nucleation phase

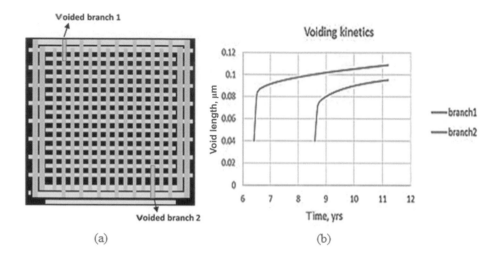

Figure 9.20 (a) An upper-level layout with voided branches; (b) void evolution kinetics. Reprinted from [56], with the permission of AIP Publishing.

durations) to reach void saturation. This would be a more conservative estimate of the grid TTF.

Additional variations of the residual stress can be expected in the junctions between line segments with polygranular and near-bamboo structures [51] and in the vicinity (above and below) of interlayer vias [57]. Such variations can be addressed by introducing a lookup table to provide different values of the effective bulk modulus B for corresponding layout segment configurations.

The physics discussed so far provides a basis for accurate calculation of the stress evolution in a multisegment tree. As an example, Figure 9.21 shows the kinetics of stress evolution at all tree junctions and the stress distribution across the three-terminal tree in Figure 9.22.

The extended Korhonen model starts out as a system of PDEs coupled to each other by the boundary conditions (9) and (10), which are then scaled and discretized to reduce the model to a system of ordinary differential equations (ODE), which is a linear time-invariant (LTI) system. The method can then proceed to numerically solve the ODE system at successive time points to track the stress evolution and find the corresponding void nucleation times.

To account for the random nature of EM degradation, Monte Carlo random sampling was used to estimate the MTTF. In each Monte Carlo iteration, new randomly generated diffusivities were assigned to all the branches in the grid. This effectively produces a new whole power grid, which was referred to as a sample grid. Then the TTF values were generated based on the mesh model, and another based on the series model for comparison purposes. With enough samples, two averages were found as the estimates of the series MTTF (MTF_S) and the mesh MTTF (MTF_M) in Table 9.2 [58]. The stress conditions used to calculate MTTF were similar to what was used for Table 9.1, except that the critical stress was taken as 600 MPa at temperature of 300 K.

Table 9.2. Comparison of power grid MTTF as estimated using the Black model and the extended Korhonen model.

Power grids				Performance metrics		
Grid name	# of nodes	# of branches	# of trees	MTF_s (years)	MTF_m (years)	t_{CPU} (mins)
ibmpg1	6K	11K	709	3.3	7.1	0.6
ibmpg2	62K	61K	462	6.7	12.0	1.2
ibmpg3	410K	401K	8.1K	4.5	6.8	4.2
ibmpg4	475K	465K	9.6K	9.0	16.8	6.6
ibmpg5	249K	496K	2K	4.5	6.4	1.8
ibmpg6	404K	798K	10.2K	5.6	11.2	14.0
ibmpgnew1	316K	698K	19.5K	34	13.3	4.8
ibmpgnew2	718K	698K	19.5K	4.7	7.4	3.6
PG1	560K	558K	2.6K	4.4	17.2	2.4
PG2	1.2M	1.2M	5.6K	3.6	10.4	7.2
PG3	1.6M	1.6M	6.9K	3.9	8.5	3.6
PG4	2.6.M	2.6M	12.2K	3.3	14.8	9.0
PG5	4.1M	4.1M	12.7K	4.4	9.2	8.4

Source: © 2019 IEEE. Reprinted, with permission, from [58].

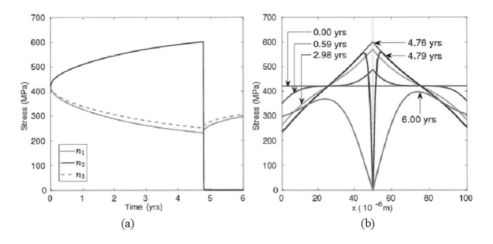

Figure 9.21 (a) Evolution of stress at junctions with time and (b) stress profile evolution with time. Here $L_1 = L_2 = 50$ μm, and $j_1 = -j_2 = 6e9$ A/m^2. © 2018 IEEE. Reprinted, with permission, from [28].

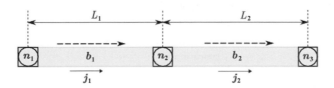

Figure 9.22 A simple three-terminal tree. © 2018 IEEE. Reprinted, with permission, from [28].

Computation speed is enhanced by using a filtering scheme that estimates upfront the set of trees that are most likely to impact the MTTF of the grid, with minimal impact on accuracy. The process also includes a predictive scheme that allows for faster MTTF estimation by extrapolating the solution (stress curve) obtained from a few initial time points. This has also been shown to have minimal impact on accuracy.

The resulting approach was tested on the same set of IBM power grid benchmarks, [47], on a quad-core 3.4 GHz Linux machine with 32 GB of RAM. The MTTFs estimated using the physics-based approach were on average $3\times$ longer than those based on the Black model, supporting the claim that the Black model is not accurate enough for modern power grids and confirming the need for physical models. Having achieved a runtime of less than seven hours for the largest assessed grid (15.5 million nodes, 12.4 million branches, and 3.2 million current sources), this approach has been demonstrated as suitable for large VLSI circuits.

In order to show the inaccuracy of the Black model, two scenarios, based on two interconnect trees T_1 and T_2 taken from ibmpg2, have been presented in [28]. Both trees are straight metal stripes with 192 branches each. T_1 has a high current density profile, with maximum branch current density being 5.31×10^9 A/m^2 (Figure 9.23). In this case, the Black model predicts the first failure time of about 6.2 years, whereas the actual failure time found using the extended Korhonen model is about 13.2 years,

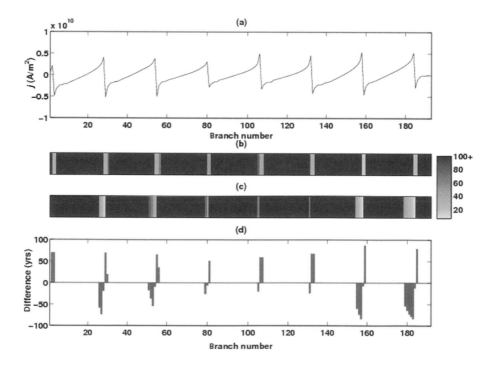

Figure 9.23 (a) Current density profile for T_1 and MTTFs estimated using (b) the extended Korhonen model (MTTF$_{ekm}$), (c) the Black model (MTTF$_{blk}$), and (d) MTTF$_{blk}$ − MTTF$_{ekm}$. © 2018 IEEE. Reprinted, with permission, from [28].

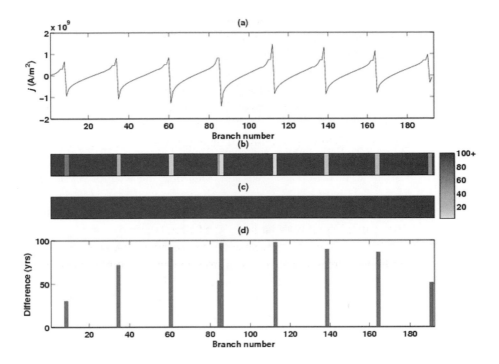

Figure 9.24 (a) Current density profile for T_2 and MTTFs estimated using (b) the extended Korhonen model (MTTF_{ekm}), (c) the Black model (MTTF_{blk}), and (d) $\text{MTTF}_{\text{blk}} - \text{MTTF}_{\text{ekm}}$. © 2018 IEEE. Reprinted, with permission, from [28].

which is ~2× longer. T_2 has a low current density profile, with maximum branch current density being 1.44×10^9 A/m² (Figure 9.24). Here, due to the Blech effect, the Black model predicts that no failure would occur. However, based on a physical model to account for material flow between the branches, it was found that the first failure would occur at about 2.44 years. Thus, the Black model was pessimistic in the first scenario and highly optimistic in the second one. This shows that lifetime estimates using the Black model can be highly inaccurate.

The developed 1D model has a built-in capability to account for an early failure, which happens if a large enough void forms below a via [28, 58] (Figure 9.25). Removal of a via, as it happens during the early failures, has a significant impact on grid reliability. In our model, once we have determined the void volume, we check for the following two conditions: (i) is the void located below a via (this is determined based on the geometry of the grid); and (ii) is the void large enough to disconnect the via. If both conditions are met, this void leads to an early failure, so that we remove the via from the power grid and update the voltage drops and current density values. In order to assess the impact of early failures on the grid lifetime, a case study using the ibmpg2 grid was presented in [28, 58]. The mesh MTTF was estimated under two settings, one where early failure detection is on and the other where early failure detection is turned off. As can be seen from Figure 9.26b, turning

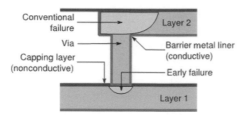

Figure 9.25 Early failures and conventional failures. © 2018 IEEE. Reprinted, with permission, from [28].

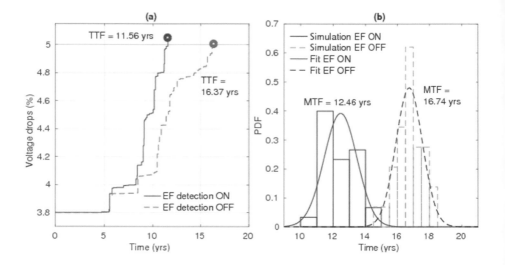

Figure 9.26 Impact of early failures (EF) on (a) the maximum voltage drop (shown for one sample grid) and (b) estimated mesh MTF for ibmpg2. © 2018 IEEE. Reprinted, with permission, from [28].

off early failures gives an optimistic MTF estimate that is 34% longer than the actual MTF. Thus, if the target product lifetime is set as 15 years, this grid will fail the EM sign-off due to the impact of early failures but would erroneously succeed if early failures are ignored. The difference in MTFs stems from the influence of early failures on node voltage drops. It is shown in Figure 9.26a how the maximum node voltage drop changes with time (for one sample grid) as the voids nucleate due to EM. Since early failures lead to removal of a via, their impact on voltage drops is more severe, which ultimately leads to shorter lifetimes. In general, the effect of early failures gets more pronounced as the difference between the maximum initial voltage drop and the threshold increases.

In order to improve model predictability and, as a result, the accuracy of MTTF predictions, a number of model enhancements should be done. For example, the methodology for finding the saturated void volume V_{SV} in interconnect trees should be improved. The formalism employed in [28] describes V_{SV} for the case of a single line [19, 26]. It should be extended to the case of a multisegment interconnect tree, and the effect of the already nucleated voids on new void nucleation and growth should be accounted for. Effect of microstructure on EM-induced stress evolution and voiding should be implemented in a way similar to that discussed by Korhonen et al. in [29]. As discussed in literature [59], the effect of the EM-induced plasticity creates new interfaces for atomic diffusion in later stages of the electromigration process in the form of dislocations and grain boundaries parallel to the direction of the current, which can be addressed in 1D modeling by adjusting the atomic diffusivity at the calibration stage. Finally, different types of void growth kinetics should be implemented in the case of initially voidless trees versus trees with preexisting saturated voids, similar to what was proposed in [54].

These model enhancements will require additional spatial discretization of the tree segments and solution of even larger systems of ODEs, which in turn will demand further improvement of the numerical techniques to provide a faster computation speed. Thus, we can conclude that the critically needed analysis of EM-induced voltage-drop degradation in on-chip power/ground grids is a justification and a strong drive for further development of accurate 1D EM models and fast simulation techniques.

9.6   Summary

In this chapter, we summarize the current results that have been achieved with a recently developed methodology for EM assessment in commercial-grade power-delivery networks. Despite the need for various additional improvements, the physics-based EM verification and checking methodology are already available, and the numerical capabilities recently developed have demonstrated that the industry-accepted EM assessment based on the Black model cannot accurately predict lifetime of modern power grids. The methodology based on the Black model calls for significant overdesign, while, today, there is very little margin left for electromigration. It has been demonstrated that pessimism, which is natural for this methodology, is very high: for example, grids that must survive 10 years are being designed to survive 40 years or more. One might think that such pessimism is not a bad strategy in VLSI. However, too much pessimism in the power grid can be a big problem. It leads to overuse of metal area, leaving little room for signal routing, which makes EM signoff extremely difficult in modern designs, thus increasing design complexity and design time. In contrast, the newly developed physics-based EM assessment approach provides the MTTF for any given power grid, which can be a DC, RC, or RLC netlist, and user-specified current sources and voltages (see Figure 9.27). If adopted, this approach can effectively relax the very conservative current density design rules to allow many improvements in power, time to market, and design cost.

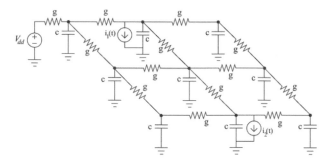

Figure 9.27 Power grid schematics with user-provided current sources and voltages. © 2018 IEEE. Reprinted, with permission, from [5].

References

1. J. Warnock, Circuit design challenges at the 14nm technology node, *Proceedings of the 48th Annual Design Automation Conference (DAC)* (San Francisco: ACM, 2011), 464–467.
2. V. Sukharev, Beyond Black's equation: full-chip EM/SM assessment in 3D IC stack, *Microelectron. Engineering* **120** (2014), 99–105.
3. J. Lienig, M. Thiele, *Fundamentals of Electromigration-Aware Integrated Circuit Design* (Switzerland: Springer, 2018).
4. M. Ohring, *Reliability and Failure of Electronic Materials and Devices* (San Diego: Academic Press, 1998).
5. V. Sukharev and F. N. Najm, Electromigration check: where the design and reliability methodologies meet, *IEEE Transactions on Device and Materials Reliability* **18** (2018), 498–507.
6. J. R. Black, Electromigration: a brief survey and some recent results, *IEEE Transactions on Electron Devices* **16** (1969), 338–347.
7. S. P. Hau-Riege and C. V. Thompson, Experimental characterization and modeling of the reliability of interconnect trees, *Journal of Applied Physics* **89** (2001), 601–609.
8. J. R. Lloyd, New models for interconnect failure in advanced IC technology. *Proceedings of the 14th International Symposium on the Physical and Failure Analysis of Integrated Circuits (IPFA)* (Piscataway: IEEE, 2008), 297–302.
9. M. Hauschildt, M. Gall, C. Hennesthal et al., Electromogration void nucleation and growth analysis using large-scale early failure statistics. *Proceedings of the 13th International Workshop on Stress-Induced Phenomena and Reliability in 3D Microelectronics*, ed. P. S. Ho, C. K. Hu, M. Nakamoto, et al. (Kyoto: AIP Conference Proceedings 1601, 2014), 89–98.
10. M. Hauschildt, C. Hennesthal, G. Talut, et al., Electromigration early failure void nucleation and growth phenomena in Cu and Cu(Mn) interconnects, *Proceedings of the 2013 IEEE International Reliability Physics Symposium (IRPS)* (Anaheim: IEEE, 2013), 2C.1.1–2C.1.1-6.
11. G. L. Gan, C. V. Thompson, K. L. Pey, and W. K. Choi, Experimental characterization and modeling of the reliability of three-terminal dual-damascene Cu interconnect trees, *Journal of Applied Physics* **94** (2003), 1222–1228.

12. M.-H. Lin and T. Oates, Electromigration failure time model of general circuit-like interconnects, *IEEE Transactions on Device and Materials Reliability* **17** (2017), 381–398.

13. C. V. Thompson, S. P. Hau-Riege, and V. K. Andleigh, Modeling and experimental characterization of electromigration in interconnect trees, *Proceedings of the 5th International Workshop on Stress-Induced Phenomena in Metallization*, ed. O. Kraft, E. Arzt, C. Volkert, and P. S. Ho. (Stuttgart: AIP Conference Proceedings 491, 1999), 150–162.

14. C. V. Thompson, C. L. Gan, S. M. Alam, and D. E. Troxel, Experiments and models for circuit-level assessment of the reliability of Cu metallization, *2010 IEEE International Interconnect Technology Conference Proceedings (IITC)* (Burlingame: IEEE, 2004), 69–72.

15. M. J. Dion, Electromigration lifetime enhancement for lines with multiple branches, *Proceedings of the 2000 IEEE International Reliability and Physics Symposium (IRPS)* (San Jose: IEEE, 2000), 324–332.

16. H.-B. Chen, S. X.-D. Tan, X. Huang, T. Kim, and V. Sukharev, Analytical modeling and characterization of electromigration effects for multibranch interconnect trees, *IEEE Transactions on Computer-Aided Design for Integrated Circuits and Systems* **35** (2016), 1811–1824.

17. I. A. Blech, Electromigration in thin aluminum films on titanium nitride, *Journal of Applied Physics* **47** (1976), 1203–1208.

18. S. Chatterjee, M. B. Fawaz, and F. N. Najm, Redundancy-aware electromigration checking for mesh power grids, *Proceedings of the 32th International Conference on Computer-Aided Design (ICCAD)* (San Jose: IEEE/ACM, 2013), 540–547.

19. X. Huang, T. Yu, V. Sukharev, and S. X.-D. Tan, Physics-based electromigration assessment for power grid networks, *Proceedings of the 51st Annual Design Automation Conference (DAC)* (San Francisco: ACM, 2014), 1–6.

20. B. Ouattara, L. Doyen, D. Ney, H. Mehrez, and P. Bazargan-Sabet, Power grid redundant path contribution in system on chip (SoC) robustness against electromigration, *Microelectronics Reliability* **54** (2014), 1702–1706.

21. B. Li, A. Kim, P. McLaughlin, B. Linder, and C. Christiansen, Electromigration characteristics of power grid like structures, *Proceedings of the 2018 IEEE International Reliability Physics Symposium (IRPS)* (San Francisco: IEEE, 2018), 4F3.1–4F3.5.

22. C. Zhou, R. Fung, S.-J. Wen, R. Wong, and C. Kim, Electromigration effects in power grids characterized from a 65 nm test chip, *IEEE Transactions on Device and Materials Reliability* 20 (2020), 74–83.

23. F. Chen, E. Mccullen, C. Christiansen, et al., Diagnostic electromigration reliability evaluation with a local sensing structure, *Proceedings of the 2015 IEEE International Reliability and Physics Symposium (IRPS)*. (Monterey: IEEE, 2015), 2D4.1–2D4.7.

24. K. Croes, M. Iofrano, C. J. Wilson, et al., Study study of void formation kinetics in Cu interconnects using local sense structures, *Proceedings of the 2011 IEEE International Reliability Physics Symposium (IRPS)* (Monterey: IEEE, 2011), 3E5.1–3E5.7.

25. F. Bana, L. Arnaud, D. Ney, R. Galand, and Y. Wouters, Effects of current density on electromigration resistance trace analysis, *Final Report of the 2011 IEEE International Integrated Reliability Workshop* (South Lake Tahoe: IEEE, 2011), 59–62.

26. S. Chatterjee, V. Sukharev, and F. N. Najm, Fast physics-based electromigration checking for on-die power grids, *Proceedings of the 35th International Conference on Computer-Aided Design (ICCAD)* (Austin: IEEE/ACM, 2016), 110.1–110.8.

27. X. Huang, A. Kteyan, S. X.-D. Tan, and V. Sukharev, Physics-based electromigration models and full-chip assessment for power grid networks, *IEEE Transactions on Computer-Aided Design on Integrated Circuits and Systems* **35** (2016), 1848–1861.

28. S. Chatterjee, V. Sukharev, and F. N. Najm, Power grid electromigration checking using physics-based models, *IEEE Transactions on Computer-Aided Design Integrated Circuits and Systems* **37** (2018), 1317–1330.

29. M. A. Korhonen, P. Borgesen, K. N. Tu, and C.-Y. Li, Stress evolution due to electromigration in confined metal lines, *Journal of Applied Physics* **73** (1993), 3790–3799.

30. S. A. Alam, C. L. Gan, C. V. Thompson, and D. E. Troxel, Reliability computer-aided design tool for full-chip electromigration analysis and comparison with different interconnect metallization, *Microelectronics Journal* **38** (2007), 463–473.

31. D.-A. Li, M. Marek-Sadowska, and S. Nassif, A method for improving power grid resilience to electromigration-caused via failures, *IEEE Transactions on Very Large Scale Integrated (VLSI) Systems* **23** (2015), 118–130.

32. M. Chew, A. Aslyan, J.-H. Choy, and X. Huang, Accurate full-chip estimation of power map, current densities and temperature for EM assessment, *Proceedings of the 33th International Conference on Computer-Aided Design (ICCAD)* (San Jose: IEEE/ACM, 2014), 440–445.

33. X. Huang, V. Sukharev, J-H Choy, M. Chew, T. Kim, and S. X.-D. Tan, Electromigration assessment for power grid networks considering temperature and thermal stress effects, *Integration, the VLSI Journal* **55** (2016), 307–315.

34. D. A. Li, M. Marek-Sadowska, and S. R. Nassif, T-VEMA: a temperature- and variation-aware electromigration power grid analysis tool, *IEEE Transactions on Very Large Scale Integration (VLSI) Systems* **23** (2015), 2327–2331.

35. V. Sukharev, X. Huang, and S. X.-D. Tan, Electromigration induced stress evolution under alternate current and pulse current loads, *Journal of Applied Physics* **118** (2015), 034504.1–034504.10.

36. J. A. Maiz, Characterization of electromigration under bidirectional (BC) and pulsed unidirectional (PDC) currents, *Proceedings of the 1989 IEEE International Reliability Physics Symposium (IRPS)* (San Diego: IEEE, 1989), 220–228.

37. K. Liew, N. W. Cheung, and C. Hu, Projecting interconnect electromigration lifetime for arbitrary current waveforms, *IEEE Transactions on Electron Devices* **37** (1990), 1343–1351.

38. K. Hatanaka, T. Noguchi, and K. Maeguchi, A generalized lifetime model for electromigration under pulsed DC/AC stress conditions, *Digest of Technical Papers of the 1989 Symposium on VLSI Technology* (Tokyo: 1989), 19–20.

39. H. Chenming, Reliability phenomena under AC stress, *Microelectronics Reliability* **38** (1998), 1–5.

40. W. R. Hunter, Self-consistent solutions for allowed interconnect current density. I. Implications for technology evolution, *IEEE Transactions on Electron Devices* **44** (1997), 304–309.

41. R. Moning, R. R. Keller, and C. A. Volkert, Thermal fatigue testing of thin metal films, *Review of Scientific Instruments* **75** (2004), 4997–5004.

42. S. Gradshteyn and I. M. Ryzhik, *Table of Integrals, Series, and Products*, 8th ed. (Waltham: Academic Press, 2014).

43. COMSOL, Inc., Burlington, MA. 8 New England Executive Park.

44. P. Jain, S. S. Sapatnekar, and J. Cortadella, A retargetable and accurate methodology for logic-IP-internal electromigration assessment, *Proceedings of the Annual Asian-Pacific Design Automation Conference (ASP-DAC)* (Chiba: ACM, 2015), 346–351.

45. L. M. Ting, J. S. May, W. R. Hunter, and J. W. McPherson, AC electro- migration characterization and modeling of multilayered interconnects, *Proceedings of the 1993 IEEE International Reliability Physics Symposium (IRPS)* (Atlanta: IEEE, 1993), 311–316.

46. Z. Suo, Reliability of interconnect structures, *Volume 8: Interfacial and Nanoscale Failure, Comprehensive Structural Integrity*, eds. W. Gerberich and W. Yang (Amsterdam: Elsevier, 2003), 265–324.

47. S. R. Nassif, Power grid analysis benchmarks, *Proceedings of the Annual Asian-Pacific Design Automation Conference (ASP-DAC)* (Seoul: ACM, 2008), 376–381.

48. S. Chatterjee, V. Sukharev, and F. N. Najm, Fast physics-based electromigration assessment by efficient solution of linear time-invariant (LTI) systems, *Proceedings of the 36th International Conference on Computer-Aided Design (ICCAD)* (Irvine: IEEE/ACM, 2017), 659–666.

49. W. Huang, M. R. Stan, and K. Skadron, Parameterized physical compact thermal modeling, *IEEE Transactions on Components and Packaging Technologies* **28** (2005), 615–622.

50. R. M. Christensen, *Mechanics of Composite Materials* (Mineola: Dover Publications, 2005).

51. S. P. Hau-Riegea and C. V. Thompson, The effects of the mechanical properties of the confinement material on electromigration in metallic interconnects, *Journal of Materials Research* **15** (2000), 1797–1802.

52. V. Sukharev, J.-H. Choy, A. Kteyan, et al., Carrier mobility shift in advanced silicon nodes due to chip-package interaction, *Journal of Electronic Packaging* **139** (2017), 020906-1-12.

53. P. Karmarkar, X. Xu, and K. El-Sayed, temperature and process dependent material characterization and multiscale stress evolution analysis for performance and reliability management under chip package interaction, *International Symposium on Microelectronics, IMAPS 2017* (Raleigh: ASME, 2017), 000013–000024.

54. J. He, Z. Suo, T. N. Marieb, and J. A. Maiz, Electromigration lifetime and critical void volume, *Applied Physics Letters* **85** (2004), 4639–4641.

55. V. Sukharev, A. Kteyan, and X. Huang, Postvoiding stress evolution in confined metal lines, *IEEE Transactions on Device and Materials Reliability* **16** (2016), 50–60.

56. S. Torosyan, A. Kteyan, V. Sukharev, J.-H. Choy, and F. N. Najm, Novel physics-based tool-prototype for electromigration assessment in commercial-grade power delivery networks, *Journal of Vacuum Science and Technology. B* **39** (2021), 013203-1-6.

57. V. Sukharev, Stress modeling for copper interconnect structures. *Materials for Information Technology*, ed. E. Zschech, C. Whelan, and T. Mikolajick (London: Springer, 2005), 251–263.

58. F. N. Najm and V. Sukharev, Efficient simulation of electromigration damage in large chip power grids using accurate physical models, *Proceedings of the 2019 IEEE International Reliability Physics Symposium (IRPS)* (Monterey: IEEE, 2019), 2A-MB.1-10.

59. A. S. Budiman, C. S. Hau-Riege, W. C. Baek, et al., Electromigration-induced plastic deformation in Cu interconnects: effects on current density exponent, n, and implications for EM reliability assessment, *Journal of Electronic Materials* 39 (2010), 2483–2488.

Index